Academic Software Installation Instructions

Arena requires Microsoft® Windows 2000 (Service Pack 3 or later), Windows Server 2003, Windows XP (Service Pack 2 or later) or Windows Vista Business Edition; under Windows 2000 and Windows XP you must have Administrator privileges to install the software.

Authorization to Copy Software

This academic software can be installed on any university computer, as well as on students' computers. It is intended for use in conjunction with this book for the purpose of learning simulation and Arena. You have the right to use and make copies of the software for academic use for teaching and research purposes only.

Commercial use of the software is prohibited.

This textbook uses the Arena version available at book release or publishing. It is expected that the Arena examples in this textbook can be compiled and run in the accompanying version as well as future Arena versions. Since new releases of Arena typically outpace updates to textbooks, academic institutions are encouraged to keep their lab and research software up to date and to make copies of their latest install CDs available to students (This is in agreement with EULA specifications), replacing what is included in this book. NOTE: Students should use the same software version as those used in the labs. Arena is forward, but not backward compatible (e.g., an Arena 11.0 model can be loaded into Arena 12.0, but an Arena 12.0 model cannot be loaded into Arena 11.0).

Installing the Arena Software

Follow this sequence to install your Arena software. Please note that you cannot merely accept all the defaults; there are some specific steps you must follow during the installation process:

1. Insert the Arena CD to initiate the autorun program, which displays the Arena installation screen. If it does not run automatically, browse the CD directory to locate autorun.exe and double-click it to start the installation.
2. From the installation dialog, select *Install Arena 12.00.00 (CPR 9)*. When prompted for a serial number, enter STUDENT to activate the academic version. Doing so customizes the install, provides access to the examples referenced in the textbook, and allows you to build larger models.
3. When choosing a location to install Arena on the PC's hard drive, please note that Arena will be placed in the Arena subfolder of the folder you specify.
4. After Arena installs, reboot your computer if requested.
5. If you have further questions, please refer to the User Zone section of our Web site at www.ArenaSimulation.com.

License activation is not supplied with, or required for, the STUDENT version of Arena. If you see an option to install the activation for Arena, this option should be cleared (unchecked). If you are installing the Arena PE Educational Lab Package or for more information on license activation or any other aspect of installation, select *Installation Notes* from the Arena installation screen.

System Requirements

The minimum requirements/recommendations for running the Arena software are:
- Windows Vista Business Edition, Windows XP Professional (SP2 or later), Windows XP Home (SP2 or later), Windows 2000 Professional (SP4 or later), Windows 2000 Server (SP4 or later), Windows Server 2003 (SP1 or later).
- Microsoft Internet Explorer 6.0 (Service Pack 1) or later.
- Adobe Acrobat Reader 7.0 or later to view documentation
- You must have Administrator privileges to install the software; however it is not necessary to have Administrator privileges to run Arena.
- Hard drive with 75-250 MB free disk space (depending on operating system and options installed)
- 256 MB RAM minimum recommendation
- Minimum Pentium processor, 500Mhz or higher
- The running and animation of simulation models can be calculation-intensive, so a faster processor with additional memory may result in significantly improved performance. In addition, a larger monitor and a screen resolution of at least 1024 x 768 is recommended for improved animation viewing.
- **Under Windows 2000 and Windows XP, you must have Administrator privileges to install the software.**

Manuel Rossetti's Supporting Files

Additional chapter files (e.g. Arena and Excel) are provided on the disk. You can access/retrieve them by browsing to the "Manuel Rossetti Supporting Files" folder when the CD is in your computer's optical drive.

Simulation Modeling and Arena

MANUEL D. ROSSETTI

University of Arkansas

WILEY

John Wiley & Sons, Inc.

VP & EXECUTIVE PUBLISHER	Don Fowley
ACQUISITIONS EDITOR	Jennifer Welter
SENIOR PRODUCTION EDITOR	Nicole Repasky
MARKETING MANAGER	Christopher Ruel
DESIGNER	Madelyn Lesure
PRODUCTION MANAGEMENT SERVICES	Thomson Digital Limited
EDITORIAL ASSISTANT	Mark Owens
MEDIA EDITOR	Lauren Sapira
COVER PHOTO	WizData/Images.com

This book was set in Times Roman by Thomson Digital Limited and printed and bound by Hamilton Printing. The cover was printed by Hamilton Printing.

To order books or for customer service please, call 1-800-CALL WILEY (225-5945).

Library of Congress Cataloging-in-Publication Data

Rossetti, Manuel D. (Manuel David), 1962-
   Simulation modeling and Arena / Manuel D. Rossetti.
      p.   cm.
   Includes bibliographical references and index.
   ISBN 978-0-470-09726-7 (cloth/cd)
1. Simulation methods. 2.   Discrete-time systems. 3.   Arena (Computer file).
I. Title.
   T57.62.R67 2008
   003'.3–dc22   2008032167

Printed in the United States of America

10 9 8 7 6 5 4 3 2 1

Preface

Intended Audience

This is an introductory textbook for a first course in discrete-event simulation modeling and analysis for upper-level undergraduate students as well as entering graduate students. While the text is focused on engineering students (primarily industrial engineering), it could also be used by advanced students in business, computer science, and other disciplines where simulation is practiced. Of course, practitioners interested in learning simulation and Arena could also use this book independently.

What is Simulation Modeling?

Discrete-event simulation is an important tool for the modeling of complex systems. It is used to represent manufacturing, transportation, and service systems in a computer program for the purpose of performing experiments. Representation of the system via a computer program enables the testing of engineering design changes without disruption to the system being modeled. Simulation modeling involves elements of system modeling, computer programming, probability and statistics, and engineering design. Because simulation modeling involves these individually challenging topics, the teaching and learning of simulation modeling can be difficult for both instructors and students. Instructors are faced with the task of presenting computer programming concepts, probability modeling, and statistical analysis, all within the context of demonstrating how to model complex systems such as factories and supply chains. In addition, because of the complexity associated with simulation modeling, specialized computer languages are needed, and thus must be taught to students for use during the model-building process. This book is intended to help instructors with this daunting task.

Approach, and Use of Arena

Traditionally, there have been two major types of simulation textbooks (1) those that emphasize the theoretical (and mostly statistical) aspects of simulation, and (2) those that emphasize the simulation language or package. The intention of this book is to merge these two aspects of simulation textbooks, while adding and emphasizing the art of model building. Thus, the book contains chapters on modeling and chapters that emphasize the statistical aspects of simulation. However, the coverage of statistical analysis is integrated with the modeling in such a way to emphasize the importance of both topics.

This book uses the Arena Simulation Environment as the primary modeling tool for teaching simulation. Arena is one of the leading simulation modeling packages in the world and has a strong and active user base. While the book uses Arena as the primary modeling tool, the book is not intended to be a user's guide to Arena. Instead, Arena is used as the vehicle for explaining important simulation concepts.

I feel strongly that simulation is best learned by doing. The book is structured to enable and encourage students to get engaged in the material. The overall approach to presenting the material is grounded in a hands-on concept of education. The style of writing is informal, tutorial, and centered around examples that students can implement while reading the chapters. The book assumes a basic knowledge of probability and statistics, and an introductory knowledge of computer programming. Even though these topics are assumed, the book provides integrated material that should help readers review the basics of these topics. Thus, instructors who use this book should not have to formally cover this material, and can be assured that students who read the book will be aware of these concepts within the context of simulation.

Organization of the Book

The first chapter is an introduction to the field of simulation modeling and an introduction to the Arena modeling environment. After Chapter 1, the student should know what simulation is and be able to put different types of simulation into context. Chapter 1 also introduces the important concept of how a discrete-event clock "ticks" and sets the stage for process modeling using activity diagramming. Finally, a simple (but comprehensive) example of Arena is presented so that students will feel comfortable with the tool.

Chapter 2 dives deeper into process-oriented modeling. The statistical aspects of simulation are downplayed within the chapter. The Basic Process template within Arena is thoroughly covered. Important concepts within process-oriented modeling (e.g., entities, attributes, activities, state variables, etc.) are discussed within the context of a number of examples. In addition, a deeper understanding of Arena is developed, including flow of control, input/output, variables, arrays, and debugging. After finishing Chapter 2, the reader should be able to model interesting systems from a process viewpoint using Arena.

Chapter 3 emphasizes the role of randomness in simulation. Specifically, the chapter presents input modeling, random number generation, and random variate generation techniques. After Chapter 3, the student should be able to model the input distributions required for simulation using such tools as EXCEL®, MINITAB™, and ARENA's™ input analyzer. In addition, the student will know why random number generators and their control are essential for simulation modeling. Finally, the primary techniques for generating random variates from probability distributions are covered (e.g., inversion, acceptance/rejection, convolution, and composition). An appendix is available for Chapter 3 that demonstrates how to perform Monte Carlo simulation within Excel.

Building on the use of stochastic elements in simulation, Chapter 4 discusses the major methods by which simulation output analysis must account for randomness. The various types of statistical quantities (observation based versus time persistent) are defined and then statistical methods are introduced for their analysis. Specifically, the chapter covers the method of replication for finite-horizon simulations, analysis of the initialization-transient period, the replication deletion method, and the batch means method. In addition, the use of simulation to make decisions based on competing alternatives is presented.

Chapter 5 returns to model building by presenting models for important classic modeling situations in queuing and inventory theory. Both analytical and simulation approaches to modeling these systems are covered. For instructors who work in a curriculum that has a separate course on these topics, this chapter presents an opportunity to concentrate on simulating these systems. The analytical material could easily be skipped without loss of continuity; however, often students learn the most about these systems through simulation. In situations where this material is not covered separately, background is presented on these topics to ensure that students can apply the basics of queuing theory and are aware of basic inventory models. The basic models are then extended so that students understand how simulation quickly becomes necessary when modeling more realistic situations.

Chapter 6 presents a thorough treatment of the entity-transfer and material-handling constructs within Arena. Students learn the fundamentals of resource-constrained transfers, free path transporters, conveyors, and fixed path transporters. The animation of models containing these elements is also emphasized.

Chapter 7 pulls together a number of miscellaneous topics that round out the use of Arena. In particular, the chapter covers Arena's activity based costing model and presents advanced aspects of modeling with resources (e.g., schedules and failure modeling). This chapter also presents a few useful modules that were not previously covered (e.g., picking stations, generic stations, and picking up and dropping off entities). An introduction to using Visual Basic and Arena is also presented.

Finally, Chapter 8 presents a detailed case study using Arena. An IIE/Rockwell Software Arena Contest problem is solved in its entirety. This chapter ensures that students will be ready to solve such a problem if assigned the same as a project for the course. The chapter wraps up with some practical advice for performing simulation projects.

Special Features

- *Learning Objectives:* Each chapter begins with specific learning objectives.
- *Output Analysis integrated with Arena:* The statistical aspects of simulation (e.g., output analysis) are integrated with the tool (e.g., Arena). More detailed discussions of the statistical aspects of simulation are presented than are found in many other simulation language–oriented textbooks.
- *Activity-based Learning:* Studies have shown that activity-based learning is critical to student retention of material. The text is organized around model building with the intention that students should be following along at the computer while working through the chapters. Instructors can perform the activities or organize computer laboratory exercises around the development of the models in the text.

- *Emphasis on Computer Programming aspect of Simulation:*   The computer programming aspects of simulation receive special emphasis. Students who take a course based on this text will be expected to have had a least one entry-level computer programming course; however, even with this background most students are woefully ill-prepared to use computers to solve problems. The theory-based textbooks do not cover this material and the simulation package textbooks attempt to downplay the programming aspects of their environment so that the modeling environment appears attractive to non–computer-oriented users. This book is intended to enable students to understand the inner workings of the simulation environment and thus demystify the "black box." The language elements of the simulation environment are compared to standard computer language elements so that students can make the appropriate analogies to already studied material.
- *Language-independent Conceptual Modeling Process:*   While Arena is the modeling tool, the conceptual modeling process presented in the text is based on language-independent methods, including but not limited to rich picturing, elementary flow charting, activity diagramming, and pseudo-code development. The emphasis is placed on developing a specification for a model that could be implemented in any simulation language environment.
- *Classic Stochastic Models:*   Classic stochastic models from operations research are covered. One chapter is dedicated to queuing and inventory models. In many curricula, if the analytical models are presented, they will be taught in a different course. In my opinion, the simulation of classic models along with their analytical treatment can provide for deeper student learning on these topics. In addition, the presentation of these classic models both analytically and through simulation provides simple systems on which to build the teaching of complex, more practical extensions.
- *Examples and Exercises:*   Comprehensive examples, exercises, questions, and problem sets were developed from the author's teaching, research, and industrial experiences.
- *Arena Software:*   The CD that accompanies the text contains the student version of Arena. In addition, the chapter files (e.g., Arena and Excel) are provided on the disk. You can access/retrieve these files by browsing to the "Manuel Rossetti Supporting Files" folder when the CD is in your computer's optical drive.

Student Resources

The following resources are available on the book website at www.wiley.com/college/rossetti:

- The chapter files (e.g., Arena and Excel).
- Appendix Files as PDF for handy reference.

Instructor Resources

The following resources are available on the instructor section of the book website at www.wiley.com/college/rossetti. Please visit the website to register for a password to access these resources:

- The chapter files (e.g., Arena and Excel) are provided on a disk.
- Lecture Slides: A comprehensive set of Powerpoint slides are for instructors to use within the classroom.

- Sample Syllabus
- Solutions Manual: Detailed or outlined solutions for the exercises in the text are available including the Arena model files where appropriate.
- Image Gallery: All illustrations from the text in electronic format, appropriate for inclusion in lecture slides.

Course Syllabus Suggestion

Early versions of the manuscript for this textbook were used over many semesters in my course at the University of Arkansas. The course that I teach is geared toward junior/senior, undergraduate industrial engineering students. In addition, graduate students who have not yet had a course in simulation enroll in the course. Graduate students are given extra homework assignments and are tested on some of the more theoretical aspects presented in the text (e.g., acceptance/rejection). I am able to cover Chapters 1 through 7 during a typical 16-week semester offering. A typical topic outline follows:

Number of lectures	Topic	Reading
1	Introduction	Chapter 1
4	Basic process modeling	Chapter 2
2	Input modeling	Chapter 3
2	Random number generation	Chapter 3
2	Random variate generation	Chapter 3
2	Finite horizon simulation	Chapter 4
3	Infinite horizon simulation	Chapter 4
2	Comparing alternatives	Chapter 4
4	Queuing and Inventory models	Chapter 5
5	Entity transfer and material handling constructs	Chapter 6
3	Miscellaneous topics in Arena modelling	Chapter 7

I use two exams and a project in the course. Exam 1 covers Chapters 1 to 3 and finite-horizon simulation from Chapter 4. Exam 2 covers the remaining portion of Chapter 4 plus Chapters 5 and 6. I do not formally test the students on the material in Chapter 7 since they will be using all previously learned material and components of Chapter 7 when preparing their final project. Students are assigned homework throughout the semester (about eight assignments). In addition to formal lectures, my course has a computer-based laboratory component that meets 1 day per week. During this time, the students are required to work on computer-based assignments that are based on the examples in the textbook.

Acknowledgements

Special thanks go to the team at Wiley, who helped me pull this project together, especially Jenny Welter, whose patience and guidance was extremely helpful. I would like to thank the following colleagues who reviewed the manuscript for this book, at various stages, and provided valuable feedback:

Michael Branson, Oklahoma State University
Todd Ebert, California State University - Long Beach
David Goldsman, Georgia Institute of Technology
Jeffrey W. Herrmann, University of Maryland
Samuel H. Huang, University of Cincinnati
James S. Noble, University of Missouri
Jeffrey S. Smith, Auburn University
Milton Smith, Clemson University
Kevin Taaffe, Clemson University

I would also like to thank the students in my classes who tested the manuscript and provided feedback, suggestions, and comments. In particular, I would like to thank Ron Walker, Roger Snelgrove, and Laura Jordan for contributing to the solutions manual for the text.

Lastly, I would like to thank my children, Joseph and Maria, who gave me their support and understanding, and my wife, Amy, who not only gave me support, but also helped with creating figures and diagrams, and with proofreading. Thanks so much!

About the Author

Manuel D. Rossetti, Ph.D., P.E., is an associate professor in the Industrial Engineering Department at the University of Arkansas. He received his doctorate in industrial and systems engineering from The Ohio State University. He has published over 70 journal and conference articles in the areas of transportation, manufacturing, health care, and simulation, and he has obtained over $3.1 million in extramural research funding. His research interests include the design, analysis, and optimization of manufacturing, health care, and transportation systems using stochastic modeling, computer simulation, and operations research techniques. He teaches courses in the areas of simulation, transportation and logistics, database systems, and inventory management. He was selected as a Lilly Teaching Fellow in 1997/1998 and has been nominated three times for outstanding teaching awards. He serves as an associate editor for the *International Journal of Modeling and Simulation*, and is active in the Institute of Industrial Engineers, the Institute for Operations Research and Management Science, and the American Society for Engineering Education. He served as co-editor for the 2004 Winter Simulation Conference, and will be co-editor for the 2009 Winter Simulation Conference.

Brief Contents

Table of Contents

4
Analyzing Simulation Output   **215**

5
Modeling Queueing and Inventory Systems   **303**

8

Application of Simulation Modeling   *502*

CHAPTER 1

Simulation Modeling

LEARNING OBJECTIVES

After completing this chapter, you should:

- Understand what computer simulation is and why it is an important analysis tool.
- Understand the various types of computer simulation.
- Be able to explain how the simulation clock works and how randomness is used in simulation models.
- Be able to develop and read an activity flow diagram.
- Understand the basic components of the Arena™ Environment.
- Be able to create, run, animate, and examine the results of an introductory Arena™ model.

In this book, you will learn how to model systems within a computer environment in order to analyze system design configurations. The models that you will build and exercise are called *simulation models*. When developing a simulation model, the modeler attempts to represent the system in such a way that the representation assumes or mimics the system's pertinent outward qualities. When you execute the simulation model, you are performing a simulation. In other words, simulation is an instantiation of the act of simulating. A simulation is often the next best thing to observing the real system. If you have confidence in your simulation, you can use it to infer how the real system will operate. You can then use your inference to understand and improve the system's performance.

In general, simulations can take on many forms. Almost everyone is familiar with the board game Life™. In this game, the players imitate life by going to college, getting a job, getting married, and so on, and finally retiring. This board game is a simulation of life. Another example is the military's war game exercises, which are simulations of battlefield conditions. Both types of simulations involve a physical representation of the thing being simulated. The board game, the rules, and the players represent the simulation model. The battlefield, the rules of engagement, and the combatants are also physical representations. No wishful thinking will make the simulations that you develop in this book real. This is the first rule to remember about simulation. A simulation is only a model (representation) of the real thing. You can make your simulations as realistic as time and technology allow, but they are not the real thing. As you would never confuse a toy airplane with a real airplane, you should never confuse a simulation of a system with the real system. You may laugh at this analogy, but as you apply simulation to the real world you will encounter analysts who forget this rule. Don't be one.

All the previous examples involved a physical representation or model (tangible or real things simulating other real things). In this book, you will develop computer models that simulate real systems. Ranvindran et al. (1987) defined computer simulation as "[a] numerical technique for conducting experiments on a digital computer which involves logical and mathematical relationships that interact to describe the behavior of a system over time." Computer simulations provide an extra layer of abstraction from reality that allows fuller control of the progression of and interaction with the simulation. In addition, even though computer simulations are one step removed from reality, they are often capable of providing constructs that cannot be incorporated into physical simulations. For example, a computer-based airplane flight simulator may contain emergency conditions that would be too dangerous or too costly in a physical simulation training scenario. This representational power of computer modeling is one of the main reasons why computer simulation is used.

1.1   Why Simulate?

Imagine trying to analyze the following situation. Patients arrive at an emergency room. The arrival of the patients to the emergency department (ED) occurs randomly and may vary with the day of the week and even the hour of the day. The hospital has a triage station, where the arriving patient's condition is monitored. If the patient's condition warrants immediate attention, the patient is expedited to an emergency room bed to be attended by a doctor and a nurse. In this case, the patient's admission information may be obtained from a relative. If the patient does not require immediate attention, the patient goes through the admission process, where the patient's information is obtained. The patient is then directed to the waiting room, to wait for allocation to a room, a doctor, and a nurse. The doctors and nurses within the ED must monitor the patients by performing tests and diagnosing symptoms on a periodic basis. As the patient receives care, the patient may be moved to and require other facilities (MRI, x-ray, etc.). Eventually, the patient is either discharged after receiving care or admitted to the main hospital.

The hospital is interested in conducting a study of the ED in order to improve the care of the patients while better utilizing the available resources. To investigate this situation, you might need to understand the behavior of certain measures of performance such as

- The average and maximum number of patients that waited during the day.
- The average patient waiting time while in the ED.
- The probability that a patient has to wait longer than three hours.
- The average and maximum number of rooms required per hour.
- The average utilization of doctors and nurses (and certain equipment).

Because of the importance of ED operations, the hospital has historical records available on the operation of the department through its patient tracking system. With these records, you might be able to estimate the current performance of the ED. Despite the availability of this information, when conducting a study of the ED, you might want to propose changes in how the department will operate (e.g., staffing levels) in the future. Thus, you are faced with trying to predict the future behavior of the system and its performance when making changes to the system. In this situation, you cannot realistically experiment with the actual system without possibly endangering the lives or care of the patients. Thus, it would be better to model the system and to test the effect of changes on the model. If the model has acceptable fidelity, then you can infer how the changes will affect the real system. This is where simulation techniques can be utilized.

Readers familiar with operations research and industrial engineering techniques may be thinking that the ED can be analyzed by using queuing models. Queuing models will be discussed later in this book; however, for the present situation, the application of queuing models will most likely be inadequate due to the complex policies for allocating nurses, doctors, and beds to patients. In addition, the dynamic nature of this system (nonstationary arrivals, changing staffing levels, etc.) cannot be successfully modeled with current analytical queuing models. Queuing models might be used to analyze portions of the system, but analyzing the dynamic behavior of the entire system is beyond the capability of these types of models. Analysis of the entire system is *not* beyond the capability of simulation modeling.

A key advantage of simulation modeling is its capability of modeling entire systems and complex interrelationships. The representational power of simulation provides the flexible modeling that is required for capturing complex processes. As a result, all important interactions among components of the system can be accounted for within the model. The modeling of these interactions is inherent in simulation modeling because simulation imitates the behavior of the real system (as closely as necessary). The prediction of the future behavior of the system is then achieved by monitoring the behavior of various modeling scenarios as a function of simulated time. Real-world systems are often too complex for analytical models and often too expensive to experiment with directly. Simulation models allow the modeling of this complexity and enable low-cost experimentation to make inferences about how the actual system might behave.

1.2   Types of Computer Simulation

The main purpose of a simulation model is to allow observations about a particular system to be collected as a function of time. So far the word "system" has been used in much of the discussion, without formally discussing what a system is. According to Blanchard and Fabrycky (1990), a system is a set of interrelated components working together toward a common objective. The standard for systems engineering provides a deeper definition: "A system is a composite of people, products, and processes that provide a capability to satisfy stated needs. A complete system includes the facilities, equipment (hardware and software), materials, services, data, skilled personnel, and techniques required to achieve, provide, and sustain system effectiveness." (Air Force Systems Command, 1991).

Figure 1-1 illustrates the fact that a system is embedded within an environment, and that typically a system requires inputs and produces output using internal components. How you model a particular system will depend on the intended use of the model and how you perceive the system. The modeler's view of the system colors how it is conceptualized. For example, in the emergency room situation, would the system boundaries encompass the ambulance dispatching and delivery process, or should the details of the operating room be modeled? Clearly, the emergency room has these components, but your conceptualization of it as a system may or may not include these items, and thus, your decisions regarding how to conceptualize the system will drive the level of abstraction within your modeling. An important point to remember is that two perfectly logical and rational people can look at the same thing and conceptualize it as two entirely different systems based on their own Weltanschauung or worldview.

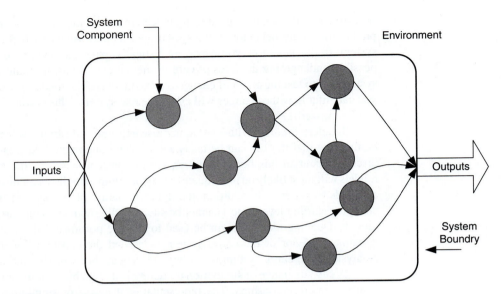

■ **FIGURE 1-1** **Conceptualization of a system**

Because how you conceptualize a system drives your modeling, describing some general system classifications is useful. Systems may be classified according to whether they are man-made (e.g., manufacturing system) or natural (e.g., solar system). A system can be physical (e.g., an airport) or conceptual (e.g., a system of equations). If stochastic or random behavior is an important component of the system, then the system is said to be stochastic; if not, then it is considered deterministic. One of the more useful ways to look at a system is whether it changes with respect to time. A system that does not change significantly with respect to time is said to be static, otherwise it is considered dynamic. If a system is dynamic, you might want to consider how it evolves over time. A dynamic system is *discrete* if the state of the system changes at discrete points in time, and *continuous* if the state of the system changes continuously over time. This dichotomy is purely a function of your level of abstraction. If conceptualizing a system as discrete serves your purposes, then you can call the system discrete. Figure 1-2 illustrates general system classifications. This book primarily examines stochastic, dynamic, and discrete systems.

The main purpose of a simulation model is to allow observations about a particular system to be gathered as a function of time. From that standpoint, the two distinct types of simulation are discrete event and continuous.

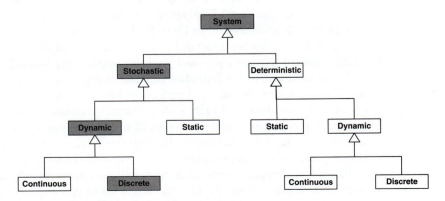

■ **FIGURE 1-2** **General types of systems**

Just as discrete systems change at discrete points in time, in discrete-event simulation observations are gathered at selected points in time when certain changes take place in the system. These selected points in time are called *events*. On the other hand, continuous simulation requires that observations be collected continuously at every point in time (or at least that the system is described for all points in time). The types of models to be examined in this book are called *discrete-event simulation* models.

To illustrate the difference between the two types of simulation, contrast a fast-food service counter with that of an oil-loading facility that is filling tankers. In the fast-food service counter system, changes in the status of the system occur when a customer either arrives to place an order or when the customer receives a food order. At these two events, measures such as queue length and waiting time will be affected. At all the other points in time, these measures remain either unchanged (e.g., queue length) or not yet ready for observation (e.g., waiting time of the customer). For this reason, the system does not need to be observed on a continuous basis. The system need only be observed at selected discrete points in time; thus, a discrete-event simulation model is applicable.

In the case of the oil tanker-loading example, one of the measures of performance is the amount of oil in each tanker. Because the oil is a liquid, it cannot be readily divided into discrete components. That is, it flows continuously into the tanker. It is not necessary (or practical) to track the molecules of oil, when you only care about the level of the oil in the tanker. In this case, a model of the system must describe the rate of flow over time and the output of the model is presented as a function of time. Systems such as these are often modeled using differential equations. The solution of these equations involves numerical methods that integrate the state of the modeled system over time. This, in essence, involves dividing time into small equal intervals and stepping through time.

Often both the discrete and continuous viewpoints are relevant in modeling a system. For example, if an oil tanker arrives at the port to be filled, there is an arrival event that changes the state of the system. This type of modeling situation is called *combined continuous discrete modeling*. Arena^{TM}, the simulation language used in this book, has modeling constructs for both continuous and discrete modeling; however, this book does not cover the modeling of continuous or combined continuous discrete systems. There are many useful references on this topic. See Pegden (1995) and Kelton et al. (2004) for an overview of how Arena^{TM} handles this topic. The focus of this book will be on discrete-event modeling. The key to understanding this type of modeling is to develop an appreciation for how the computer keeps track of time.

1.3   How the Discrete-Event Clock Works

This section introduces the concept of how time evolves in a discrete-event simulation. This topic will also be revisited in future chapters after you have developed a deeper understanding for a number of underlying elements of simulation modeling.

In discrete-event dynamic systems, an event is something that happens at an instant in time that corresponds to a change in system state. An event can be conceptualized as a transmission of information that causes an action resulting in a change in system state. Let's consider a simple bank that has two tellers serving customers from a single waiting line. In this situation, the system of interest is the bank tellers (whether they are idle or busy) and any customers waiting in the line. Assume that the bank opens at 9 a.m., which can be used to designate time zero for the simulation. It should be clear that if the bank does not have any customers arrive during the day that the state of the bank will not change. In addition, when a customer arrives, the number of customers in the bank will increase by one. In other words, the arrival of a customer changes the state of the bank. Thus, the arrival of a customer can be considered an event.

Customer	Time of arrival	Time between arrival
1	2	2
2	5	3
3	7	2
4	15	8

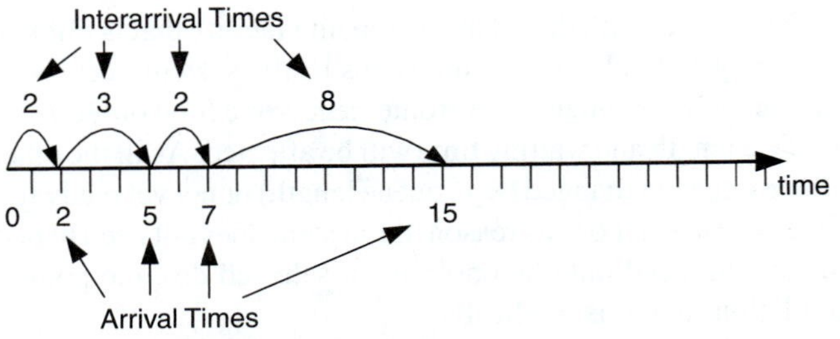

■ **FIGURE 1-3**
Customer arrival process

Figure 1-3 illustrates a timeline of customer arrival events. The values for the arrival times have been rounded to whole numbers, but in general the arrival times can be any real valued numbers greater than zero. As seen in the figure, the first customer arrives at time 2, and 3 minutes elapses before customer 2 arrives at time 5. From the discrete-event perspective, nothing is happening in the bank from time [0,2); however, at time 2 an arrival event occurs and the subsequent actions associated with this event need to be accounted for with respect to the state of the system. Since an event causes actions that result in a change of system state, the modeler asks, "What are the actions that occur when a customer arrives?"

- The customer enters the waiting line.
- If there is an available teller, the customer will immediately exit the line and the available teller will begin to provide service.
- If there are no tellers available, the customer will remain waiting in the line until a teller becomes available.

Now, consider the arrival of the first customer. The bank opens at 9 a.m. with no customers and all the tellers idle. The first customer enters and immediately exits the queue at 9:02 a.m. (or time 2) and service begins. After the customer completes service, the customer exits the bank. When a customer completes service (and leaves the bank), the number of customers in the bank decreases by 1. It should be clear that certain actions occur when a customer completes service. These actions correspond to the second event associated with this system, the customer service completion event. What actions occur at this event?

- The customer departs the bank.
- If there are waiting customers, the teller indicates to the next customer to approach the teller window for service. The customer exits the waiting line and begins service with the teller.
- If there are no waiting customers, the teller becomes idle.

Figure 1-4 shows the service times for each of the four customers that arrived in Figure 1-3. Customer 1 enters service at time 2 because there were no other customers present in the bank. Suppose that it is now 9:02 a.m. (time 2) and that the service time of customer 1 is known *in advance* to be 8 minutes. The time that customer 1 is going to complete service can thus be precomputed. Based on the information in the figure, customer 1 will complete service at time 10 (current time + service time = 2 + 8 = 10). Thus, it should be apparent that for a

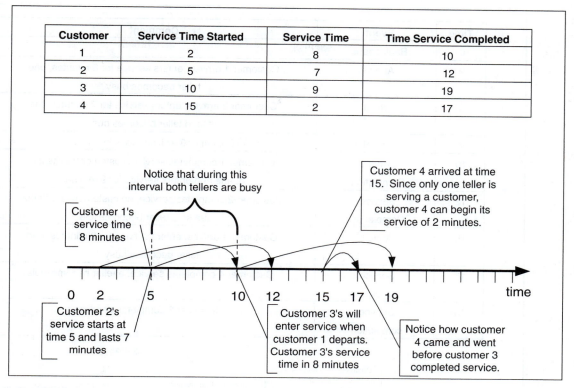

Customer	Service Time Started	Service Time	Time Service Completed
1	2	8	10
2	5	7	12
3	10	9	19
4	15	2	17

Notice that during this interval both tellers are busy

Customer 1's service time 8 minutes

Customer 4 arrived at time 15. Since only one teller is serving a customer, customer 4 can begin its service of 2 minutes.

0 2 5 10 12 15 17 19   time

Customer 2's service starts at time 5 and lasts 7 minutes

Customer 3's will enter service when customer 1 departs. Customer 3's service time in 8 minutes

Notice how customer 4 came and went before customer 3 completed service.

■ **FIGURE 1-4** Customer service process

modeler to re-create the bank's behavior over a time period, knowledge of customer service times is necessary. The service time of each customer coupled with the knowledge of when the customer began service allows advance computation of the time that the customer will complete service and exit the bank. A careful inspection of Figure 1-4 and knowledge of how banks operate should indicate that a customer will begin service either immediately upon arrival (when a teller is available) or coinciding with when another customer departs the bank after being served. This latter situation occurs when the queue is not empty after a customer completes service. The times that service completions and arrivals occur constitute the pertinent events for simulating this banking system.

If the arrival and service completion events are combined as in Figures 1-3 and 1-4, then the time-ordered sequence of events for the system can be determined. Figure 1-5 illustrates the events ordered by time from smallest event time to largest event time. Looking forward from time 2, the next event to occur will be at time 5 when the second customer arrives. When you simulate a system, you must be able to generate a sequence of events so that when each event occurs, the appropriate actions to change the system state are invoked.

The real system will simply evolve over time; however, a simulation of the system must re-create the events. In simulation, events are created by adding additional logic to the normal state-changing actions. This additional logic is responsible for scheduling future events that are implied by the actions in the current event. For example, when a customer arrives, the time to the next arrival can be generated and scheduled to occur at some future time. This can be done by generating the time until the next arrival and adding it to the current time to determine the actual arrival time. Thus, all arrival times do not need to be precomputed prior to beginning the simulation. For example, at time 2, customer 1 arrived. Customer 2 arrives at time 5. Thus, the time between arrivals (3 minutes) is added to the current time and customer 2's arrival at

Time	Event	Comment
0	Bank opens	
2	Arrival	Customer 1 arrives, enters service for 8 minutes, one teller becomes busy
5	Arrival	Customer 2 arrives, enters service for 7 minutes, the second teller becomes busy
7	Arrival	Customer 3 arrives, waits in queue
10	Service Completion	Customer 1 completes service, customer 3 exits the queue and enters service for 9 minutes
12	Service Completion	Customer 2 completes service, no customers are in the queue so a teller becomes idle
15	Arrival	Customer 4 arrives, enters service for 2 minutes, one teller becomes busy
17	Service Completion	Customer 4 completes service, a teller becomes idle
19	Service completion	Customer 3 completes service

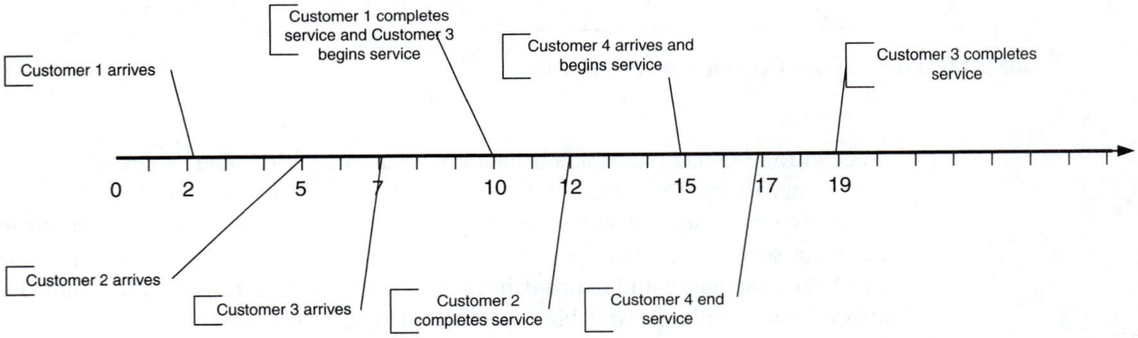

■ **FIGURE 1-5** Order in which events are processed

time 5 is scheduled to occur. Every time an arrival occurs, this additional logic is invoked to ensure that more arrivals will continue within the simulation.

Adding scheduling logic also occurs for service completion events. For example, since customer 1 immediately enters service at time 2, the service completion of customer 1 can be scheduled to occur at time 12 (current time + service time = 2 + 10 = 12). Note that other events may have already been scheduled for times previous to time 12. In addition, other events may be inserted before the service completion at time 12 actually occurs. From this scenario, you should begin to get a feeling for how a computer program can implement some of these ideas.

This intuitive notion for how a computer simulation executes leads to the realization that computer logic processing need only occur at event times. In brief, the state of the system does not change between events. Thus, it is *not necessary* to step incrementally through time checking to see if something happens at time 0.1, 0.2, 0.3, and so on (or whatever time scale you envision that is fine enough to prevent missing any events). Thus, the discrete-event simulation clock does not "tick" at regular intervals. Instead, the simulation clock jumps from event time to event time. As the simulation moves from the current event to the next event, the current simulation time is updated to the time of the next event, and any changes to the system associated with the next event are executed. This allows the simulation to evolve over time.

1.4   Randomness in Simulation

The discrete-event processing scheme outlined in the previous section for the bank system depends on having two pieces of data: the time between arrivals and the service times of the customers. You may be wondering how such information can be secured during the execution of a simulation model. In most real-life situations, the arrival process and the service process occur in a random fashion. Even though the processes may be random, it does not mean that you cannot describe or model the randomness. To have any hope of simulating the situation, you must be able to model the randomness. One of the ways to model this randomness is to describe the phenomenon as a random variable governed by a particular probability distribution. For example, if customer arrivals occur according to a Poisson process, then based on probability theory, it is known that the distribution of interarrival times is an exponential distribution. In general, information about how the customers arrive must be secured either through direct observation of the system or by using historical data. If neither source of information is available, then some plausible assumptions must be made to describe the random process via a probability model.

If historical data are available, there are two basic choices for how to handle the modeling. The first choice is to develop a probability model based on the data. The second choice is to try to drive the simulation directly from the historical data. The latter approach is not recommended. First, it is extremely unlikely that the captured data will be in a directly useable form. Second, it is even more unlikely that the data will be able to adequately represent all the modeling scenarios that you will need through the course of experimenting with the model. For example, suppose that you only have 1 day's worth of arrival data, but you need to simulate a month's worth of system operation. If you simply re-drive your simulation using the 1 day's worth of data, you cannot possibly simulate different days! Developing probability models from historical data encompassing all desired scenarios through time or from data that you capture while developing your model is preferable. Chapter 3 will discuss some of the tools and techniques for modeling probability distributions.

Once a probability model has been developed, statistical theory provides the means for obtaining random samples based on the use of uniformly distributed random numbers on the interval (0,1). These random samples are then used to map the future occurrence of an event on the time scale. For example, if the interarrival time is exponential, then a random sample drawn from that distribution would represent the time interval until the occurrence of the next arrival. The process of generating random numbers and random variables in simulation models will also be discussed in Chapter 3.

1.5   Simulation Languages

Discrete event simulation normally involves a tremendous amount of computation. Consequently, the use of computers to carry out these computations is essential. The sheer quantity of computation, however, is not the only obstacle in simulation. Upon considering the previous bank teller example, you will discover that it involves a complex logical structure that requires special expertise before it can be translated into a computer model. Attempting to implement the simulation model from scratch in a general purpose language such as FORTRAN, Visual Basic, C/C++, or Java will require above-average programming skills. In the absence of specialized libraries for these languages that lift some of the burden from the user, simulation as a tool would be relegated to "elite" programmers. Luckily, the repetitive nature of computations in simulation allows the development of computer libraries that are applicable to simulation modeling situations. For example, libraries or packages must be available for ordering and processing events chronologically, as well as generating random numbers and automatically collecting statistics. Such a library for simulating discrete-event systems in Java is available from the author (see Rossetti, 2008).

The computational power and storage capacity of computers have driven the development of specialized simulation languages. Some languages have been developed for continuous and discrete simulations, and combined continuous and discrete modeling. All simulation languages provide certain standard programming facilities and differ in how the user takes advantage of these facilities. There is normally a trade-off between how flexible the language is in representing certain modeling situations. Usually, languages that are highly flexible in representing complex situations require more work (and care) by the user to account for how the model logic is developed. Some languages are more programming oriented (e.g., SIMSCRIPT™) and others are more "drag and drop" (e.g., ProModel™, Arena™).

The choice of a simulation language is a difficult one. There are many competing languages, each with its own advantages and disadvantages. The Institute for Operations Research and Management Science (INFORMS) often publishes a yearly product review covering commercial simulation languages (see, for example, http://lionhrtpub.com/orms/). In addition to this useful comparison, you should examine the Winter Simulation Conference (www.wintersim.org). The conference has hands-on exhibits of simulation software, and the conference proceedings often have tutorials for the various software packages. Past proceedings have been made available electronically through the generous support of the INFORMS Society for Simulation (http://www.informs-sim.org/wscpapers. html).

Arena™ was chosen for this textbook because of the author's experience with the software, its ease of use, and the availability of student versions of the software. While all languages have flaws, using a simulation language is essential in performing high-performance simulation studies. Most, if not all simulation companies, have strong support to assist the user in learning their software. Arena™ has a strong academic and industrial user base, and is very competitive in the simulation marketplace. Once you learn one simulation language well, it is much easier to switch to other languages and to understand which languages will be more appropriate for certain modeling situations.

Arena™ is fundamentally a process description—based language. That is, when using Arena™, the modeler describes the process that an "entity" experiences while flowing through or using the elements of the system. You will learn about how Arena™ facilitates process modeling throughout this textbook.

1.6   Getting Started with Arena™

This section presents a simple model to introduce and provide an overview of the modeling capabilities in the Arena™ Environment. Figure 1-6 illustrates the Arena™ Environment with the View, Draw, Animate, and Animate Transfer toolbars detached (floating) in the environment. In addition, the Project Bar contains the project templates (Advanced Transfer, Advanced Process, Basic Process, Flow Process), the report panel, and the navigate panels. By right-clicking in the project template area, you can access the context menu for attaching/detaching project templates. The project templates, indicated by the diamond flowchart symbol are used to build models in the model window. Normally, the Advanced Transfer and Flow Process panels are not attached. Additional modeling constructs can be accessed by attaching

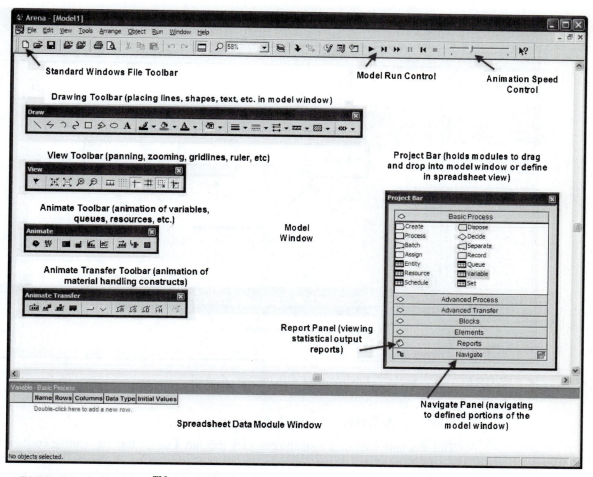

■ FIGURE 1-6 Arena™ environment

these templates to the project bar. In addition, you can control the size and view options for the modules. For example, you can change the view of the modules to see them as large icons, small icons, or as text only.

The report panel allows you to drill down into the reports that are generated after a simulation run. The navigation panel allows the definition and use of links (bookmarks) to prespecified areas of the model window. The Arena™ Environment (Figure 1-6) normally has the toolbars docked, allowing unobstructed access to the model window and the spreadsheet (data) window. The flow chart—oriented symbols from the project templates are "dragged and dropped" into the model window. The flow chart symbols are called *modules*. The spreadsheet view presents a row/column format for the currently selected module in the model window. In addition, the spreadsheet view allows data entry for modules (e.g., VARIABLE) used in the model, for which there is not a flow chart oriented representation. You will learn about the various characteristics of flow chart modules and data modules as you proceed through this text.

To hide or view the toolbars, you can right-click in the gray toolbar area and check the desired toolbars in the contextual pop-up menu. In addition, you can also drag the toolbars to the desired location in the environment. Figure 1-7 illustrates the Arena™ Environment with the toolbars and project bar docked in standard locations. Modules (CREATE, PROCESS, DISPOSE) can be accessed from the basic process panel in the model window, and the spreadsheet (data) window is showing the spreadsheet view for the CREATE module.

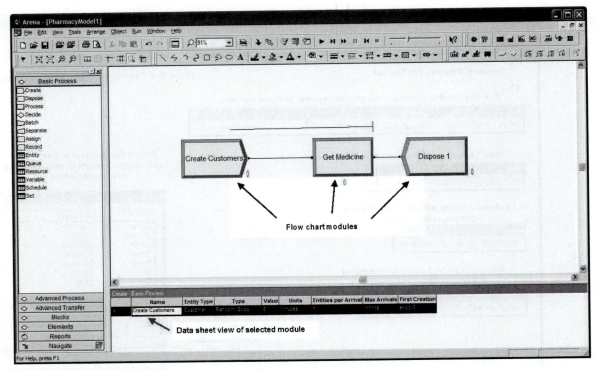

■ **FIGURE 1-7**   Arena™ Environment with toolbars docked

1.6.1   A Drive-Through Pharmacy

This example considers a small pharmacy that has a single line for waiting customers and a single pharmacist. Assume that customers arrive at a drive-through pharmacy window according to a Poisson distribution with a mean of 10 per hour. The time that it takes the pharmacist to serve the customer is random, and an analysis of the service times has indicated that the time is well modeled with an exponential distribution with a mean of 3 minutes. Customers who arrive at the pharmacy are served in the order of arrival and enough space is available in the parking area of the adjacent grocery store to accommodate any waiting customers.

The drive-through pharmacy system can be conceptualized as a single-server, waiting-line system, where the server is the pharmacist. An idealized representation of this system is shown in Figure 1-8. If the pharmacist is busy serving a customer, then additional customers will wait in line. In such a situation, management might be interested in how long customers wait in line before being served by the pharmacist. In addition, management might want to predict if the number of waiting cars will be large. Finally, managers might want to estimate pharmacist utilization in order to ensure that he or she is not too busy.

1.6.2   Modeling the System

Example Arena™ models used throughout this text are included on the companion compact disk. At the very least, you should explore the completed models as they are discussed in the text; however, in order to get the most out of these examples, you should try to follow along using Arena™ to build the models. In some cases, you can start from scratch. In other cases, a starting model is provided so that you can perform the enhancements. Working through these examples is very important for you to develop a good understanding of the material in the text. Let's get started working with Arena™.

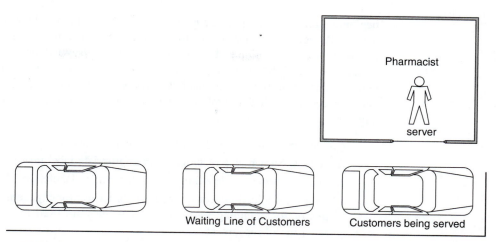

■ **FIGURE 1-8**   **Drive-through pharmacy**

Arena™ is predicated on modeling the process flow of "entities" through a system. Thus, the first question to ask is, *What is the system?* In this situation, the system is the pharmacist and the potential customers as idealized in Figure 1-8. Now you should consider the entities of the system. An entity is a conceptual thing of importance that flows through a system potentially using the resources of the system. Therefore, one of the first questions to ask when developing an Arena™ model is, *What are the entities?* In this situation, the entities are the customers that need to use the pharmacy. This is because customers are discrete things that enter the system, flow through the system, and then depart the system.

Because entities often use things as they flow through the system, a natural question is to ask: *What resources are used by the entities?* A resource is something used by the entities that may constrain the flow of the entities in the system. Another way to think of resources is to think of the components that provide service in the system. In this situation, the entities "use" the pharmacist in order to get their medicine. Thus, the pharmacist can be modeled as a resource.

Since Arena™ facilitates the modeling of the process flows of the entities through a system, it is natural to ask, *What are the process flows?* In answering this question, it is conceptually useful to pretend that you are the entity and to ask, *What do I do?* In this situation, a pharmacy customer arrives, gets served, and leaves.

As you can see, initial modeling involves identifying the elements of the system and what those elements do. The next step in building an Arena™ model is to enhance your understanding of the system through conceptual modeling. For this simple system, a very useful conceptual model has already been given in Figure 1-8. As you proceed through this text, you will learn other conceptual model-building techniques. From the conceptual modeling, you might proceed to more logical modeling in preparation for using Arena™. A useful logical modeling tool is to develop pseudocode for the situation. Here is some potential pseudocode for this situation.

- Create customers according to the Poisson arrival process.
- Process the customers through the pharmacy.
- Dispose of the customers as they leave the system.

The pseudocode represents the logical flow of an entity (customer) through the drive-through pharmacy: arrive (create), get served (process), leave (dispose).

Another useful conceptual modeling tool is the *activity diagram*. An *activity* is an operation that takes time to complete. An activity is associated with the state of an object over a time interval. Activities are defined by the occurrence of two events that represent the activity's beginning time and ending time and mark the entrance and exit of the state associated with the activity. An activity diagram is a pictorial representation of the process (steps of activities) for an entity and its interaction with resources while in the system. If the entity is a temporary entity (i.e., it flows through the system), the activity diagram is called an activity flow diagram. If the entity is permanent (i.e., it remains in the system throughout its life), the activity diagram is called an activity cycle diagram. The notation of an activity diagram is very simple, and can be augmented as needed to explain additional concepts.

Queues: Shown as a circle with queue labeled inside

Activities: Shown as a rectangle with appropriate label inside

Resources: Shown as small circles with resource labeled inside

Lines/arcs: Indicating flow (precedence ordering) for engagement of entities in activities or for obtaining resources, dotted lines are used to indicate the seizing and releasing of resources

Zigzag lines: Indicate the creation or destruction of entities

Activity diagrams are especially useful for illustrating how entities interact with resources. Activity diagrams are easy to build by hand and serve as a useful communication mechanism. Since they have a simple set of symbols, it is easy to use an activity diagram to communicate with people who have little simulation background. Activity diagrams are an excellent mechanism to document your conceptual model of the system before building the model in Arena^{TM}.

Figure 1-9 shows the activity diagram for the pharmacy situation. This diagram was built using Visio software, and the drawing is available with the supplemental files for this chapter. You can use the drawing to copy and paste from, in order to form other activity diagrams. The diagram describes the life of an entity in the system. The zigzag lines at the top of the diagram indicate the creation of an entity. While not necessary, the diagram has been augmented with Arena^{TM} like pseudocode to represent the CREATE statement. Consider following the life of the customer through the pharmacy. Following the direction of the arrows, the customers are first created and then enter the queue. Note that the diagram clearly shows that there is a queue for the drive-through customers. Think of the entity flowing through the diagram. As customers flow through the queue, each attempts to start an activity. In this case, the activity requires a resource. The pharmacist is shown as a resource (circle) next to the rectangle that represents the service activity.

The customer requires the resource in order to start its service activity. This is indicated by the dashed arrow from the pharmacist (resource) to the top of the service activity rectangle. If the customer does not obtain the resource, the customer waits in the queue; once they receive the number of units of the resource requested, they proceed with the activity. The activity represents a delay of time, and in this case the resource is used throughout the delay. After the activity is completed, the customer releases the pharmacist (resource). This is indicated by another dashed arrow, with the direction indicating that the resource is "being put back" or released. After the service activity is completed, the customer leaves the system. This is indicated with the zigzag lines going nowhere and augmented with the keyword DISPOSE. The dashed arrows of a typical activity diagram have also been augmented with the Arena^{TM} like pseudocode of SEIZE and RELEASE. The conceptual model of this system is summarized in the following.

System: The system has a pharmacist that acts as a resource, customers that act as entities, and a queue to hold the waiting customers. The state of the system includes the number of customers in the system, in the queue, and in service.

Events: Arrivals of customers in the system occur with an inter-event time that is exponentially distributed with a mean of 6 minutes.

Activities: The service time of the customers is exponentially distributed with a mean of 3 minutes.

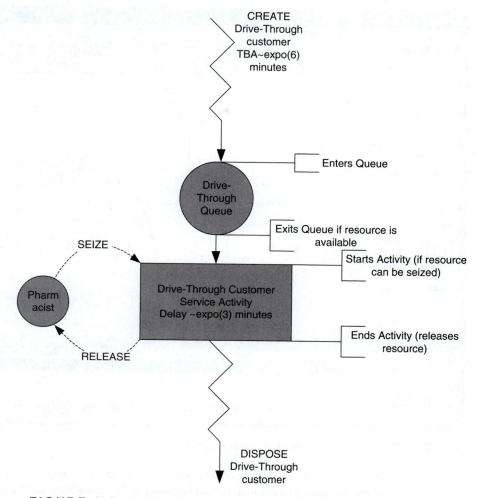

■ **FIGURE 1-9**   Activity diagram of drive-through pharmacy

Conditional delays: A conditional delay occurs when an entity has to wait for a condition to occur in order to proceed. In this system, the customer may have to wait in a queue until the pharmacist becomes available.

With an activity diagram and pseudocode such as these to represent a solid conceptual understanding of the system, you can begin the model development process in Arena™.

1.6.3   Implementing the Model in Arena™

If you haven't already done so, open the Arena™ Environment. Using the Basic Process Panel template, drag the CREATE, PROCESS, and DISPOSE modules into the model window and make sure that they are connected as shown in Figure 1-10. In order to drag and drop the modules, simply select the module in the Basic Process template and drag the module into the model window, letting go of the mouse when you have positioned the module in the desired location. If the module in the model window is highlighted (selected), the next dragged module will automatically be connected. If you placed the modules and they were not automatically connected, then you can use the connect toolbar icon. Select the icon and then select the "from" module near the connection exit point, and then holding the mouse

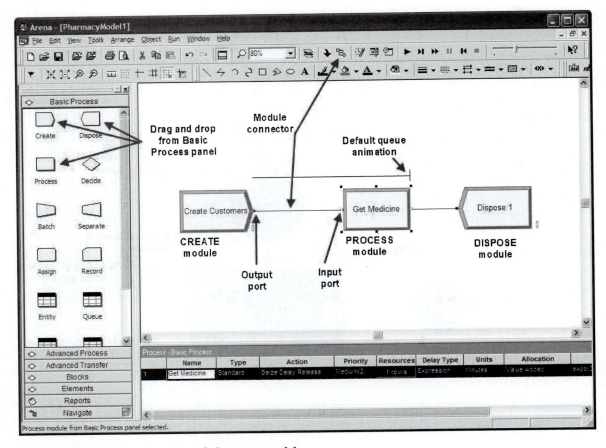

■ **FIGURE 1-10** Overview of pharmacy model

button down, drag to form a connection line to the entry point of the "to" module. The names of the modules will not match what is shown in the figure. You will name the modules as you fill out their dialog boxes.

1.6.4 Specify the Arrival Process

Within Arena™, nothing happens unless entities enter the model. This is done through the use of the CREATE module. In the current example, pharmacy customers arrive according to a Poisson process with a mean of $\lambda = 10$ per hour. According to probability theory, this implies that the time between arrivals is exponentially distributed with a mean of $(1 / \lambda)$. Thus, for this situation, the mean time between arrivals is 6 minutes.

$$\frac{1}{\lambda} = \frac{1 \text{ hour}}{10 \text{ customers}} \times \frac{60 \text{ minutes}}{1 \text{ hour}} = \frac{6 \text{ minutes}}{\text{customer}}$$

Open up the CREATE module (by double-clicking on it) and fill it in as shown in Figure 1-11. The distribution of the "Time Between Arrivals" textbox is by default Exponential, Random (expo) in the figure. In this instance, the "Value" textbox refers to the mean time between arrivals. "Entities per Arrival" specifies how many customers arrive at each arrival. Since more than one customer does not arrive at a time, the value is set to 1. "Max Arrivals" specifies the total number of arrival events that will occur in the CREATE module. This is set at the default, "Infinite", since a fixed number of customer arrivals is not specified for this example. The "First Creation" textbox specifies the time of the first arrival event. Technically, the time to

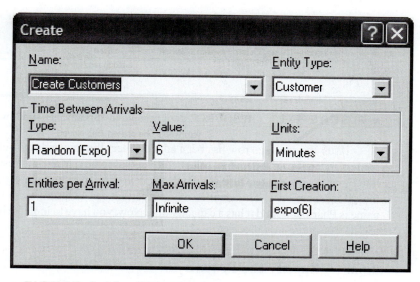

■ **FIGURE 1-11   CREATE module**

the first event for a Poisson arrival process is exponentially distributed. Hence, expo(6) has been used with the appropriate mean time, where *expo(mean)* is a function in Arena™ that generates random variables according to the exponential distribution function with the supplied mean.

Make sure that you specify the units for time as minutes and be sure to press OK; otherwise, your work in the module will be lost. In addition, as you build models you never know what could go wrong; therefore, you should save your models often as you proceed through the model building process. Take the opportunity to save your model before proceeding.

There is one last thing that you should do related to the entities before proceeding. Since the customers drive cars, you will change the picture associated with the customer entity that was defined in the CREATE module. To do this you will use the ENTITY module in the Basic Process panel. The ENTITY module is a data module. A data module cannot be dragged and dropped into the model window. Data modules require the model builder to enter information in either the spreadsheet window or in a dialog box. To see the dialog box, select the row from the spreadsheet view, and right-click and choose the edit via dialog box option. You will use the spreadsheet view here. Select the ENTITY module in the Basic Process panel and use the corresponding spreadsheet view to select the picture for the entity as shown in Figure 1-12.

1.6.5   Specify the Resources

Go to the Basic Process Panel and select the RESOURCE module. This module is also a data module. As you selected the RESOURCE module, the spreadsheet window should have changed to reflect this selection. Double-click on the row in the spreadsheet view of the RESOURCE module to add a resource to the model (Figure 1-13).

Entity - Basic Process					
	Entity Type	Initial Picture	Holding Cost / Hour	Initial VA Cost	Initial NVA Cost
1	Customer	Picture.Van	0.0	0.0	0.0
	Double-click here to add a new row.				

■ **FIGURE 1-12   ENTITY module**

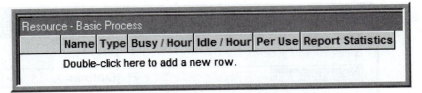

■ **FIGURE 1-13   RESOURCE data module in spreadsheet window**

After the resource row has been added, select the row and right-click, which accesses a context menu. From this menu, select edit via dialog box. Make the resulting dialog box look like Figure 1-14. You can also type in the same information that you typed in the dialog box in the spreadsheet view. This defines a resource that can be used in the model. The resource's name is Pharmacist and it has a capacity of 1. Now, you have to indicate how the customers will use this resource in a process.

1.6.6   Specify the Process

A process can be thought of as a set of activities experienced by an entity. There are two basic ways in which processing can occur: resource constrained and resource unconstrained. In the current situation, because there is only one pharmacist and the customer will need to wait when the pharmacist is busy, the situation is resource constrained. A PROCESS module provides for the basic modeling of processes in an Arena™ model. You specify this in a PROCESS module by having the entity "seize" the resource for the specified usage time. After the usage time has

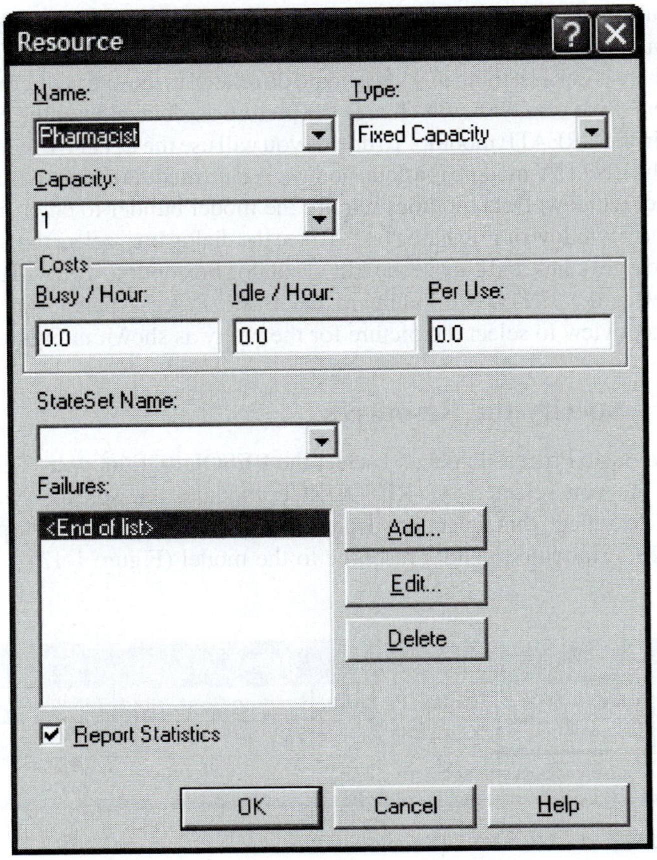

■ **FIGURE 1-14**
RESOURCE dialog box

elapsed, you specify that the resource is "released." The basic pseudocode can now be modified to capture this concept.

1. Create customers according to Poisson arrival process.
2. Process the customers through the pharmacy.

Seize the pharmacist.

Delay for the service.

Release the pharmacist.

3. Dispose of the customers as they leave the system.

Open up the PROCESS module and fill it out as indicated in Figure 1-15. Change the "Action" drop-down dialog box to the SEIZE, DELAY, RELEASE option. Use the "Add" button in the "Resources" area to indicate which resources the customer will use. In the pop up Resources dialog box, indicate the name of the resource and the number of units desired by the customer. Within the "Delay Type" area, choose Expression as the type of delay and type in expo(3) as the expression. This indicates that the delay, which represents the service time, will be randomly distributed according to an exponential distribution with a mean of 3 minutes. Make sure to change the units accordingly. Now you are ready to specify how to run the model.

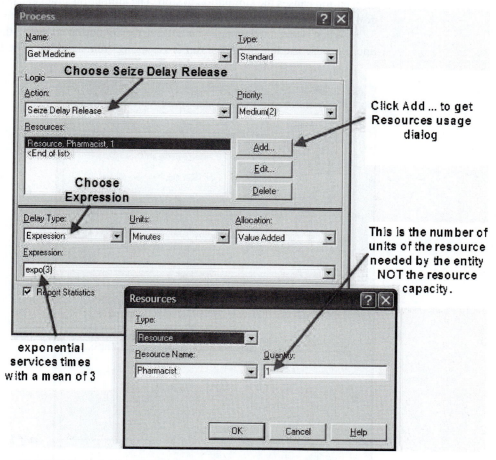

■ **FIGURE 1-15** **PROCESS module SEIZE, DELAY, RELEASE option**

1.6.7   Specify Run Parameters

Let's assume that the pharmacy is open 24 hours a day, 7 days a week. In other words, it is always open. In addition, assume that the arrival process does not vary over time. Finally, assume that management is interested in understanding the long-term behavior of this system in terms of the average waiting time of customers, average number of customers, and utilization of the pharmacist.

To simulate this situation over time, you must specify how long to run the model. Ideally, since management is interested in long-run performance, you should run the model for an infinite amount of time to get long-term performance; however, you probably don't want to wait that long! For the sake of simplicity, assume that 10,000 hours of operation is long enough. Within the Arena™ environment, go to the Run menu item and choose Setup. After the Setup dialog box appears, select the Replication parameters tab and fill it out as shown in Figure 1-16. The "Replication Length" textbox specifies how long the simulation will run. Note that the base time units were changed to minutes. This ensures that information reported by the model is converted to minutes in the output reports.

Now the model is ready to be executed. You can use the Run menu to do this or you can use the convenient VCR-like run toolbar (Figure 1-17). The Run button causes the simulation to run until the stopping condition is met. The Fast Forward button runs the simulation without animation until the stopping condition is met. The Pause button suspends the simulation run. The Start Over button stops the run and starts the simulation again. The Stop button causes the simulation to stop. The animation slider causes the animation to slow down (to the left) or speed up (to the right).

When you run the model, you will see the animation related to the customers waiting in line. Because the length of this run is very long, you should use the fast forward button on

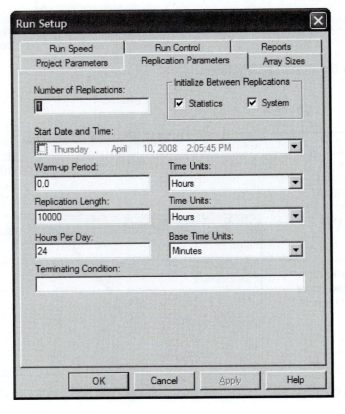

■ **FIGURE 1-16   Run setup replication parameters tab**

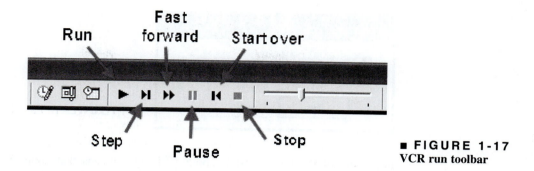

■ **FIGURE 1-17**
VCR run toolbar

the VCR run toolbar to execute the model to completion without the animation. Instead of using the fast forward button, you can significantly speed up the execution of your model by running the model in batch mode without animation. To do this, you can use the Run menu as shown in Figure 1-18; when you use the VCR run button thereafter, the model will run much faster without animation.

1.6.8   Analyze the Results

After running the model, you will see a dialog box indicating that the simulation run is completed and that the simulation results are ready (Figure 1-19). Answer "yes" to open up the report viewer. Arena™ writes the simulation output to a database and then uses Crystal Reports™ to prepare standardized reports.

In the reports preview area, you can drill down to see the statistics that you need. Go ahead and do this so that you see the report window as indicated in Figure 1-20. The reports indicate that customers wait about 3 minutes on average in the line. Arena™ reports the average of the waiting times in the queue as well as the associated 95% confidence interval half-width. In addition, Arena™ reports the minimum and maximum of the observed values. These results indicate that the maximum a customer waited to get service was 16 minutes. The

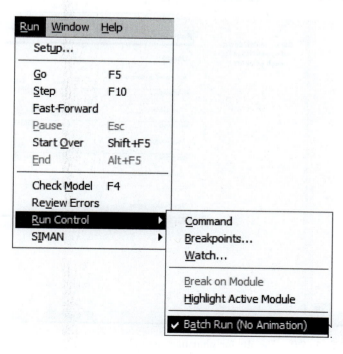

■ **FIGURE 1-18**   **Batch run no animation option**

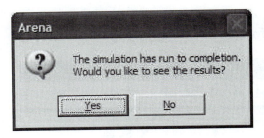

■ **FIGURE 1-19** Run completion dialog box

utilization of the pharmacist is about 50%. This means that about 50% of the time the pharmacist was busy. For this type of system, this is probably not a bad utilization, considering that the pharmacist probably has other in-store duties. The reports also indicate that there was less than one customer on average waiting for service. Using the Reports panel in the Project Bar, you can also select specific reports.

Arena™ also produces a text-based version of these reports in the directory associated with the model. In the Windows File Explorer, select the *modelname.out* file (see Figure 1-21). This file can be read with any text editor such as Notepad (Figure 1-22). Also, as indicated in Figure 1-21, Arena™ generates additional files when you run the model. The *modelname.mdb* file is a Microsoft Access™ database that holds the information displayed by Crystal Reports™. The *modelname.p* file is generated when the model is checked or run. If you have errors or

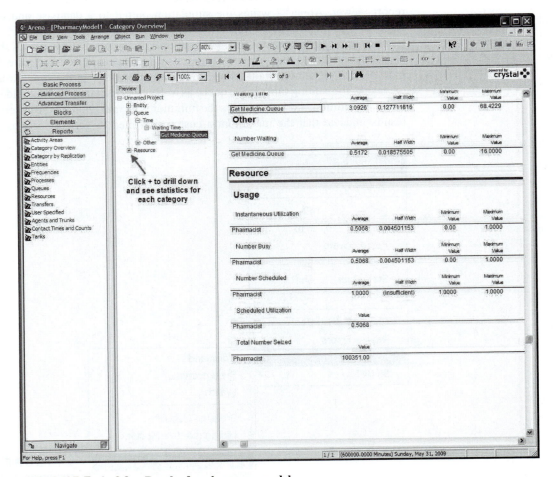

■ FIGURE 1-20   Results for pharmacy model

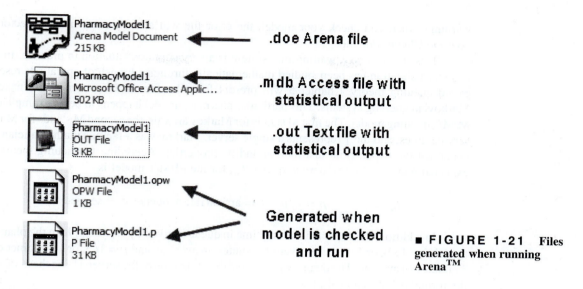

■ FIGURE 1-21 Files generated when running Arena™

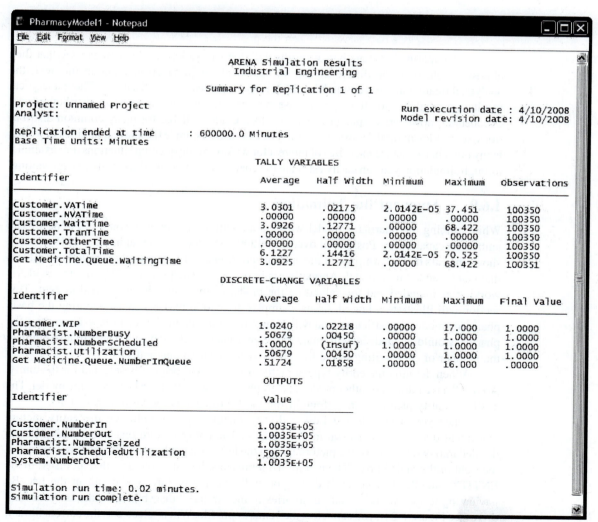

■ FIGURE 1-22   Text file summary results

warnings when you check your model, the error file will also show up in the directory of your *modelname.doe* file

This single-server, waiting-line system is a very common situation in practice. In fact, this exact situation has been studied mathematically through a branch of operations research called queuing theory. Chapter 5 will present formulas for the long-term performance of Markovian queuing systems. This particular pharmacy model happens to be an example of an M/M/1 queuing model. The first M stands for Markov arrivals, the second M stands for Markov service times, and the 1 represents a single server. Markov was a famous mathematician who examined the exponential distribution and its properties. According to queuing theory, the expected number of customers in queue, L_q, for the M/M/1 model is

$$L_q = \frac{\rho^2}{1 - \rho} \quad \rho = \lambda/\mu \quad \lambda = \text{arrival rate to queue}; \ \mu = \text{service rate}$$

In addition, the expected waiting time in queue is given by $W_q = L_q/\lambda$. In the pharmacy model, $\lambda = 1/6$, or 1 customer every 6 minutes on average, and $\mu = 1/3$, or 1 customer every 3 minutes on average. The quantity ρ, is called the utilization of the server. Using these values in the formulas for L_q and W_q results in

$$\rho = 0.5 \quad L_q = \frac{0.5 \times 0.5}{1 - 0.5} = 0.5 \quad W_q = \frac{0.5}{1/6} = 3 \text{ minutes}$$

In comparing these analytical results with the simulation results, you can see that they almost match, with the difference due to statistical sampling error. Later in this text, the analytical treatment of queues and the simulation of queues will be discussed. These analytical results are available for this special case because the arrival and service distributions are exponential; however, simple analytical results are not available for many common distributions, such as log normal. With simulation, you can easily estimate the above quantities as well as many other performance measures of interest for wide-ranging queuing situations. For example, through simulation you can easily estimate the chance that there are three or more cars waiting.

1.6.9 Augment the Animation

While running the pharmacy model with the animation on, you saw some of Arena's basic animation capabilities. Positioned over the PROCESS module is an animation queue, which shows the cars that are waiting for the pharmacist. Arena™ has many other animation features that can be added to a model. Animation is important for helping to verify that the model is working as intended and for validating the model's representation of the real system. This section will illustrate a few of Arena's simpler animation features by augmenting the basic pharmacy model. In particular, you will change the entity picture from a van to a car, draw the pharmacy building, and show the state of the pharmacist (idle or busy). You will also animate the number of cars waiting in the line (both visually and numerically).

When defining the entity type, a picture of a van was originally selected for the entity. Arena™ has a number of other predefined graphical pictures that can be used in the model. The available entity pictures can be found from the Edit menu as shown in Figure 1-23.

Once you have selected Edit > Entity Pictures, you should see the entity picture placement dialog box as shown in Figure 1-24. The entity picture placement dialog box is divided into two functions: the picture list (on the left of dialog box) and the picture library (on the right of the dialog box). The picture list represents the entity pictures that are listed in the ENTITY module. If you scroll down you will find the picture of the van in the list. By navigating to the Arena™ picture libraries in the Arena™ distribution folder on your hard drive (see Figure 1-24), you can select the *Vehicles.plb* file. When selected as the current library, your entity picture placement dialog box should look as shown in Figure 1-25.

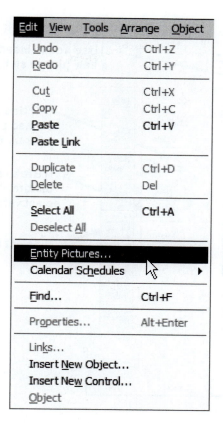

■ **FIGURE 1-23**   Accessing entity pictures

Scroll down in the vehicle library and select the red car, and then select the << button to move the picture to the picture list. Finally, you should give the picture a name in the list (e.g., Red Car). Now, go back to the ENTITY module and associate the picture with your entity. The name does not appear on the Entity module drop-down list, but the picture is available. Just type in the correct name in the Entity module to have the entity picture represented by the red car.

Now you will use Arena's drawing tools to draw the outline of the pharmacy building. Arena™ has standard drawing tools (e.g., lines, rectangles, polygons, etc.) as well as ways to add text, set fill options, and so on. These tools are available through the Drawing toolbar as shown in Figure 1-26. You can also turn on the drawing grid and the ruler, which are useful in placing the drawing objects. In this situation, drawing will be kept to a rather simple/crude representation of the pharmacy. Essentially, the pharmacy will be represented as a box, the drive-through lane as a line with the road line pattern, and the pass-through service window as two lines with 3-point thickness.

Figure 1-27 illustrates the background drawing for the pharmacy. The Arena™ drawing editor works like most computer-based drawing editors. You should just drag and drop the items, and set the fill options, thickness, and line pattern as you go. This portion of this example is available in the *PharmacyModel1WithAnimationS1.doe* file.

In the next part of the example, you will animate the resource so that you can see when the pharmacist is busy or idle, provide a location to show the customer currently being served, and better represent the waiting cars in the drawing.

In Arena™ a resource can be in one of four default states (idle, busy, inactive, and failed). The inactive and failed states will be discussed later in the text. The idle state represents the situation that at least 1 unit of the resource is available. In the pharmacy model, there is only one pharmacist, so if the pharmacist is not serving the customer, the pharmacist is considered idle. The busy state represents the situation where all available units of the resource are currently

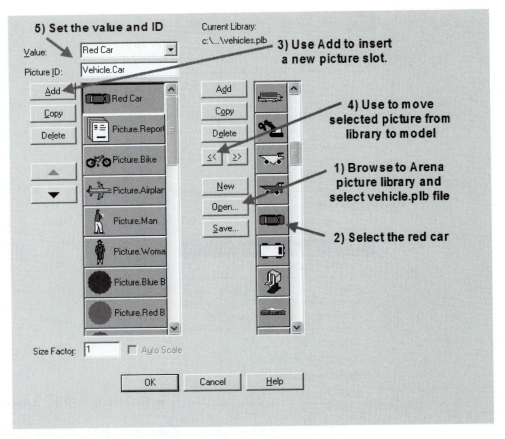

■ **FIGURE 1-24** Entity picture placement dialog

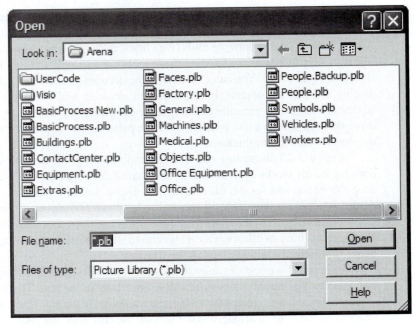

■ **FIGURE 1-25** Arena picture library files

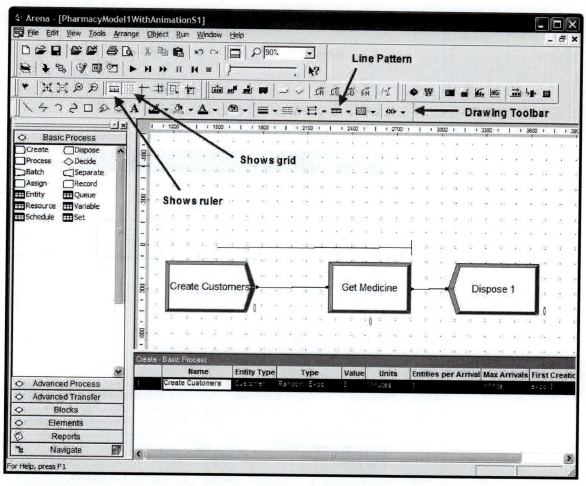

■ **FIGURE 1-26**   Basic drawing tools in Arena™

seized. In the case of the pharmacy model, since there is only 1 resource unit (the pharmacist), the resource is busy when the sole pharmacist is seized. In the animation, you will visibly represent the state of the pharmacist. This can be done through Arena's animate toolbar and in particular with the animate resource button.

1. Click the Resource button on the Animate toolbar.
2. The Resource Placement dialog box appears. Use the Open button to navigate to the *people.plb* picture library (e.g., *C:\Program Files\Rockwell Software\Arena*).
3. Select the picture icon of the overhead view of a person from the library's list of pictures (see Figure 1-28), and then select the Copy button. This will cause a copy of the picture to be made in the library.
4. After copying the picture, double-click on the picture. This will put you in the resource editor, which is similar to a drawing editor. For simplicity's sake, just select the picture and change the fill color to green. If you want to make it look just like the picture in the text, you need to ungroup the picture elements and change their appearance as necessary.
5. Assign the white version of the person icon to the idle state and the green picture that was just made to the busy state. To change the idle picture, click the Idle button in the table on

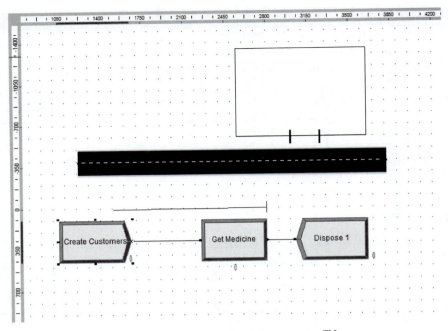

■ **FIGURE 1-27** Simple pharmacy drawing in Arena™

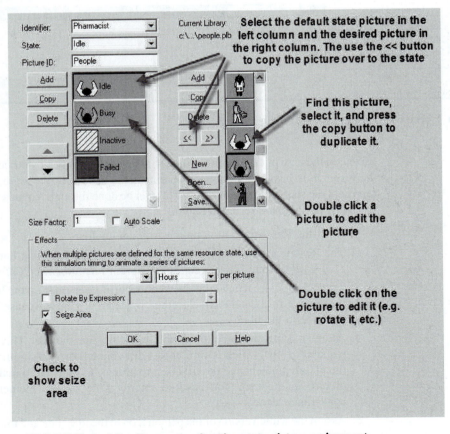

■ **FIGURE 1-28** Resource animation state picture assignment

the left. From the picture library table on the right, select the picture of the white overhead-view person picture. Click the Transfer button between the tables to use the picture for the Idle resource state. To change the busy picture, click the Busy button in the table on the left. Select from the picture library table on the right, the picture of the green worker that was just made. Click the Transfer button between the tables to use the selected picture to represent the pharmacist's busy picture icon.

6. Now you need to rotate the picture so that the person is facing down. Double-click on the new Idle resource picture to open the resource editor. Then, select the whole picture and choose Arrange > Rotate in order to rotate the picture so that the pharmacist is facing down. Close the editor and do this for the Busy state as well.

7. Click on the Seize Area check box. This will be discussed shortly.

8. Click OK to close the dialog box. (All other fields can be left with default values.) Arena™ will ask about whether you want to save the changes to the picture library. If you want to keep the green overhead-view person for use in later models, answer "yes" during the save process. The cursor will appear as a cross-hair. Move it to the model window and click to place the pharmacist resource animation picture in the pharmacy.

Now that the pharmacist is placed in the pharmacy, it would be nice to show the pharmacist serving the current customer. This can be done by using the seize area for a resource. The seize area animates the entities that have seized the resource. By default, the seize area is not shown on the animation of a resource. The checking of the Seize Area check box allows the seize area to be visible. After placing the resource animation icon, zoom in on the icon and note a small double circle, which is the seize area. Select the seize area and drag it so that it is positioned in the road adjacent to the pharmacy. Now you need to place the queue of waiting cars on the road. Select the animation queue on top of the Get Medicine PROCESS module and delete it. Go to the Queue button on the Animate toolbar and select it. This will allow you to create a new queue animation element that is not physically attached to the Get Medicine PROCESS module and place it in the road. Fill in the Queue animation dialog box as shown in Figure 1-29. Once you press OK, the cursor will turn in to a cross-hair and allow you to place the animation queue.

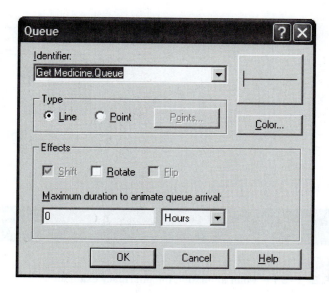

■ **FIGURE 1-29** Queue animation dialog box

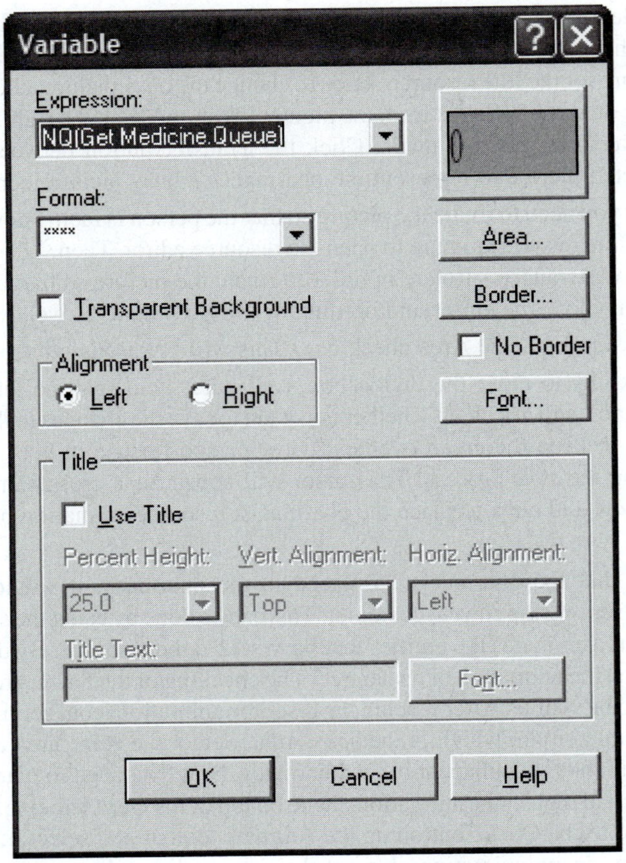

■ FIGURE 1-30
Variable animation dialog box

The final piece of animation that you will perform is to show the current number of waiting cars in the pharmacy. Select the Variable button on the Animate toolbar and fill out the resulting dialog box as per Figure 1-30. The cursor should turn to cross-hairs, and you should place the variable animation inside the pharmacy as shown in Figure 1-31.

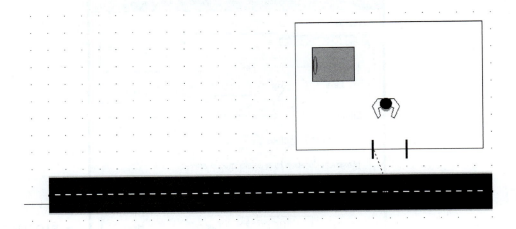

■ FIGURE 1-31   Final animation for pharmacy example

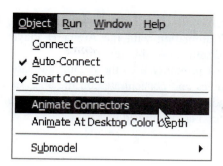

■ **FIGURE 1-32**   Disabling connector animation

The last thing to do before running the animation will be to turn off the animation associated with the flow chart connectors. This will stop the animation of the customers in the flow chart modules, and thus reduce distractions. The Object menu contains the option to disable/enable the animation of the connectors as shown in Figure 1-32.

The final version of the animated pharmacy model is available in the file, *Pharmacy-Modle1WithAnimationS2.doe*. If you run the model with animation on, you will see the pharmacist indicated as busy or idle, the car currently in service, any cars lined up waiting, and the current number of cars waiting as part of the animation.

Animation is an important part of the simulation process. It is extremely useful for demonstrating the model to decision makers and experts. With this example, you have actually learned a great deal about animating an Arena™ model. One of the standard practices is to rely on the default animation that is available with the flow chart modules. Then, when animation is important for validation and for convincing decision makers, the simulation analyst can devote time to making the animation more compelling. Often, the animation is developed in an area of the model that can be easily accessed via the Navigate bar. Many detailed examples of what is possible with Arena™ animation come as demo models with the Arena™ distribution. For example, consider exploring the models in the Arena™ Examples folder (e.g., *Flexible Manufacturing.doe*) (Figure 1-33). If you find that the Arena™ drawing tools are limiting, then you can use programs such as Visio™ or AutoCAD™ to make the drawings and import them as background for your Arena™ model. For more information on this process, conduct a search on importing Visio

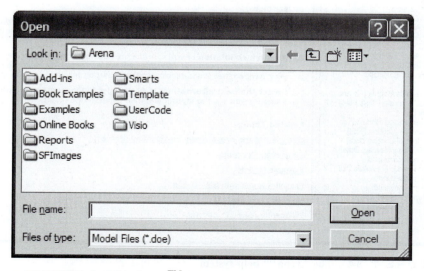

■ **FIGURE 1-33**   Arena™ 9.0 folders

drawings or importing AutoCAD drawings in Arena's help system. Arena™ also has the capability of developing 3D animation with the Arena3D Player. The 3D player is a separate product and is not discussed in this text.

The first part of this text primarily uses the default animation that comes with the placement of flow chart modules and concentrates on the analysis of simulation models; however, Chapter 6 presents more on the topic of animation, when material handling constructs (e.g., conveyors and transporters) are examined.

1.7   Getting Help in Arena™

Arena™ has an extensive help system. Each module's dialog box has a Help button, which will bring up help specific to that module. The help files use hyperlinks to important information. In fact, as can be seen in Figure 1-34, Arena™ has an overview of the modeling process as part of the help system. The example files are especially useful for getting a feel for what is possible with

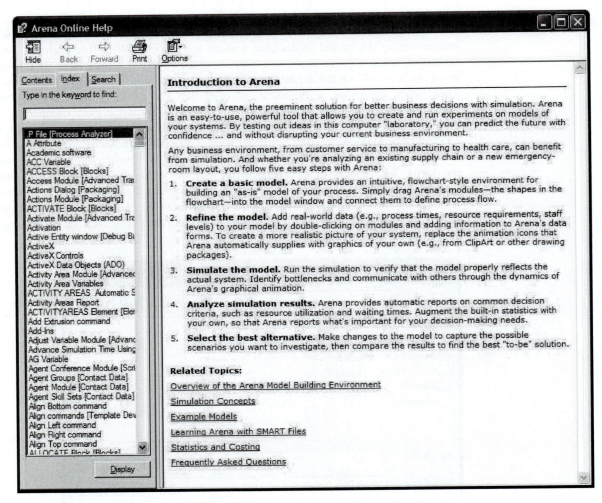

■ **FIGURE 1-34**   The Arena™ help system

Arena™. This text will use a number of Arena's example models to illustrate various concepts. For example, the Smarts file folder (see Figure 1-33) has small Arena™ models that illustrate the use of particular modules. You are encouraged to explore and study the Arena™ Help system. In addition to the online help system, the Arena™ Environment comes with user manuals in the form of PDF documents, which include an introductory tutorial in the "Arena™ User's Guide" in the Online Books folder.

This section introduced the Arena™ Environment. Using this environment, you can build simulation models using a drag-and-drop construction process. The Arena™ Environment facilitates the model building process, the model running process, and the output analysis process. Arena™ has many other facets that have yet to be touched upon. Not only does Arena™ allow the user to build and analyze simulation models, it also makes experimentation easier with the Process Analyzer, which will automatically run multiple simulations in a single batch run, and allow the comparison of these scenarios. In addition, Arena's Input Analyzer helps the user to build models for the probability distributions used in a model. Finally, through its integration with OptQuest™, Arena™ can assist in finding optimal design configurations via simulation. When using these tools, you will need to follow a methodology, which is examined in the next section.

1.8   Simulation Methodology

This section presents a brief overview of the steps of simulation modeling by discussing the process within the context of a methodology. A methodology is simply a series of steps to follow. Since simulation involves systems modeling, a simulation methodology based on the general precepts of solving a problem through systems analysis is presented here. A general methodology for solving problems can be stated as follows.

1. **D**efine the problem.
2. **E**stablish measures of performance for evaluation.
3. **G**enerate alternative solutions.
4. **R**ank alternative solutions.
5. **E**valuate and iterate as necessary.
6. **E**xecute and evaluate the solution.

The first step in the DEGREE methodology helps to ensure that you are solving the right problem. The second step helps to ensure that you are solving the problem for the right reason; that is, your metrics must be coherent with your problem. The next two steps ensure that you look at and evaluate multiple solutions to the problem. In other words, these steps help to ensure that you develop the right solution to the problem. A good methodology recognizes that the analyst needs to evaluate the method itself. In step 5, the analyst evaluates how the process is proceeding and allows for iteration. Although an important concept, iteration is alien to many modelers. Iteration recognizes that the problem-solving process can be repeated until the desired *degree* of modeling fidelity has been achieved. Start the modeling at a level that allows it to be initiated and do not try to address the entire situation in each of the steps. Start with small models that work and build them up until you have reached your desired goals. It is important to get started and get something established at each step, and then continually repeat the steps until the model is representing reality in the way that you intended. The final step is often overlooked. Simulation is often used to recommend a solution to a problem. This step indicates that if you have the opportunity, you should execute the solution by implementing the decisions. Finally, you should always follow up to ensure that the projected benefits of the solution were obtained.

The DEGREE problem-solving methodology should serve you well; however, simulation involves certain unique actions that must be performed during the general overall problem-solving process. When applying DEGREE to a problem that may require simulation, the general DEGREE approach needs to be modified to explicitly consider how simulation will interact with the overall problem-solving process.

Figure 1-35 represents a refined general methodology for applying simulation to problem solving:

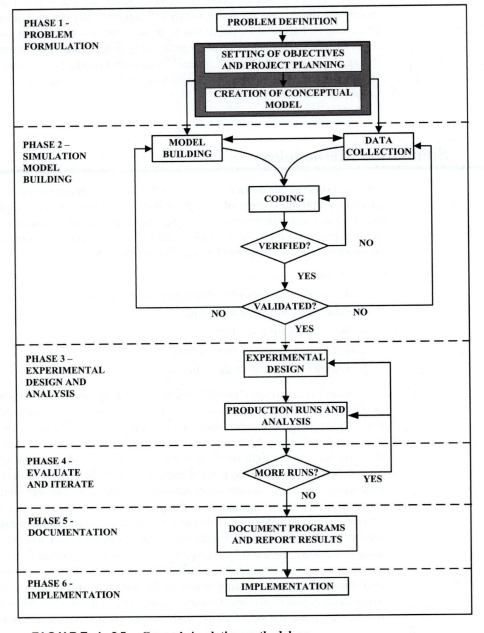

■ **FIGURE 1-35**   **General simulation methodology**

The first phase, *problem formulation*, captures the essence of the first two steps in the DEGREE process. The second phase, *model building*, captures the essence of step 3 of the DEGREE process. When building models, you are either explicitly or implicitly developing certain design alternatives. The third phase, *experimental design and analysis*, encapsulates some of steps 3 and 4 of the DEGREE process. In designing experiments, design alternatives are specified, and when analyzing experiments their worth is being evaluated with respect to problem objectives. The fourth phase, *evaluate and iterate*, captures the notion of iteration. Finally, the fifth and sixth phases, *documentation* and *implementation* complete the simulation process. Documentation is essential to ensure the ongoing and future use of the simulation model, and implementation recognizes that simulation projects often fail in the absence of follow-through on the recommended solutions.

The problem formulation phase of the study consists of five primary activities.

1. Defining the problem.
2. Defining the system.
3. Establishing performance metrics.
4. Building conceptual models.
5. Documenting modeling assumptions.

A problem starts with a perceived need. These activities are useful in developing an appreciation for and an understanding of the problem requirements. The basic output of the problem definition activity is a *problem definition statement*. A problem definition statement is a narrative discussion of the problem. A problem definition statement is necessary to accurately and concisely represent the problem for the analyst and for the problem stakeholders. The problem definition statement should include all the required assumptions made during the modeling process. It is important to document your assumptions so that you can examine their effect on the model during the verification, validation, and experimental analysis steps of the methodology. A well-written problem definition statement ensures that the problem is well understood and that all parties agree on the nature of the problem and the goals of the study.

The general goals of a simulation study often include

Comparison of system alternatives and their performance measures across various factors (decision variables) with respect to some objectives.

Optimization, a special case of comparison in which you try to find the system configuration that optimizes performance subject to constraints.

Prediction of system behavior at some future point in time.

Investigation to learn about and gain insight into the behavior of the system given various inputs.

These general goals will need to be specialized to the problem under study. The problem definition should include a detailed description of the objectives of the study, the desired outputs from the model, and the types of scenarios to be examined or decisions to be made.

The second activity of this phase produces a definition of the system. A *system definition statement* is necessary to accurately and concisely define the system, particularly its boundaries. The system definition statement is a narrative and often contains a pictorial representation of the major elements of the system. This ensures that the simulation study is focused on the appropriate areas of interest to the stakeholders and that the scope of the project is well understood.

When defining the problem and the system, one should naturally begin to develop an understanding of how to measure system performance. The third activity of problem formulation makes this explicit by encouraging the analyst to define the required performance measures for the model. To meaningfully compare alternative scenarios, objective and measurable metrics describing the performance of the system are necessary. The performance metrics should include quantitative statistical measures from any models used in the analysis (e.g., simulation models), quantitative measures from the systems analysis, (e.g., cost/benefits), and qualitative assessments (e.g., technical feasibility, human operational feasibility). The focus should be placed on the performance measures that are considered to be the most important to system decision makers and tied directly to the objectives of the simulation study. Evaluation of alternatives can then proceed in an objective and unbiased manner to determine which system scenario performs the best according to decision-maker preferences.

The problem definition statement, the system definition statement, and explicit performance metrics set the stage for more detailed modeling. These activities should be captured in written form. Within this text, you will develop models of certain "ready-made" book problems. One way to accomplish the problem formulation phase of a simulation study is to consider writing yourself a "book problem." You will need enough detail in these documents that a simulation analyst (you) can develop a model in *any* simulation language for the given situation. The example problem in Chapter 8 represents an excellent sample of problem and system definition statements. If you have the opportunity to do a "real-life" project as part of your study of simulation, you might want to use the book problems in this text and the example in Chapter 8 to learn how to write reasonable problem/system definition statements.

With a good understanding of the problem and of the system under study, you should be ready to begin your detailed model formulations. Model formulation does not mean an Arena™ program. You should instead use conceptual modeling tools, including conceptual diagrams, flow charts, and so on, before using software to implement a model. The purpose of conceptual modeling tools is to convey a more detailed system description so that the model may be translated into a computer representation. General descriptions help to highlight the areas and processes of the system that the model will simulate. Detailed descriptions assist in simulation model development and coding efforts. Some relevant diagramming constructs include

Context diagrams: A context diagram assists in conveying the general system description. The diagram is a pictorial representation of the system that often includes typically encountered flow patterns. Context diagrams are often part of the system description document. Figure 1-8 shows a simple example of a context diagram. There are no rules for developing context diagrams. If you have an artistic side, here is your opportunity to shine!

Activity diagrams: An activity diagram is a pictorial representation of the process for an entity and its interaction with resources while in the system. If the entity is temporary (i.e., it flows through the system), the activity diagram is called an activity flow diagram. If the entity is permanent (i.e., it remains in the system throughout its life), the activity diagram is called an activity cycle diagram. Activity diagrams will be used extensively in this text.

Software engineering diagrams: Because simulation entails software development, the wide variety of software engineering diagramming techniques can be utilized to provide information for the model builder. Diagrams such as flow charts, database diagrams, IDEF (ICAM DEFinition language) diagrams, UML (unified modeling language) diagrams, and state charts are useful in documenting complex modeling situations. These techniques assist development and coding efforts by focusing attention on describing, and thus understanding, the elements in the system. In this text, activity diagrams will be augmented with some simple flow chart symbols, and some simple state diagrams will be used to illustrate various concepts.

In your modeling, you should start with an easy conceptual model that captures the basic aspects and behaviors of the system. Then, you should begin to add details and consider additional functionality. Finally, you should always remember that the complexity of the model has to remain proportional to the quality of the available data and the degree of validity necessary to meet the objectives of the study. In other words, don't try to model the world!

After developing a solid conceptual model of the situation, simulation model building can begin. During the simulation model-building phase, alternative system design configurations are developed based on the previously developed conceptual models. Additional project planning is also performed to yield specifications for the equipment, resources, and timing required for the development of the simulation models. The simulation models used to evaluate the alternative solutions are then developed, verified, validated, and prepared for analysis. Within the context of a simulation project this process includes

Input data preparation: Input data is analyzed to determine the nature of the data and to determine further data collection needs. Necessary data are also classified by area. This classification establishes the different aspects of the model that are used in model development.

Model Translation: The act of implementing the model in computer code, including timing and general procedures and the translation of the conceptual models into computer simulation program representations.

Verification: Verification of the computer simulation model is performed to determine whether the program performs as intended. Verification consists of model debugging to locate any errors in the simulation code. Errors of particular importance include improper flow control or entity creation, failure to release resources, and logical/ arithmetic errors or incorrectly observed statistics. Model debugging also includes scenario repetition using identical random number seeds, "stressing" the model through a sensitivity analysis (varying factors and their levels) to ensure compliance with anticipated behavior, and testing of individual modules in the simulation code.

Validation: Validation of the simulation model is performed to determine whether the simulation model adequately represents the real system. The simulation model is shown to personnel (of various levels) associated with the system in question. Their input concerning model realism is critical in establishing the validity of the simulation. In addition, further observations of the system are performed to ensure model validity with respect to actual system performance. A simple technique is to statistically compare the output of the simulation model to the output from the real system and to analyze whether there is a significant (and practical) difference between the two.

Modeling input parameters will be illustrated in Chapter 3. Model translation will be a large component of each chapter as you learn how to develop Arena™ models. Verification techniques will be discussed briefly in Chapter 2 and again in Chapter 8. Model validation will not be a major component of this text, primarily because the models will be examples made for educational purposes. This does not mean that you should ignore this important topic. You are encouraged to examine references on validation, such as Balci (1997, 1998).

After you are confident that your model has been verified and validated to suit your purposes, you can begin to use the model to perform experiments that investigate the goals and objectives of the project. Preliminary simulation experiments should be performed to set the statistical parameters associated with the main experimental study. These techniques are discussed in Chapter 4. The experimental method should use the simulation model to generate benchmark statistics of current system operations. The simulation model is then altered to conform to a potential scenario and is re-run to generate comparative statistics. This process is continued, cycling through suggested scenarios and generating comparative statistics to allow evaluation of alternative solutions. In this manner, objective assessments of alternative scenarios can be made.

For a small set of alternatives, this "one at a time" approach is reasonable; however, often there are a significant number of design factors that can affect the performance of the model. In this situation, the analyst should consider utilizing formal experimental design techniques. This step should include a detailed specification of the experimental design (e.g., factorial) and any advanced output analysis techniques (e.g., batching, initialization bias prevention, variance reduction techniques, multiple comparison procedures) that may be required during the execution of the experiments. During this step of the process, any quantitative models developed during the previous steps are exercised. Within the context of a simulation project, the computer simulation model is exercised at each of the design points in the stipulated experimental design.

Using the criteria specified by system decision makers, and using the simulation model's statistical results, alternative scenarios should then be analyzed and ranked. A methodology should be used to allow the comparison of the scenarios that have multiple performance measures that trade off against each other.

If you are satisfied that the simulation has achieved your objectives, then you should document and implement the recommended solutions. If not, you can iterate as necessary and determine whether any additional data, models, experimentation, or analysis is needed to achieve your modeling objectives. Good documentation should consist of at least two parts: a technical manual, which can be used by the same analyst or by other analysts, and a user manual. A good technical manual is very useful when the project has to be modified, and it can be a very important contribution to software reusability and portability. The approach to documenting the example models in this text can be used as an example for how to document your models. In addition to good model development documentation, often the simulation model will be used by nonanalysts. In this situation, a good user manual for how to use and exercise the model is imperative. The user manual is a product for the user who may not be an expert in programming or simulation issues; therefore, clearness and simplicity should be its main characteristics.

Given the goals of the study, the analyst should develop plans to implement the recommended solutions and follow through with the installation and integration of the solutions. After implementation, the project should be evaluated as to whether the proposed solution met the intended objectives.

1.9   Organization of the Book

This chapter introduced basic concepts of simulation. Chapter 2 will further your study of Arena™ as a tool for developing simulation programs by discussing some of the basic modeling constructs available in Arena™. With certain basic simulation and programming concepts mastered in Chapters 1 and 2, the next step is to investigate how randomness is incorporated into simulation. Chapter 3 presents general methods for random number and random variable generation, and discusses how Arena™ enables the incorporation of probabilistic models into simulations. Whenever a simulation uses probability models for input parameters, it will also produce output that is random. Chapter 4 examines some of the statistical issues involved in analyzing simulation output data. You will also see how Arena™ makes this analysis easier for the user.

The first four chapters comprise a foundation for developing system modeling skills. Chapter 5 then examines some common modeling situations involving queues and inventory systems. The modeling of situations involving material handling systems is presented in Chapter 6. Next, Chapter 7 discusses some miscellaneous topics in simulation modeling and examines some of Arena's programming capabilities. The final chapter illustrates the use of simulation on a practical case study. Through this study you will have a solid foundation for understanding what it takes to model more realistic systems found in practice.

Simulation is a tool that can assist analysts in improving system performance. There are many other aspects of simulation in addition to those in Arena^{TM} that will be considered in this text. I hope that you will find this a useful and interesting experience.

Exercises

1.1 Using the *Arena^{TM} User's Guide*, build an activity diagram of the home mortgage application model.

1.2 Using the *Arena^{TM} User's Guide*, build the home mortgage application model and perform the two modeling extensions suggested in the tutorial. Prepare in snapshots of your model building process (using the screen capture functionality of your computer (ctl-alt-printscreen)) for answering the questions posed by the tutorial's extensions. Summarize the differences between the model results in the form of a table. Describe the statistical significance of the results.

1.3 Using the formulas for the M/M/1 queuing model, compute the expected waiting time in queue and the expected number of customers in queue if the mean of the service distribution is 5 minutes rather than 3 minutes. What do you think of the performance in this situation? Is the utilization reasonable? Why or why not?

1.4 Using the drive-through pharmacy Arena^{TM} model, estimate the expected waiting time in the queue and the expected number of customers in the queue if the mean of the service distribution is 5 minutes rather than 3 minutes. How do your results compare to exercise 1.3?

1.5 Redo exercise 1.4, but instead of using an exponential distribution for the service time, assume that the service time is a constant 5 minutes. How do your results compare to those of exercises 1.3 and 1.4?

1.6 Recall the banking situation of Section 1.1. A simulation analyst observed the operation of the bank and recorded the information given in Table 1-1.

■ **TABLE 1-1**
Customer Arrival and Service Times

Customer number	Time of arrival	Service time
1	3	4
2	11	4
3	13	4
4	14	3
5	17	2
6	19	4
7	21	3
8	27	2
9	32	2
10	35	4
11	38	3
12	45	2
13	50	3
14	53	4
15	55	4

■ **TABLE 1-2**
Bank Re-Creation Table

(1)	(2)	(3)	(4)	(5)	(6)	(7)	(8)	(9)	(10)
Event time	Customer number	Event type	Arrival time	Start service time	Depart time	Time in system	Number in queue	Teller status	Teller idle time
0	—	—	—	—	—	—	0	I	—
3	1	A	3	3	7		0	B	3
7	1	D				4	0	I	
11	2	A	11	11	15		0	B	4
13	3	A	13	15	19		1	B	
14	4	A	14	19	22		2	B	

The information was recorded right after the bank opened during a period of time when only *one* teller was working. From this information, you would like to re-create the operation of the system. Table 1-2 presents the operation of the system for the first four customers. The row for time 0 indicates that there were no customers present and that the teller was idle. The second row marks the time of the first event, which happens to be the arrival of the first customer at time 3.0.

The columns of the table denote specifics for what happened at the given time. For example, from Table 1-1 we know that customer 1 arrived at time 3.0 and that the customer started service at time 3 (because the teller was idle). From Table 1-1 we also know that the first customer had a service time of 4 minutes. Thus, we can compute column (6), which indicates when the customer will depart. The time in the system, column (7), should be computed when the customer actually leaves the system. When the customer arrived at time 3, there were 0 customers in the queue, and the teller became busy as shown in columns (8) and (9). Since the teller became busy, we can compute how long the teller had been idle as shown in column (10). According to Table 1-1, there were no arrivals before time 7, which is when the first customer should depart. Thus, the third row indicates the next event as customer 1 departing. At this event, the system time in column (7) can be computed (time of departure − time of arrival = 7 − 3 = 4). In addition, since there are no customers in the queue, the teller's status becomes idle in column (9). Verify that you understand how rows for the event times of 11, 13, and 15 are tabulated. Complete the table and compute the average system times for the customers. What percentage of the total time was the teller idle?

1.7 Complete a table like Table 1-2 for all the data in Table 1-1.

(a) Compute the average time that customers spent in the system.

(b) Compute the percentage of time that the teller was idle.

(c) Compute the percentage of time that there were 0, 1, 2, and 3 customers in the queue.

1.8 Using the resources at http://www.informs-sim.org/wscpapers.html, find an application of simulation to a real system and discuss why simulation was important to the analysis.

1.9 Read Balci, O., Verification, validation, and accreditation, in Medeiros DJ, Watson EF, Carson JS, Manivannan MS, editors, Proceedings of the 1998 Winter Simulation Conference; IEEE Piscataway New York; 1998. p. 41.48. (http://www.informs-cs.org/wsc98papers/ 006.PDF), and discuss the difference between verification and validation. Why is the verification and validation of a simulation model important?

1.10 Customers arrive to a gas station with two pumps. Each pump can reasonably accommodate a total of two cars. If all the space for the cars is full, potential customers will balk (leave without getting gas). What measures of performance will be useful in evaluating the effectiveness of the gas station? Describe how you would collect the necessary interarrival and service times of the customers to simulate this system.

1.11 Categorize the following situations as either discrete or continuous (or a combination of both). In each situation, specify the objective of developing a simulation model.

(a) Calls come into a computer support center randomly over time. Each call is handled in first-come, first-served order. If the problem cannot be solved on the first contact, follow-up contacts are scheduled.

(b) A computer manufacturer supports on-line sales. A customer configures and places his/her order via a secured web portal.

(c) A native fish population is being threatened by a foreign predator species that was accidentally introduced. We are interested in understanding the dynamic interactions between the populations of the two species.

(d) The progress of a viral infection in the human body is being modeled to understand the time that it takes to cause specific symptoms.

CHAPTER 2

Basic Process Modeling

LEARNING OBJECTIVES

After completing this chapter, you should be able to:

- Define and explain the key elements of process-oriented simulations
- Identify entities and their attributes and use them within Arena[TM]
- Identify, define, and use variables within Arena[TM]
- Use Arena[TM] to perform basic input/output operations
- Program flow of control within an Arena[TM] model
- Understand the programming aspects of Arena[TM], including debugging
- Understand and model shared resources
- Understand and model the batching and separation of entities

Chapter 1 described a system as a set of interrelated components that work together to achieve common objectives. In this chapter, a method for modeling the operation of a system by describing its processes is presented. In the simplest sense, a process can be thought of as a sequence of activities. An activity is an element of the system that takes a particular time interval to complete. In the pharmacy example, the service of the customer by the pharmacist was an activity. The representation of the dynamic behavior of the system by describing the process flows of the entities moving through the system is called process-oriented modeling. When developing a simulation model using the process view, there are a number of terms and concepts that are often used. Before learning some of these concepts in more detail, it is important that you begin with an understanding of some of the vocabulary used in simulation. The following terms will be used throughout the text:

System: A set of interrelated components that act together over time to achieve common objectives.

Parameters: Model inputs that are unchanging properties of the system. These are typically quantities (variables) that are part of the environment which the modeler feels cannot be controlled or changed.

Variables: Quantities that are properties of the system (as a whole) that change or are determined by the relationships among system components as it evolves through time.

System State: Snapshot of the system at a particular point in time characterized by the values of the variables that are necessary for determining the future evolution of the system from the present time. The minimum set of variables considered necessary to describe the future evolution of the system is known as the system's state variables.

Entity: Object of interest in the system whose movement or operation within the system may cause events to occur.

Attribute: Property or variable associated with an entity.

Event: Instantaneous occurrence or action that changes the state of the system at a particular point in time.

Activity: Interval of time bounded by two events (start event and end event).

Resource: Limited quantity of items used (e.g., seized and released) by entities as they proceed through the system. A resource has a capacity that governs the total quantity of items that may be available. All the items in the resource are homogeneous, meaning that they are indistinguishable. If an entity attempts to seize a resource that does not have any units available, it may wait in a queue.

Queue: Location that holds entities when their movement is constrained within the system.

Future Event List: A list that contains the time-ordered sequence of events for the simulation.

When developing models, it will be useful to identify system elements that fit some of these definitions. Entities constitute an excellent place to develop an understanding of these concepts because process-oriented modeling is predicated on describing the life of an entity as it moves through the system.

2.1   Entities, Attributes, and Variables

When modeling a system, there are often many types of entities. For example, consider a retail store. Besides customers, the products might also be considered entities. The products are received by the store and "wait" on the shelves until customers select them for purchase. Entities may come in groups and then are processed individually or they might start out as individual units that are formed into groups. For instance, a truck arriving at the store may be an entity that consists of many pallets that contain products. The customers select the products from the shelves and during the check-out process the products are placed in bags. The customers then carry their bags to their cars. Entities are uniquely identifiable within the system. If there are two customers in the store, they can be distinguished by the values of their attributes. For example, considering a product as an entity, it may have serial number, weight, category, and price attributes. The set of attributes for a type of entity is called its "attribute set." While all products might have these attributes, the attributes do not necessarily have the same values. For example, consider the following two products:

(serial number = 12345, weight = 8 ounces, category = green beans, price = $0.87)

(serial number = 98765, weight = 8 ounces, category = corn, price = $1.12)

The products carry or retain these attributes and their values as they move through the system. In other words, attributes are attached to or associated with entities. The values of the attributes might change during the operation of the system. For example, a price mark-down for green beans might occur after a given period of time. Attributes can be thought of as variables that are attached to entities.

Not all information in a system is local to the entities. For example, the number of customers in the store, the number of carts, and the number of check-out lanes are all characteristics of the system. These types of data are called "system attributes." In simulation models, this information can be modeled with global variables or other data modules (e.g., resources) to which all entities have access. By making these quantities visible at the system level, the information can be shared between the entities within the model and between the components of the system.

Figure 2-1 illustrates the difference between global (system) variables and entities with their attributes in the context of a warehouse. In the figure, the trucks are entities with attributes: arrival time, type of product, amount of product, and load tracking number. Note that both of the trucks have these attributes, but each truck has different values for their attributes. The figure also illustrates examples of global variables, such as, number of trucks loading, number of trucks unloading, number of busy forklifts, and so on. This type of information belongs to the entire system.

Once a basic appreciation of system components is accomplished through identifying system variables, entities, and attributes, you must begin to understand the processes in the system. Developing the process description for the various types of entities in a system and connecting the flow of entities to the changes in state variables of the system is the essence of process-oriented modeling. In order for entities to flow through the model and experience processes, you must be able to create (and dispose of) the entities. The following section describes how Arena^{TM} allows the modeler to create and dispose of entities.

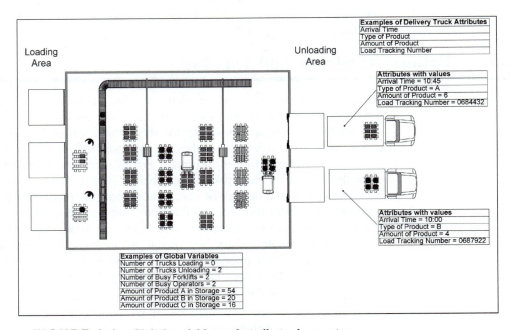

■ **FIGURE 2-1**   **Global variables and attributes in a system**

2.2   Creating and Disposing of Entities

The basic mechanism by which entities are introduced into an Arena™ model is the CREATE module. An entity is an object that flows in the model. As an entity flows through the model it causes the execution of each module through which it flows. Because of this, nothing will happen in an Arena™ model unless entities are created.[1] Entities can be used to represent instances of actual physical objects that move through the system. For example, in a model of a manufacturing system, entities that represent the parts that are produced by the system will need to be created. Sometimes entities do not have a physical representation in the system and are simply used to represent some sort of logical use. For example, an entity, whose sole purpose is to generate random numbers or to read data from an input file, may be created in a model. You will learn various ways to model with entities as you proceed through the text.

Figure 2-2 illustrates the CREATE module dialog box that opens when double-clicking on the flowchart symbol in the model window. Understanding what the dialog box entries mean is essential to writing Arena™ models. Each dialog box has a Help button associated with it. Clicking on this button will reveal help files that explain the basic functionality of the module. In addition, the text field prompts are explained in the help system.

According to the Arena™ help files, the basic dialog box entries are as follows:

Type: Type of arrival stream to be generated. Types include Random (uses an Exponential distribution, user specifies the mean of the distribution), Schedule (specifies a nonhomogeneous Poisson process with rates determined from the specified Schedule module), Constant (user specifies a constant value, e.g., 100), and Expression (pull-down list of various distributions or general expressions written by the user).

Value: Determines the mean of the exponential distribution (if Random is used) or the constant value (if Constant is used) for the time between arrivals. Applies only when Type is Random or Constant. Can be a general expression when the type is specified as Expression.

Entities per Arrival: Number of entities that will enter the system at a given time with each arrival. This allows for the modeling of batch arrivals.

Max Arrivals: Maximum number of arrival events permitted by the module. When this value is reached, the creation of new entities by the module ceases.

First Creation: Starting time for the first entity to arrive into the system. Does not apply when Type is Schedule.

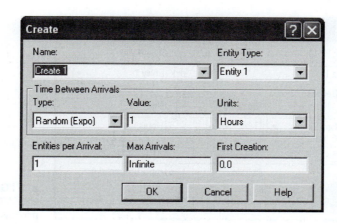

■ **FIGURE 2-2**
CREATE module flowchart symbol and dialog box

---

[1]Actually, this statement is not quite true and not quite false. You don't necessarily have to have a CREATE module in the model to get things to happen, but this requires the use of techniques that will be discussed in Chapter 7.

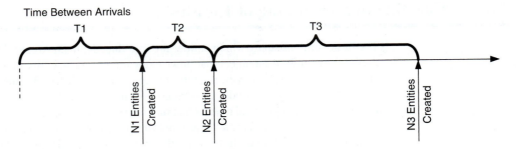

■ **FIGURE 2-3** Example arrival process

The CREATE module defines a repeating pattern for an arrival process. The time between arrivals is governed by the specification of the type of arrival stream. The time between arrivals specifies an ordered sequence of events in time at which entities are created and introduced to the model. At each event, a number of entities can be created. The number of entities created is governed by the Entities per Arrival field, which may be stochastic. The first arrival is governed by the specification of the First Creation time field in the dialog box, which also may be stochastic. The maximum number of arrival events is governed by the Max Arrivals entry. Figure 2-3 illustrates an arrival process where the time of the first arrival is given by T1, the time of the second arrival is given by T1 + T2, and the time of the third arrival is given by T1 + T2 + T3. In the figure, the number of arriving entities at each arriving event is given by N1, N2, and N3 respectively.

For example, to specify a Poisson arrival process with mean rate λ, you can use a CREATE module as shown in Figure 2-4. Why does this specify a Poisson arrival process? Because the time between arrivals for a Poisson process with rate λ is exponentially distributed with the mean of the exponential distribution being $1/\lambda$.

To specify a compound arrival process, use the "Entities per Arrival" field. For example, suppose you have a compound Poisson process[2] where the distribution for the number created at each arrival is governed by a discrete distribution:

$$P(X = x) = \begin{cases} 0.2 & x = 1 \\ 0.3 & x = 2 \\ 0.5 & x = 3 \end{cases}$$

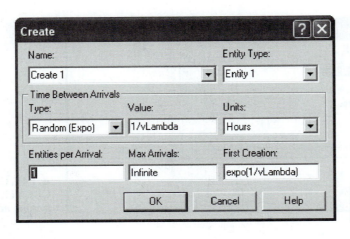

■ **FIGURE 2-4**
CREATE module for Poisson process

[2]See Ross (1997) for more information on the theory of compound Poisson processes.

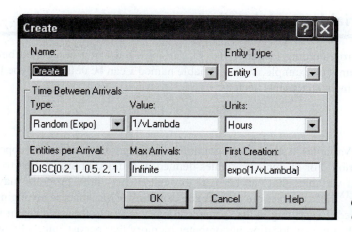

■ **FIGURE 2-5**
CREATE module for compound Poisson process

Figure 2-5 shows the CREATE module for this compound Poisson process with the entities per arrival field using the discrete empirical distribution function, DISC(0.2, 1, 0.5, 2, 1.0, 3), in Arena™.

When developing models, it is often useful to create an entity and use it to trigger other actions in the model. For example, to read in information from a file, a "logical" entity can be created at time zero and used to execute a READ module. To create one and only one entity at time zero, specify the *Max Arrivals* field as one, the "First Creation" field as 0.0, the Entities per Arrival as 1, and the Type field as Constant. The value specified for the constant is really immaterial since only one entity will be created.

Each entity that is created allocates memory for the entity. Once the entity has traveled through its process in the model, the entity should be disposed. The DISPOSE module acts as a "sink" to dispose of entities when they are no longer needed in the model.

Figure 2-6 illustrates the dialog box for the DISPOSE module. The DISPOSE module indicates the number of entities that are disposed by the module in the animation. In addition, the default action is to record the entity statistics prior to disposing the entity. If the Entities Statistics Collection field is checked in the Project Parameters tab of the Run/Setup dialog box, the entity statistics will include VA Time (value added time), NVA Time (non-value added time), Wait Time, Transfer Time, Other Time and entity Total Time. If the Costing Statistics Collection field is also checked in the Project Parameters tab of the Run/Setup dialog box, additional entity statistics include VA Cost (value added cost), NVA Cost (non-value added cost), Wait Cost, Transfer Cost, Other Cost, and Total Cost. Additionally, the number of entities leaving the system (for a given entity type) and currently in the system (WIP, work in process) will be tabulated. Chapter 7 will describe the use of some of these attributes in the context of Arena's cost modeling system.

You now know how to introduce entities into a model and how to dispose of entities after you are done with them. To make use of the entities in the model, you must first understand how to define and use variables and entity attributes in an Arena™ model.

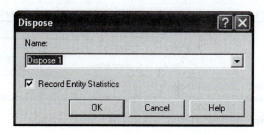

■ **FIGURE 2-6** **DISPOSE module dialog box**

2.3   Defining Variables and Attributes

In a programming language like C, variables must first be defined before using them in a program. For example, in C, a variable named x can be defined as type float and then used in assignment statements such as:

$$float\ x;$$
$$x = x + 2;$$

Variables in standard programming languages have a specific scope associated with their execution. For example, variables defined in a function are limited to use in that function. Global variables that can be used throughout the entire program may also be defined. In Arena™, all variables are global. You can think of variables as belonging to the system being modeled. Everything in the system can have access to the variables of the system. Variables provide a named location in computer memory that persists until changed during execution. The naming conventions for declaring variables in Arena™ are quite liberal. This can be both a curse and a blessing. Because of this, you should adopt a standard naming convention when defining variables. The models in this text try to start the name of all variables with the letter v. For instance, in the previously discussed CREATE module examples, the variable *vLambda* was used to represent the arrival rate of the Poisson process. This makes it easy to distinguish variables from other Arena™ constructs when reviewing, debugging, and documenting the model. In addition, the naming rules allow the use of spaces in a variable name, such as, "This is a legal name" (without the quotations) is a legally named variable in Arena™. Suppose you needed to count the number of parts and decided to name your variable *Part Count*. The use of spaces in this manner should be discouraged because woe unto you if you have this problem: *Part Count* and *Part Count*. Can you see the problem? The second instance of *Part Count* has two spaces between *Part* and *Count*. Try searching through every dialog box to find that in a model! According to the naming convention recommended here, you should name this variable *vPartCount*, concatenating and capitalizing the individual components of the name. Of course you are free to name your variables whatever you desire as long as the names do not conflict with already existing variables or keywords in Arena™. To learn more about the specialized variables available in Arena™, you should refer to the portable document file (PDF) document *Arena™ Variables Guide*, which comes with the Arena™ installation. Reading this document should give you a better feeling for the functional capabilities of Arena™.

To declare variables, you use the VARIABLE module found in the Basic Process template. The VARIABLE module is a data module and is accessed either through the spreadsheet view (Figure 2-7) or through a dialog box.

In Arena™, variables can be scalars or can be defined as arrays. The arrays can be either one-dimensional (1D) or two-dimensional (2D). For 1D arrays, you specify the number of rows. The net effect is that there will be three memory locations defined which can be indexed by the numbers 1, 2, and 3. When defining a 2D array of variables, you specify the number of

Variable - Basic Process						
	Name	**Rows**	**Columns**	**Data Type**	**Clear Option**	**Initial Values**	**Report Statistics**
1	vScalar ▼			Real	System	0 rows	☐
2	v1DArray	3		Real	System	3 rows	☐
3	v2DArray	2	5	Real	System	10 rows	☐
	Double-click here to add a new row.						

■ **FIGURE 2-7**   Spreadsheet view of VARIABLE data module

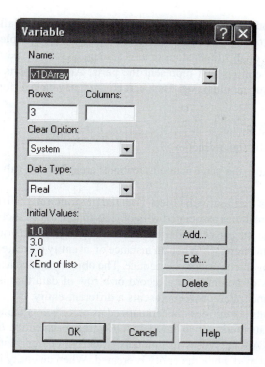

■ **FIGURE 2-8**   **Dialog box for defining variables**

rows and columns. Array indexes start at 1 and run through the size of the specified dimension. A runtime error will occur if the index used to access an array is not in the bounds of the array's dimensions.

When defining a variable or an array of variables, Arena^{TM} allows you to specify the initial value(s) (Figure 2-8). The default initial value for all variables is zero. By using the spreadsheet view in the Data window, see Figure 2-9, you can easily enter the initial values for variables and arrays. All variables are treated as real numbers in Arena^{TM}. If you need to represent an integer, such as *vPartCount*, you simply represent it in Arena^{TM} as a variable.

It is important to remember that all variables or arrays are global in Arena^{TM}. Variables are used to share information across all modules in the model. You should think of variables as characteristics or properties of the model as a whole. Variables belong to the entire system being modeled.

Initial Values

	1	2	3	4	5
1	1.1	1.2	3.1	9.0	99.9
2	10.5	4.3	0.0	2.9	27.4

Variable - Basic Process

	Name	Rows	Columns	Data Type	Clear Option	Initial Values	Report Statistics
1	vScalar			Real	System	0 rows	☐ **Click to**
2	v1DArray	3		Real	System	3 rows	☐ **enter**
3	v2DArray	2	5	Real	System	10 rows	☐ **values**

Double-click here to add a new row.

■ **FIGURE 2-9**   **Spreadsheet view for 2-D array**

Arena^{TM} has another data type that is capable of holding values called an "attribute." An attribute is a named property or characteristic of an entity. For example, suppose the time to produce a part depends on the dimensions or area associated with the part. A CREATE module can be used to create entities that represent the parts. After the parts are created, the area associated with each part should be set. Each part created may have different values for its area attribute.

Part 1: area = 5 square inches

Part 2: area = 10 square inches

Parts 1 and 2 have the same named attribute, area, but the value for this attribute is different.

For those readers familiar with object-oriented programming, attributes in Arena^{TM} are similar in concept to attributes associated with objects in languages such as VB.Net, Java, and C++. If you understand the use of attributes in those languages, then you have a basis for how attributes can be used in Arena^{TM}. When an instance of an entity is created in Arena^{TM}, it is like creating an object in an object-oriented language. The object instance has attributes associated with it. You can think of an entity as a record or a row of data that is associated with that particular object instance. Each row represents a different entity.

Table 2-1 presents six entities conceptualized as records in a table. The column *IDENT* represents the *IDENT* attribute, which uniquely *IDENT*ifies each entity. *IDENT* is a special pre-defined attribute that Arena^{TM} uses to uniquely track entities currently in the model. The *IDENT* attribute is assigned when the entity is created. When an entity is created in Arena^{TM}, memory is allocated for the entity and all of its attributes. A different value for *IDENT* is assigned to each entity that currently exists in the model. When an entity is disposed, the memory associated with that entity is released and the values for *IDENT* will be reused for newly created entities. Thus, the *IDENT* attribute uniquely identifies entities currently in the model. No two entities currently in the model have the same value for the *IDENT* attribute. You can think of the creating and disposing of entities as adding and deleting records in the model's "entity table." The *Entity.SerialNumber* attribute is also a unique number assigned to an entity when it is created; however, if the entity is ever duplicated (cloned) in the model, the clones will have the same value for the *Entity.SerialNumber* attribute. In the table, there are two entities (1 and 4) that are duplicates. The duplication of entities will be discussed later in this chapter. The size, weight, and processing time attributes are three user defined attributes associated with each of the entities. To find a description of the full list of predefined entity attributes, perform a search on "attributes and entity-related variables" in the Arena^{TM} Help system.

When you define a user-defined attribute, you are adding another "column" to the "entity table." The implication of this statement is very important. The defining of a user-defined attribute associates that attribute with every entity type. This is quite different than what

■ **TABLE 2-1**

Entities Conceptualized as Records

IDENT	Entity.SerialNumber	Size	Weight	Processing Time
1	1001	2	33	20
2	1002	3	22	55
3	1003	1	11	44
4	1001	2	33	20
5	1004	5	10	14
6	1005	4	14	10

■ **TABLE 2-2**
Different Types of Entities Conceptualized as Records

IDENT	Type	Size	Weight	MoveTime	Processing Time
1	1	2	33	——————	20
2	1	3	22	——————	55
3	2	1	11	23	——————

you see in object-oriented languages where you can associate specific attributes with specific classes of objects. In Arena™, an entity is an entity. You can specify attributes that can allow you to conceptualize them as different types. This is conceptually defining the attribute set for the entity type that was previously mentioned. For example, you might create entities and think of them as parts flowing in a manufacturing system. In a manufacturing system, there might also be entities that represent pallets that assist in material handling in the system. A part may have a processing time attribute and a pallet may have a move time attribute. In Table 2-2, the type attribute indicates the type of entity (part = 1, pallet = 2). Note that all entities have the move time and processing time attributes *defined*. These attributes can be used regardless of the actual type of entity. In using attributes in Arena™, it is up to the modeler to provide the appropriate context (understanding of the attribute set) for an entity and then to use the attributes appropriately in that context.

There are two ways to distinguish between different types of entities. Arena™ has a pre-defined attribute, called *Entity.Type*, which can be used to set the type of the entity. In addition, you can specify a user defined attribute to indicate the type of the entity. Arena's on-line help has this to say about the *Entity.Type* attribute:

> *Entity.Type [(Entity Number)]—Entity type attribute.* This attribute refers to one of the types (or names) of entities defined in the Entities element. Entity type is used to set initial values for the entity picture and the cost attributes. It is also used for grouping entity statistics (e.g., each entity's statistics will be reported as part of all statistics for entities of the same type).

The Entity module in the Basic process panel allows you to define entity types. The dialog box shown in Figure 2-10 shows the basic text fields for defining an entity type. The most important field is the Entity Type field, which gives the name of the entity type. The Initial Picture field allows a picture to be assigned to the *Entity.Picture* attribute when entities of this type are created. The other fields refer to how costs are tabulated for the entity when applying Arena's activity based costing (ABC) functionality. This functionality will be discussed in Chapter 7.

The advantage of using the *Entity.Type* attribute is that Arena™ then makes other functions easier, such as collecting statistics by entity type (via the DISPOSE module), activity-based costing, and displaying the entity picture. Using the *Entity.Type* attribute makes it a little more difficult to randomly assign an entity type to an entity, although this can still be done using Arena's Set concept, which will be discussed later. By using a user-defined attribute, you can change and interpret the attribute very easily by mapping a number to the type as was done in Table 2-2.

Now that a basic understanding of the concept of an entity attribute has been developed, you still need to know how to declare and use the attributes. The declaration and use of attributes in Arena™ is similar to how variables can be declared and used in Visual Basic. Unless you have the Option Explicit keyword specified for Visual Basic, any variables can be

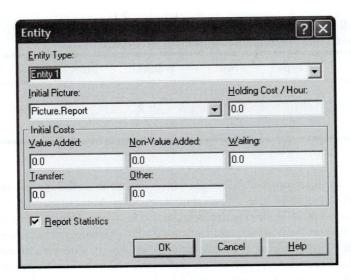

■ **FIGURE 2-10**
Entity module dialog box

declared at their first use. For attributes in Arena™ this also occurs. For example, to make an attribute called *myArea* available, simply use it in a module, such as an ASSIGN module. This does cause difficulties in debugging. It is also recommended that you adopt a naming convention for your attributes. This text uses the convention of placing *my* in front of the name of the attribute, as in *myArea*. This little mnemonic indicates that the attribute belongs to the entity. In addition, spaces in the name are not used, and the convention of the concatenation and the capitalization of the words in the name is used. Of course, you are free to label your attributes whatever you desire as long as the names do not conflict with already existing attributes or keywords in Arena™. Attributes can also be formally declared using the ATTRIBUTES element available on the ELEMENTS template panel. This also allows the user to define attributes that are arrays.

2.4   Processing Entities

This section overviews the types of operations facilitated by the Basic Process template panel and some of the programming constructs in Arena™. This is not meant to be a detailed discussion on these constructs, but rather a conceptual overview of the capabilities in order to set up the examples later in this and other chapters.

Entities move between modules via a direct connection or, as will be seen later, via an entity transfer mechanism. When an entity moves via a direct connection, the movement is instantaneous. That is, the entity completes the movement at the current time without any advance of the simulation clock. Some modules in Arena™ execute at the current event time and others may cause the flow of the entity to stop its movement until the operation is completed. The entity that is currently moving is called the "active entity." The active entity will continue to move until it is delayed or blocked by some module. To implement a simple time duration delay, you can use the PROCESS module on the Basic Process template with the delay option selected, or you can use the DELAY module from the Advanced Process template. This allows the modeling of entities that experience activities in the system.

An activity can take place with or without a resource. An activity that takes place without a resource is called "unconstrained." For example, the walking of customers from one location to another in the grocery store can be modeled with an unconstrained activity (e.g., a process delay). Entities that enter the DELAY module schedule an event representing the end of the activity and then wait until the end of the activity event occurs before continuing their motion in the model. In some instances the activity will require a resource. In which case, the entity must be allocated the required number of units of the resource before starting the activity. These types of activities are called "resource constrained." For example, suppose a part needs to move from a drilling station to a grinding station and suppose the part requires a forklift to be able to move. This type of movement is resource (forklift) constrained. Resource constrained activities in a system are often modeled using the RESOURCE module in Arena™. Resource constrained movement, as in the part/forklift example, can be modeled either using a RESOURCE or a TRANSPORTER module (found on the Advanced Transfer panel). When the entity's movement is constrained, the entity will need a place to wait. The QUEUE module defines the characteristics of the ordering of the waiting line. In most instances, if a module may constrain the movement of an entity, a queue will be automatically attached to the module. For example, the HOLD module allows the entity to be held in a queue until a specific signal is activated by another entity or until a specific condition is met.

In most practical situations, an entity may follow alternative paths through the system. In this case, the modeler needs to be able to direct the entity through the system. Two fundamental mechanisms are available for directing entities: probabilistic and conditional. In the case of probabilistic routing, the entity selects from a set of paths, randomly according to a distribution. In the case of conditional routing, the entity selects a path based on a set of conditions in the system. Both of these mechanisms are available via the DECIDE module in Arena™. Besides the DECIDE module, Arena™ allows the selection of actions via various rules. For example, a resource can be selected based on the least number of busy units. Arena™ also has a number of programming based mechanisms (e.g., IF-THEN-ELSE, WHILE, etc.) available in the BLOCKS template that facilitate entity flow. Finally, Arena™ has many advanced ways to define routes through the system via the Advanced Transfer panel. These include various material handling constructs such as conveyors and transporters.

Besides the CREATE module, the SEPARATE module allows entities to duplicate themselves in a model. As a complement to the SEPARATE module, the BATCH module allows a set of entities to be combined into either a permanent or temporary group. In the case of a temporary group, the entities are represented as one entity that can be separated via the SEPARATE module back into the original entities. Batching allows the entities to move together as one unit (e.g., a pallet of items). Entities can also pickup and drop off other entities as they move via the PICKUP and DROPOFF modules on the Advanced Process panel. Finally, entities can wait in queues until a matching condition occurs via the MATCH module. While this does not group the entities, it allows their movements to be synchronized when the matching condition is met.

In the previously mentioned process modeling concepts, most of the concepts correspond to some physical phenomenon (e.g., using a resource, traveling from one location to another, etc.). These constructs assist in representing the physical system in the simulation model; however, these constructs must be augmented by programming constructs that facilitate other aspects of simulation modeling. In particular, there must be ways to get input into the model and to get output from the model, to set up and run the model, to collect statistics, to assign variables/attributes, to animate the model, and to trace/debug the simulation model.

Input and output are available in a number of different formats (e.g., text files, spreadsheets, databases, VBA forms, etc.). The basic way to handle input and output in an Arena™ model comes from the READWRITE module and the FILES module. In order to set up and run the model, the Run/Setup dialog box illustrated in Chapter 1's pharmacy example is used. To collect statistics in a module, the RECORD and STATISTIC modules are used. The ASSIGN module can be used to change the value of various attributes and variables in a model. Finally, to trace and debug a model, you can use some of Arena's built-in animation and also Arena's Run Controller.

You should now have an overview of the basic concepts available for modeling in Arena™. In the rest of this chapter and other chapters will examine the application of these concepts in a number of examples.

2.5   Attributes, Variables, and Some I/O

In this section, you will be introduced to how to use attributes and variables. In addition, statistical collection and writing data to a file will be introduced by expanding on the pharmacy model of Chapter 1.

2.5.1   Modifying the Pharmacy Model

Suppose that customers who arrive at the pharmacy have 1, 2, or 3 prescriptions to be filled. There is a 50% chance that they have 1 prescription to fill, a 30% chance that they have 2 prescriptions to fill, and a 20% chance that they have 3 prescriptions to fill. The time to service a customer is still exponentially distributed but the mean varies according to how many prescriptions that they need filled as show in Table 2-3.

For illustrative purposes, the current number of customers having 1, 2, or 3 prescriptions in the system is needed in the animation of the model. In addition, the total number of prescriptions in the system should also be displayed. Statistics should be collected on the average number of prescriptions in the system and on the average time a customer spends in the system. For diagnostic purposes, the following information should be captured to a file:

- When a customer arrives: The current simulation time, the entity number (*IDENT*), the entity's serial number, the number of prescriptions for the customer
- When a customer departs: The current simulation time, the entity number (*IDENT*), the entity's serial number, the number of prescriptions for the customer, and the time of arrival, and the total time spent in the system

In addition to the above information, each of the above two cases should be denoted in the output. The simulation should be run for only 30 customers.

Until you get comfortable with modeling, you should follow a basic modeling recipe and develop answers to the following questions:

- What is the system? What information is known by the system?
- What are the required performance measures?
- What are the entities? What information must be recorded or remembered for each entity? How are entities introduced into the system?
- What are the resources that are used by the entities? Which entities use which resources and how?

■ TABLE 2-3
Prescription Related Data

#Prescriptions	Chance	Service Time Mean
1	50%	3 minutes
2	30%	4 minutes
3	20%	5 minutes

Exhibit 2-1   Pseudo-Code for Multiple Prescription Model

```
CREATE customer according to Poisson process
ASSIGN number of prescriptions for customer and arriving time
  increment number of prescriptions in system and by type
WRITE the customer arrival information to a file
PROCESS the customer with the pharmacist
ASSIGN decrement the number of prescriptions in the system and by type
RECORD statistics on system time
WRITE the customer departing information to a file
DISPOSE the customer leaving the system
```

- What are the process flows? Sketch the process or make an activity flow diagram.
- Develop pseudo-code for the situation.
- Implement the model in Arena™.

Upon considering each of these steps in turn, you should have a good start in modeling the system.

In this example, the system is again the pharmacy and its waiting line. The system knows about the chance for each prescription amount, the interarrival time distribution, and the service time distributions with their means by prescription amount. The system must also keep track of how many prescriptions of each type are in the pharmacy and the total number of prescriptions in the pharmacy. The performance of the system will be measured by using the average number of prescriptions in the system and the average time a customer spends in the system. As in the previous model, the entity will be the customers. Each customer must know the number of prescriptions needed. As before, each customer requires a pharmacist to have their prescription filled. The activity diagram will be exactly the same as in Chapter 1 because the activity of the customer has not changed. The basic pseudo-code for this situation is given in Exhibit 2-1.

In implementing this model in Arena™, the following modules will be used:

CREATE: Creates the 30 customers

ASSIGN: Assigns values to variables and attributes

READ/WRITE: Writes out the necessary information from the customers

PROCESS: Implements the prescription-filling activity with the required pharmacist

RECORD: Records statistics on customers' system time

DISPOSE: Disposes of the entities that were created.

FILE: Defines the characteristics of the operating system file used by the READWRITE module within a data module.

VARIABLE: Defines variables to track the number of customers having different numbers of prescriptions and to track the total number of prescriptions.

To begin modeling, the variables and attributes necessary for the problem need to be understood. In the problem, the system needs to keep track of the total number of prescriptions and the number of customers that need one, two, or three prescriptions filled. Therefore, variables to keep track of each of these items are required. A scalar variable can be used to track the number of prescriptions in the system and a 1D variable with three elements can be used to track the number of customers requiring each of the three prescription amounts.

In the example, each customer needs to know how many prescriptions that he/she needs to have filled. Therefore, the number of prescriptions needs to be an entity attribute. In addition, the arrival time of each customer and the customer's total time in the system must be written out to a file. Thus, each entity representing a customer must remember when it arrived at the pharmacy. Do real customers need to remember this information? No! However, the *modeling* requires this information. Thus, attributes (and variables) can be additional characteristics required by the modeler that are not necessarily part of the real system. Thus, the arrival time of each customer should be modeled as an entity attribute.

To understand why this information *must* be stored in an attribute consider what would happen if you tried to use a variable to store the number of prescriptions of the incoming customer. Entities move from one block to another during the simulation. In general, entities may delay their progress due to simulation logic. When this happens, other entities are allowed to move. If the number of prescriptions for the customer was stored in a (global) variable, then during the movement cycle of the other entities, the other entities might change the value of the variable. Thus, the information concerning the current customer's number of prescriptions will be over written by the next customer that enters the system. Thus, this information *must be carried with* the entity. Global variables are very useful for communicating information between entities; however, as in any programming language, care must be taken to use them correctly.

The information in Table 2-3 also needs to be modeled. Does this information belong to the system or to each customer? While the information is about customers, each customer does not need to "carry" this information. The information is really about how to create the customers. The system must know how to create the customers, thus this information belongs to the system. The number of prescriptions per customer can be modeled as a random variable with a discrete distribution. The information about the mean of the service time distributions can be kept in a 1D variable with three elements representing the mean of each distribution.

In what follows, the already developed pharmacy model will be used as a basis for making the necessary changes for this situation. If you want to follow along with the construction of the model, then make a copy of the pharmacy model of Chapter 1. You should start by defining the variables for the example. With your copy of the model open, go to the Basic Panel and select the VARIABLE data module and define the following variables:

> *vNum Prescriptions:* This scalar variable should count the total number of prescriptions in the system. This variable should be initialized at the value 0.0.
>
> *vNP:* This 1D variable should have 3 rows and should count the number of customers having 1, 2, 3 prescriptions currently in the system. The index into the array is the prescription quantity. Each element of this variable should be initialized at the value 0.0.
>
> *vService Mean:* This 1D variable should have 3 rows representing the mean of the service distributions for the 3 prescription quantities. The index into the array is the prescription quantity. The elements (1, 2, and 3) of this variable should be initialized at the values 3, 4, and 5 respectively.

Given the information concerning the variables, the variables can be defined as shown in Figure 2-11. For scalar variables, you can tell Arena^{TM} to automatically report statistics on the time average value of the variable. Chapter 4 will more formally define time-average statistics, but for now you can think of them as computing the average value of the variable over time. The initial values can be easily defined using the spreadsheet view.

Now, you are ready to modify the modules from the previous model in Arena's model window. The CREATE module is already in the model. In what follows, you will be inserting some modules between the CREATE module and the PROCESS module and inserting some modules between the PROCESS module and the DISPOSE module. You should select and delete the connector links between these modules. You should then drag an ASSIGN module from the Basic Process panel to the model window so that you can connect it to the CREATE module.

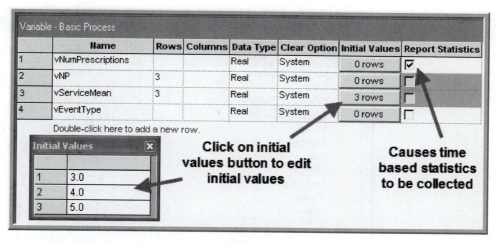

■ **FIGURE 2-11**   Defining the variables

2.5.2   Using the ASSIGN module

Because the ASSIGN module is directly connected to the CREATE module, each entity that is created will pass through the ASSIGN module causing each assignment statement to be executed in the order in which the assignments are listed in the module. The ASSIGN module simply represents a series of logical assignments, such as *Let X* = 2, as you would have in any common programming language. Figure 2-12 illustrates the dialog boxes associated with the ASSIGN module. Using the Add button, you can add a new assignment to the list of assignments for the module. Add an attribute named *myNP*, and give it the value DISC(0.5, 1, 0.8, 2, 1.0, 3).

The function DISC($cp_1, v_1, cp_2, v_2, \cdots, cp_n, v_n$) will supply a random number according to a discrete probability function, where cp_i are cumulative probabilities and v_i is the value—that is, $P(X = v_i) = cp_{i+1} - cp_i$ with $cp_0 = 0$ and $cp_n = 1$. The last cumulative value must be 1.0 since it must define a probability distribution. For this situation, there is a 50% chance of having one prescription, a $(80 - 50 = 30\%)$ chance for two prescriptions, and $(100 - 80 = 20\%)$

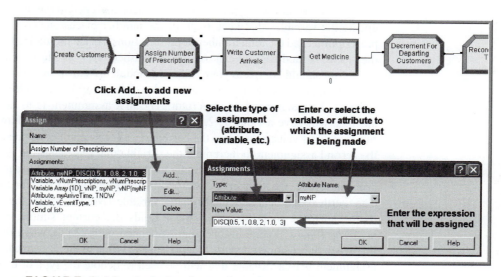

■ **FIGURE 2-12**   Assigning the number of prescriptions

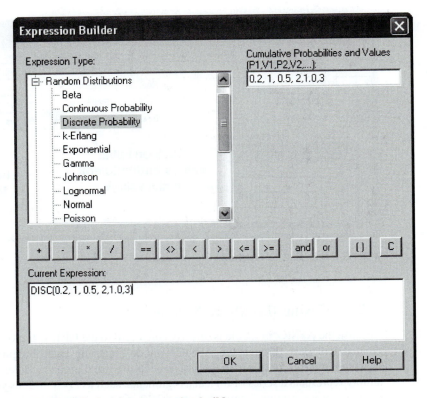

■ **FIGURE 2-13**   **Expression builder**

chance for three prescriptions. The generation and use of random variables will be discussed in detail in Chapter 3.

The Expression Builder can be using to build complicated expressions in Arena™ by right-clicking in the text field and choosing Build Expression. Figure 2-13 illustrates the Expression Builder with the DISC distribution function and shows a list of available probability distributions in Arena™.

Once the number of prescriptions has been assigned to the incoming customer, you can update the variables that keep track of the number of customers with the given number of prescriptions and the total number of prescriptions. As an alternative to the dialog box view of the ASSIGN module, you can use the spreadsheet view as illustrated in Figure 2-14.

You should complete your ASSIGN module as shown in Figure 2-14. The spreadsheet view of Figure 2-14 clearly shows the order of the assignments in the ASSIGN module. The order of the assignments is important! The first assignment is to the attribute *myNP*. This attribute will have a value 1, 2, or 3, depending on the random draw caused by the DISC function. This value represents the current customer's number of prescriptions and is used in the second assignment to increment the variable that represents the total number of prescriptions in the system. In the third assignment the attribute *myNP* is used to index into the variable array that counts the number of customers in the system with the given number of prescriptions. The fourth assignment uses the Arena™ variable *TNOW* to assign the current simulation time to the attribute *myArriveTime*. *TNOW* holds the current simulation. Thus, the attribute, *myArriveTime* will contain the value associated with *TNOW* when the customer arrived, so that when the customer departs the total elapsed time in the system can be computed. The last assignment is to a variable that will be used to denote the type of event {1 for arrival, 2 for departure} when writing out the customer's information.

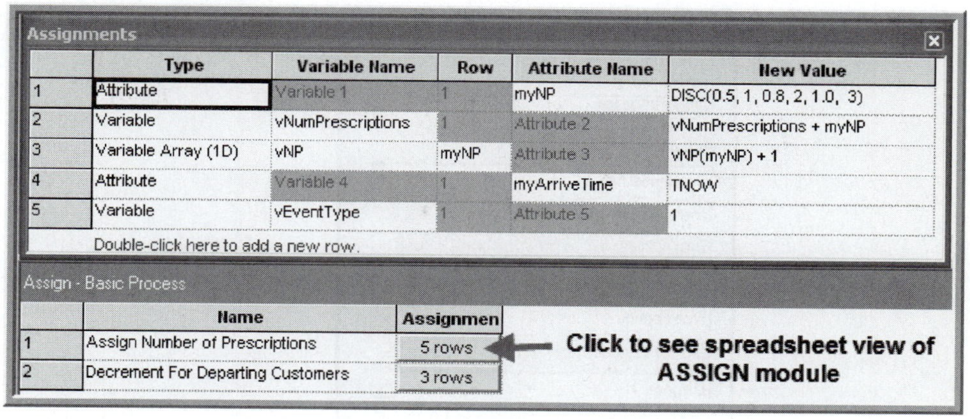
Assignments
Type | Variable Name | Row | Attribute Name | New Value
1 Attribute | Variable 1 | 1 | myNP | DISC(0.5, 1, 0.8, 2, 1.0, 3)
2 Variable | vNumPrescriptions | 1 | Attribute 2 | vNumPrescriptions + myNP
3 Variable Array (1D) | vNP | myNP | Attribute 3 | vNP(myNP) + 1
4 Attribute | Variable 4 | 1 | myArriveTime | TNOW
5 Variable | vEventType | 1 | Attribute 5 | 1
Double-click here to add a new row.
Assign - Basic Process
Name | Assignmen
1 Assign Number of Prescriptions | 5 rows
2 Decrement For Departing Customers | 3 rows
Click to see spreadsheet view of ASSIGN module

■ **FIGURE 2-14** Spreadsheet view of ASSIGN module

2.5.3 Using the READWRITE Module

As with any programming language, Arena™ has the ability to read and write information from and to files. Often the READWRITE module is left to the coverage of miscellaneous topics in Arena™; however, in this section, the READWRITE module is introduced so that you can be aware that it is available and possibly use it when debugging your models. The READWRITE module is very useful for writing information from a simulation and for capturing values from the simulation for post-processing and other analysis.

This example uses the READWRITE module to write out the arriving and departing customer's information to a text file. The information can be used to trace the customer's actions through time. There are better ways to trace entities with Arena™, but this example uses the READWRITE module so that you can gain an understanding of how Arena™ processes entities. In order to use the READWRITE module, you must first define the file to which the data will be written. You do this by using the FILE data module in the Advanced Process panel. To get the FILE dialog box, insert a row into the file area and then choose edit via the dialog box from the right-click context menu. Open the FILE dialog box and make it look like the FILE module shown in Figure 2-15. This will define a file in the operating system to which data can be written or from which data can be read. Arena's FILE module allows the user to work with text files, Excel™ files, Access™ files, as well as the other file types available in the Access Type drop-down menu dialog box. In this case, the Sequential option has been chosen, which indicates to Arena™ that the file is a text file for which the records will be in the same sequence that they were written. In addition, the Free Format option has been selected, which essentially writes each value to the file, separated by a space. You should review the help file on the FILE module for additional information on the other access file types.

Now, open the READWRITE module and make it look like the READWRITE module given in Figure 2-16. Use the drop-down menu to select the file that you just defined and use the Add button to tell the module what data to write out. After selecting the Add button, you should use the Type drop-down to select the data type of the item that you want to write out to the file and the item to write out. For the data type Other, you must type in the name of the item to be written out (it will not be available in a drop-down context). In this case, the following is being written to the file:

Tnow: Current simulation time

vEventType: Type of event {1 for arrival, 2 for departure}

Ident: Entity identity number of the customer

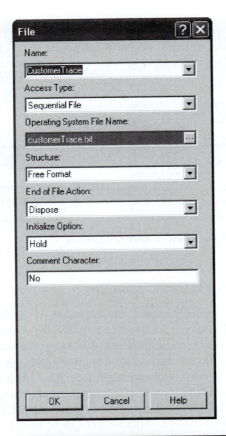

■ FIGURE 2-15 FILE module

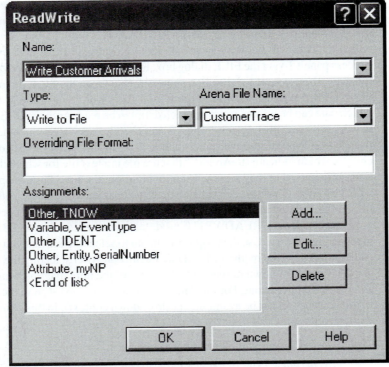

■ FIGURE 2-16 READWRITE module

Entity.SerialNumber: Serial number of the customer entity

myNP: The number of prescriptions for the customer

This information is primarily about the current (active) entity. The active entity is the entity that is currently moving through the model's modules. In addition to writing out information about the active entity or the state of the system, the READWRITE module is quite useful for reading in parameter values at the beginning of the simulation and for writing out statistical values at the end of a simulation. These options will be explored in Chapter 7 when illustrating how to read/write to and from Excel™ and Access™.

After writing out the values to the file, the customer entity proceeds to the PROCESS module. Because the mean of the service time distribution depends on the number of prescriptions, the PROCESS module needs to be changed. In Figure 2-17, use the 1D variable, *vServiceMean*, and the attribute, *myNP*, to select the appropriate mean for the exponential distribution in the PROCESS module. Remember that the attribute *myNP* has a value {1, 2, or 3} depending on the number of prescriptions for the customer. This value is used as the index into the vServiceMean 1D variable to select the appropriate mean. Since Arena™ has limited data structures, the ability using arrays in this manner is quite common.

Now, that the customer has received service, you can update the variables that keep track of the number of prescriptions in the system and the number of customers having one, two, and three prescriptions. This can be done with an ASSIGN module placed after the PROCESS module. Figure 2-18 shows the assignments for after a customer completes service. The first assignment decrements the global variable, *vNumPrescriptions*, by the amount of prescriptions, *myNP*, associated with the current customer. Then, this same attribute is used to index

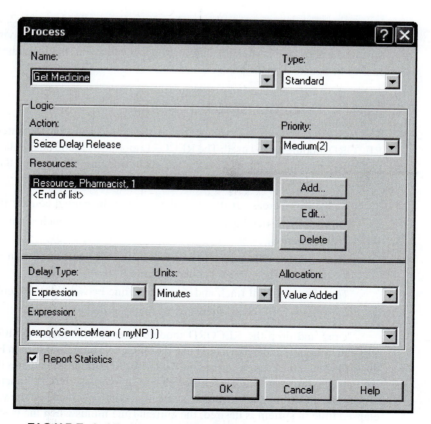

■ **FIGURE 2-17**   **Changed PROCESS module**

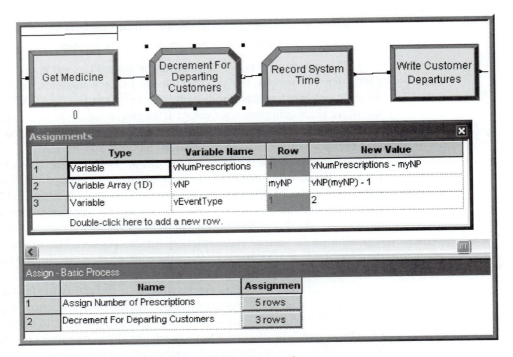

■ **FIGURE 2-18**   Updating variables after service

into the array that is counting the number of customers that have {1, 2, or 3} prescriptions to reduce the number by 1 since a customer of the designated type is leaving. In preparation for writing the departing customer's information to the file, you should set the variable, *vEventType*, to 2 to designate a departure.

2.5.4   Using the RECORD Module

The problem states that the average time in the system must be computed. To do this, Arena's RECORD module can be used. The RECORD module is found on the Basic Process panel. The RECORD module tabulates information each time an entity passes through it. The options include:

Count will increase or decrease the value of the named counter by the specified value.

Entity Statistics will generate general entity statistics, such as time and costing/duration information.

Time Interval will calculate and record the difference between a specified attribute's value and the current simulation time.

Time Between will track and record the time between entities entering the module.

Expression will record the value of the specified expression.

Drag and drop the RECORD module after the previous ASSIGN module. Open the RECORD module and make it look like the RECORD module shown in Figure 2-19. Make sure that you choose the Expression option. The RECORD module evaluates the expression given in the Value textbox field and passes the value to an internal Arena™ function that automatically tallies statistics (min, max, average, standard deviation, count) on the value passed. Since the elapsed time in system is desired, the current time minus when the customer arrived must be computed. The results of the statistics will be shown on the default Arena™ reports labeled

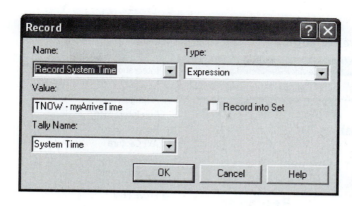

■ **FIGURE 2-19**
Record module

with the name of the tally, such as *SystemTime*. The Time Interval option could also have been used by supplying the *myArriveTime* attribute. You should explore and try out this option to see that the two options produce the same results.

The final model change is to include another READWRITE module to write out the information about the departing customer. The READWRITE module also has a spreadsheet view for assigning the information that is to be read in or written out. Figure 2-20 shows the spreadsheet view for the departing customer's READWRITE module.

The module is set to first write out the simulation time via the variable *TNOW*, then the type of event, *vEventType*, which was set to 2 in the previous ASSIGN module. Then, it writes out the information about the departing entity (*IDENT, Entity.SerialNumber, myNP, myArriveTime*). Finally, the system time for the customer is written out to the file.

2.5.5   Animating a Variable

The problem asks to have the value of the number of prescriptions currently in the system displayed in the model window. This can be done by using the variable animation features in

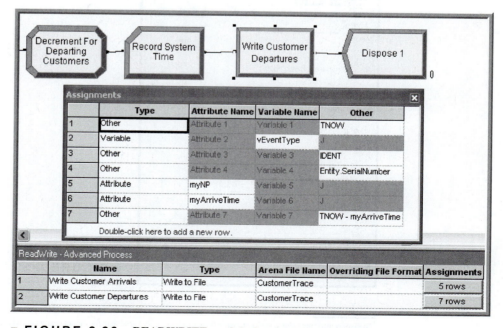

■ **FIGURE 2-20**   READWRITE module for departing customers

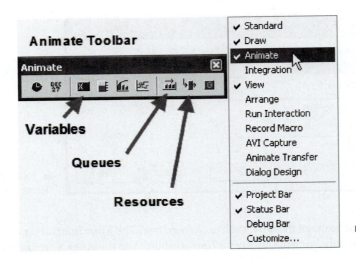

■ **FIGURE 2-21**
Animate toolbar

Arena[TM]. Figure 2-21 shows the Animate toolbar in Arena[TM]. If the toolbar is not showing in your Arena[TM] Environment, then you can right-click in the toolbar area and make sure that the Animate option is checked as shown in Figure 2-21.

You will use the Animate Variables option of the Animate toolbar to display the current number of prescriptions in the system. Click on the button with the 0.0 on the toolbar. The Animate variable dialog box will appear (see Figure 2-22). By using the drop-down box labeled Expression, you can select the appropriate variable to display. In Arena[TM], only the values of variables can be displayed[3] in the animation. After filling out the dialog box and selecting OK,

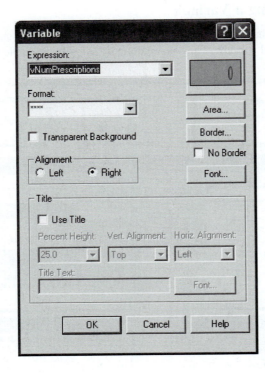

■ **FIGURE 2-22** **Animate variable dialog**

---

[3]To display attribute values, you first assign them to a temporary variable in the model near when/where you want the information. Then use the Animate Variables option of the Animate toolbar to show the value of the temporary variable.

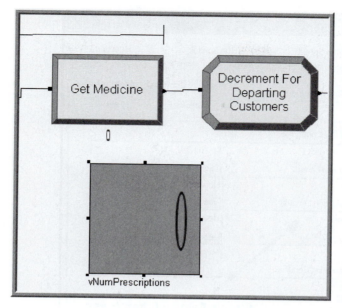

■ **FIGURE 2-23**
**Animating the number of
prescriptions in the system**

your cursor will change from the arrow selection form to the cross-hairs form. By placing your cursor in the model window and clicking you can indicate where you want the animation to appear in the model window. After the first click, drag over the area where you want the animation in order to size the animation and then click again. After completing this process, you should have something that looks like Figure 2-23. During the simulation run, this area will display the variable *vNumPrescriptions* as it changes with respect to time.

2.5.6   Running the Model

Two final changes to the model are necessary to prepare the model to run only 30 customers. First, the maximum number of arrivals for the CREATE module must be set to 30. This will ensure that only 30 customers are created by the CREATE module. Second, the Run Setup > Replication Parameters dialog box needs to indicate an infinite replication length.

These two changes (shown in Figures 2-24 and 2-25) ensure that the simulation will stop after 30 customers are created. If you do not specify a replication length for the simulation,

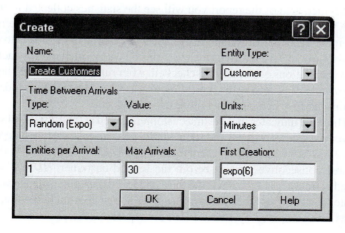

■ **FIGURE 2-24**
Creating only 30 customers

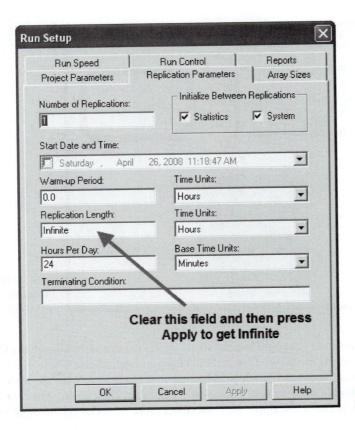

■ **FIGURE 2-25** No replication length specified

Arena^{TM} will process whatever entities are created in the model until there are no more entities to process. If a maximum number of arrivals for the CREATE module is not specified and the replication length was infinite, then the simulation will never end!

After completing all these changes to the model, use Run Setup > Check Model to check for syntax errors. If there are errors, you should carefully go back through the dialog boxes to ensure that you typed and specified everything correctly. The final Arena^{TM} file, *Pharmacy-ModelEx2-1.doe*, is available in the supporting files for this chapter. To see the animation, you should make sure that you uncheck the Batch Run (No animation) option in the Run > Run Control menu. You should run your model and agree to see the simulation report. When you do, you should see that 30 entities passed through the system. By clicking on the User Specified area of the report you will see the statistics collected for the RECORD module and the VARIABLE module that was specified when constructing the model.

Figure 2-26 indicates that the average system time of the 30 customers that departed the system was about 7.1096. Since the system time was written out for each departing customer, Arena's calculations can be double-checked. After running the model, there should be a text file called *customerTrace.txt* in the same directory as where you ran the model. Table 2-4 shows the data from the first five customers through the system.

The table indicates that the first customer arrived at time 2.076914 with 1 prescription and was given the *IDENT* value 2 and serial number 1. The customer then departed, event type = 2, at time 2.257428 with a total system time of (2.257428 − 2.076914 = 0.180514). There were then three consecutive arrivals before the second customer, (*IDENT* = 3, serial number 2) departed at time 13.31427. Thus, the third and fourth customers had to wait in the queue. If you compute the average of the values of the system time column in the file, you will get the same value as that computed by Arena^{TM} on the reports.

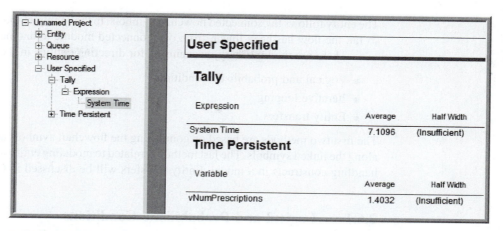

■ **FIGURE 2-26**   System time output report

■ **TABLE 2-4**
Data from Output File in Tabular Form

TNOW	Event Type	IDENT	Serial Number	Number Prescriptions	Arrive Time	System Time
2.076914	1	2	1	1		
2.257428	2	2	1	1	2.076914	0.180514
4.74824	1	3	2	1		
5.592303	1	4	3	3		
10.34295	1	5	4	1		
13.31427	2	3	2	1	4.74824	8.566026
21.39914	1	6	5	1		
21.89796	2	4	3	3	5.592303	16.30566
23.79078	2	5	4	1	10.342949	13.447826
26.0986	2	6	5	1	21.399139	4.699463

In this example, you have learned more about the CREATE module and you have seen how you can define variables and attributes in your models. With the use of the ASSIGN module you can easily change the values of variables and attributes as entities move through the model. In addition, the RECORD module allows you to target specific quantities in the model for statistical analysis. Finally, the READWRITE module was used to write out information concerning the state of the system and the current entity when specific events occurred in the model. This is useful when diagnosing problems with the model or for capturing specific values to files for further analysis. The simple variable animation feature of Arena™ was also demonstrated.

This example had a simple linear flow, with each module directly connected to the previous module. Complex systems often have more complicated flow patterns. The next section introduces how to better control the flow of entities through the model.

2.6   Flow of Control in Arena™

When developing computer programs in standard languages, programmers have the ability to direct the flow of the program so that certain statements are executed, perhaps repeatedly. In Arena™, the flowchart programming paradigm makes this flow of control somewhat obvious.

The entity follows the connected flowchart symbols. In the examples that have been considered so far, the flow has been linear, from one connected module to another.

There are three primary mechanisms for directing entities in the model:

- Logical and probabilistic conditions
- Iterative looping
- Entity transfers

The first two methods are based on connecting the flowchart symbols and directing the entity along the linked symbols. The last method is related to modeling entity movement and material handling constructs in a model. Entity transfers will be discussed in Chapter 6.

2.6.1   Logical and Probabilistic Conditions

A programming language such as C has constructs like if-then-else that can direct the execution of the program given specific conditions. Arena™ has three constructs for modeling conditional logical tests—DECIDE, BRANCH, and IF-ELSE-ENDIF. The DECIDE module is the more commonly used form and is found on the Basic Process template panel. Figure 2-27 illustrates the DECIDE module flowchart symbol and its dialog box. As can be seen in the figure, a DECIDE module has four types (2-way by Chance, 2-way by Condition, N-way by Chance, and N-way by Condition). The "by condition" options will be discussed first.

The 2-way by condition option implements a basic if-then-else construct, with the entity being directed along either of the two indicated paths based on whether or not the condition is true or false. In pseudo-code form, the 2-way by condition DECIDE module acts as follows:

```
If condition is true
        Send entity out true exit point
Else
        Send entity out false exit point
```

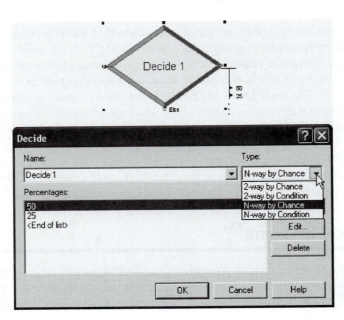

■ **FIGURE 2-27**
DECIDE flowchart symbol and module dialog box

The N-way by condition option, implements something similar to a switch statement in C or a case statement in Visual Basic. In pseudo-code form, the N-way by condition DECIDE module acts as follows:

```
If condition 1 is true, go out condition 1's exit point.
Else If condition 2 is true, go out condition 2's exit point.
Else If condition 3 is true, go out condition 3's exit point.

  And so on...

Else If condition N is true, go out condition N's exit point.
Else If all previous conditions are false, go out ''else'' last exit point.
```

As soon as a condition tests to true, the entity is sent out that exit point. No further tests are performed. The order of the tests is clearly important. By placing certain conditions before other conditions, the modeler can provide a simplistic priority over the conditions.

Conditions can be any legal Arena™ expression that evaluates to a logical (Boolean) value. The DECIDE module's dialog box, see Figure 2-28, facilitates the building of conditions. The user can write expressions for testing attributes, variables, array elements, and entity types. In addition, the user can write a general expression. The use of Arena's Expression Builder is very useful in assisting with the development of expressions and helps with preventing syntax and spelling mistakes.

Table 2-5 lists the mathematical and logical operators that can be used in Arena™ expressions. The logical operators have two forms. For example, equality can be tested either by using .EQ. or two equals signs (==). The precedence for the operators is also listed in the table. Parentheses can be used to group the sections of an expression so that the desired evaluation will be performed in the appropriate order.

The "by chance" option for the DECIDE module allows the entity to randomly pick from the list of exit points. In the 2-way by Chance option, the user specifies the probability associated with the true exit point. The remaining probability is associated with the false exit point. In the N-way by Chance option, the user specifies the percent true for each of the exit points. The Else exit point uses the remaining probability for its chance. The percent true for each exit point can be easily specified using the spreadsheet view as shown in Figure 2-29. When an entity enters the by chance DECIDE module, Arena™ randomly generates a number and selects the exit point based on the specified probability distribution across the exit points.

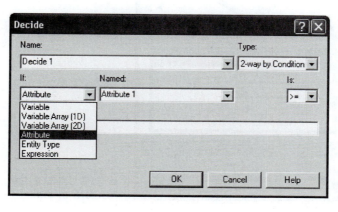

■ **FIGURE 2-28** 2-way DECIDE module dialog box

■ **TABLE 2-5**
Mathematical and Logical *Operators*

Operator	Operation	Priority
Math Operators		
**	Exponentiation	1 (highest)
/	Division	2
*	Multiplication	2
−	Subtraction	3
+	Addition	3
Logical Operators		
.EQ. , ==	Equality comparison	4
.NE. , < >	Non-equality comparison	4
.LT. , <	Less than comparison	4
.GT. , >	Greater than comparison	4
.LE. , <=	Less than or equal to comparison	4
.GE. , > =	Greater than or equal to comparison	4
.AND. , &&	Conjunction (and)	5
.OR. , ‖	Inclusive disjunction (or)	5

Arena's BRANCH block found on the Blocks Template panel also performs very similarly to the DECIDE module. The Blocks Template contains the original SIMAN commands from which the other templates are derived, see Figure 2-30. SIMAN is the underlying text-based programming language for Arena™. SIMAN's relationship with Arena™ is discussed further, later in this chapter. The Blocks template panel can be attached by right-clicking in the project bar and choosing *Blocks.tpo* from Arena's Template folder. The

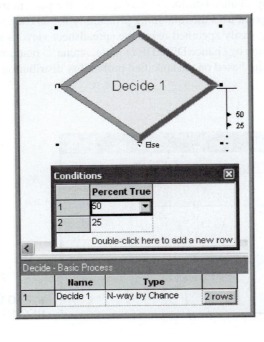

■ **FIGURE 2-29** **N-way chance DECIDE module**

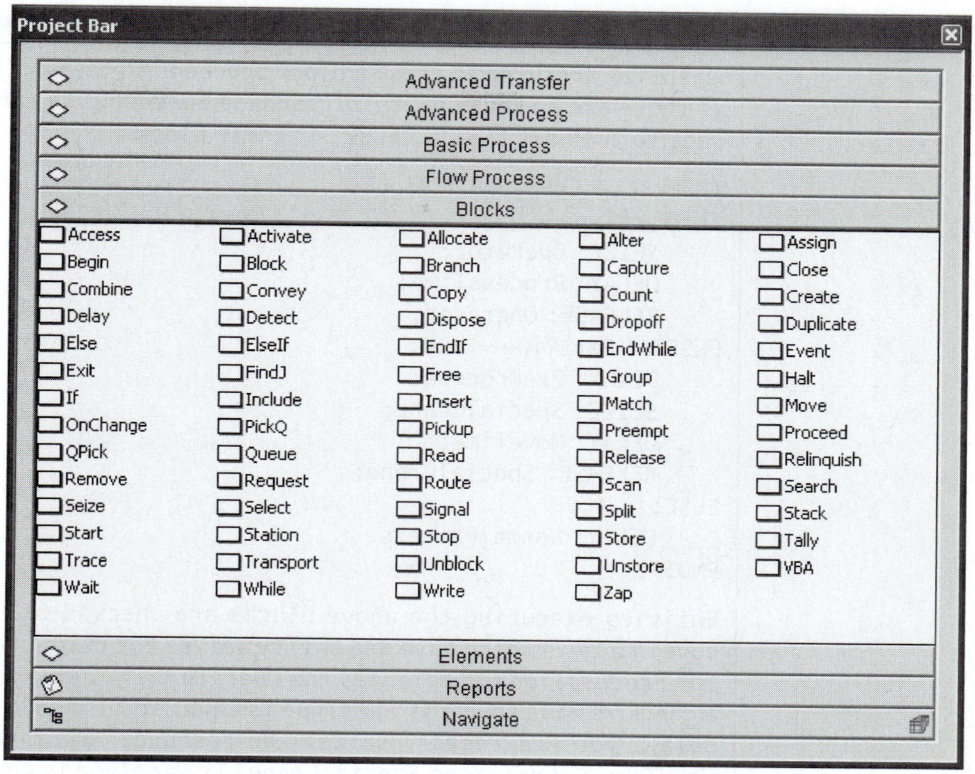

■ **FIGURE 2-30**   **Blocks template panel**

BRANCH block works in a similar fashion as the DECIDE module except that you specify the maximum number of branches, the branch type, and the random number stream. The use of random number streams is discussed in the next chapter.

The BRANCH block directs the arriving entity along one of the branches according to the type of branching. The key difference is the use of the maximum number of branches. If the maximum number of branches is one, then this block works essentially the same as the DECIDE module; however, if the maximum number of branches is greater than one, duplicate entities will be created and allowed to flow along the branches for any of the conditions that are met. The BRANCH block has been largely superseded by the DECIDE module. Unless you are certain that you need to utilize this duplicating entity functionality, you should stick with the DECIDE module. Further details on how the BRANCH block handles duplicates and the order that they exit the block are available in the Arena™ Help system by selecting Help on the BRANCH block dialog box.

The final logical flow construct that will be discussed is Arena's IF-ELSEIF-ELSE-ENDIF blocks. These blocks are also found on the Blocks Template. These blocks work in a similar fashion as the standard if-then-else constructs available in major languages. Exhibit 2-2 shows an example use of the IF-ELSEIF-ELSE-ENDIF block combination from Arena's help files. If the logical condition associated with the IF block is true, the entity will only execute the statements between the testing block and the next ELSEIF, ELSE, or ENDIF block combination. An ENDIF block must be used to terminate the use of combinations of these blocks. When using these blocks, the sets of statements are mutually exclusive. Only one set will execute based on the evaluation of the conditions tested. For example, in the exhibit, if *PartType* equals 1,

Exhibit 2-2 IF-ELSEIF-ELSE-ENDIF Example from Arena's Help File

```
An entity arriving at this IF block proceeds to the next block in the model
if there are any busy units of resource Sweeper. Otherwise, the entity is
sent to the next ELSEIF, ELSE, or ENDIF block.

IF-ELSEIF-ELSE-ENDIF Blocks
IF: PartType == 1;
     SEIZE: Operator;
     DELAY: ProcessTime;
     RELEASE: Operator;
ELSEIF: PartType == 14;
     DELAY: PreProcess;
     SEIZE: SpecialCrane;
     DELAY: MoveTime;
     RELEASE: SpecialCrane;
ELSE;
     DELAY: NormalProcess;
ENDIF;

Entities executing the above blocks are checked to see if PartType is
equal to 1, in which case the entity seizes the resource Operator, delays
for ProcessTime and releases the Operator. If PartType is not equal to 1,
a check is made to see if PartType is equal to 14, in which case the entity
delays for PreProcess, seizes the resource SpecialCrane, delays for
MoveTime and releases SpecialCrane. If PartType is neither 1 nor 14, the
entity delays for NormalProcess.
```

then only the statements

    SEIZE: *Operator;*
    DELAY: *ProcessTime;*
    RELEASE: *Operator;*

will be executed. Unfortunately, the use of these blocks requires the user to place and connect each of the modules, that is, you cannot write this sort of text directly in the Arena™ Environment. This often becomes tedious and takes up much screen real estate. The use of multiple DECIDE modules can largely take the place of the use of IF-ELSEIF-ELSE-ENDIF blocks; however, their use can be convenient for complicated model logic and if you prefer a more structured programming style.

2.6.2 Iterative Looping

Iterative looping is often performed in computer programming. Languages such as C have constructs like for-loop, do-while, and so on, to allow a set of statements to be iteratively executed. Arena™ has the WHILE/ENDWHILE blocks to perform this task as well as flowchart mechanisms to accomplish iteration, essentially by sending the entity "back" through a series of statements. The use of the WHILE-ENDWHILE combination is similar to that for the IF-ELSEIF-ELSE-ENDIF blocks. Exhibit 2-3 illustrates the use of the blocks in SIMAN code. Again, this text format cannot be directly developed in the Arena™ Environment. The user must drag and drop and hook up the blocks in the model. An error will occur if a WHILE block is not matched up with a corresponding ENDWHILE block.

Exhibit 2-3   WHILE-ENDWHILE Arena Help Example

```
Syntax
WHILE: Condition or Expression;
Basic Use
ASSIGN: J=0;
WHILE: J<10;
    ASSIGN: J=J+1:
    ProcessTime(J)=PTime(1,J);
ENDWHILE;
```

Promt	Entry
Condition or Expression	J < 10

The entity first assigns J equal to 0. Then, while J is less than 10, the entity assigns *ProcessTime(J)* equal to *PTime(1,J)*. This effectively assigns *ProcessTime(1)* through *ProcessTime(10)* equal to *PTime(1,1)* through *PTime(1,10)*. As soon as J is incremented to 10, the loop is terminated and the entity is transferred to the ENDWHILE block.

In the following example, you will examine iterative looping in Arena™ through the use of an example. In addition, you will learn more about variables, attributes, and a new construct call EXPRESSIONS.

2.6.3   Example: Iterative Looping, Expressions, and Submodels

In this example, you will examine how to define and use attributes, variables, and expressions in Arena™. Also, you will learn how to delay an entity along its flow path. In addition, you will learn some basic flow of control mechanisms for looping or iterating in Arena™. This model will use the following modules:

CREATE: Two instances of this module will be used to have two different arrival processes into the model.

ASSIGN: This module will be used to assign values to variables and attributes.

WHILE & ENDWHILE: These modules will be used to loop the entities until a condition is true.

DECIDE: This module will also be used to loop entities until a condition is true

PROCESS: This module will be used to simulate simple time delays in the model.

DISPOSE: This module will be used to dispose of the entities that were created.

VARIABLE: This data module will be used to define variables and arrays for the model.

EXPRESSION: This data module will be used to define named expression to be used in the model.

SUBMODEL: Submodels are areas of the model window that contain modules that have been aggregated into a single module.

This example is based partly on Arena's SMARTS file 183. This model will be considerably larger than the previous models. The final model will look something like the one shown in Figure 2-31.

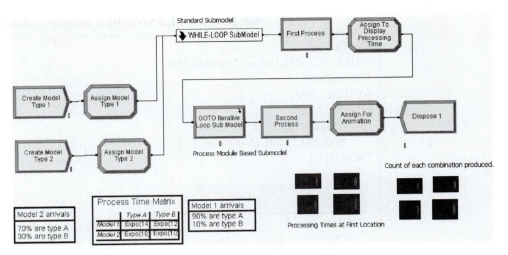

■ **FIGURE 2-31** Completed model for iterative looping example

This system produces products. The products have different model configurations (models 1 and 2) that are being produced in a small manufacturing system. Model 1 arrives according to a Poisson process with a mean rate of one model every 12 minutes. The second model type also arrives according to a Poisson arrival process but with a mean arrival rate of one arrival every 22 minutes. Furthermore, in a model configuration, there are two types of products produced, type A and type B.

Table 2-6 shows the data for this example. In the table, 90% of model configuration 1 is of type A. In addition, the base process time for type A's for model configuration 1 is exponentially distributed with a mean of 14 minutes. Clearly, the base process time for the products depends on the model configuration and the type of product A or B. The actual processing time is the sum of 10 random draws from the base distribution. Processing occurs at two identical sequential locations. After the processing is complete, the product leaves the system.

When building this model, you will display the final processing time for each product at each of the two locations and display a count of the number of each configuration and type that was produced.

2.6.3.1 Building the Model.
Following the basic modeling recipe, consider the question, "What are the entities?" By definition, entities are things that flow through the system. They are transient. They are created and disposed. For this simple model, the products are produced by the system. The products flow through the system. These are definite candidates for entities. In this problem, there are actually four types of entities: Two main types Model 1 and Model 2, which are further classified into product type A or product type B. Now, consider how to represent the entity types. As was previously mentioned, this can be done in two ways: (1) by using the ENTITY module to define an entity type, or (2) by using user defined attributes. Either method can be used in this situation, but by using user defined attributes, you can more easily facilitate some of the answers to some of the other modeling recipe questions.

■ **TABLE 2-6**

Model	Mean Rate	Type A (%, process time)	Type B (%, process time)
1	1 every 12 minutes	90%, expo(14) minutes	10%, expo(12) minutes
2	1 every 22 minutes	70%, expo(16) minutes	30%, expo(10) minutes

Next, consider the question, "What are the attributes of the entities?" This question is partly answered already by the decision to use user defined attributes to distinguish between the types of entities. Let's define two attributes, *myModel* and *myType* to represent the classification by model and by product type for the entities. For example, if *myModel* = 1, then the entity is of the configuration 1 type model. To continue addressing the required attributes, you need to consider the answers to the following questions:

- What data do the entities need as they move through the model? Can they carry the data with them as they move or can the information be shared across entities?
- What data belong to the system as a whole?

The answer to the first question is that the parts need their processing times. The answer to the second question can have multiple answers, but for this model, the distributions associated with the processing times need to be known and this information doesn't change in the model. This should be a hint that it can be stored globally. In addition, the processing time needs to be determined based on different distributions at each of the two locations.

While you might first think of using variables to represent the processing time at each location, you will soon realize that there is a problem with this approach. Assume that a variable holds the processing time for the entity that is currently processing at location 1. While that entity is processing, another entity can come along and over write the variable. This is clearly a problem. The better approach is to realize that the entities can "carry" their processing times with them after the processing times have been computed. This should push you toward the use of an attribute to hold the processing time. Let's decide to have an attribute called *myProcessingTime* that can be used to hold the computed processing time for the entity before it begins processing. If information is not to be carried by specific entities, but it still needs to be accessed, then this is a hint that it belongs to the system as a whole. In this case, the processing time distributions need to be accessed by any potential entity.

How can the entities be created? The problem description gives information about two Poisson arrival processes. While there are other ways to proceed, the most straightforward approach is to use a CREATE module for each arrival process. If this is done, one CREATE module will create the entities having model configuration 1 and the other CREATE module will create the entities having model configuration 2. How do you assign data to the entities? An ASSIGN module can be used for this task, but more importantly, the model configuration, the product type, and the processing time all will need to be assigned. In particular, the product type can be randomly assigned according to a given distribution based on the model configuration. In addition, the processing time is the sum of 10 draws from the processing time distribution. Thus, a way to compute the final processing time for the entity before it starts processing must be determined.

Exhibit 2-4 shows the basic pseudo-code for the example. After creation, the entities will have their model and type assigned. Then, they will proceed for processing. Prior to processing at the first station, the processing time is determined. In the pseudo-code, a WHILE-ENDWHILE construct is used to sum up the processing time. After the processing time has been determined, the entity then will experience a delay for the specified time. When the processing is done at the first station, the processing time for the second station is used. Here the processing time is determined in an iterative fashion by using a *goto* construct coupled with an if statement. After the processing time has been determined, the entity again delays for the processing before being disposed.

The model will follow the outline presented in the pseudo-code. To support the modeling in the Arena™ Environment, you must first define a number of variables and other data.

2.6.3.2 VARIABLE Module. Begin by defining the variables to be used in the model (Figure 2-32). Open the Arena™ Environment and define the following variables:

Exhibit 2-4   Pseudo-Code for Example

```
CREATE 10 entities with TBA EXPO(12)
ASSIGN myModel=1
    myType=DISC(.9,1,1,2)
    GOTO label A

CREATE 10 entities with TBA EXPO(22)
ASSIGN myModel=2
    myType=DISC(.7,1,1,2)
    GOTO label A

A: // determine the processing time for first station
  ASSIGN vCounter=1
    vNumIterations=10
    myProcessingTime=0
  WHILE vCounter≤vNumIterations
    ASSIGN myProcessingTime=myProcessingTime + ePTime(myModel, myType)
    vCounter=vCounter+1
  ENDWHILE

DELAY for myProcessingTime
// determine the processing time for the second station
  ASSIGN vCounter=0
    vNumIterations=10
    myProcessingTime=0
B: ASSIGN myProcessingTime=myProcessingTime+ePTime(myModel, myType)
    vCounter=vCounter+1
DECIDE if vCounter≤vNumIterations
    If true, GOTO B

DELAY for myProcessingTime

DISPOSE
```

	Name	Rows	Columns	Data Type	Clear Option	Initial Values	Report Statistics
1	vCounter			Real	System	0 rows	☐
2	vNumIterations			Real	System	1 rows	☐
3	vMTCount	2	2	Real	System	0 rows	☐
4	vPTLocation1	2	2	Real	System	0 rows	☐
5	vPTLocation2	2	2	Real	System	0 rows	☐
	Double-click here to add a new row.						

■ **FIGURE 2-32**   Defining the variables

■ **TABLE 2-7**
Arena's Distributions

Beta distribution	BETA(Alpha1,Alpha2[,Stream])
Normal distribution	NORM(Mean,SD[,Stream])
Empirical Continuous distribution	CONT(Prob1,Value1,Prob2,Value2, . . . [,Stream])
NSExpo distribution	Non-homogeneous Poisson process
Empirical DISCRETE distribution	DISC(Prob1,Value1,Prob2,Value2, . . . [,Stream])
Poisson distribution	POIS(Mean[,Stream])
k-Erlang distribution	ERLA(Mean, k[,Stream])
Lognormal distribution	LOGN(LogMean,LogStd[,Stream])
Random Number Between 0 and 1	RA
Exponential distribution	EXPO(Mean[,Stream])
Triangular distribution	TRIA(Min,Mode,Max[,Stream])
Gamma distribution	GAMM(scale,shape[,Stream])
Uniform distribution	UNIF(Min,Max[,Stream])
Johnson distribution	JOHN(shape1,shape2,scale,location[,Stream])
Weibull distribution	WEIBull(scale,shape[,Stream])

vCounter: Used to count the number of times the entity goes through the WHILE-ENDWHILE loop, a scalar variable. This WHILE-ENDWHILE loop will be used to compute the processing time at each location.

vNumIterations: Used to indicate the total number of times to go through the WHILE-ENDWHILE loop, a scalar variable. This should be initialized to the value 10.

vMTCount: Used to assist with counting and displaying the number of entities of each model/type combination, a 2D variable (2 rows, 2 columns)

vPTLocation1: Used to assist with displaying the processing time of each model/type combination, a two2D variable (2 rows, 2 columns)

vPTLocation2: Used to assist with displaying the processing time of each model/type combination, a 2D variable (2 rows, 2 columns)

At this juncture, a way is needed to store the distributions associated with each model/product type. A distribution in Arena™ is simply a mathematical expression. Table 2-7 presents a listing of Arena's built-in probability distributions. In the table, the function name is given along with the parameters required by the function. The [,Stream] indicates an optional parameter which helps in controlling the randomness of the function.

In addition to using distributions in expressions, you can use logical constructs and mathematical operators to build expressions. Arena™ also has a list of mathematical functions available to the modeler as show in Table 2-8.

To represent the model/product type distributions, the EXPRESSION data module can be used.

2.6.3.3 EXPRESSION Module. While you can build expressions using the previously mentioned functions and operators, a mechanism is needed to "remember" the expressions, and in particular, for this problem, the appropriate distribution must be looked up based on the model type and the product type. The EXPRESSION module found on the Advanced Process template allows just this functionality. Expressions as defined in an EXPRESSION module are simply names assigned to expressions. An expression defined in an EXPRESSION module is substituted by Arena™ wherever the named expression appears in the model. The EXPRESSION module allows for the modeler to define an array of expressions.

■ **TABLE 2-8**
Arena's Mathematical Functions

Function	Description
ABS(a)	Absolute value
ACOS(a)	Arc cosine
AINT(a)	Truncate
AMOD(a1, a2)	Real remainder, returns (a1-(AINT(a1/a2)*a2))
ANINT(a)	Round to nearest integer
ASIN(a)	Arc sine
ATAN(a)	Arc tangent
COS(a)	Cosine
EP(a)	Exponential (e^a)
HCOS(a)	Hyperbolic cosine
HSIN(a)	Hyperbolic sine
HTAN(a)	Hyperbolic tangent
MN(a1, a2, . . .)	Minimum value
MOD(a1, a2)	Integer remainder, same as AMOD except the arguments are truncated to integer values first
MX(a1, a2, . . .)	Maximum value
LN(a)	Natural logarithm
LOG(a)	Common logarithm
SIN(a)	Sine
SQRT(a)	Square root
TAN(a)	Tangent

Go to the Advanced Process panel and define the expressions used in the model by clicking on the EXPRESSION module in the project bar, defining the name of the expression, ePTime, to have two rows and two columns, where the row designates the model type and the column designates the product type. Use the spreadsheet view to enter the exponential distribution with the appropriate mean values as indicated in Figure 2-33.

This creates an arrayed expression. This is similar to an arrayed variable, except that the value stored in the array element can be any valid Arena™ expression. Now that you have the basic data predefined for the model, you can continue the model building process by starting with the CREATE modules. Since the model is large, you will begin by laying down the first part of the model. You should place the two CREATE modules and the two ASSIGN modules into the model window as shown in Figure 2-31. Fill in the create modules as indicated in Figure 2-34. This example model only requires 10 products of each model configuration to be created.

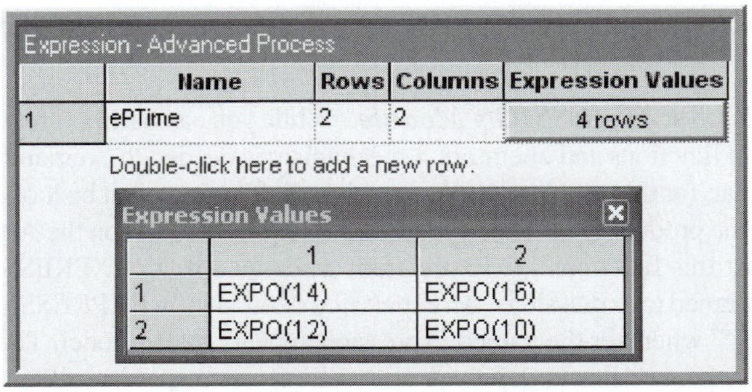

■ **FIGURE 2-33**
Expression module spreadsheet view

	Name	Entity Type	Type	Value	Units	Entities per Arrival	Max Arrivals	First Creation
1	Create Model Type 1	Entity 1	Random (Expo)	12	Minutes	1	10	0.0
2	Create Model Type 2	Entity 1	Random (Expo)	22	Minutes	1	10	0.0

■ **FIGURE 2-34**   **CREATE modules spreadsheet view**

The ASSIGN modules can be used to define the attributes and to assign their values. In the ASSIGN modules, the values of the *myModel* and *myType* attributes are assigned. There are two different models—1 and 2. Also, each model is assigned a type, dependent on a specific probability distribution for that model. The DISC() function can be used to represent the model type. Open the first ASSIGN module for model type 1 and make it look like the ASSIGN module show in Figure 2-35.

For variety, you can use the spreadsheet view to fill out the second ASSIGN module. Click on the ASSIGN module in the model window. The corresponding ASSIGN module will be selected in the spreadsheet view. Click on the assignment rows to get the assignments window and fill it in as indicated in Figure 2-36.

2.6.3.4 Hierarchical Submodels. In this section, the middle portion of the model as indicated in Figure 2-37, which includes a submodel, will be built. A submodel in Arena™ is a named collection of modules that is used to organize the main model. Go to the Object menu and choose Submodel > Add Submodel. Place the cross-haired cursor that appears in the

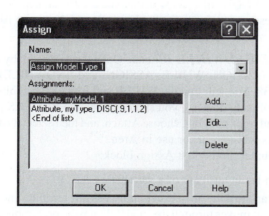

■ **FIGURE 2-35**   **ASSIGN module dialog box for model type 1**

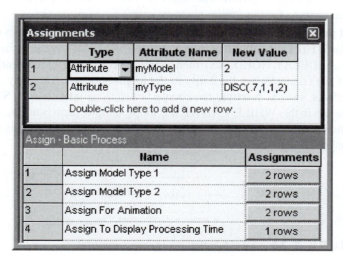

■ **FIGURE 2-36**   **Second ASSIGN module in spreadsheet view**

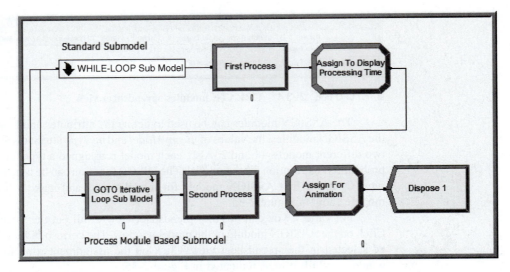

■ **FIGURE 2-37** First location processing

model window at the desired location and connect your ASSIGN modules to it as indicated in the figure. By right-clicking on the submodel and selecting the properties menu item, you can give the submodel a name. This submodel will demonstrate the use of a WHILE-ENDWHILE loop in Arena^{TM}. The WHILE- ENDWHILE construct will be used as an iterative looping technique to compute the total processing time according to the number of iterations (10) for the given model/ type combination.

2.6.3.5 WHILE-ENDWHILE Blocks.
Open the submodel by double-clicking on it and add the modules (ASSIGN, WHILE, ASSIGN, and ENDWHILE) to the submodel as indicated in Figure 2-38. You will find the WHILE and ENDWHILE blocks on the BLOCKS panel. If you haven't already done so, you should attach the BLOCKS panel by going to the Basic Process panel and right-clicking, choose Attach, and then select the file called *Blocks.tpo*. A new panel of blocks will appear for use in Arena^{TM}.

Initialize the counter, in the first Assign block:

1. The counter can be an attribute or a variable.
2. If the counter is an attribute, then every entity has this attribute and the value will persist with the entity as it moves through the model. It may be preferable to have the counter as an attribute for the reason given in item 3.
3. If the counter is a (global) variable, then it is accessible from anywhere in the model. If the WHILE-ENDWHILE loop contains a block that can cause the entity to stop moving (e.g., DELAY, HOLD, etc.), then it may be possible for another entity to change the value of the counter when the entity in the loop is stopped. This may cause unintended side effects, such as resetting the counter, causing an infinite loop. Since there are no time delays in this WHILE-ENDWHILE loop example, there is no danger of this. Remember, there can only be one entity moving (being processed by the simulation engine) at anytime while the simulation is running.

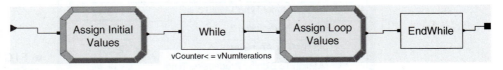

■ **FIGURE 2-38** First location processing submodel

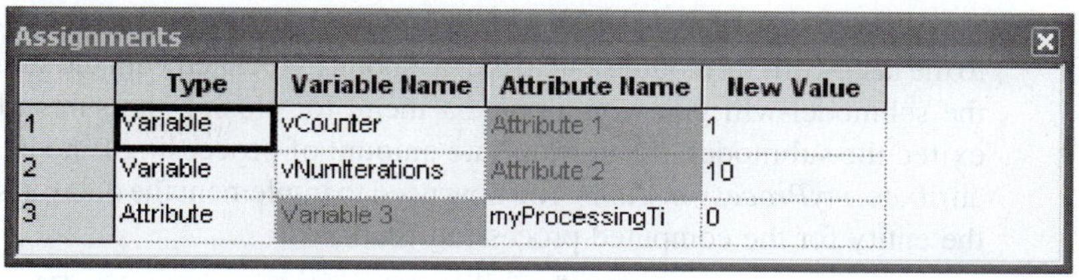

Assignments				
	Type	Variable Name	Attribute Name	New Value
1	Variable	vCounter	Attribute 1	1
2	Variable	vNumIterations	Attribute 2	10
3	Attribute	Variable 3	myProcessingTi	0

■ **FIGURE 2-39**   ASSIGN in first submodel

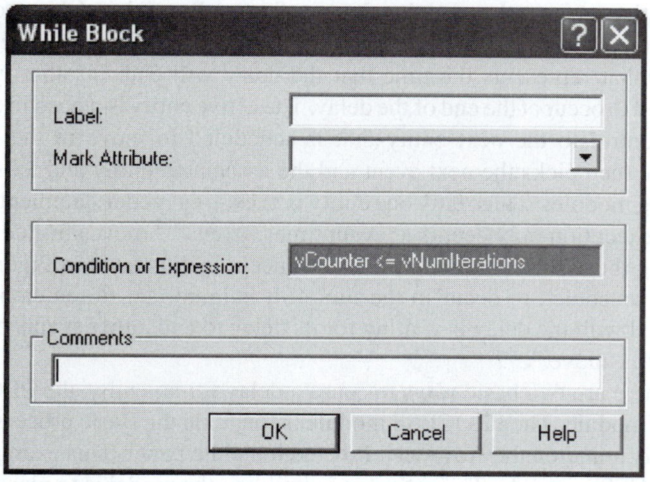

While Block

Label:

Mark Attribute:

Condition or Expression: vCounter <= vNumIterations

Comments

OK   Cancel   Help

■ **FIGURE 2-40**
WHILE block in first submodel

Open the first ASSIGN module and make it look like that shown in Figure 2-39. Note how the variables are initialized prior to the WHILE. In addition, the *myProcessingTime* attribute has been defined and initialized to zero.

Now the counter must be checked in the condition expression of the WHILE block. The WHILE and ENDWHILE blocks can be found on the Blocks Panel. Open the WHILE block and make it look like that shown in Figure 2-40. When the variable *vCounter* is greater than *vNumIterations*, the WHILE condition will evaluate to false and the entity will exit the WHILE-ENDWHILE loop.

Make sure to update the value of the counter variable or the evaluation expression in the body of the WHILE and ENDWHILE loop; otherwise, you may end up with an infinite loop. Open the second ASSIGN module and make the assignments as shown in Figure 2-41. In the first assignment, the attributes *myModel* and *myType* are used to index into the ePTime array of expressions. Each time the entity hits this assignment statement, the value of the *myProcessingTime* attribute is updated based on the previous value, resulting in a summation. The variable *vCounter* is incremented each time to ensure that the WHILE loop will eventually end.

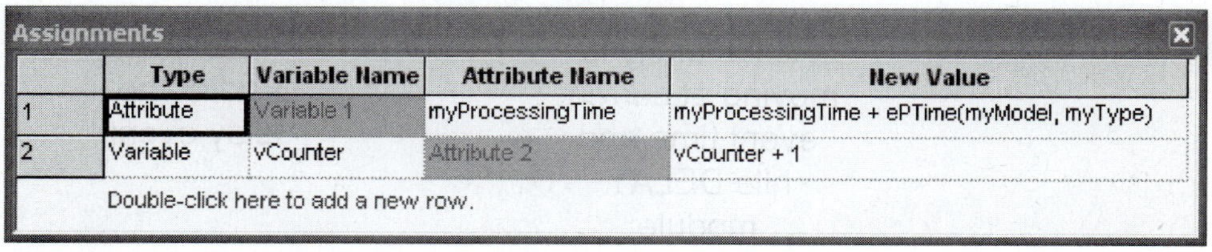

Assignments				
	Type	Variable Name	Attribute Name	New Value
1	Attribute	Variable 1	myProcessingTime	myProcessingTime + ePTime(myModel, myType)
2	Variable	vCounter	Attribute 2	vCounter + 1
	Double-click here to add a new row.			

■ **FIGURE 2-41**   Second ASSIGN module in first submodel

There is nothing to fill in for the ENDWHILE block. Just make sure that it is connected to the last ASSIGN module and to the exit point associated with the submodel. Right clicking in the submodel will give you a context menu for closing the submodel. After the entity has exited the submodel, the appropriate amount of processing time has been computed in the attribute, *myProcessingTime*. You now need to implement the delay associated with processing the entity for the computed processing time.

2.6.3.6 PROCESS Module Delay Option. In Arena^{TM}, only the active entity is moving through the modules at any time. The active entity is the entity that the simulation executive has determined is associated with the current event. At the current event time, the active entity executes the modules along its flow path until it is delayed or cannot proceed due to a condition in the model. When the active entity is time delayed, an event is placed on the event calendar that represents the time that the delay will end. In other words, a future event is scheduled to occur at the end of the delay. The active entity is placed in the future events list and yields control to the next entity that is scheduled to move in the model. The simulation executive then picks the next event and the associated entity and tells that entity to continue executing modules. Thus, only one entity is executing model statements at the current time and module execution only occurs at event times. Arena^{TM} moves from event time to event time, executing the associated modules. This concept is illustrated in Figure 2-42, where a simple delay is scheduled to occur in the future. It is important to remember that while the entity associated with the delay is waiting for its delay to end, other events may occur causing other entities to "move."

There are two basic ways to cause a delay for an entity, the PROCESS module and the DELAY module. The PROCESS module is found on the Basic process template. The DELAY module is found on the Advanced Process template panel. This example illustrates the use of the PROCESS module. The next step in building the model is to place the PROCESS module after the submodel as shown in Figure 2-37. Open the PROCESS model and select the Delay option for the Action text drop-down box in the logic area of the dialog box. Now you can indicate the length of the delay. Arena^{TM} allows the user to select the type of delay. In this case, you want to delay by the amount of time computed for the processing time. This value is stored in the *myProcessingTime* attribute. In the Delay Type area, choose expression, and in the expression text field, fill in *myProcessingTime* as shown in Figure 2-43. Be careful to

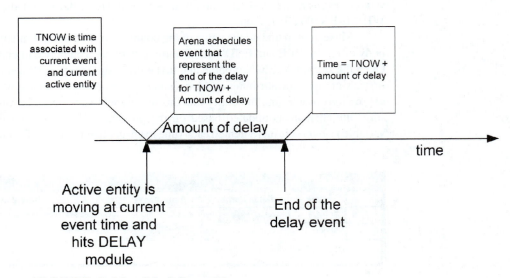

■ **FIGURE 2-42** Scheduling a delay

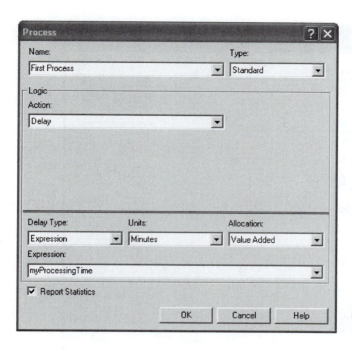

■ **FIGURE 2-43**
PROCESS module

appropriately set the time units (minutes) associated with the delay. A common mistake is to not set the units and have too long or too short a delay. The too long a delay is often caught during debugging because the entities take such a long time to leave the module.

Now you will add an ASSIGN module to facilitate the displaying of the values of variables in the model window during the simulation. This will also illustrate how arrays can be used in a model. Using the spreadsheet view, make the ASSIGN module look like that show in Figure 2-44.

The values of attributes cannot be displayed using the variable animation constructs in Arena™. So, in this ASSIGN module, an array variable is used to temporarily assign the value of the attribute to the proper location of the array variable so that it can be displayed. Go to the toolbar area in Arena™ and right-click. Make sure that the Animate toolbar is available. From the Animate toolbar, choose the 0.0 button. You will get a dialog box for entering the variable to display. Fill out the dialog box so that it looks like Figure 2-45. After filling out the dialog box and choosing OK, you will need to place the cross-hair somewhere near the ASSIGN module in the model window. In fact, the element can be placed anywhere in the model window. Repeat this process for the other elements of the *vPTLocation1* array.

The model can now be completed. Lay down the modules indicated in Figure 2-37 to complete the model. In Figure 2-37, the second submodel uses a PROCESS based submodel. It looks like a PROCESS module, and is labeled GOTO Iterative Loop Sub Model (see Figure 2-46). To create this submodel, lay down a PROCESS module and change its type to submodel. In what follows, you will examine a different method for looping in the second submodel. Other

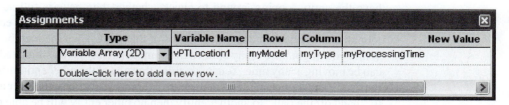

■ **FIGURE 2-44   ASSIGN module with arrays**

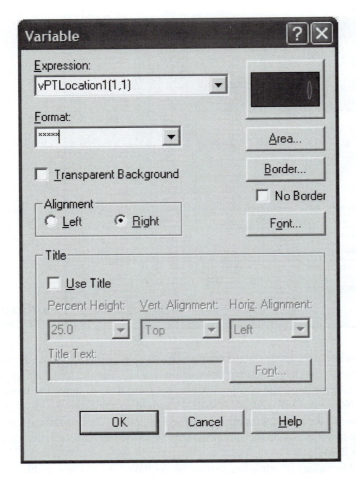

■ **FIGURE 2-45**
Animate variable dialog box

than that, the rest of the model is similar to what has already been done. The second submodel contains the modules shown in Figure 2-47. Place and connect the ASSIGN, DECIDE, and ASSIGN modules as indicated in Figures 2-48 and 2-49. In this logic, the entity initializes a counter for counting the number of times through the loop and updates the counter. An If/Then DECIDE module is used to redirect the entity back through the loop the appropriate number of times. This is GOTO programming!

Again, be sure to use logic that will change the tested condition (in this case *vCounter*) so that you do not get an infinite loop. I prefer the use of the WHILE-ENDWHILE blocks instead of this "go-to" oriented implementation, but it is ultimately a matter of taste.

In general, you should be careful when implementing logic in which the entity simply loops around to check a condition. As mentioned, you must ensure that the condition can change to prevent an infinite loop. A common error is to have the entity check a condition that can be changed from somewhere else in the model, such as a queue becoming full. Suppose you have the entity looping around to check whether the queue is full so that when the queue becomes full the entity can perform other logic. This seems harmless enough; however, if the looping entity does not change the status of the queue, then the loop becomes infinite! Why? While you might have many entities in the model at any given simulated time, only one entity can be moving at any time. You might think that one of the other entities in the model may enter the queue and thus change its status. If the looping entity does not enter a module that causes it to "hand off the ball" to another entity then no other entity will be able to move to change the queue condition. If this occurs, you will have an infinite loop.

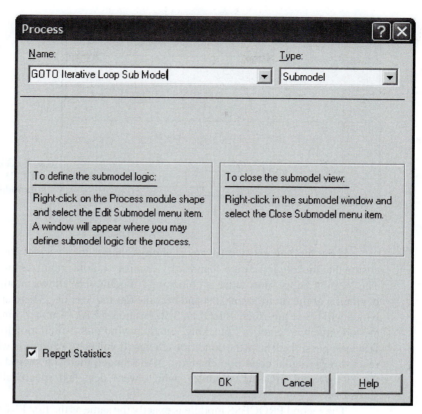

■ **FIGURE 2-46** PROCESS-based submodel

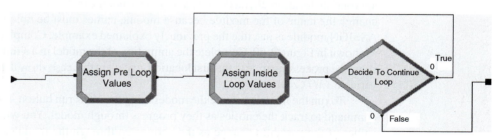

■ **FIGURE 2-47** Second submodel

Assignments — Assign Pre-Loop Values

	Type	Variable Name	Attribute Name	New Value
1	Variable	vCounter	Attribute 1	0
2	Variable	vNumIterations	Attribute 2	10
3	Attribute	Variable 3	myProcessingTime	0

Assignments — Assign Inside Loop Values

	Type	Variable Name	Attribute Name	New Value
1	Attribute	Variable 1	myProcessingTime	myProcessingTime + ePTime(myModel, myType)
2	Variable	vCounter	Attribute 2	vCounter + 1

■ **FIGURE 2-48** ASSIGN modules in second submodel

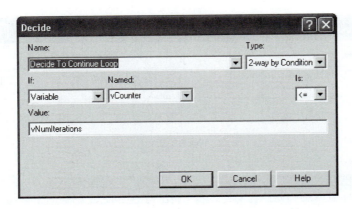

■ **FIGURE 2-49**
DECIDE module in second submodel

If the looping entity does not explicitly change the checked condition, then you must ensure that the looping entity "hands off" control to another entity. How can this be achieved? The looping entity must enter a "blocking" module that allows another entity to get to the beginning of the future events list and become the active entity. There are a variety of modules which will cause this, such as HOLD, SEIZE, and DELAY. Another thing to remember, even if the looping entity enters a "blocking" module, your other simulation logic must ensure that it is at least possible to have the condition change. If you have a situation where an entity needs to react to a particular condition changing in the model, then you should consider investigating the HOLD and SIGNAL modules, which were designed specifically for these types of situations.

The second PROCESS module is exactly the same as the first PROCESS module. In fact, you can just copy the first PROCESS module and paste the copy into the model at the appropriate location. If you do copy and paste a module in this fashion, you should make sure to change the name of the module because module names must be unique in Arena™. The last ASSIGN module is just like the previously explained example. Complete the ASSIGN module as shown in Figure 2-50. Complete the animation of the model in a similar fashion as you did to show the processing time at the first location, except in this case show the counts captured by the variable *vMTCount*.

To run the model, execute the model using the VCR run button. It is useful to use the step command to track the entities as they progress through model. You will be able to see that the values of the variables change as the entities go through the modules.

To recap, in this example you learned about attributes and how they are attached to entities. In addition, you saw how to share global information through the use of arrays of variables and arrays of expressions. The concept of scheduling events via the use of the delay option in the PROCESS panel was also introduced. Finally, your memory of basic programming constructs such as decision logic and iterative processing was also refreshed.

Assignments						
	Type	**Variable Name**	**Row**	**Column**	**New Value**	
1	Variable Array (2D)	vPTLocation2	myModel	myType	myProcessingTime	
2	Variable Array (2D)	vMTCount	myModel	myType	vMTCount(myModel,myType) + 1	

■ **FIGURE 2-50**   **Last ASSIGN module—assigned for animation**

2.7   Batching and Separating Entities

Of the flowchart modules on the Basic Process panel, there are two remaining modules to discuss: BATCH and SEPARATE. The BATCH module allows entities to be grouped together into a temporary or permanent representative entity. A representative entity is an entity that consists of a group of entities that can travel together and be processed as if there was only one entity. A temporary representative entity can be split apart, by a SEPARATE module, so that the individual entities can again be processed. For a permanent representative entity, the batched individual entities cannot be separated. You can use a permanent entity when you no longer need the information associated with the individual entities in the group. You can think of a temporary entity as an "open bag" that holds the batched members in the group, and a permanent entity as a "closed bag." Temporary entities are useful for modeling a manufacturing system when items are assembled and disassembled during processing. A permanent entity is useful in modeling a situation where the assembly is complete.

The batching of entities can be performed based on a number of criteria. For example, the batch might form when at least 10 entities are available. Thus, an important consideration in batching is how to determine the size of the batch. In addition, batches might be formed based on entities that have similar attributes. For example, suppose red and blue paper folders are being manufactured in batches of five of a specific color for sale by color. The batching of entities is also useful in synchronizing the movement of entities. As previously indicated, when potential members of a batch enter the BATCH module, they wait until the batching conditions have occurred before they continue their movement with the representative entity in the model. Because of this, the BATCH module has a queue that holds the entities that are waiting to be batched. In the case of a temporary representative entity, the SEPARATE module provides the functionality to split apart the representative entity into the individual entities. The SEPARATE module also can *duplicate or clone* the incoming entity. This provides an additional mechanism to introduce new entities into the model, that is, another way of creating entities. In the following examples, the BATCH and SEPARATE modules will be examined to investigate some of the rules involved in the use of these modules.

EXAMPLE 2-1   TIE-DYE T-SHIRTS

Suppose production orders for tie-dye T-shirts arrive at a production facility according to a Poisson process with a mean rate of one per hour. There are two basic psychedelic designs involving either red or blue dye. For some reason the blue shirts are a little more popular than the red shirts so that when an order arrives about 70% of the time it is for the blue dye designs. In addition, there are two different package sizes for the shirts, three and five units. There is a 25% chance that the order will be for a package size of 5 and a 75% chance that the order will be for a package size of three. Each of the shirts must be individually hand made to the customer's order design specifications. The time to produce a shirt (of either color) is uniformly distributed in the range of 15 to 25 minutes. There are currently two workers who are set up to make either shirt. When an order arrives to the facility, its type (red or blue) is determined and the pack size is determined. Then, the appropriate number of white (undyed) shirts are sent to the shirt makers with a note pinned to the shirt indicating the customer order, its basic design, and the pack size for the order. Meanwhile, the paperwork for the order is processed and a customized packaging letter and box is prepared to hold the order. It takes another worker between 8 to 10 minutes to make the box and print a custom thank-you note. After the packaging is made, it waits prior to final inspection for the shirts associated with the order. After the shirts are combined with the packaging, they are inspected by the packaging worker, which has a packaging time distributed according to a triangular distribution with a minimum of 5 minutes, a most likely value of 10 minutes, and a maximum value of 15 minutes. Finally, the boxed customer order is sent to shipping.

2.7.1 Conceptualizing the Model

Before proceeding, you might want to jot down your answers to the modeling recipe questions and then you can compare how you are doing with respect to what is presented in this section. The modeling recipe questions follow:

- What is the system? What information is known by the system?
- What are the required performance measures?
- What are the entities? What information must be recorded or remembered for each entity? How are entities introduced into the system?
- What are the resources that are used by the entities? Which entities use which resources and how?
- What are the process flows? Sketch the process or make an activity flow diagram.
- Develop pseudo-code for the situation.
- Implement the model in Arena™.

The entities can be conceptualized as the arriving orders. Since the shirts are processed individually, they should also be considered entities. In addition, the type of order (red or blue) and the size of the order (three or five) must be tracked. Since the type of the order and the size of the order are properties of the order, attributes can be used to model this information. The resources are the two shirt makers and the packager. The flow is described in the scenario statement: orders arrive, shirts are made, and meanwhile packaging is made. Then, orders are assembled, inspected, and finally shipped.

It should be clear that a CREATE module, set up to generate Poisson arrivals can create the orders, but if shirts are entities, how should they be created? To do this, a SEPARATE module can be used to make the number of shirts required based on the size of the order. After this, there will be two types of entities in the model, the orders and the shirts. The shirts can be made and meanwhile the order can be processed. When the shirts for an order are made, they need to be combined together and then matched with the order. This implies that a method is required to uniquely identify the order. This is another piece of information that both the order and the shirt require. Thus, an attribute will be used to note the order number.

The activity diagram for this situation is given in Figure 2-51. After the order is created, the process separates into the order making process and the shirt-making process. Note that the orders and shirts must be synchronized together after each of these processes. In addition, the order making process and the final packaging process share the packager as a resource.

Exhibit 2-5 represents the activity diagram in pseudo-code. In the exhibit, a variable is incremented each time an order is created and then the value of the variable is assigned to the attribute representing the order. In addition, the order type and the order size are both assigned based on the probability distribution information that was given in the problem. The two pseudo-code segments labeled A and B represent the parallel processing.

2.7.2 Building the Model

The completed Arena™ model is given in Figure 2-52. Note how there is a parallel structure in the middle of the model. This is where the orders are processed separately from the shirts. This is quite a common pattern in simulation modeling.

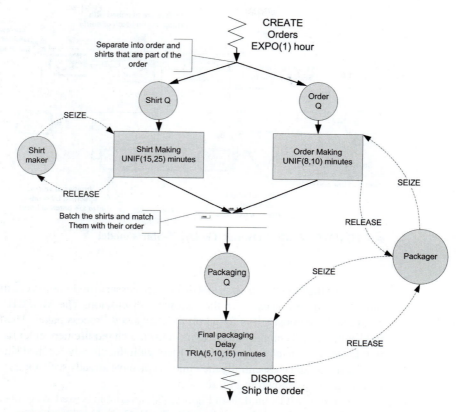

■ FIGURE 2-51   Activity diagram for tie-dye T-shirts example

Exhibit 2-5   **Pseudo-Code for Tie-Dye T-Shirts Example**

```
CREATE orders every hour exponentially

ASSIGN vOrderNumber=vOrderNumber+1
    myOrderNumber=vOrderNumber
    myOrderType=DISC(0.7, 1, 1.0, 2)
    myOrderSize=DISC(0.75, 3, 1.0, 5)
SEPARATE into order and shirts based on myOrderSize
    Shirts GOTO Label A shirt making
    Order GOTO Label B order making

A: PROCESS order using packager for UNIF(8,10) minutes
Wait until shirts for matching order are completed
    When match occurs, GOTO Label C, order forming

B: PROCESS each shirt using shirt maker for UNIF(15,20) minutes
BATCH shirts on order together
Wait until order for the matching shirts are completed
    When match occurs, GOTO Label C, order forming

C: BATCH order and shirts together
PROCESS completed order by packager for TRIA(5,10,15) minutes
Ship order out of the system
```

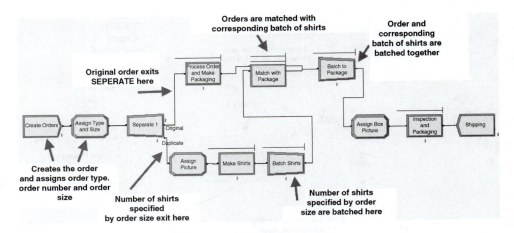

■ **FIGURE 2-52** **Overall Tie-Dye T-Shirts model**

Take the time now to drag and drop the necessary modules to your model area. Each of the modules will be discussed in the following discussion. The MATCH module has not been mentioned previously. It is found on the Advanced Process panel. The MATCH module will hold entities until enough entities to meet the match requirement enter the queues. This module will be used to match the processed orders with the already batched shirts for the orders. The CREATE module is essentially the same as you have already seen. Open your CREATE module and fill it out as shown in Figure 2-53.

In the ASSIGN module of Figure 2-54, a variable is used to count each order as it arrives. This unique number is then assigned to the *myOrderNumber* attribute.[4] This attribute will be used to uniquely identify the order in the model. Then, the DISC() random distribution function is used to assign the type of order (Blue = 1, Red = 2) according the given probabilities. Note the concept of blue and red are mapped to the numbers 1 and 2, respectively. This is often the case in Arena™, since attributes and variables can only be real numbers. The DISC() distribution is used to randomly assign the size of the order. This will be remembered by the entity in the *myOrderSize* attribute.

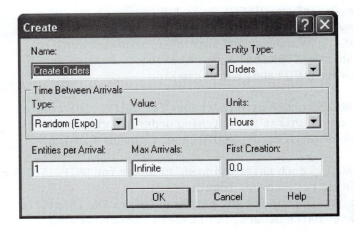

■ **FIGURE 2-53**
CREATE model for Tie-Dye
T-Shirt model

---

[4]The Entity.SerialNumber attribute can also be used for this purpose. A user defined attribute was used here to illustrate how to "do it yourself."

Assignments

	Type	Variable Name	Attribute Name	
1	Variable	vOrderNumber	Attribute 1	vOrderNumber + 1
2	Attribute	Variable 2	myOrderNumber	vOrderNumber
3	Attribute	Variable 3	myOrderType	DISC(0.7, 1, 1.0, 2)
4	Attribute	Variable 4	myOrderSize	DISC(0.75, 3, 1.0, 5)

■ **FIGURE 2-54**   Assigning the order number, type, and size

After you have filled in the dialog boxes as shown, you can proceed to the SEPARATE module. The SEPARATE module has two options, Split Existing Batch and Duplicate Original. The Duplicate Original option is used here. The entity that enters the SEPARATE module is an order; however, after proceeding through the module both orders and shirts will depart from the module. Open the SEPARTE module and note the textbox labeled # of Duplicates as shown in Figure 2-55. This field indicates the number of entities that will be created by the SEPARATE module. It can be any valid Arena^{TM} expression. The size of the order should be used to indicate how many entities to create. For example, if the *myOrderSize* attribute was set to 3, then three additional entities will be cloned or duplicated from the original entering entity. For modeling purposes, these entities will be conceptualized as the shirts. Don't worry about the text field in the dialog box relating to cost attributes at this time.

When an entity enters the SEPARATE module, the original will exit along the original exit point, and the duplicate (or split off) entities will exit along the duplicate exit point. Since the created entities are duplicates, they will have the *same* values for all attributes as the original. Thus, each shirt will know its order type (*myOrderType*) and its order size (*myOrderSize*). In addition, each shirt will also know which order it belongs to through the (*myOrderNumber*) attribute. These shirt attributes will be used when combining the orders with their shirts after all the processing is complete.

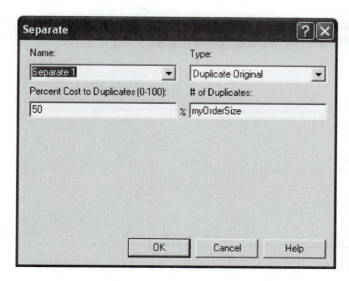

■ **FIGURE 2-55**   Using the SEPARATE module to create the shirts

Resource - Basic Process						
	Name	**Type**	**Capacity**	**Busy / Hour**	**Idle / Hour**	**Per Use**
1	ShirtMakers	Fixed Capacity	2	0.0	0.0	0.0
2	Packager	Fixed Capacity	1	0.0	0.0	0.0
	Double-click here to add a new row.					

■ **FIGURE 2-56** Defining the resources

Next, you should define and use the resources in the model. First you should add the resources to the model using the resource data sheet view as shown in Figure 2-56. Note here that there are two units of capacity for the shirt maker resource to represent the two workers involved in this process.

The packager is used to process the orders along the original entity path. Figure 2-57 shows the dialog box for using the packager. This is similar to how the pharmacist was implemented in the drive through pharmacy model. Fill out both PROCESS modules as shown in Figures 2-57 and 2-58.

After the order's packaging is complete, it must wait until the shirts associated with the order are made. As seen in Figure 2-52, the orders go to a MATCH module, where they will be matched by the attribute *myOrderNumber* with the associated group of shirts that exit the BATCH module named *Batch Shirts*. Open the BATCH module that you previously placed, and fill it out as shown in Figure 2-59. This BATCH module creates a permanent entity because the shirts do not need to be processed individually after being combined into the size of the order. The entities

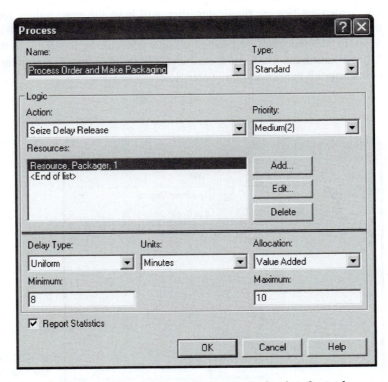

■ **FIGURE 2-57** Seizing, delaying, and releasing the packager

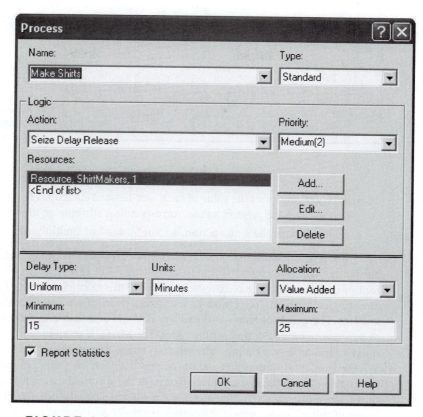

■ **FIGURE 2-58**   Seizing, delaying, and releasing the shirt makers

that enter this module are shirts. They can be red or blue. The shirts enter the module and wait until the number of entities (shirts) in the batch queue with the indicated attributes reaches *myOrderSize*. Then these entities (shirts) are combined and leave as a permanent entity.

The attributes of the representative entity are determined by the selection indicated in the Save Criterion textbox field. All the user-defined attributes of the representative entity are assigned based on the Save Criterion.

■ *First* or *Last* assigns the user-defined attributes based on the first/last entity forming the batch.

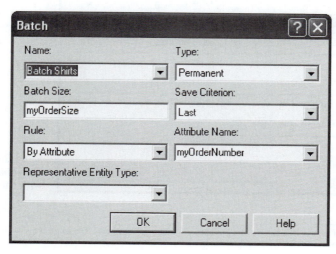

■ **FIGURE 2-59**
Batching the shirts together

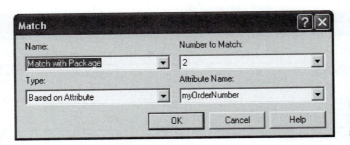

■ **FIGURE 2-60**
**Matching the order with the
shirt group**

- *Product* multiplies the value of each user-defined attribute among all entities in the batch
  and assigns the product to the corresponding attribute of the representative entity.
- *Sum* performs the same action, adding instead of multiplying.

For a detailed discussion of how Arena™ handles the assigning of its special-purpose
attributes, you should refer to the help files on the BATCH module. In fact, reading the help files
for this important module is extremely useful. In this example, *First* or *Last* can be used. When
an entity leaves the BATCH module, it will be a group of *myOrderSize* shirts for specified order
number. The group of shirts will exit the BATCH and then enter the MATCH module to be
matched with the packaging for the order. The MATCH module in Figure 2-60 has two match
queues. The order/packaging enters the top queue and groups of shirts enter the lower queue.
Whenever an entity enters the module, Arena™ will check if the appropriate number of entities
to match are available in each queue that meet the specified matching criteria. In this example,
the order number is used as the matching attribute so that the order/packaging will wait for the
group of shirts (or vice versa). After they are matched, the matching entities are released from
the MATCH[5] module.

The group of shirts and the order/packaging are now synchronized. It is very common to use
a BATCH module, as done here, to make the synchronization permanent by batching the entities
together as per Figure 2-61. Since the order number is unique and the batch size is 2, the result of
this BATCH module will be to combine the group of shirts and the order/packaging together into a

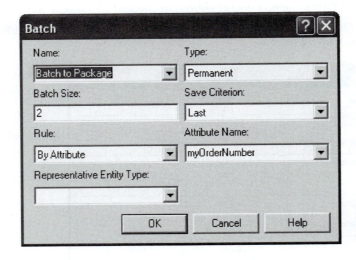

■ **FIGURE 2-61**
**Batching the matched order
with the shirt group**

---

[5]The use of the MATCH module here is primarily for illustrative purposes. The same results can be obtained without the MATCH
module. Try it!

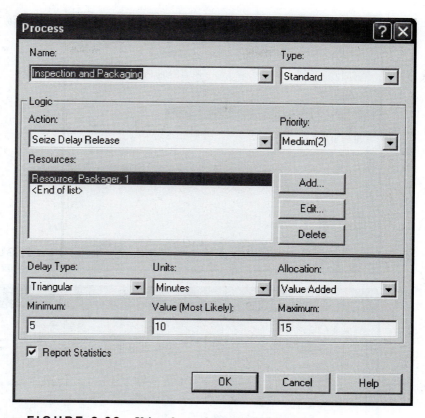

■ **FIGURE 2-62**   Using the packager to do inspection

permanent entity. After the order is completed, it is sent to a PROCESS module (Figure 2-62) where it is inspected and then sent to a DISPOSE module, which represents shipping.

The only items that have not been discussed are the two ASSIGN modules labeled *Assign 2* and *Assign 3*. In these modules, the animation picture of the entities is changed so that the operation of the SEPARATE, MATCH, and BATCH modules will be more visible in the animation. To do this, a SET is used to hold the pictures for each type of entity.

The SET module is found on the Basic Process panel as a data module. A set is a group of related (similar) objects that are held in a list in Arena™. In this instance, a set is defined to hold a list of animation pictures. Click on the SET module and use the edit via dialog box option to add the picture of the blue ball and the red ball to a set with its type set to Entity Picture as shown in Figure 2-63. The ordering of the list of the entries in the set is important. The first location in the set is associated with the blue ball and the second location in the set is associated with the red ball. This is done because the numbers 1 and 2 have been mapped to the type of order, Blue = 1 and Red = 2. The order type attribute can be used to index into the set to assign the appropriate picture to the entity.

In Figure 2-64, an assignment of type Other has been made to the *Entity.Picture* special purpose attribute. The expression builder was used to build the appropriate expression for indexing into the set as shown in Figure 2-65. Figure 2-66 indicates the resulting ASSIGN module for directly assigning a picture to an entity. In this case, since the shirts and packaging have been combined, the combined entity is shown as a box.

You should run the model for 8 hours and set the base time unit to minutes. When you run the model use the animation slider bar to slow down the speed of the animation. You will see in the animation how the SEPARATE, MATCH, and BATCH modules operate. For this short

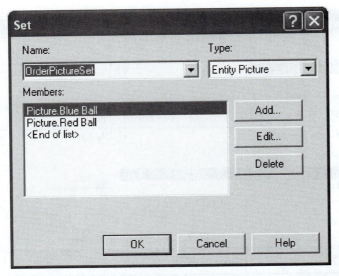

■ **FIGURE 2-63**
Defining a picture set

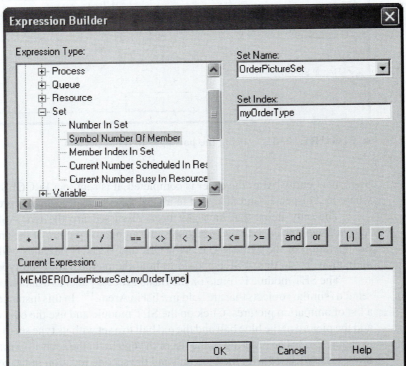

■ **FIGURE 2-64** Assign picture: assigning an animation picture based on a set index

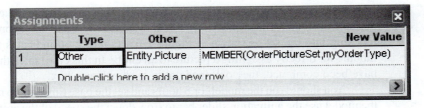

■ **FIGURE 2-65** Using the expression builder to index into a set

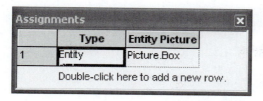

■ **FIGURE 2-66** Assign box picture: directly assigning an animation picture

period of time, it took about 75 minutes on average to produce an order. You are asked to further explore the operation of this system in the exercises.

In this example, you learned that parallel processing can be easily accomplished in Arena^{TM} by using a SEPARATE module. This facilitates the creation of other entity (types) in the model and allows them to proceed "in parallel." The BATCH module shows how you can combine (and synchronize) the movement of a group of entities. In addition, the MATCH module also facilitates the synchronization of entity movement. Finally, you saw that Arena^{TM} also has another data structure for holding information (besides variables and attributes) called a "set." A set acts as a list of similar items and can be indexed to return the items in the set. The use of the set allowed an index to be used in an animation picture set to assign different pictures to the entities as they move through the model. This is especially useful in debugging and interpreting the actions of modules in Arena^{TM}.

2.8   SIMAN and Arena's Run Controller

In a programming language such as Visual Basic or Java, programmers must first define the data types that they will use (int, float, double, etc.) and declare the variables, arrays, classes, and other characteristics for those data types to be used in the program. The programmer then uses flow of control (if-then, while, etc.) statements to use the data types to perform the desired task. Programming in Arena^{TM} is quite different than programming in a language like C. An Arena^{TM} simulation program consists of the flowchart and data modules that have been used or defined in the model. By dragging and dropping flowchart modules and filling in the appropriate dialog box entries you are writing a program. Along with programming comes debugging and making sure that the program is working as it was intended. To be able to better debug your Arena^{TM} programs, you must have, at the very least, a rudimentary understanding of SIMAN, Arena's underlying language. SIMAN code is produced by the Arena^{TM} Environment, compiled, and then executed. This section provides an overview of SIMAN and Arena's debugger called the Run Controller. To better understand the programming aspects of Arena^{TM}, it is useful to first dissect an Arena^{TM} program.

2.8.1   An Arena^{TM} Program

This section reexamines the pharmacy model that was built in Chapter 1. Using the Run > SIMAN > View menu (Figure 2-67) will generate two files associated with the SIMAN program, the *mod* (model) file and the *exp* (experiment) file. The files can be viewed using the Window menu in the Arena^{TM} Environment.

The mod file contains the SIMAN representation for the flowchart modules that were laid out in the model window. The *exp* file contains the SIMAN representation for the data modules and simulation run control parameters that are used during the execution of the simulation (*mod* stands for model, and *exp* stands for experiment).

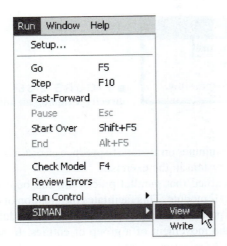

■ **FIGURE 2-67**   **Run** > **SIMAN** > **View menu**

Exhibit 2-6 presents the experiment (*exp*) file for the pharmacy model. For readers who are familiar with C or C++ programming, the *exp* file is similar to the header files used in those languages. The experiment file defines the elements that are to be used by the model during the execution of the simulation. As indicated for this file, the major elements are categorized as PROJECT, VARIABLES, QUEUES, RESOURCES, PICTURES, REPLI-CATE, and ENTITIES. Additional categories that are not used (and thus not shown) in this model include TALLIES and OUTPUTS. Each of these constructs will be discussed as you learn more about the modules in the Arena™ Environment. The experiment module declares and defines that certain elements will be used in the model. For example, the RESOURCES element indicates that there is a resource called *Pharmacist* that can be used in the model. Understanding the exact syntax of the SIMAN elements is not necessarily important, but being able to look for and interpret these elements can be useful in diagnosing certain errors that occur during the model debugging process. It is important to understand that these elements are defined during the model building process, especially when data modules are filled out in the Arena™ Environment.

Exhibit 2-7 presents the contents of the SIMAN model (mod) file for the Pharmacy model. The model file is similar in spirit to the *.c or *.cpp files in C/C++ programming. It represents the flowchart portion of the model. In this code, the semicolon (;) indicates a comment line or an end of a statement. The capitalized commands represent special SIMAN language keywords. For example, the ASSIGN keyword is used to indicate an assignment statement, that is, *variable = expression*. There are two keywords that you should immediately recognize from the model window, CREATE and DISPOSE. Each non-commented line has a line number that identifies the SIMAN statement. For example, 57$ identifies the DISPOSE statement. These line numbers are especially useful in following the execution of the code. For example, in the code, note the use of the NEXT() keyword. For example, on line number 54$, the keyword NEXT(1$) redirects the flow of control to the line statement 1$. You should also note that many lines of code are generated from the placement of the three modules (CREATE, PROCESS, DISPOSE). Much of this SIMAN code is added to the model to enable additional statistical collection. The exact syntax associated with this generated SIMAN code will not be the focus of the discussion; however, the ability to review this code and interpret its basic functionality is essential when developing and debugging complex models. The code generated in the experiment and model files is not directly editable. The only way to change this code is to change the model in the data modules or in the model window. Finally, the SIMAN code is not directly executed on the computer. Before running the model, the SIMAN code is checked for

Exhibit 2-6 SIMAN exp File Contents for Pharmacy Model

```
PROJECT, "Unnamed Project"," ",,,No,Yes,Yes,Yes,No,No,No,No,No,No;

VARIABLES: Get Medicine.WIP,CLEAR(System),CATEGORY("Exclude-Exclu-
de"),DATATYPE(Real):
    Dispose 1.NumberOut,CLEAR(Statistics),CATEGORY("Exclude"):
    Create Customers.NumberOut,CLEAR(Statistics),CATEGORY("Ex-
clude"):
    Get Medicine.NumberIn,CLEAR(Statistics),CATEGORY("Exclude"):
    Get Medicine.NumberOut,CLEAR(Statistics),CATEGORY("Exclude");

QUEUES: Get Medicine.Queue,FIFO,,AUTOSTATS(Yes,,);

PICTURES:     Picture.Airplane:
              Picture.Green Ball:
              Picture.Blue Page:
              Picture.Telephone:
              Picture.Blue Ball:
              Picture.Yellow Page:
              Picture.EMail:
              Picture.Yellow Ball:
              Picture.Bike:
              Picture.Report:
              Picture.Van:
              Picture.Widgets:
              Picture.Envelope:
              Picture.Fax:
              Picture.Truck:
              Picture.Person:
              Picture.Letter:
              Picture.Box:
              Picture.Woman:
              Picture.Package:
              Picture.Man:
```

syntax or other errors and then is translated to an executable form via a C/C++ compiler translation.

In a programming language like C, the program starts at a clearly defined starting point, such as the main() function. Arena^{TM} has no such main() function. Examining the actual SIMAN code that is generated from the model building process does not yield a specific starting point either. So where does Arena^{TM} start executing? To fully answer this question involves many aspects of simulation that have not yet been examined. Thus, at this point, a simplified answer to the question will be presented.

A simulation program models the changes in a system at specific events in time. These events are scheduled and executed by the simulation executive ordered by time. Arena^{TM} starts executing with the first scheduled event and then processes each event sequentially until there are no more events to be executed. Of course, this begs the question of how the first event gets

Exhibit 2-7   SIMAN mod File Contents for Pharmacist Model

```
;
; Model statements for module: BasicProcess.Create 1 (Create Customers)
;
2$         CREATE,
1,MinutesToBaseTime(expo(6)),Customer:MinutesToBaseTime(EXPO(6)):
NEXT(3$);

3$   ASSIGN:      Create Customers.NumberOut=Create Customers.NumberOut
+ 1:NEXT(0$);
;
;
; Model statements for module: BasicProcess.Process 1 (Get Medicine)
;
0$        ASSIGN:       Get Medicine.NumberIn=Get Medicine.NumberIn + 1:
                        Get Medicine.WIP=Get Medicine.WIP+1;
9$        QUEUE,        Get Medicine.Queue;
8$        SEIZE,        2,VA:
                        Pharmacist,1:NEXT(7$);

7$        DELAY:        expo(3),,VA;
6$        RELEASE:      Pharmacist,1;
54$       ASSIGN:       Get Medicine.NumberOut=Get Medicine.NumberOut + 1:
                        Get Medicine.WIP=Get Medicine.WIP-1:NEXT(1$);
;
;
; Model statements for module: BasicProcess.Dispose 1 (Dispose 1)
;
1$        ASSIGN:       Dispose 1.NumberOut=Dispose 1.NumberOut + 1;
57$       DISPOSE:      Yes;
```

scheduled. In Arena[TM], the first event is determined[6] by the CREATE module with the smallest First Creation time. Figure 2-68 illustrates the dialog box for the CREATE module. The text field labeled First Creation accepts an expression that must evaluate to a real number. In the figure, the expo() function is used to generate a random variable from the exponential distribution with a mean of 6 minutes to be used as the time of the event associated with the creation of an entity from this CREATE module.

A natural question at this point is, "What if there are ties?" In other words, what if there are two or more CREATE modules with a First Creation time set to the same value. For example, it is quite common to have multiple CREATE modules with their First Creation time set at zero. Which CREATE module will go first? This is where looking at the SIMAN model output is useful. The CREATE module with the smallest First Creation time that is listed first in the SIMAN model file will be the first to execute. Thus, if there are multiple CREATE modules, each with a First Creation time set to zero, then the CREATE module that is listed first will create its entity first.

---

[6]Again, this is a simplified discussion. There are other ways to get an initial event scheduled and Arena[TM] uses some internally generated events.

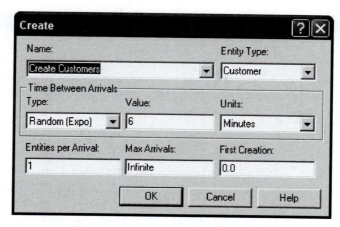

■ **FIGURE 2-68**   CREATE module dialog box

2.8.2   Using the Run Controller

An integral part of any programming effort is debugging the program. Simulation program development in Arena[TM] is no different. You have already seen how to do some rudimentary tracing via the READWRITE module; however, this section discusses some of the debugging and tracing capabilities that are built in the Arena[TM] Environment. Arena's debugging and tracing facilities come in two major forms: the Run Controller and animation. Some of the basic capabilities of Arena's animation have already been demonstrated. Using animation to assist in checking your model logic is highly recommended. Aspects of animation will be discussed throughout this text; however, this section concentrates on using Arena's Run Controller, which is located in the Run menu as shown in Figure 2-69.

Figure 2-70 illustrates how the Arena[TM] Environment appears after the Run Controller has been invoked on the original pharmacy model. The Run Controller allows tracing and debugging actions to be performed by the user through the use of various commands. Let's first learn about the various commands and capabilities of the Run Controller. Go to Arena[TM] Help, and in the index search, type in "command-driven Run Controller introduction." You should see a screen that resembles the one shown in Figure 2-71.

The Run Controller allows the Arena[TM] system to be executed in a debugging mode. With the Run Controller, you can set trace options, watch variables and attributes, run the model until certain time or conditions are true, and examine the values of entities attributes, and so on. In the help system, use the hyperlink that says Run Controller commands. You will see a

■ **FIGURE 2-69**   Invoking the Run Controller

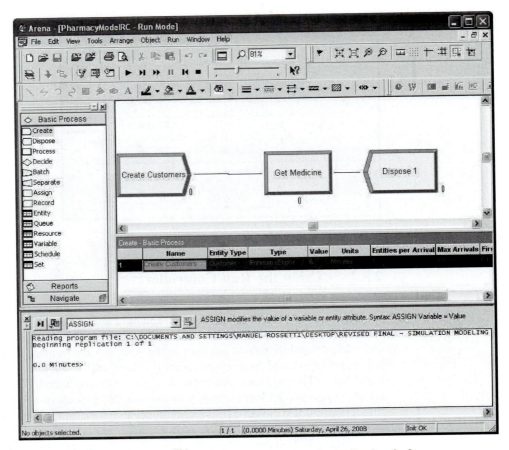

■ **FIGURE 2-70** Arena™ Environment with Run Controller invoked

list of commands for the Run Controller (Figure 2-72). Selecting any of these provides detailed instructions on how to use the command in the Run Controller environment.

In what follows, some simple commands will be used to examine the drive through pharmacy model. Open the pharmacy model in Arena™ and then open up the CREATE module and make sure that the time of the first arrival is set to 0.0. In addition, open up the PROCESS module and make sure that the delay expression is EXPO(5). This will help in ensuring that your output looks like what is shown in the figures. The first arrival will be discussed in what follows.

Now, you should invoke the Run Controller as indicated in Figure 2-69. This will start the run of the model. Arena™ will immediately pause the execution of the model at the first execution step. You should see a window as illustrated in Figure 2-70. Let's first turn the tracing on for the model. By default, Arena™ embeds tracing output into the SIMAN code that is generated. Information about the current module and the active entity will be displayed. To turn on the tracing, enter "set trace" at the prompt or select the command from the drop-down menu of commands, then press return to enter the command. You can also press the toggle trace button.

Now select the VCR run button for single stepping through the model. The button is indicated in Figure 2-74. You should single step again. Your command window should look like Figure 2-74. Remember that the CREATE module was set to create a single arrival at time 0.0. This is exactly what is displayed. In addition, the trace output is telling you that the next arrival will be coming at time 2.0769136. The asterisk marking the output indicates the SIMAN

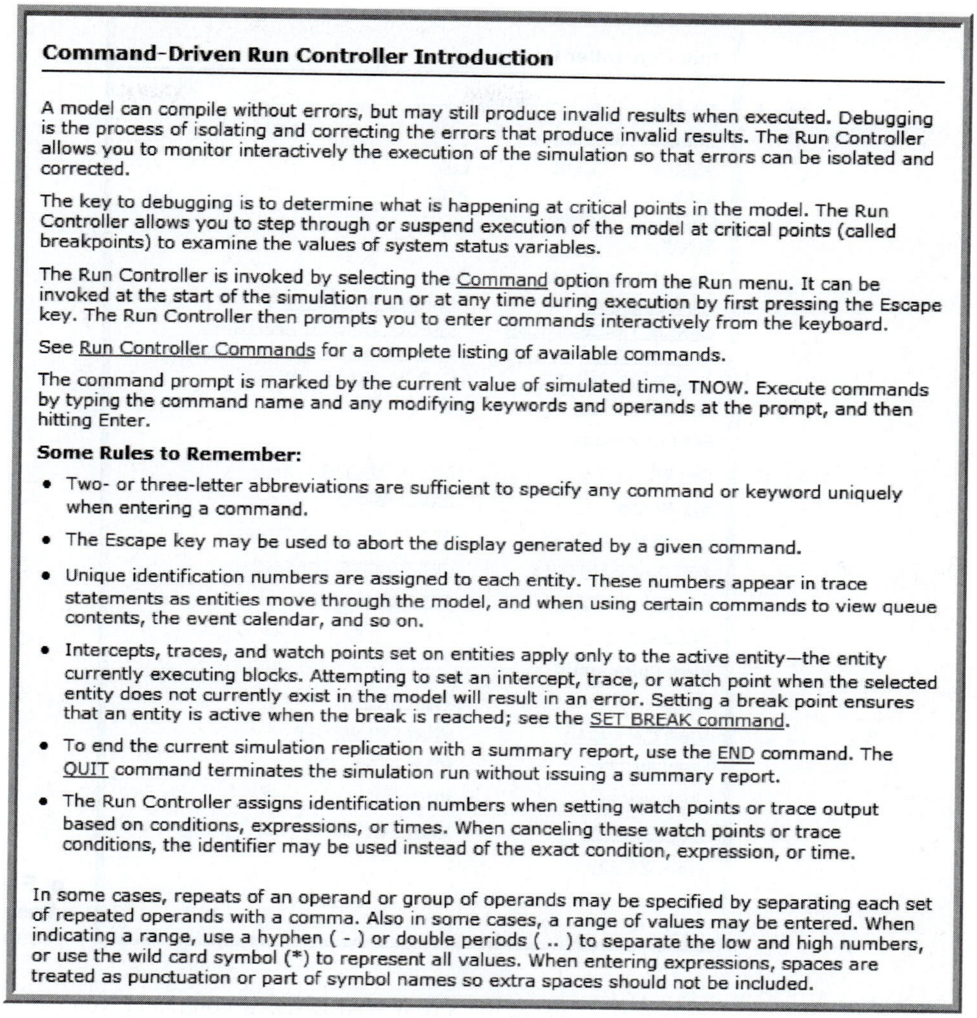

Command-Driven Run Controller Introduction

A model can compile without errors, but may still produce invalid results when executed. Debugging is the process of isolating and correcting the errors that produce invalid results. The Run Controller allows you to monitor interactively the execution of the simulation so that errors can be isolated and corrected.

The key to debugging is to determine what is happening at critical points in the model. The Run Controller allows you to step through or suspend execution of the model at critical points (called breakpoints) to examine the values of system status variables.

The Run Controller is invoked by selecting the <u>Command</u> option from the Run menu. It can be invoked at the start of the simulation run or at any time during execution by first pressing the Escape key. The Run Controller then prompts you to enter commands interactively from the keyboard.

See <u>Run Controller Commands</u> for a complete listing of available commands.

The command prompt is marked by the current value of simulated time, TNOW. Execute commands by typing the command name and any modifying keywords and operands at the prompt, and then hitting Enter.

Some Rules to Remember:

- Two- or three-letter abbreviations are sufficient to specify any command or keyword uniquely when entering a command.

- The Escape key may be used to abort the display generated by a given command.

- Unique identification numbers are assigned to each entity. These numbers appear in trace statements as entities move through the model, and when using certain commands to view queue contents, the event calendar, and so on.

- Intercepts, traces, and watch points set on entities apply only to the active entity—the entity currently executing blocks. Attempting to set an intercept, trace, or watch point when the selected entity does not currently exist in the model will result in an error. Setting a break point ensures that an entity is active when the break is reached; see the <u>SET BREAK command</u>.

- To end the current simulation replication with a summary report, use the <u>END</u> command. The <u>QUIT</u> command terminates the simulation run without issuing a summary report.

- The Run Controller assigns identification numbers when setting watch points or trace output based on conditions, expressions, or times. When canceling these watch points or trace conditions, the identifier may be used instead of the exact condition, expression, or time.

In some cases, repeats of an operand or group of operands may be specified by separating each set of repeated operands with a comma. Also in some cases, a range of values may be entered. When indicating a range, use a hyphen (-) or double periods (..) to separate the low and high numbers, or use the wild card symbol (*) to represent all values. When entering expressions, spaces are treated as punctuation or part of symbol names so extra spaces should not be included.

■ **FIGURE 2-71**   **Arena™ Help on Run Controller**

statement that will be executed on the next step. Then, when the step executes the trace results are given and the next statement to execute is indicated.

If you press the single step button twice, you will see that ASSIGN statements are executed and that the current entity is about to execute a QUEUE statement. After seven steps, the Run Controller output should appear as shown in Figure 2-75.

From this trace output, it is apparent that the entity was sent directly to the SEIZE block from the QUEUE block where it seized 1 unit of the pharmacist. Then, the entity proceeded to the DELAY block where it was delayed by 2.2261054 time units. This is the service time of the entity. Since the entity is entering service at time 0.0, it will thus leave at time 2.2261054. In summary, Entity 2 was created at time zero, arrived at the system, and attempted to get a unit of the pharmacist. Since the pharmacist was idle at the start of the simulation, the entity did not have to wait in the queue and was able to immediately start service.

Something new has happened. The time has jumped to 2.0769136. As you might recall, this was the time indicated by the CREATE module for the next entity to arrive. This entity is denoted by the number 3. Entity 3 is now the active entity. What happened to Entity 2? It was placed on the future events list by the DELAY module and is scheduled to "wake up" at time

Run Controller Commands

The following Run Controller Commands are available:

General Commands

ASSIGN	CLEAR	END
EVENT	GO	STEP
QUIT	SIGNAL	
SHOW		

Cancel Commands

CANCEL BREAK	CANCEL INTERCEPT
CANCEL TRACE BLOCKS	CANCEL TRACE CONDITIONS
CANCEL TRACE ENTITIES	CANCEL TRACE EXPRESSIONS
CANCEL TRACE FILE	CANCEL TRACE TIMES
CANCEL WATCH	

Set Commands

SET BREAK	SET INTERCEPT
SET MODEL	SET TRACE
SET TRACE BLOCKS	SET TRACE CONDITIONS
SET TRACE ENTITIES	SET TRACE EXPRESSIONS
SET TRACE FILE	SET TRACE TIMES
SET WATCH	

View Commands

VIEW	VIEW BREAK
VIEW CALENDAR	VIEW CONVEYORS
VIEW ENTITY	VIEW INTERCEPT
VIEW MODEL	VIEW QUEUE
VIEW SOURCE	VIEW TRACE
VIEW WATCH	

■ **FIGURE 2-72**
Arena™ Run Controller commands

2.2261054. By using the VIEW CALENDAR command, you can see all the events that are scheduled on the event calendar. Figure 2-76 illustrates the output of the VIEW CALENDAR command. There are two data structures in Arena™ to hold entities that are scheduled. The "current events chain" represents the entities that are scheduled to occur at the current time. The "future events heap" represents those entities that are scheduled in the future.

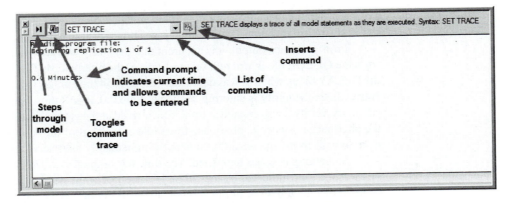

■ **FIGURE 2-73**   Using commands in the Run Controller

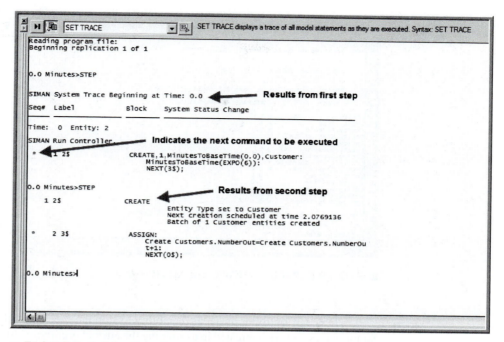

■ **FIGURE 2-74**   Run Controller output window after two steps

This output indicates that no other entities are scheduled for the current time and that Entity 2 is scheduled to activate at time 2.2261054. The entity is not going to "arrive" as indicated. This is a typo in the Arena™ output. The actual word should be "activate." In addition, the calendar indicates that an entity numbered 1 is scheduled to cause the end of the simulation at time 600000.0. Arena™ uses an internal entity to schedule the event that represents the end of a replication. If there are a large number of entities on the calendar, then the VIEW CALENDAR command can be very verbose.

Single stepping the model allows the next entity to be created and arrive at the model. This was foretold in the last trace of the CREATE module.

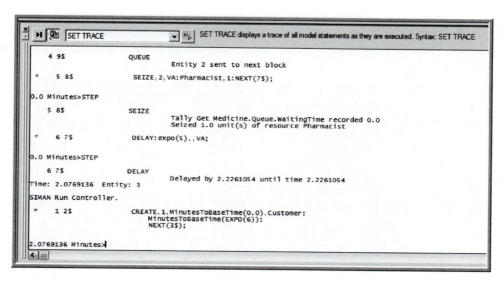

■ **FIGURE 2-75**   Run Controller output window after seven steps

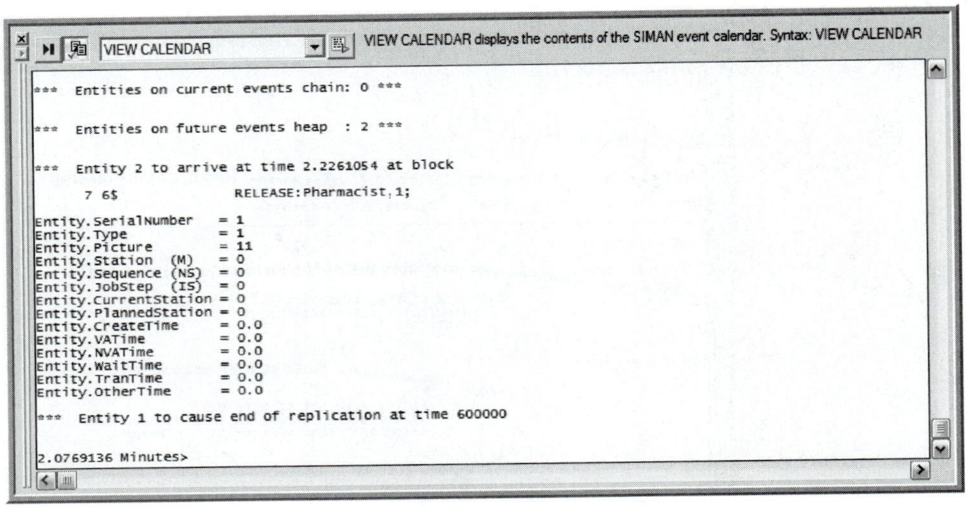

■ **FIGURE 2-76** **VIEW CALENDAR output**

The Run Controller provides the ability to stop the execution when a condition occurs. Suppose that you are interested in examining the attributes of the customers in the queue when at least 2 are waiting. You can set a "watch" on the number of customers in the queue and this will cause Arena™ to stop whenever this condition becomes true. Then, by using the VIEW QUEUE command you can view the contents of the customer waiting queue. Carefully type in SET WATCH NQ(Get Medicine.Queue) == 2 at the Run Controller prompt and hit return. Your Run Controller should resemble Figure 2-77.

When you type GO at the Run Controller prompt, Arena™ will run until the watch condition is met, see Figure 2-78. The VIEW QUEUE command will show all the entities in all the queues, see Figure 2-79. You can also select only certain queues and only certain entities. See the Arena's help for how to specialize the commands.

Arena™ also has a debugging bar in the run command controller that can be accessed from the Run > Run Control > Breakpoints menu (see Figure 2-69). This allows the user to easily set breakpoints, view the calendar, view the attributes of the active entity, and set watches on the model. The debug bar is illustrated for this execution after the next step in Figure 2-80. In the figure, the active entity tab is open to allow the attributes of the entity to be seen. The functionality of the debug bar is similar to many debuggers. The Arena™ help system has a

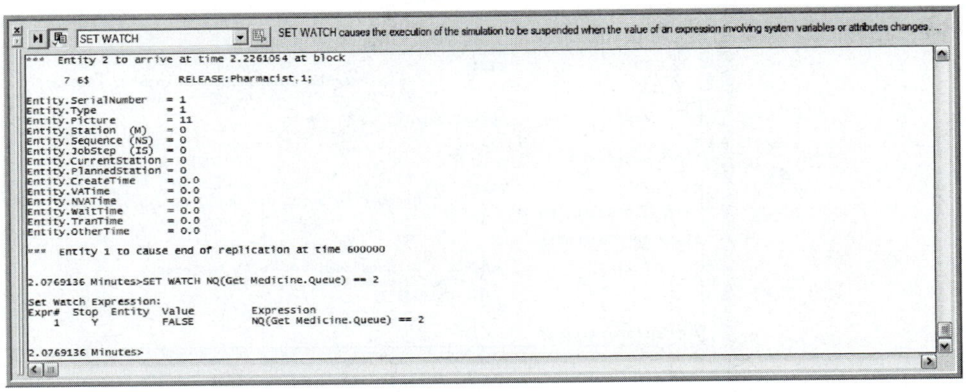

■ **FIGURE 2-77** **Setting a WATCH**

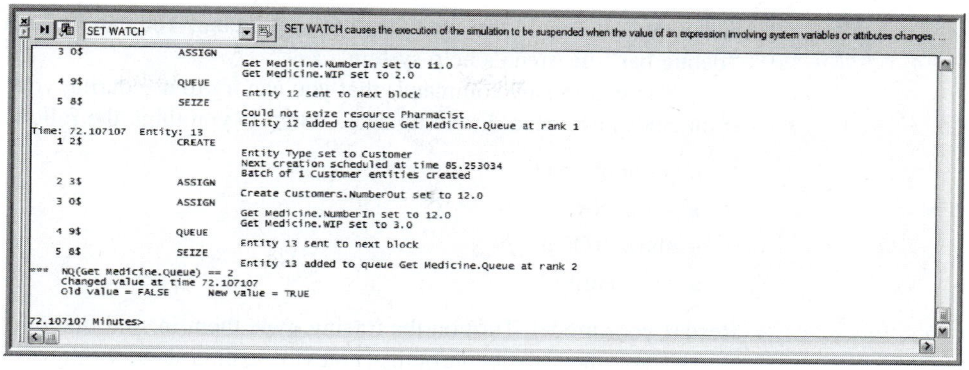

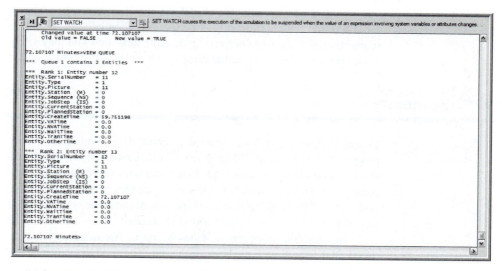

■ FIGURE 2-78   Output after WATCH condition break

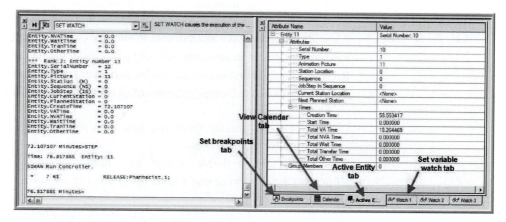

■ FIGURE 2-79   Output after VIEW QUEUE command

■ FIGURE 2-80   Debug window in Run Controller

discussion on how to use the tabs in the debug bar. You can find out more by searching for "debug bar" in Arena's help system.

There are many commands that you can try to use during your debugging. Only a few commands have been discussed here. What do you think the following commands do?

- go until 10.0
- show NR(*)
- show NQ(*)
- view entity

Restart your model. Turn on the tracing, type them in, and find out!

There are just a couple of final notes to mention about the Run Controller. When the model runs to completion or if you stop it during the run, the Run Controller window will show the text based statistical results for your model, which can be quite useful. In addition, after you are done tracing your model, be sure to turn the tracing off (CANCEL TRACE) and to save the model with the tracing off. After you turn the tracing on for a model, Arena™ remembers this even though you might not be running the model via the Run Controller. The trace output will still be produced and will significantly slow the execution of your model. In addition, the memory requirements of the trace output can be quite large and you may get out of memory errors for what appears to be inexplicable reasons. Don't forget to turn the tracing off and resave!

2.9   Summary

In this chapter, you have learned a great deal about how Arena™ can function as a programming language. While Arena™ provides an environment to build simulation models, this model building process requires the user to have a basic knowledge of programming to be productive. This chapter has made some analogies between common programming languages and Arena™ by showing you how typical programming aspects (defining variables, attributes, I/O, iteration, etc.) can be performed in Arena™.

In addition, all of the modules found in Arena's Basic Process panel, except for the SCHEDULE module have been discussed. The modules covered include the following:

CREATE: Used to create and introduce entities into the model according to a pattern.

DISPOSE: Used to dispose of entities once they have completed their activities in the model.

PROCESS: Used to allow an entity to experience an activity with the possible use of a resource.

ASSIGN: Used to make assignments to variables and attributes in the model

RECORD: Used to capture and tabulate statistics in the model.

BATCH: Used to combine entities into a permanent or temporary representative entity.

SEPARATE: Used to create duplicates of an existing entity or to split a batched group of entities.

DECIDE: Used to provide alternative flow paths for an entity based on probabilistic or condition based branching.

VARIABLE: Used to define variables to be used in the model.

RESOURCE: Used to define a quantity of units of a resource that can be seized and released by entities.

QUEUE: Used to define a waiting line for entities whose flow is currently stopped in the model.

ENTITY: Used to define different entity types for use in the model.

SET: Used to define a list of elements in Arena™ that can be indexed by the location in the list.

In addition to Arena's Basic Process panel, the following constructs have been introduced:

READWRITE: From the Advanced Process panel, this module allows input and output to occur in the model.

FILE: From the Advanced Process panel, this module defines the characteristics of the operating system file used in a READWRITE module.

EXPRESSION: From the Advanced Process panel, this module allows the user to define named logical/mathematical expressions that can be used throughout the model.

DELAY: From the Advanced Process panel, this module allows an entity to experience a delay in movement via the scheduling of an event.

MATCH: From the Advanced Process panel, this module allows entities to wait in queues until a user specified matching criteria occurs.

IFELSEIFELSEENDIF: From the Blocks panel, these modules allow standard logic based flow of control.

WHILEENDWHILE: From the Blocks panel, these modules allow for iterative looping.

BRANCH: From the Blocks panel, this module allows probabilistic and condition based path determination along with cloning of entities.

You also learned that Arena™ has a number of variables (e.g., *TNOW*), attributes (e.g., *Entity.Type*), and mathematical functions (e.g., NORM(), ABS()) that can be used in a model.

To develop a program in Arena™, you used data modules to define the elements to be used in the model and flowchart modules to specify the logical flow of the model. The flowchart style that Arena™ has its advantages and disadvantages. The primary advantage is that you can quickly build useful models in the environment without really knowing how to program. This is a great boon to the use of simulation technology. The primary disadvantage is that the flowchart paradigm makes it difficult to organize and develop code that is well structured.

Because of this disadvantage, I strongly encourage you to plan your simulation model carefully (on paper) prior to entering it into the Arena™ Environment. If you just sit down and try to program at the computer, the final result may be confusing "spaghetti code." You should have a plan for defining your variables and use a naming convention for things that you use in the model. For example, attach an R to the end of your resource names or add a *v* to the beginning of your variables. Also, you should fill in the description property for your modules and use good common sense module names. In addition, you should list out in some sort of pseudo-code the logic of your model. Additional examples of this will be given in future chapters of this text. Finally, you should use the submodel feature in Arena™ to organize your code into manageable and logically consistent pieces. You should treat simulation model development in Arena™ more like a programming effort than you might have first thought.

The next two chapters will build on the modeling foundations learned in this chapter. Chapter 3 will concentrate on building models that incorporate randomness and how Arena™ facilitates modeling random processes. Then, in Chapter 4, you will learn how to analyze models that have randomness, through the use of proper statistical techniques. Along the way, more of the modules and concepts needed to build more realistic simulation models in Arena™ will be presented.

Exercises

2.1 Suppose that X is a random variable with a $N(\mu = 2, \sigma = 1.5)$ normal distribution. Generate 100 observations of X using Arena^{TM}.

(a) Estimate the mean and the variance from your observations using Arena^{TM}.

(b) Estimate the $P\{X \geq 3\}$ using Arena^{TM}.

(c) Write out your observations to a file and check your Arena^{TM} results via your favorite statistical package.

2.2 Suppose that customers arriving at the drive-through pharmacy can decide to enter the store instead of entering the drive-through lane. Assume a 90% chance that the arriving customer decides to use the drive-through pharmacy and a 10% chance that the customer decides to use the store. Model this situation with Arena^{TM} and discuss the effect on the performance of the drive-through lane.

2.3 Suppose that a customer arriving at the drive-through pharmacy will decide to *balk* if the number of cars waiting in line is four or more. A customer is said to balk if he or she refuses to enter the system and simply departs without receiving service. Model this situation using Arena^{TM} and estimate the probability that a customer will balk because the line is too long.

2.4 Incoming phone calls arrive according to a Poisson process with a rate of three calls per hour. Each call is either for the accounting department or the customer service department. There is a 30% chance that a call is for the accounting department and a 70% chance the call is for the customer service department. The accounting department has one accountant available to answer the call, which typically lasts uniformly between 30 to 90 minutes. The customer service department has three operators who handle incoming calls. Each operator has their own queue. An incoming call designated for customer service is routed to operator 1, operator 2, or operator 3 with a 25%, 45%, and 30% chance, respectively. *Operator* 1 typically takes uniformly between 30 and 90 minutes to answer a call. The call-answering time of *Operator* 2 is distributed according to a triangular distribution with a minimum of 30 minutes, a mode of 60 minutes, and a maximum of 90 minutes. *Operator* 3 typically takes exponential 60 minutes to answer a call. Run your simulation for 8 hours and estimate the average queue length for calls at the accountant, operator 1, operator 2, and operator 3.

2.5 Write a program in Arena^{TM} using iterative looping to compute the value of 7 factorial.

2.6 Samples of 20 parts from a metal-grinding process are selected every hour. Typically, 2% of the parts need reworking. Let X denote the number of parts in the sample of 20 that require rework. A process problem is suspected if X exceeds its mean by more than three standard deviations. Using Arena^{TM}, simulate 3000 hours of the process, that is, 3000 samples of size 20, and estimate the chance that X exceeds its expected value by more than three standard deviations.

2.7 Samples of 20 parts from a metal-grinding process are selected every hour. Typically 2% of the parts need rework. Let X denote the number of parts in the sample of 20 that require rework. A process problem is suspected if X exceeds its mean by more than one standard deviation. Each time X exceeds its mean by more than one standard deviation, all X of the parts requiring rework are sent to a rework station. Each part consists of two subcomponents, which are split off and repaired separately. The splitting process takes 1 worker and lasts Uniform(1, 2) minutes per part. After the subcomponents have been split, they are repaired in different processes. Subcomponent 1 takes Uniform(5, 10) minutes to repair with 1 worker at its repair process and subcomponent 2 takes expo(7.5) minutes to repair with 1 worker at its repair process. Once both of the subcomponents have been repaired, they are joined back together to form the original part. The joining process takes 5 minutes with 1 worker. The part is then sent back to the main production area, which is outside the scope of this problem. Simulate 3000 hours of production and estimate the average time that it takes a part to be repaired.

2.8 In the tie-dye T-shirt model, the owner is expecting the business to grow during the summer season. The owner is interested in estimating the average time to produce an order and the utilization of the workers if the arrival rate for orders increases. Rerun the model for 30 8-hour days with the arrival rate increased by 20, 40, 60, and 80%. Will the system have trouble meeting the demand? Use the statistics produced by Arena™ to answer this question.

2.9 Suppose that the inspection and packaging process has been split into two processes for the tie-dye T-shirt system and assume that there is an additional worker to perform inspection. The inspection process is uniformly distributed between 2 and 5 minutes. After inspection there is a 4% chance that the entire order will have to be scrapped (and redone). If the order fails inspection, the scrapped order should be counted and a new order should be initiated into the system. Hint: Consider redirecting the order back to the original SEPARATE module. If the order passes inspection, it goes to packaging where the packaging time is distributed according to a triangular distribution, with parameters (2, 4, 10) all in minutes. Rerun the model for 30 8-hour days, with the arrival rate increased by 20, 40, 60, and 80%. Will the system have trouble meeting the demand? In other words, how does the throughput (number of shirts produced per day) change in response to the increasing demand rate?

2.10 Two-piece suits are processed by a dry cleaner, as follows. Suits arrive with exponential time between arrival times having a mean of 10 minutes, and are all initially served by server 1, perhaps after a wait in a FIFO queue (see Figure 2-81). Upon completion of service at server 1, one piece of the suit (the jacket) goes to server 2, and the other part (the pants) to server 3. During service at server 2, the jacket has a probability of 0.05 of being damaged, and while at server 3 the probability of a pair of pants being damaged is 0.10. Upon leaving server 2, the jackets go into a queue for server 4. Server 4 matches and reassembles suit parts, initiating this when the server is idle, and two parts from the same suit are available. If both parts of the reassembled suit are undamaged, the suit is returned to the customer. If either (or both) of the parts is (are) damaged, the suit goes to customer relations (server 5). Assume that all service times are exponentially distributed with the means (in minutes) in Table 2-9.

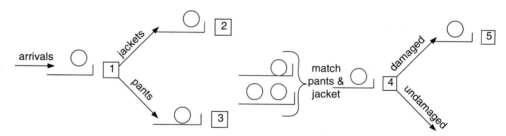

■ **FIGURE 2-81**   **Dry cleaning shop**

■ **TABLE 2-9**
Service Distribution Data for Problem 2.11

Server#	Mean service time
1	6
2	4
3	5
4	5 (if undamaged)
4	8 (if damaged)
5	12

The system is initially empty and idle. Assume that the dry cleaning shop opens at 7 a.m. and operates in this manner until 7 p.m. After 7 p.m., a skeleton crew completes any unfinished orders. In this problem, management is only interested in the system's performance during the 7 a.m. to 7 p.m. time period.

Develop an Arena™ simulation model for this system. Hints:

See Smarts file, *Smarts134.doe*, or the tie-dye T-shirt example in this chapter.

Use an attribute to indicate whether the jacket is damaged.

Use an attribute to indicate whether the pair of pants is damaged.

Use the attributes to properly decide on the processing time for the fourth server.

Use the attributes to determine whether the order goes to customer relations.

The model can be completed using fewer than 20 modules.

Collect statistics on the total time in the system for each type of outcome (either damaged or not), separately. In addition, report the statistics for each queue and the utilization of each server.[7]

---

[7]Material based on A. Law, *Simulation Modeling and Analysis*, 4th ed. (New York: McGraw-Hill, 2007), Problem 2.26, p. 174, and Figure 2.67. Used with permission of The McGraw-Hill Companies.

CHAPTER 3

Modeling Randomness in Simulation

LEARNING OBJECTIVES

After completing this chapter, you should be able to:

- Model probability distributions based on simulation input data
- Describe and use the basic pseudo-random-number generation methods
- Describe and use common random variate generation methods
- Use Arena™ to model systems that contain randomness
- Perform basic spreadsheet simulations in Excel

Simulation is often used to depict the behaviors or characteristics of an existing system. Thus, simulation is a form of descriptive modeling. However, a key use of simulation is to convey the required behaviors or properties of a proposed system. In this situation, simulation is used to prescribe a solution. In other words, simulation can also be a form of prescriptive modeling. Figure 3-1 illustrates the concept of using simulation to recommend a solution. In the figure, a simulation model is used for predicting the behavior of the system. Input models are used to characterize the system and its environment. An evaluative model is used to evaluate the output of the simulation model to understand how the output compares to desired goals. The alternative generator is used to generate different scenarios to be feed into the simulation model for evaluation. Through a feedback mechanism the inputs can be changed based on the evaluation of the outputs and eventually a recommended solution can be achieved.

For example, in the pharmacy system, suppose that the probability of a customer waiting longer than 3 minutes in line had to be less than 10%. To form design alternatives, the inputs (e.g., number of pharmacists, possibly the service process) can be varied. Each alternative can then be evaluated to see if the waiting time criterion is met. In this simple situation, you might

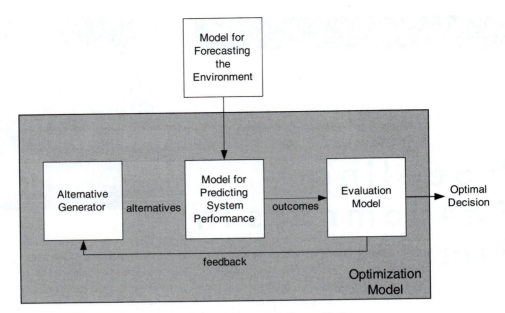

■ **F I G U R E 3 - 1**   **Using simulation for prescriptive analysis**

act as your own alternative generator and the evaluative model is as simple as meeting a criteria; however, in more complex models, there will often be hundreds of inputs to vary and multiple competing objectives.

One of the most important advantages of using simulation models when prescribing solutions to problems is simulation's ability to include randomness when characterizing the system. As stated in Chapter 1, randomness in simulation is often represented by using appropriate probability distributions. For example, in the pharmacy model, the service time distribution was governed by an exponential distribution. In the model, the service time distribution is a key *input* for driving the behavior of the simulation. Modeling the inputs and outputs associated with a simulation model using probability and statistics is a principal topic of Chapters 3 and 4.

In this chapter, you will learn how to model the input processes of a simulation using stochastic processes and how these models are implemented in order to generate the random variables necessary to drive a stochastic simulation. If the inputs represent randomness in the model, the outputs of the simulation will also be random. Thus, the observations from simulation models must be characterized in the form of statistics. The statistical aspects of simulation modeling will be covered in the next chapter. The following section reviews the basics of probabilistic modeling. You should use it as a refresher on this material so that you are ready to apply these techniques in your simulation modeling.

3.1   Random Variables and Probability Distributions

This section discusses some concepts in probability and statistics that are especially relevant to simulation. These will serve you well as you model randomness in the inputs of your simulation models. When an input process for a simulation is stochastic, you must develop a probabilistic model to characterize the process's behavior over time. Suppose that you are modeling the service times in the pharmacy example. Let X_i be a random variable that represents the service

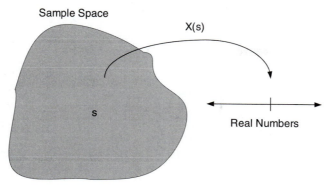

Sample Space

X(s)

s

Real Numbers

■ **FIGURE 3-2** **Random variables map outcomes to real numbers**

time of the ith customer. A random variable is simply a function that assigns a real number to each outcome, s, in a random process that has a set of possible outcomes, S. Figure 3-2 illustrates this concept.

In this case, the random variable is the service time of the customer, and the outcomes are the possible values that the service times can take on, that is, the range of possible values for the service times. The determination of the range of the random variable is part of the modeling process. For example, if the range of the service time random variable is the set of all possible positive real numbers, that is, $X_i \in \Re^+$, or in other words, $X_i \geq 0$, then the service time should be modeled as a *continuous* random variable. Suppose instead that the service time can only take on one of five discrete values $\{2, 5, 7, 8, 10\}$, then the service time random variable should be modeled as a *discrete* random variable. Thus, the first decision in modeling a stochastic input process is to appropriately define a random variable and its possible range of values.

The next decision is to characterize the probability distribution for the random variable. A probability distribution for a random variable is a function that maps from the range of the random variable to a real number, $p \in [0, 1]$. The value of p should be interpreted as the probability associated with the event represented by the random variable (Figure 3-3).

For a discrete random variable, X, with possible values $x_1, x_2, \ldots, x_n$ (n may be infinite), the probability distribution, $f(x)$, is called the probability mass function (PMF) and is denoted

$$f(x_i) = P(X = x_i)$$

where $f(x_i) \geq 0$ for all x_i and $\sum_{i=1}^{n} f(x_i) = 1$. The PMF describes the probability value associated with each discrete value of the random variable. For a continuous random variable,

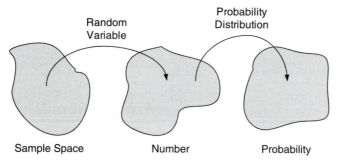

Random Variable

Probability Distribution

Sample Space   Number   Probability

■ **FIGURE 3-3** **Probability distributions map random variables to probabilities**

X, the probability distribution, $f(x)$, is called a probability density function (PDF) and has the properties:

$$f(x) \geq 0$$

$$\int_{-\infty}^{\infty} f(x)dx = 1 \quad \text{(The area must sum to 1.)}$$

$$P(a \leq x \leq b) = \int_{a}^{b} f(x)dx \quad \text{(The area under } f(x) \text{ between } a \text{ and } b.)$$

The probability density function describes the probability associated with a range of possible values for the random variable. A cumulative distribution function (CDF) for the random variable can also be defined. For a discrete random variable X, the cumulative distribution function is defined as

$$F(x) = P(X \leq x) = \sum_{x_i \leq x} f(x_i)$$

and satisfies $0 \leq F(x) \leq 1$, and if $x \leq y$ then $F(x) \leq F(y)$. The cumulative distribution function of a continuous random variable X is

$$F(x) = P(X \leq x) = \int_{-\infty}^{x} f(u)du \quad \text{for } -\infty < x < \infty$$

In addition to describing the probability distribution or equivalently the cumulative distribution function of the random variable it is useful to have a characterization of the first two moments of the distribution in the form of the expected value and the variance of the random variable. The *expected value* of a discrete random variable X, is denoted by $E[X]$ and is defined as

$$E[X] = \sum_{x} x f(x)$$

where the sum is defined through all possible values of x. The *variance* of X is denoted by $V(X)$ and is defined as

$$V[X] = E[(X - E[X])^2] = \sum_{x} (x - E[X])^2 f(x) = \sum_{x} x^2 f(x) - (E[X])^2$$

which is equivalent to $V[X] = E[X^2] - (E[X])^2$ where $E[X^2] = \sum_{x} x^2 f(x)$. Suppose X is a continuous random variable with PDF, $f(x)$, then the expected value of X is

$$E[X] = \int_{-\infty}^{\infty} x f(x)dx$$

and the variance of X is

$$V[X] = \int_{-\infty}^{\infty} (x - E[X])^2 f(x)dx = \int_{-\infty}^{\infty} x^2 f(x)dx - (E[X])^2$$

which is equivalent to $V[X] = E[X^2] - (E[X])^2$, where $E[X^2] = \int_{-\infty}^{\infty} x^2 f(x)dx$.

■ **TABLE 3-1**
Common Discrete Distributions

Distribution, random variable X	PMF	$E[X]$ and $V[X]$
Bernoulli(p) The number of successes in one trial	$P(X = 1) = p$ $P(X = 0) = 1 - p$	$E[X] = p$ $V[X] = p(1 - p)$
Binomial(n, p) The number of successes in n Bernoulli trials	$P(X = x) = \binom{n}{x} p^x (1 - p)^{n-x}$ $0 \le p \le 1 \quad x = 0, 1, \ldots, n$	$E[X] = np$ $V[X] = np(1 - p)$
Geometric(p) The number of trials until the first success in a sequence of Bernoulli trials	$P(X = x) = p(1 - p)^{x-1}$ $0 \le p \le 1 \quad x = 1, 2, \ldots$	$E[X] = 1/p$ $V[X] = (1 - p)/p^2$
Negative Binomial(r, p), Defn. 1 The number of trials until the rth success in a sequence of Bernoulli trials	$P(X = x) = \binom{x-1}{r-1} p^r (1 - p)^{x-r}$ $0 \le p \le 1 \quad x = r, r+1, \ldots,$	$E[X] = r/p$ $V[X] = r(1 - p)/p^2$
Negative Binomial, Defn. 2 The number of failures prior to the rth success in a sequence of Bernoulli trials	$P(Y = y) = \binom{y+r-1}{r-1} p^r (1 - p)^y$ $0 \le p \le 1 \quad y = 0, 1, \ldots,$	$E[Y] = r(1 - p)/p$ $V[Y] = r(1 - p)/p^2$
Poisson(λ) The number of event occurrences during a specified period of time	$P(X = x) = \dfrac{e^{-\lambda}\lambda^x}{x!} \quad x = 0, 1, \ldots,$ $\lambda > 0$	$E[X] = \lambda$ $V[X] = \lambda$
Discrete uniform(a, b)	$P(X = x) = \dfrac{1}{b - a + 1}$ $a \le b \quad x = a, a+1, \ldots, b$	$E[X] = (b + a)/2$ $V[X] = ((b - a + 1)^2 - 1)/12$
Discrete uniform	$P(X = x) = \dfrac{1}{n}$ $x = x_1, x_2, x_3, \ldots, x_n; n \in \aleph$	$E[X] = \dfrac{1}{n}\sum_{i=1}^{n} x_i$ $V[X] = \dfrac{1}{n}\sum_{i=1}^{n} x_i^2 - (E[X])^2$

Note: $\binom{n}{x} = \dfrac{n!}{(n-x)!x!}.$

There are a wide variety of discrete and continuous random variables that often occur in simulation modeling. Table 3-1 summarizes some common discrete distributions.

A Bernoulli distribution models the behavior of a random variable that follows a Bernoulli trial. A Bernoulli trial has the following characteristics:

- The result of each trial may be considered either a success or a failure, that is, only two outcomes are possible.
- The probability, p, of success is the same in every trial.
- The trials are independent. That is, trials have no influence over each other or memory of each other.

A Bernoulli distribution is often used to model a flow process in which an entity can take two possible paths at a decision point. For example, if a part is inspected and the chance of that the part is defective is p and each part is independent of any other part, then whether the part is defective can be characterized by a Bernoulli distribution.

A binomial distribution models a random variable that represents the number of successes in n Bernoulli trials where Y_i represents the outcome of the ith trial. Since success is indicated with 1, the number of successes is simply the sum of the Y's, $X = \sum_{i=1}^{n} Y_i$. The random variable, X, will have a binomial distribution with parameters n for the number of trials and p for the probability of success for each trial. For example, suppose 10 parts are to be produced and assume that whether each part is defective is a Bernoulli trial, then the number of defective parts out of the 10 parts that are defective is characterized by a binomial distribution with parameters $n = 10$ and p. The simplest way to simulate a binomial random variable is to first simulate each Bernoulli trial and then count the number of successes.

A geometric distribution models a random variable that represents the number of trials until the first success in a sequence of Bernoulli trials. Suppose that a repair machine is available to rework parts from a machining center. In addition, suppose that the repair machine is only set up after the first defective part is produced. Then, the number of parts until the repair machine is set up can be modeled by a geometric distribution. As seen in Table 3-1, the geometric distribution is a special case of the negative binomial distribution. Suppose now that instead of setting up the repair machine when the first defective part is produced that the system waits until four defective parts are produced. Then, the number of parts produced until the machine is set up can be modeled by a negative binomial distribution. That is, if X denotes a random variable that represents the number of Bernoulli trials until r success occur, then X has a negative binomial distribution.

The discrete uniform distributions given in the table can be used to model situations where there are a finite discrete set of outcomes that have equal probability of occurring. The first definition is a special case of the second definition where the set of possible values is defined over a consecutive set of integers on the range $[a, b]$. Suppose the parts are numbered in a queue waiting to be inspected. Each part is denoted by its rank in the queue, such as 1 = first in queue, . . . , 10 = last in queue. Then, a part can be randomly selected for inspection by using a discrete uniform distribution over the range from 1 to 10.

The Poisson distribution is often used to model the number of events in an interval of time, such as the number of phone calls to a call center in an hour, the number of persons arriving at a bank, the number of cars arriving at an intersection in an hour, the number of demands for inventory in a month, and so on. In addition, it is used to model the number of defects that occur in a length (or volume) of an item, such as the number of typos in a book or the number of potholes in a mile of road. Consider an interval of real numbers and assume that incidents occur at random throughout the interval. If the interval can be partitioned into subintervals of small enough length such that

- the probability of more than one incident in a subinterval is zero
- the probability of one incident in a subinterval is the same for all intervals and it is proportional to the length of the subinterval
- the number of incidents in each subinterval is independent of other subintervals

then the number of incidents in the interval is governed by a Poisson distribution. If a Poisson random variable represents the number of incidents in some interval, then the mean of the random variable must equal the expected number of incidents in the same length of interval. In other words, the units must match. When examining the number of incidents in a unit of time,

■ **TABLE 3-2**
Common Continuous Distributions

Distribution	$F(x)$	$E[X]$	$V[X]$
Uniform(a, b)	$\dfrac{x-a}{b-a} \quad a < x < b$	$\dfrac{a+b}{2}$	$\dfrac{(b-a)^2}{12}$
Normal(μ, σ^2)	No closed form	μ	σ^2
Exponential(θ)	$1 - e^{-x/\theta}$	θ	θ^2
Erlang(θ, r)	$1 - \sum_{n=0}^{r-1} \dfrac{e^{-x/\theta}(x/\theta)^n}{n!}$	$r\theta$	$r\theta^2$
Gamma(β, α)	If r is a positive integer, see Erlang; otherwise, no closed form	$\alpha\beta$	$\alpha\beta^2$
Weibull(β, α) $\beta = scale$ $\alpha = shape$	$1 - e^{-(x/\beta)^\alpha}$	$\dfrac{\beta}{\alpha}\Gamma\left(\dfrac{1}{\alpha}\right)$	$\dfrac{\beta^2}{\alpha}\left\{2\Gamma\left(\dfrac{2}{\alpha}\right) - \dfrac{1}{\alpha}\left(\Gamma\left(\dfrac{1}{\alpha}\right)\right)^2\right\}$
Lognormal(μ_l, σ_l^2)	No closed form $\mu = \ln\left(\mu_l/\sqrt{\sigma_l^2 + \mu_l^2}\right)$ $\sigma^2 = \ln\left((\sigma_l^2 + \mu_l^2)/\mu_l^2\right)$	$e^{\mu + \frac{\sigma^2}{2}}$	$e^{2\mu+\sigma^2}\left(e^{\sigma^2} - 1\right)$
Beta(α_1, α_2)	No closed form	$\dfrac{\alpha_1}{\alpha_1 + \alpha_2}$	$\dfrac{\alpha_1\alpha_2}{(\alpha_1 + \alpha_2)^2(\alpha_1 + \alpha_2 + 1)}$
Triangular(a, m, b) $a = minimum$ $b = maximum$ $m = mode$	$\begin{cases} \dfrac{(x-a)^2}{(b-a)(m-a)} & a \leq x \leq m \\ 1 - \dfrac{(b-x)^2}{(b-a)(b-m)} & m \leq x \leq b \end{cases}$	$\dfrac{a+b+m}{3}$	$\dfrac{a^2 + b^2 + m^2 - ab - am - bm}{18}$

$X(t)$, the Poisson distribution is often written as follows:

$$P(X(t) = x) = \frac{e^{-\lambda t}(\lambda t)^x}{x!} \qquad \lambda > 0 \quad x = 0, 1, \ldots,$$

The Poisson distribution is intimately related to the exponential distribution, a common continuous distribution shown in Table 3-2. The *Arena*^{TM} *Users Guide* contains an appendix that describes the distributions available in Arena^{TM}. These can also be found by searching "distributions" in Arena's Help system.

The continuous uniform distribution can be used to model situations in which your data set is small or nonexistent, and it is reasonable to assume that everything is equally likely in an interval. This distribution will also play a key role in the generation of random samples from other distributions. This point is discussed further in Section 3.3.2. The triangular distribution is also useful in situations characterized by no or little data if you can characterize a most

■ **TABLE 3-3**
Common Continuous Distribution Modeling Situations

Distribution	Common modeling situations
Uniform	When you have no data Everything is equally likely to occur within an interval Task times
Normal	Modeling errors Modeling measurements, length, etc. Modeling the sum of a large number of other random variables
Exponential	Time to perform a task Time between failures Distance between defects
Erlang	Service times Multiple phases of service with each phase exponential
Weibull	Time to failure Time to complete a task
Gamma	Repair times Time to complete a task
Lognormal	Time to perform a task Quantities that are the product of a large number of other quantities
Triangular	Rough model in the absence of data Assume a minimum, a maximum, and a most likely value
Beta	Useful for modeling task times in bounded range with little data Modeling probability as a random variable

likely value for the random variable in addition to its range (minimum and maximum). Table 3-3 lists these modeling situations as well as others for various common continuous distributions.

While the normal distribution is a mainstay of probability and statistics, you need to be careful when using it as a distribution for input models because it is defined over the entire range of real numbers. For example, in simulation, the time to perform a task is often required; however, time must be a positive real number. Clearly, since a normal distribution can have negative values, using a normal distribution to model task times can be problematic.

3.2   Input Distribution Modeling

When performing a simulation study, there is no substitution for actually observing the system and collecting the data required for the modeling effort. As outlined in Section 8 of Chapter 1, a good simulation methodology recognizes that modeling and data collection often occur in parallel. That is, observing the system allows conceptual modeling, which in turn allows for understanding the input models needed for the simulation. Data collection for the input models allows further observation of the system and further refinement of the conceptual model, including the identification of additional input models. Eventually, this cycle converges to the point where the modeler has a well-defined understanding of the input data requirements. The data for the input model must be collected and modeled.

Input modeling begins with data collection, probability, statistics, and analysis. There are many methods available for collecting data, including time study analysis, work sampling, historical records, and automatically collected data. Time study and work sampling methods are covered in a standard industrial engineering curriculum. Observing the time that an operator takes to perform a task via a time study results in a set of observations of the task times. Hopefully, there will be sufficient observations for applying the techniques discussed in this section. Work sampling is useful for developing the proportion of time associated with various activities. This sort of study can be useful in identifying probabilities associated with performing tasks and for validating the output from the simulation models. Historical records and automatically collected data hold promise for allowing more data to be collected, but also pose difficulties related to the quality of the data collected. In any of the abovementioned methods, the input models will only be as good as the data and processes used to collect the data.

One especially important caveat for new simulation practitioners: Do not rely on the people in the system you are modeling to correctly collect the data for you. If you do rely on them to collect the data, you must develop documents that clearly define what data are needed and how to collect the data. In addition, you should train them to collect the data using the methods that you have documented. Only through careful instruction and control of data collection processes will you have confidence in your input modeling.

A typical input modeling process includes the following procedures:

1. Documenting the process being modeled: Describe the process being modeled and define the random variable to be collected. When collecting task times, you should pay careful attention to clearly defining when the task starts and when the task ends. You should also document what triggers the task.

2. Developing a plan for collecting the data and then collecting the data: Develop a sampling plan, describe how to collect the data, perform a pilot run of your plan, and then collect the data.

3. Graphical and statistical analysis of the data: Using standard statistical analysis tools, you should visually examine your data. This should include such plots as a histogram, a time series plot, and an autocorrelation plot. Again, using statistical analysis tools you should summarize the basic statistical properties of the data, e.g., sample average, sample variance, minimum, maximum, quartiles, and so on.

4. Hypothesizing distributions: Using what you have learned from Steps 1 through 3 above, you should hypothesize possible distributions for the data.

5. Estimating parameters: Once you have possible distributions in mind you need to estimate the parameters of those distributions so that you can analyze whether the distribution provides a good model for the data. With current software such as the Arena™ Input Analyzer, BestFit™, and ExpertFit™, this step, as well as Steps 3, 4, and 6, have been largely automated.

6. Checking goodness of fit for hypothesized distributions: In this step, you should assess whether the hypothesized probability distributions provide a good fit for the data. This should be done both graphically (e.g., P-P plots and Q-Q plots) and via statistical tests (e.g., chi-square test, Kolmogorov-Smirnov test). As part of this step you should perform some sensitivity analysis on your fitted model.

During the input modeling process and after it is complete, you should document your process. This is important for two reasons. First, much can be learned about a system simply by collecting and analyzing data. Second, in order to have your simulation model accepted as useful by decision makers, they must *believe* in the input models. Even decision makers who are not simulation savvy understand the old adage "garbage in = garbage out." The following section illustrates the input modeling process with a concrete example.

3.2.1   Input Modeling Example

This example starts with Step 3 of the input modeling process. That is, the data have already been collected. This explanation of the model-fitting process will review some basic statistical analysis techniques, and perform much of the analysis using only Excel[TM] and Arena's Input Analyzer. While you may have other tools available to you, such as MINITAB[TM], ExpertFit[TM], or BestFit[TM], it is useful to know how to do the analysis "from scratch." This should allow you to better understand what the more sophisticated tools are doing and make you a better modeler. For the sake of truth in advertising, the example has been constructed so as to facilitate the use of Excel. For the vast majority of modeling situations, Excel is woefully inadequate for the task at hand. The shortcomings and difficulties of using Excel will be pointed out throughout the example. In light of this, a short introduction to using MINITAB and BestFit for input modeling is provided in the appendix to this chapter. A more rigorous discussion of the topic can be found in chapter 6 of Law (2007), including the use of ExpertFit to facilitate the modeling process.

Let's continue with the pharmacy model and develop a model for the service time of the customers. Let X_i be the service time of the ith customer, where the ith service time is defined as starting when the $(i-1)st$ customer begins to drive off, and ending when the ith customer drives off after interacting with the pharmacist. In the case where there is no customer already in line when the ith customer arrives, the start of the service can be defined as the point where the customer's car arrives to the beginning of the space in front of the pharmacist's window. Note that in this definition, the time that it takes the car to pull up to the pharmacy window is being included. An alternative definition of service time might simply be the time between when the pharmacist asks the customer what they need until the time in which the customer gets the receipt. Both of these definitions are reasonable interpretations of service times, and it is up to you to decide what sort of definition fits best with the overall modeling objectives. As you can see, input modeling is as much an art as it is a science.

One hundred observations of the service time were collected using a portable digital assistant and are shown in Table 3-4 where the first observation is in row 1 column 1, the second observation is in row 2 column 1, the 21st observation is in row 1 column 2, and so forth. The observations were collected to the nearest hundredth of a second. These data are available in the Excel spreadsheet called *PharmacyInputModelingExampleData.xls* or the text file *PharmacyInputModelingExampleData.txt* that accompanies this chapter.

3.2.2   Graphical and Statistical Analysis of the Data

In the following, you will first explore what can be done easily in Excel and then you will examine alternative methods using both Arena's Input Analyzer and Output Analyzers. You should first place your data in a spreadsheet and a text file for analysis as shown in Figure 3-4. The analysis will start by summarizing the data with descriptive statistics. In Excel, first make a copy of the sheet labeled "ServiceTimes". You will work with this copy as you proceed through the example. In your copy of the *ServiceTimes* sheet, select your columns containing the data (columns A and B in Figure 3-4).

Next, select the Data menu, Sort, and sort by column B in ascending order. After adding heading labels, your spreadsheet should look like Figure 3-5. Congratulations! You have just estimated what statisticians call the *order statistics*. Let $(x_1, x_2, \ldots x_n)$ represent a sample of data. If the data are sorted from smallest to largest, then the ith ordered element can be denoted as $x_{(i)}$. For example, $x_{(1)}$ is the smallest element, and $x_{(n)}$ is the largest, so that $(x_{(1)}, x_{(2)}, \ldots x_{(n)})$

■ **TABLE 3-4**
Pharmacy Service Times

61	278.73	194.68	55.33	398.39
59.09	70.55	151.65	58.45	86.88
374.89	782.22	185.45	640.59	137.64
195.45	46.23	120.42	409.49	171.39
185.76	126.49	367.76	87.19	135.6
268.61	110.05	146.81	59	291.63
257.5	294.19	73.79	71.64	187.02
475.51	433.89	440.7	121.69	174.11
77.3	211.38	330.09	96.96	911.19
88.71	266.5	97.99	301.43	201.53
108.17	71.77	53.46	68.98	149.96
94.68	65.52	279.9	276.55	163.27
244.09	71.61	122.81	497.87	677.92
230.68	155.5	42.93	232.75	255.64
371.02	83.51	515.66	52.2	396.21
160.39	148.43	56.11	144.24	181.76
104.98	46.23	74.79	86.43	554.05
102.98	77.65	188.15	106.6	123.22
140.19	104.15	278.06	183.82	89.12
193.65	351.78	95.53	219.18	546.57

represents the ordered data, and these data are called the *order statistics*. From the order statistics, a variety of other statistics can be computed:

$$\text{Min}\,(x_i) = x_{(1)}$$

$$\text{Max}\,(x_i) = x_{(n)}$$

$$\text{Range } r = x_{(n)} - x_{(1)} \quad (r \text{ is called the sample range})$$

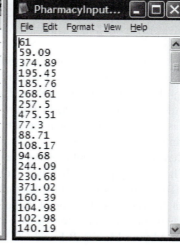

■ **FIGURE 3-4** Excel data and data in text file

	A	B	
1	**Obs#**	**Value**	
2	81	36.84	
3	84	38.5	
4	54	42.93	
5	24	46.23	
6	37	46.23	
7	85	48.3	
8	75	52.2	
9	51	53.46	
10	61	55.33	
11	56	56.11	
12	62	58.45	
13	90	58.92	
14	66	59	
15	2	59.09	
16	86	59.97	
17	1	61	

■ **FIGURE 3-5** Sorted data from smallest to largest

The median, $\tilde{x}$, is a measure of central tendency such that one-half of the data are above it and one-half of the data are below it. When n is odd, the median is $x_{\left(\frac{n+1}{2}\right)}$:

$$e.g. \quad x_{(1)} = 3, \; x_{(2)} = 5, \; x_{(3)} = 7, \; x_{(4)} = 7, \; x_{(5)} = 38$$

$$\frac{(n+1)}{2} = \frac{5+1}{2} = 3$$

$$\tilde{x} = x_{(3)}$$

When n is even, the median is $\dfrac{x_{\left(\frac{n}{2}\right)} + x_{\left(\frac{n}{2}+1\right)}}{2}$:

$$e.g. \quad x_{(1)} = 3, \quad x_{(2)} = 5, \quad x_{(3)} = 7, \quad x_{(4)} = 7$$

$$n = 4$$

$$x_{(n/2)} = x_{(2)}$$

$$x_{(n/2+1)} = x_{(3)}$$

$$\tilde{x} = \frac{x_{(2)} + x_{(3)}}{2} = \frac{5+7}{2} = 6$$

As you can see in Figure 3-5, the minimum value occurred on observation number 81 and had a value of 36.84.

Before using Excel to compute additional statistics, you should first define a *named range* for the data. In Excel, select the data. According to Figure 3-5, the data starts in cell B2 and goes down to cell B101. With the data selected, go to the named range area of Excel and type in a name for the range as indicated in Figure 3-6. In this figure, the highly descriptive name, "Data," has been used to represent the data. You can use any name that does not conflict with Excel's naming conventions.

With a named range defined, the use of many of the functions in Excel becomes easier. Instead of having to type B2:B101 over and over you can simply use "Data" as a substitute for the range. To get the summary statistics for the data, use Excel's descriptive statistics function in the Data Analysis ToolPak found in the Tools menu. If you cannot find the Data Analysis ToolPak, then you should use the Tools > Add-Ins menu to select it as shown in Figure 3-7. You may need your Microsoft Excel CD to complete this procedure.

	A	B	C
	Obs#	Value	
1			
2	81	36.84	
3	84	38.5	
4	54	42.93	
5	24	46.23	
6	37	46.23	
7	85	48.3	
8	75	52.2	
9	51	53.46	
10	61	55.33	
11	56	56.11	
12	62	58.45	
13	90	58.92	
14	66	59	
15	2	59.09	

Data ▼   *fx* 36.8

■ **FIGURE 3-6**   Defining a named range

After selecting the descriptive statistics option from the Data Analysis menu as shown in Figure 3-8, select your data for the input range (or in this case you can simply type in the named range) and specify the output range for the computed statistical quantities. Make sure that you select summary statistics and confidence interval for the mean as shown in Figure 3-9. At this point, Excel will use its built-in functions to compute the descriptive statistics for the data as shown in Figure 3-10. Rather than use the descriptive statistics option to perform this work, you

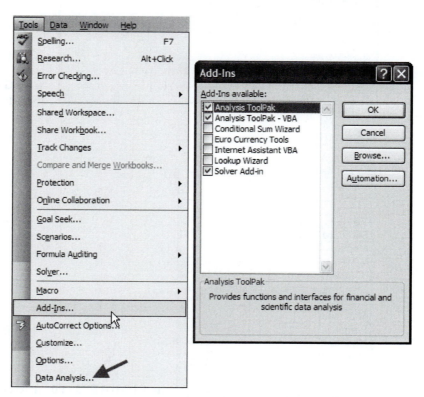

■ **FIGURE 3-7**   Getting to the Analysis ToolPak

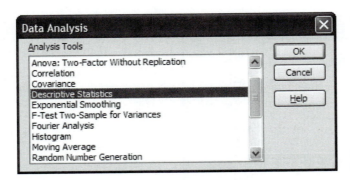

■ **FIGURE 3-8** Data analysis options

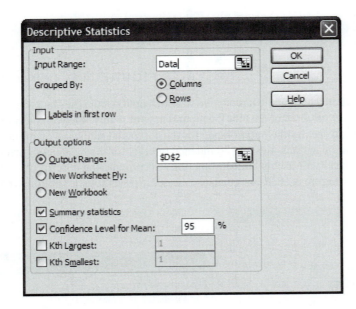

■ **FIGURE 3-9** Getting descriptive statistics

D	E
Column1	
Mean	182.7778
Standard Error	14.16111971
Median	139.11
Mode	46.23
Standard Deviation	141.6111971
Sample Variance	20053.73115
Kurtosis	3.198965798
Skewness	1.663956772
Range	745.38
Minimum	36.84
Maximum	782.22
Sum	18277.78
Count	100
Confidence Level(95.0%	28.09873306

■ **FIGURE 3-10** Excel output for descriptive statistics

G	H		G	H
Sample Average	182.7778		Sample Average	=AVERAGE(Data)
Standard Error	14.16112		Standard Error	=H6/SQRT(H14)
Median	139.11		Median	=MEDIAN(Data)
Mode	46.23		Mode	=MODE(Data)
Standard Deviation	141.6112		Standard Deviation	=STDEV(Data)
Sample Variance	20053.73		Sample Variance	=VAR(Data)
Kurtosis	3.198966		Kurtosis	=KURT(Data)
Skewness	1.663957		Skewness	=SKEW(Data)
Range	745.38		Range	=H12-H11
Minimum	36.84		Minimum	=MIN(Data)
Maximum	782.22		Maximum	=MAX(Data)
Sum	18277.78		Sum	=SUM(Data)
Count	100		Count	=COUNT(Data)
Alpha	0.05		Alpha	0.05
T-Value(Alpha/2,n-1)	1.984217		T-Value(Alpha/2,n-1)	=TINV(H15,H14-1)
CI Half-Width	28.09873		CI Half-Width	=H16*H3
UL-Confidence Interval	210.8765		UL-Confidence Interval	=H2+H17
LL-Confidence Interval	154.6791		LL-Confidence Interval	=H2-H17

■ **FIGURE 3-11**   **Useful Excel statistical functions**

can directly use the built-in Excel functions for statistical analysis. The advantage of using Excel's built-in functions is that the computation becomes "live." The Data Analysis ToolPak computes the values from Excel's statistical functions and only places the values in the output range. By directly using Excel's functions yourself you can do what-if calculations by simply changing or removing data points from the data. The statistical functions used to duplicate Excel's descriptive statistics are shown in Figure 3-11.

In Figure 3-11, the sample average and 95% confidence interval on the true expected value have been computed. As a review, consider how this confidence interval has been formed. Let $(x_1, x_2, \ldots x_n)$ represent the observations in a random sample of size n, and let $(X_1, X_2, \ldots X_n)$ represent the corresponding random variables for the sample. The random variables $(X_1, X_2, \ldots X_n)$ are a random sample of size n if (1) the X's are independent random variables, and (2) every X has the same probability distribution. These two assumptions have not been established, but let's proceed anyway. Denote the unknown cumulative distribution function of X as $F(x)$ and define the unknown expected value and variance of X with $E[X] = \theta$ and $V[X] = \sigma^2$, respectively.

A *statistic* is any function of the random variables in a sample. Any statistic that is used to estimate an unknown value such as $E[X] = \theta$ is called an "estimator." For the current situation, an estimate of $E[X] = \theta$ is desired. What would be a good estimator for this quantity? Without going into the details of the meaning of statistical goodness, you should intuitively consider the sample average as a possible estimator for θ. Define

$$\overline{X}(n) = h(X_1, X_2, \ldots, X_n) = \frac{\sum_{i=1}^{n} X_i}{n}$$

as the sample average statistic on the random sample. The function, h, is the "average function." Thus, $\overline{X}$ is a function of random variables and therefore, it is also a random variable. This makes $\overline{X}$ a statistic, since it is a function of the random variables in the sample. Any random variable has a corresponding probability distribution. The probability distribution associated with a statistic is called its *sampling distribution*, which can be used to form a confidence interval on the point estimate associated with the statistic. The point estimate is simply the value obtained from the statistic once the data have been realized. In this case, the estimate for the sample average is

$$\bar{x} = \frac{\sum_{i=1}^{100} x_i}{100} = 182.7778$$

A confidence interval expresses the degree of certainty associated with a point estimate. A specific confidence interval does not imply that the parameter θ is inside that particular interval with probability $1 - \alpha$. In fact, the true parameter is either in the interval or it is not in the interval. Instead, think of the confidence level $1 - \alpha$ as an assurance about the procedure used to compute the interval. That is, a confidence-interval procedure ensures that if a large number of confidence intervals are computed, each based on n samples, then the proportion of the confidence intervals that actually contain θ should be close to $1 - \alpha$. Any particular confidence interval will either contain the true parameter of interest or it will not. Since you do not know the true value, you can use the confidence interval to assess the risk of making a bad decision based on the point estimate. You can be $1 - \alpha$ confident in your decision making, or conversely, know that you are taking a risk of α of making a bad decision.

Under the assumption that the sample size is large enough such that the distribution of $\overline{X}$ is normally distributed, you can form an approximate confidence interval on the point estimate. Assuming that the sample variance

$$S^2(n) = \frac{\sum_{i=1}^{n} (X_i - \overline{X})^2}{n - 1}$$

is a good estimator for $V[X] = \sigma^2$, then a $(1 - \alpha) \times 100\%$ confidence interval estimate for θ is

$$\bar{x} \pm t_{\alpha/2, n-1} \frac{s}{\sqrt{n}}$$

where $t_{\alpha/2, n-1}$ is the upper $100(\alpha/2)$ percentage point of the Student's t-distribution with $n - 1$ degrees of freedom. Appendix A.4 contains the percentage points for the Student t-distribution. Excel's *TINV(p, degrees freedom)* function computes the corresponding two tailed Student's t-distribution value for a given probability and degrees of freedom. Thus, $t_{\alpha/2, n-1} = \text{TINV}(\alpha, n - 1)$ within Excel. For this problem, the 95% confidence interval is [154.68, 210.88].

Two interesting aspects emerge from these results. The first is that the distribution is positively skewed. Skewness is 0 for a symmetric distribution, positive for a right-skewed distribution, and negative for a left-skewed distribution. In addition, the sample average is greater than the median which is greater than the mode (Mode < Median < Mean). The *mode* is also used as an indicator for the central tendency of the data. The mode is commonly defined as the most likely value in the data. This implies that most of the observations are in a large group toward the left of the distribution. The longer tail is to the right. The second is that the minimum observation is significantly larger than zero. This is often true when modeling tasks that take a minimum time to perform. In this situation, it seems reasonable that a finite minimum amount of time is necessary to process any customer. This can pose some difficulty when trying to directly fit a distribution such as the gamma, Weibull, lognormal, and exponential, among others, that have a range of possible values beginning at zero.

You can continue the modeling with graphical analysis. A very useful plot for hypothesizing possible distributions is the histogram or frequency plot. Divide the range of the data into class intervals. Let $b_0, b_1, \ldots, b_k$ be the breakpoints of the class intervals, such that $[b_0, b_1), [b_1, b_2), \ldots, [b_{k-1}, b_k)$ form k disjoint and adjacent intervals where the width of the class interval is $\Delta b = b_j - b_{j-1}$. Let c_j be the frequency count of the x_i's in the jth interval and define

$$c(x) = \begin{cases} 0 & \text{if} \quad x < b_o \\ c_j & \text{if} \quad b_{j-1} \le x < b_j \\ 0 & \text{if} \quad b_x \le x \end{cases} \quad \text{for} \quad j = 1, 2, \ldots k$$

Let $h_j(x)$ be the relative frequency of x_i's in the jth interval,

$$h_j(x) = c_j(x)/n$$

Note that $\sum_{i=1}^{k} h_j(x) = 1$ over the x's. A plot of the relative frequency is called a histogram. A plot of the cumulative relative frequency, $\sum_{i=1}^{j} h_j(x)$, for each j is called a cumulative distribution plot. $h_j(x)$ resembles the true probability distribution in shape because

$$P(b_{j-1} \le X \le b_j) = \int_{b_{j-1}}^{b_j} f(x)dx = \Delta b \times f(y) \quad \text{for } y \in (b_{j-1}, b_j)$$

according to the mean value theorem of calculus, and the fact that $h(y) = h_j(x) \approx \Delta b f(y)$. Thus, $h(y)$ is proportional to $f(y)$.

The Histogram Analysis tool in Excel can be used to make histograms (frequency diagrams). This tool, which is a part of the Analysis ToolPak, calculates individual and cumulative frequencies for a cell range of data and data bins. For more information, look up "statistical analysis tools" or "About statistical analysis tools" in the Excel answer wizard. Excel allows the user to specify the break points for the cell ranges or you can accept the defaults supplied by Excel. You can enter the cell reference to a range that contains an optional set of boundary values that define bin ranges. These values should be in ascending order. Microsoft Excel counts the number of data points between the current bin number and the adjoining higher bin. An observation is counted in a particular bin if it is equal to or less than the bin breakpoint down to the last bin. All values below the first bin value are counted together, as are the values above the last bin value. If you omit the bin range, Microsoft Excel creates a set of evenly distributed bins between the data's minimum and maximum values. Excel defines bins as follows: $(-\infty, b_0], (b_0, b_1], (b_1, b_2], \ldots, (b_{k-1}, b_k], (b_k, \infty)$. In general, the class intervals used in the frequency distribution should be of equal width, although this is not an absolute requirement. A simple way to determine the class interval size is to take range/(number of desired classes). Thus, the number of intervals is the key decision parameter. In general, the visual display of the histogram is highly dependent on the number of class intervals. If Δb is too small, the histogram will tend to have a ragged shape. If Δb is too large, the resulting histogram will be very block-like. Two common rules for setting the number of intervals follow:

-choose $k \cong \sqrt{n}$
-Sturges rule $k = \lfloor 1 + \log_2 n \rfloor$

Experimentation with various values for the number of intervals is recommended. In particular, some statistical tests are sensitive to this value. Using Tools > Data Analysis > Histogram and accepting the default breakpoints results in the frequency plot as shown in Figure 3-12. In this figure, you can see the effect of the minimum value and the positive skew of the data. It also appears that the data are unimodal. A *modal bar* on a frequency histogram is a bar with a height greater than or equal to the bar(s) adjacent to it. A frequency histogram with a single modal bar is called *unimodal. Multimodal* implies multiple modal bars. *Bimodal* implies two clear modal bars.

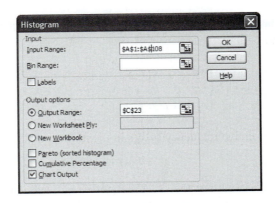

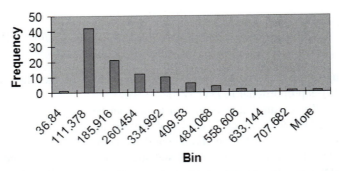

■ **FIGURE 3-12** Histogram dialog and resulting frequency plot of service times

Before proceeding to the next step in the analysis, there are two additional plots that can be made relatively easily in Excel: time series and scatterplot. The time series plot simply plots the values of the observations versus their corresponding observation number. The purpose of this plot is to determine whether some type of dependence on the order (or time) exists. For example, if you are observing a new pharmacist, service times might be progressively smaller as the pharmacist learns how to more efficiently serve customers. A time series plot can also give you an idea of the dependence in the data; for example, if you see that each high observation is often followed by a low observation, there might be negative correlation in the data. The scatterplot is designed to more readily assess dependence between pairs of data points. Both of these visual assessments are important because the fundamental assumption of many of the following statistical analysis techniques assumes a random sample. As mentioned earlier, a random sample requires that the observations are independent and identically distributed. If there is a trend in the data or if there is dependence in the data, then these vital assumptions may be violated.

The time series plot is quite easy to perform in Excel, as the user simply follows the Chart Wizard. The resulting chart is given in Figure 3-13 and can be investigated further in the spreadsheet containing the service time data.

The scatterplot is also quite easy to perform once the data have been "lagged." That is, a column next to the original data needs to be created with the resulting data shifted by the amount of the lag. In this case, the lag is 1. To check for lag-1 correlation, plot pairs (X_i, X_{i+1}), for $i = 1, 2, \ldots, n - 1$. If there is positive correlation, then the points will fall around a line with positive slope; if there is negative correlation the points will fall around a line with negative slope. Table 3-5 shows an example of this for the first 10 observations.

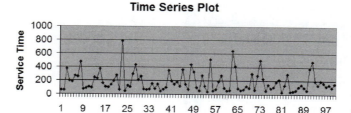

Time Series Plot

■ **FIGURE 3-13** Time series plot of service times

■ **TABLE 3-5**
Data Lagged by One Observation

$X(i)$	$X(i + 1)$
61	59.09
59.09	374.89
374.89	195.45
195.45	185.76
185.76	268.61
268.61	257.5
257.5	475.51
475.51	77.3
77.3	88.71
88.71	108.17

The resulting scatterplot can be found in Figure 3-14. In this plot, you are looking for a clear linear trend, either positive or negative. In both the time series plot and the scatterplot, indicators of potential problems do not appear.

In the following, the analysis will be continued using Arena's Input Analyzer and Arena's Output Analyzer. Arena's Input Analyzer is designed to fit a given set of distributions to a data set. The Output Analyzer is useful for some of the plotting that it facilitates. The Input Analyzer and the Output Analyzer are separate programs that can be executed outside the Arena™ Environment.

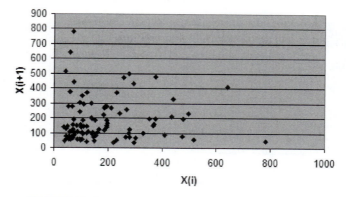

Scatter Plot

■ **FIGURE 3-14** Scatter plot of $X(i)$ vs $X(i + 1)$

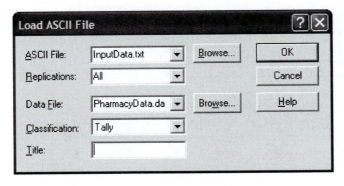

InputData - Notepad
File Edit Format View Help
1   61
2   59.09
3   374.89
4   195.45
5   185.76
6   268.61
7   257.5
8   475.51
9   77.3
10   88.71
11   108.17
12   94.68
13   244.09
14   230.68
15   371.02
16   160.39
17   104.98
18   102.98
19   140.19
20   193.65
21   278.73
22   70.55

■ **FIGURE 3-15**   Text file ready for importing into Output Analyzer

Arena's Input and Output Analyzers can be accessed in the Start Menu in the Arena™ folder. Before performing a fit of the data using the Input Analyzer, a couple of additional plots will be made using the Output Analyzer. To work with the data in the Output Analyzer, it first needs to be imported into a data group. Create a text file that has a column counting the observations and a column representing the observations. This can be easily done by cutting and pasting the data from Excel in to a text file as shown in Figure 3-15. Open up the Output Analyzer and start a new data group. Use the File > Data File > Load ASCII File menu and fill in the ASCII File dialog box as shown in Figure 3-16. Use the browse feature to select your input file. The importing action creates a file, in this case called *PharmacyData.dat*. The .dat file extension is used by files that the Output Analyzer can process. If the importing process was successful, you should see an output window that indicates the file was created and that shows the number of observations processed.

Using the Add button on the Output data group window, select the (.dat) file that you just created and add it to the data group. The Output Analyzer can work with a number of (.dat) files in what is termed a "data group." Figure 3-17 shows the Output Analyzer ready for processing. From this state, you can graph and analyze the information in the data file. The Graph > Plot feature allows you to plot the data versus time (observation number) as shown in Figure 3-18. These results confirm what was illustrated previously using Excel. Use the Graph > Histogram

Load ASCII File	? X
ASCII File: InputData.txt ▾	Browse...	OK
Replications: All ▾		Cancel
Data File: PharmacyData.da ▾	Browse...	Help
Classification: Tally ▾		
Title:		

■ **FIGURE 3-16**   Importing into Output Analyzer

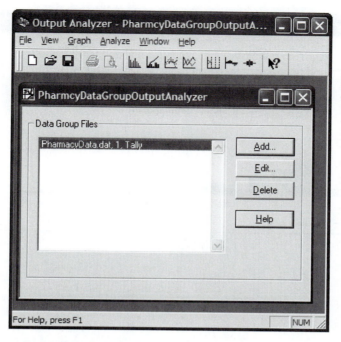

■ **FIGURE 3-17** Data group ready for processing

menu to create a histogram of the data. The dialog box is fairly self-explanatory; use the help button if you need explanation of the input fields. The resulting histogram for the pharmacy data using the default settings is shown in Figure 3-19.

The true usefulness of the Output Analyzer is its ability to easily construct an auto-correlation plot. An autocorrelation plot allows dependence in the data to be quickly examined. An autocorrelation plot is a time-series assessment tool that plots the lag-k correlation versus the lag number. Consider the sample as a sequence of observations ordered by observation number, $(X_1, X_2, \ldots X_n)$. A time series, $(X_1, X_2, \ldots X_n)$, is said to be *covariance stationary* if:

- The mean exists and $\theta = E[X_i]$, for $i = 1, 2, \ldots, n$.
- The variance exists and $Var[X_i] = \sigma^2 > 0$, for $i = 1, 2, \ldots, n$.
- The lag-k autocorrelation, $\rho_k = cor(X_i, X_{i+k})$, is not a function of i, that is, the correlation between any two points in the series does not depend on where the points are in the series, it depends only on the distance between them in the series.

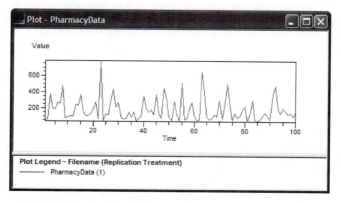

■ **FIGURE 3-18** Output Analyzer time-series plot

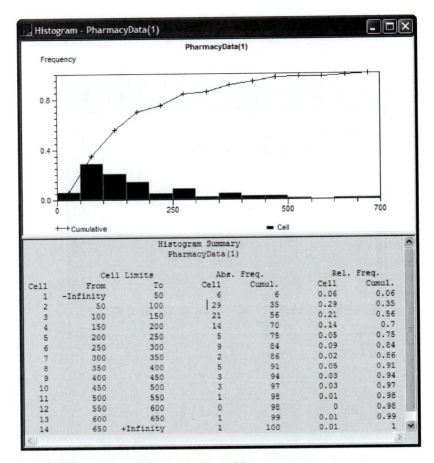

■ **FIGURE 3-19**   **Output Analyzer histogram**

Recall that the correlation between two random variables is defined as

$$cor[X, Y] = \frac{\text{cov}[X, Y]}{\sqrt{V[X] \times V[Y]}}$$

$$\text{cov}(X, Y) = E[(X - E[X])(Y - E[Y])] = E[XY] - E[X]E[Y]$$

The correlation between two random variables X and Y measures the strength of linear association. The correlation is unitless and has a range of $-1 \leq cor(X, Y) \leq 1$. If the correlation between the random variables is less than zero, then the random variables are said to be negatively correlated. This implies that if X tends to be high, then Y will tend to be low, or alternatively, if X tends to be low, then Y will tend to be high. If the correlation between the random variables is positive, the random variables are said to be positively correlated. This implies that if X tends to be high, then Y will tend to be high, or alternatively, if X tends to be low, then Y will tend to be low. If the correlation is zero, then the random variables are said to be uncorrelated. If X and Y are independent random variables, then the correlation between them will be zero. The converse of this is not necessarily true, but an assessment of the correlation should tell you something about the *linear* dependence between random variables.

The autocorrelation between two random variables that are k time points apart in the covariance stationary time series is given by

$$\rho_k = cor[X_i, X_{i+k}] = \frac{\text{cov}[X_i, X_{i+k}]}{\sqrt{V[X_i] \times V[X_{i+k}]}} = \frac{\text{cov}[X_i, X_{i+k}]}{\sigma^2} \qquad k = 1, 2, \ldots$$

A plot of ρ_k for increasing values of k is called an "autocorrelation plot." The autocorrelation function as defined above is the theoretical function. When you have data, you must estimate the values of ρ_k from the actual times series. This involves forming an estimator for ρ_k. Law (2007) suggests plotting

$$\hat{\rho}_k = \frac{\hat{C}_k}{S^2(n)} \qquad for \quad k = 1, 2, 3, \ldots, l.$$

with

$$\hat{C}_k = \frac{\sum_{i-1}^{n-k}(X_i - \overline{X}(n))(X_{i+k} - \overline{X}(n))}{n - k}$$

Most time series analysis books (e.g., Box et al. 1994), have a slightly different definition of the sample autocorrelation function:

$$r_k = \frac{c_k}{c_0} = \frac{\sum_{i-1}^{n-k}(X_i - \overline{X}(n))(X_{i+k} - \overline{X}(n))}{\sum_{i-1}^{n}(X_i - \overline{X}(n))^2}$$

where

$$c_k = \frac{1}{n}\sum_{i-1}^{n-k}(X_i - \overline{X}(n))(X_{i+k} - \overline{X}(n))$$

Note that the numerator has $(n - k)$ terms and the denominator has (n) terms. A plot of r_k versus (k) is called a "correlogram" or "sample autocorrelation plot." For the data to be uncorrelated, r_k should be approximately zero. The problem is that estimators of r_k are often not very good, especially for large (k); however, a plot should give you some idea of independence.

Formal tests can be performed using various time series techniques and their assumptions. Under the assumption that the series is white noise, $N(0, 1)$ with all $\rho_k = 0$, then for large n (Box et al. 1994),

$$Var(r_k) \cong \frac{1}{n}$$

Thus, a quick test of dependence is to check whether sampled correlations fall within a reasonable sigma band around zero. For example,

$$n = 100$$
$$Var(r_k) \cong \frac{1}{100}$$
$$s.e.(r_k) = \sqrt{\frac{1}{n}} = 0.1$$
$$\pm 2s.e.(r_k) = \pm 0.2$$

Using the Output Analyzer, a sample autocorrelation plot can be easily developed by using the menu Analyze > Correlogram and filling in the resulting dialog box, as seen in Figure 3-20. Generally, the maximum lag should be set to no larger than one-tenth of the size of the data set because the estimation of higher lags is unreliable.

The resulting autocorrelation plot is shown in Figure 3-21. It is clear that no lag-k correlation exceeds ± 0.2, and thus you can conclude that evidence is insufficient to suggest that the data are significantly correlated. The correlations appear roughly as white noise around zero. If there had been correlation, it would have been readily apparent especially in the lower

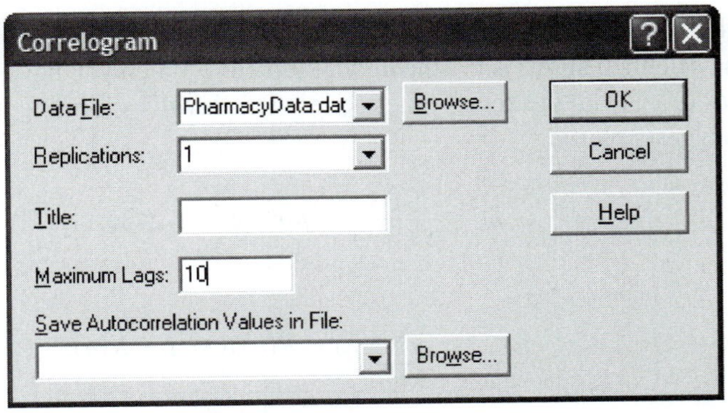

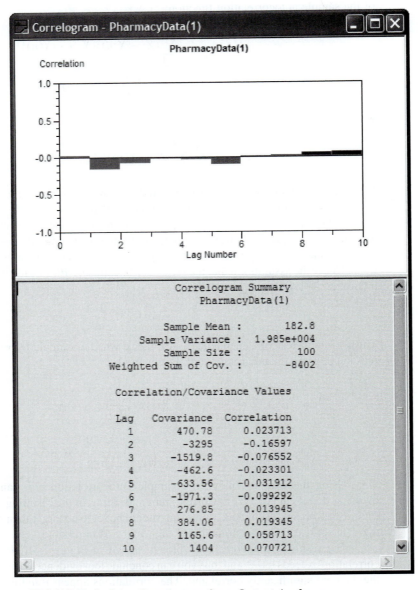

■ **FIGURE 3-20** Output Analyzer correlogram dialog

■ **FIGURE 3-21** Correlogram from Output Analyzer

lag values. The Excel spreadsheet, *PharmacyInputModelingExampleData.xls*, also shows how to create a correlogram "by hand" using Excel's correlation function.

Before continuing with the analysis, let's recap what has been learned so far:

- The data appear to be stationary. This conclusion is based on the time series plot where no discernable trend with respect to time is found in the data.
- The data appear to be independent. This conclusion is from both the scatterplot and the autocorrelation plot.
- The data have a clear minimum value significantly greater than zero.
- The distribution of the data is positively skewed and unimodal. This conclusion is readily apparent from the histogram and from the statistical summary.

The next steps involve the model-fitting processes of hypothesizing distributions, estimating the parameters, and checking for goodness of fit. Each of these steps can be done relatively easily using Arena's Input Analyzer.

3.2.3   Using Arena's Input Analyzer

After opening the Input Analyzer, choose New from the File menu to start a new input analyzer data set. Then, using File > Data File > Use Existing, you can import the text file containing the data for analysis. The resulting import should leave the Input Analyzer looking like Figure 3-22.

Save the session, which will create a (.dft) file. Note how the Input Analyzer automatically makes a histogram of the data and performs a basic statistical summary of the data. The Input Analyzer will fit many of the common distributions available in Arena^{TM}:

- Beta
- Erlang
- Exponential
- Gamma
- Lognormal
- Normal

- Triangular
- Uniform
- Weibull
- Empirical (continuous and discrete)
- Poisson
- Johnson

In addition, it will provide the Arena^{TM} expression to be used in the Arena^{TM} model. The fitting process in the Input Analyzer is highly dependent on the intervals chosen for the data histogram. Thus, it is very important that you vary the number of intervals and check the sensitivity of the fitting process to the number of intervals in the histogram. There are two basic methods by which you can perform the fitting process: (1) individually for a specific distribution, and (2) by fitting all of the possible distributions. Given the interval specification, the Input Analyzer will compute a chi-square goodness of fit statistic, Kolmogorov-Smirnov test, and squared error criteria, all of which will be discussed in what follows.

The basic approach is to compare the hypothesized distribution function in the form of the PDF, PMF, or the CDF to a fit of the data. This implies that you have hypothesized a distribution *and* estimated the parameters of the distribution in order to compare the hypothesized distribution to the data. For example, suppose you hypothesize that the gamma distribution will be a good model for the service time data. Then you would need a method to estimate the shape and scale parameters of the gamma distribution. Estimation of the

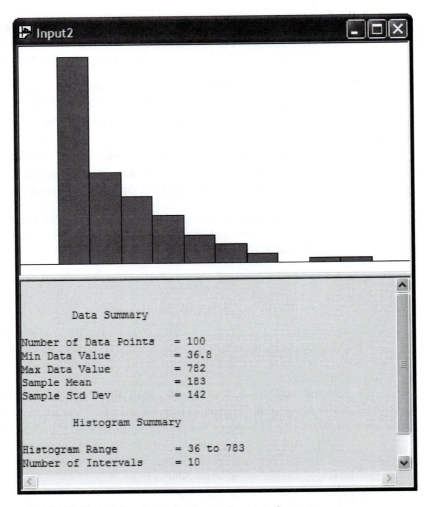

■ **FIGURE 3-22** Input Analyzer after data import

parameters from sample data is based on important statistical theory that requires the estimators for the parameters to satisfy statistical properties (e.g., unique, unbiased, invariant, and consistency). There are two main methods for estimating the parameters of distribution functions: 1) the method of moments and 2) the maximum likelihood method. The method of moments matches the empirical moments to the theoretical moments of the distribution and attempts to solve the resulting system of equations. The maximum likelihood method attempts to find the parameter values that maximize the joint probability distribution function of the sample. It is beyond the scope of this book to cover the properties of these techniques, so the interested reader is referred to Law (2007) and Casella and Berger (1990) for more detail on the theory of these methods.

To make concrete the challenges associated with fitting the parameters of the hypothesized distribution, the maximum-likelihood method for fitting the exponential distribution will be used on the service time data. Suppose you hypothesize that the distribution of the service time for the ith customer can be modeled with an exponential distribution:

$$f(x_i) = \begin{cases} \lambda e^{-\lambda x_i} & 0 \le x_i \\ 0 & otherwise \end{cases}$$

Thus, the joint probability density function of the sample is

$$g(x_1, x_2, \ldots, x_{100}) = f(x_1) f(x_2) \cdots f(x_{100}) = \prod_{i=1}^{100} f(x_i)$$

At this juncture, you must find the value of λ that maximizes the function $g(\cdot)$. Luckily, this can be easily accomplished for the exponential distribution because of its simple form for any size sample. Let $(X_1, X_2, \ldots X_n)$ represent a random sample and $(x_1, x_2, \ldots x_n)$ represent the actual observed values of the random sample. Then, the function $g(\cdot)$ as a function of λ can be written as

$$L(\lambda) = f(x_1; \lambda) f(x_2; \lambda) \cdots f(x_n; \lambda) = \prod_{i=1}^{n} f(x_i; \lambda)$$

$L(\lambda)$ is called the "likelihood function." Given the data, $(x_1, x_2, \ldots x_n)$, $L(\lambda)$ is simply a function of λ. The maximum likelihood method is predicated on finding the value of λ that maximizes the likelihood that the data came from a particular distribution, $f(x; \lambda)$. Substituting the definition of the exponential distribution in the likelihood function yields

$$L(\lambda) = \prod_{i=1}^{n} \lambda e^{-\lambda x_i} = \lambda^n e^{-\lambda \sum_{i=1}^{n} x_i}$$

It can be shown that maximizing $L(\lambda)$ is the same as maximizing $\ln(L(\lambda))$. Thus,

$$\ln L(\lambda) = n \ln \lambda - \lambda \sum_{i=1}^{n} x_i$$

Differentiating

$$\frac{d \ln L(\lambda)}{d\lambda} = \frac{n}{\lambda} - \sum_{i=1}^{n} x_i$$

and solving for the root at zero yields,

$$0 = \frac{n}{\lambda} - \sum_{i=1}^{n} x_i$$

$$\frac{n}{\lambda} = \sum_{i=1}^{n} x_i$$

$$\lambda = \frac{n}{\sum_{i=1}^{n} x_i}$$

$$\hat{\lambda} = \frac{1}{\bar{X}}$$

If the second derivative, $\dfrac{d^2 \ln L(\lambda)}{d\lambda^2} < 0$, then a maximum is obtained.

$$\frac{d^2 \ln L(\lambda)}{d\lambda^2} = \frac{-n}{\lambda^2}$$

because n is positive, the x's are positive, and λ is positive, the second derivative must be negative; therefore, the maximum likelihood estimator for the parameter of the exponential distribution is $\hat{\lambda} = \frac{1}{\bar{X}}$. As seen in this simple example, estimating the parameters of distributions involves nonlinear optimization, which in general must be performed numerically. This is the value of using software such as the Input Analyzer, BestFit, or ExpertFit

Exhibit 3-1 Exponential Distribution Summary from Input Analyzer

```
Distribution Summary
      Distribution:        Exponential
      Expression:          36 + EXPO(147)
      Square Error:        0.003955
Chi Square Test
      Number of intervals = 4
      Degrees of freedom = 2
      Test Statistic = 2.01
      Corresponding p-value = 0.387
Kolmogorov-Smirnov Test
      Test Statistic = 0.0445
      Corresponding p-value > 0.15
Data Summary
      Number of Data Points = 100
      Min Data Value = 36.8
      Max Data Value = 782
      Sample Mean = 183
      Sample Std Dev = 142
Histogram Summary
      Histogram Range = 36 to 783
      Number of Intervals = 10
```

when performing input modeling. These software tools will perform this estimation process with little difficulty.

Given that you have hypothesized the exponential distribution as a model for the data, you can use the Input Analyzer to perform a fit using the exponential distribution. With the formerly imported data in an input window in the Input Analyzer, go to the Fit menu and select the exponential distribution. The resulting analysis is shown in Exhibit 3-1 and Figure 3-23. The Input Analyzer has made a fit to the data and has recommended the Arena^{TM} expression (36+ EXPO(147)). What is this value 36? As was noted during the statistical analysis, the data suggested that the minimum value for the service time was around 36. The value 36 is called the offset or location parameter. Any distribution can have this additional parameter, which further complicates parameter estimation procedures. The Input Analyzer has an algorithm that will attempt to estimate this parameter. Is the fit a reasonable one for the service time data? From the histogram with the exponential distribution overlaid, it appears to be a reasonable fit. This can be contrasted with Figure 3-24, which show the results of fitting a uniform distribution to the data.

To understand the results of the fit, you must understand how to interpret the results from the chi-square test and the Kolmogorov-Smirnov test. The chi-square test divides the range of the data into k intervals and tests whether the number of observations that fell in each interval is close the expected number that should fall in the interval given the hypothesized distribution. The chi-square test statistic is:

$$\chi^2 = \sum_{j=1}^{k} \frac{(N_j - np_j)}{np_j}$$

where N_i is the observed number of observations that fell in the ith interval and p_i is the theoretical probability of being in the ith interval. For large n, an approximate $1 - \alpha$ level test can be performed based on a chi-square test statistic that rejects the null hypothesis if

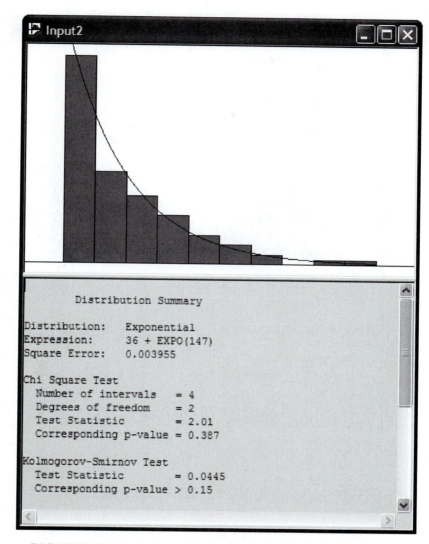

■ **FIGURE 3-23**   Histogram for exponential fit to service times

$\chi^2 > \chi^2_{\alpha,k-s-1}$, where ($s$) is the number of estimated parameters for the distribution. Appendix A.5 contains the percentage points for the chi-squared distribution. The null hypothesis is that the data are distributed according to the hypothesized distribution versus the alternative hypothesis that the data do not come from the hypothesized distribution. The Input Analyzer also shows the p-value of the test.

Accepting or rejecting H_0 based on a prespecified α does not convey a full sense of the significance of the result. A more useful mechanism for reporting significance is to report the smallest α-level at which the observed test statistic is significant. This smallest α-level is called the observed level of significance or p-value. The smaller the p-value, the more statistically significant will be the result. Thus, the p-value can be compared to the desired significance level, α. Thus, an alternative testing criterion is:

If the $p\,value > \alpha$, then do not reject H_0.

If the $p\,value \leq \alpha$, then reject H_0.

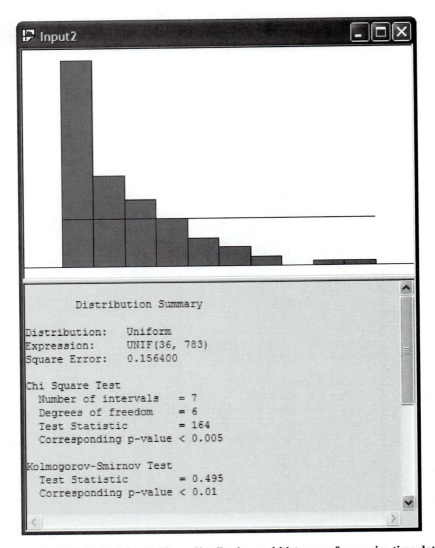

■ **FIGURE 3-24**   Uniform distribution and histogram for service time data

An alternate interpretation for the p-value is that it is the probability under H_0 of obtaining a test statistic at least as "extreme" as the observed value. Assuming that H_0 is true, then the test statistic will follow a known distributional model. The p-value can be interpreted as the chance, assuming that H_0 is true, of obtaining a more extreme value of the test statistic than the value actually obtained. Computing a p-value for a hypothesis test allows for a different mechanism for accepting or rejecting the null hypothesis. Remember that a type I error is

$$\alpha = P(\text{Type I error}) = P(\text{rejecting the null when it is in fact true})$$

This represents the chance you are willing to take to make a mistake in your conclusion. The p-value represents the actual chance associated with the test statistic being more extreme under the assumption that the null is true. A small p-value indicates that the observed result is rare under the assumption of H_0. If the p-value is "small," it indicates that an outcome as extreme as observed is possible, but not probable under H_0. In other words, chance by itself does not adequately explain the result. Thus, for a "small" p-value, you can reject H_0 as a plausible explanation of the observed outcome. If the p-value is "large," it indicates that an outcome as

Exhibit 3-2   Uniform Distribution Summary from Input Analyzer

```
Distribution Summary
     Distribution:      Uniform
     Expression:        UNIF(36, 783)
     Square Error:      0.156400
Chi Square Test
     Number of intervals = 7
     Degrees of freedom = 6
     Test Statistic = 164
     Corresponding p-value < 0.005
Kolmogorov-Smirnov Test
     Test Statistic = 0.495
     Corresponding p-value < 0.01
Data Summary
     Number of Data Points = 100
     Min Data Value = 36.8
     Max Data Value = 782
     Sample Mean = 183
     Sample Std Dev = 142
Histogram Summary
     Histogram Range = 36 to 783
     Number of Intervals = 10
```

extreme as that observed can occur with high probability. For the case of the hypothesis that the exponential distribution fits the service time data, H_0 cannot be rejected as a plausible distribution for the data. Based on Exhibit 3-2 for the uniform distribution, the corresponding p-value is less than 0.005, which leads to a rejection of the uniform distribution as a reasonable model for these observations.

The Kolmogorov-Smirnov (K-S) test compares the hypothesized distribution, $\hat{F}(x)$, to the empirical distribution and does not depend on specifying intervals for tabulating the test statistic. The K-S test is described in detail in Law (2007), who also discusses the advantages and disadvantages of the test. The empirical distribution is defined as

$$\tilde{F}_n(x_{(i)}) = \frac{i}{n}$$

and represents the proportion of the X's that are less than or equal to the ith order statistic. Therefore, the empirical distribution for each order statistic can be described as

$$\tilde{F}_n(x_{(1)}) = \frac{1}{n}$$

$$\tilde{F}_n(x_{(2)}) = \frac{2}{n} \quad \leftarrow x_{(1)} \,\&\, x_{(2)}$$

$$\vdots$$

$$\tilde{F}_n(x_{(n)}) = \frac{n}{n} = 1$$

The following continuity correction is often used when defining the empirical distribution:

$$\tilde{F}_n(x_{(i)}) = \frac{i - 0.5}{n}$$

■ **TABLE 3-6**
Empirical Distribution

Observation #	Value	Order	$\tilde{F}_n\left(x_{(i)}\right) = \frac{i}{n}$	$\tilde{F}_n\left(x_{(i)}\right) = \frac{i-0.5}{n}$
81	36.84	1	0.01	0.005
84	38.5	2	0.02	0.015
54	42.93	3	0.03	0.025
24	46.23	4	0.04	0.035
37	46.23	5	0.05	0.045
85	48.3	6	0.06	0.055
75	52.2	7	0.07	0.065
51	53.46	8	0.08	0.075
61	55.33	9	0.09	0.085
56	56.11	10	0.1	0.095

Table 3-6 shows the empirical distribution for the first 10 points in the service time data set. This is quite easy to develop using Excel.

The K-S test statistic, D_n, is defined as

$$D_n = \max\{D_n^+, D_n^-\}$$

$$D_n^+ = \max_{1 \le i \le n}\left\{\frac{i}{n} - \hat{F}(x_{(i)})\right\}$$

$$D_n^- = \max_{1 \le i \le n}\left\{\hat{F}(x_{(i)}) - \frac{i-1}{n}\right\}$$

and represents the largest vertical distance between the hypothesized distribution and the empirical distribution over the range of the distribution. This value is also relatively easy to compute using Excel, provided that the CDF of the hypothesized distribution is available. Intuitively, a large value for the K-S test statistic indicates a poor fit between the empirical and the hypothesized distributions. The null hypothesis is that the observations come from the hypothesized distribution.

Exhibit 3-1 indicates that the p-value for the K-S test is greater than 0.15, which does not suggest a serious lack of fit for the exponential distribution. The results for the uniform distribution show that the p-value for the K-S test is smaller than 0.01, which indicates that the uniform distribution is probably not a good fit for these observations.

The chi-square test has more general applicability than the K-S test. Specifically, the chi-square test applies to both continuous and discrete data; however, it suffers from dependence on the interval specification. In addition, it has a number of other shortcomings, which are discussed in Law (2007). While the K-S test can also be applied to discrete data, special tables must be used for getting the critical values. Appendix B contains critical values for the K-S test, where you reject the null hypothesis if "D_n" is greater than the critical value "D_α". Additionally, the K-S test in its original form assumes that the parameters of the hypothesized distribution are known, that is, given without being estimated from the data. Research on the effect of using the K-S test with estimated parameters has indicated that it will be conservative in the sense that the actual type I error will be less than specified. Additional advantages and disadvantages of the K-S test are given in Law (2007). Another statistical test that has been devised for testing the goodness of fit for distributions is the Anderson-Darling test, also described by Law (2007). The latter test detects tail differences and has higher power than the K-S test for many popular distributions. It can be found as standard output in software programs such as MINITAB, BestFit, and ExpertFit.

In general, you should be cautious of goodness-of-fit tests because they are unlikely to reject any distribution when you have a small number of observations, and they are likely to reject every distribution when you have many observations. The point is that for whatever software that you use for model fitting, you will need to correctly interpret the results of the statistical tests performed. Be sure to understand how these tests are computed and how sensitive the tests are to various assumptions in the model-fitting process.

The final result of interest in the Input Analyzer's distribution summary output is the value labeled *Square Error*. This is the criterion that the Input Analyzer uses to recommend a particular distribution when fitting multiple distributions to the observations. Let $b_0, b_1, \ldots, b_k$ be the breakpoints of the class intervals such that $[b_0, b_1), [b_1, b_2), \ldots, [b_{k-1}, b_k)$ form k disjoint and adjacent intervals where the width of the class interval is $\Delta b = b_j - b_{j-1}$. Let c_j be the frequency count of the x_i's in the jth interval, h_j the relative frequency of x_i's in the interval jth interval, and $\hat{p}_j$ the probability associated with the interval using the hypothesized distribution. Then, the relative frequency and probability for the jth interval are, respectively,

$$h_j = c_j/n \qquad \hat{p}_j = \int_{b_{j-1}}^{b_j} \hat{f}(x)dx$$

The squared error is defined as the sum over the intervals of the squared difference between the relative frequency and the probability associated with each interval:

$$Square\ Error = \sum_{j=1}^{k}(h_j - \hat{p}_j)^2$$

Table 3-7 shows the square error calculation for the fit of the exponential distribution to the service time data. The computed square error closely matches the value computed in the Input Analyzer, with the difference attributed to round-off errors.

When you choose the Fit All option in the Input Analyzer, each of the possible distributions are fit in turn and the summary results computed. Then, the Input Analyzer ranks the distributions from smallest to largest according to the square error criteria. As you can see from the definition of the square error criteria, the metric is dependent on the defining intervals. Thus, it is highly recommended that you test the sensitivity of the results to different values for the number of intervals.

■ **TABLE 3-7**

Square Error Calculation

j	c_j	b_j	h_j	$\hat{p}_j$	$(h_j - \hat{p}_j)^2$
1	43	111	0.43	0.399	0.000961
2	19	185	0.19	0.24	0.0025
3	14	260	0.14	0.144	1.6E-05
4	10	335	0.1	0.0866	0.00018
5	6	410	0.06	0.0521	6.24E-05
6	4	484	0.04	0.0313	7.57E-05
7	2	559	0.02	0.0188	1.44E-06
8	0	634	0	0.0113	0.000128
9	1	708	0.01	0.0068	1.02E-05
10	1	783	0.01	0.00409	3.49E-05
				Square Error	0.003969

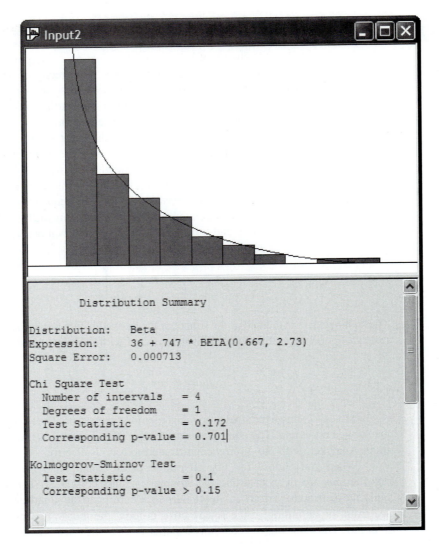

■ **FIGURE 3-25**   **Fit All recommendation**

Using the Fit All function results in the Input Analyzer suggesting that $36 +$ 747^*BETA$(0.667, 2.73)$ expression is a good fit of the model. The Window > Fit All Summary menu option will show the squared error criteria for all the distributions that were fit. See Figure 3-26 for the squared error values. The summary indicates that the Exponential distribution is second in the fitting process according to the squared-error criteria. By using Options > Parameters > Histogram, the Histogram Parameters dialog can be used to change the parameters associated with the histogram as shown in Figure 3-27.

Changing the number of intervals to 12 results in the output provided in Figures 3-28 and 3-29, which indicates that the exponential distribution is a reasonable model based on the chi-square test, the K-S test, and the squared error criteria. You are encouraged to check other fits with differing number of intervals. In most of the fits, the exponential distribution will be recommended. It is beginning to look like the exponential distribution is a reasonable model for the service time data.

Once you have narrowed your choice of distributions to a few top candidates, it is useful to perform additional visual assessments of the models. Two such assessments are

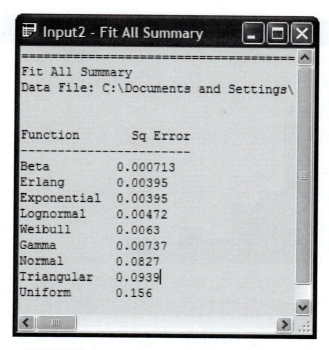

■ **FIGURE 3-26**   Summary of squared error

the P-P plot (probability-probability plot) and the Q-Q plot (quantile-quantile plot). Unfortunately, the Input Analyzer does not compute these plots. Distribution-fitting software such as BestFit and ExpertFit will easily make these plots for a wide variety of distributions. These plots can also be easily made with other general-purpose software such as MINITAB, Maple, and Mathematica, which have functions for the CDFs and inverse CDFs for many common distributions. With some effort Excel can also be used to develop these plots. In what follows, Excel will be used to illustrate how to make P-P plots and Q-Q plots for the service time data under the assumption that the exponential distribution is a good model for the data. This will also make concrete exactly what these plots represent and aid in your understanding of how to interpret the plots.

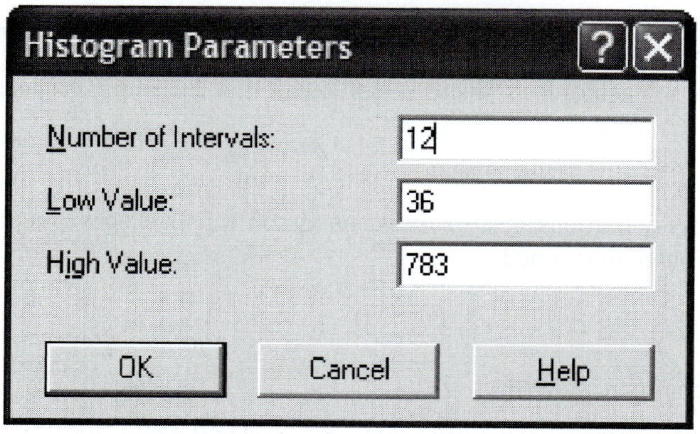

■ **FIGURE 3-27**   Changing histogram parameters

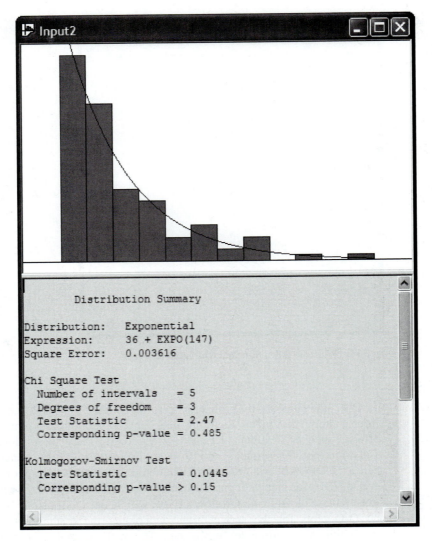

A P-P plot plots the empirical distribution function versus the theoretical distribution evaluated at each order statistic value. Recall that the empirical distribution is defined as

$$\tilde{F}_n(x_{(i)}) = \frac{i}{n}$$

Alternative definitions are also used in many software packages to account for continuous data. As previously mentioned,

$$\tilde{F}_n(x_{(i)}) = \frac{i - 0.5}{n}$$

is very common, as well as

$$\tilde{F}_n(x_{(i)}) = \frac{i - 0.375}{n + 0.25}$$

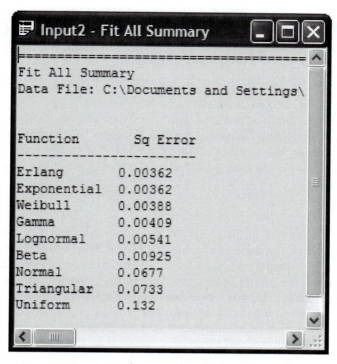

■ **FIGURE 3-29** Fit All summary with 12 intervals

which is used in MINITAB. To create a P-P plot, perform the following steps:

1. Sort the data to obtain the order statistics $(x_{(1)}, x_{(2)}, \ldots x_{(n)})$.
2. Compute $\tilde{F}_n(x_{(i)}) = \frac{i - 0.5}{n} = q_i$ for $i = 1, 2, \ldots .n$.
3. Compute $\hat{F}(x_{(i)})$ for $i = 1, 2, \ldots .n$ where $\hat{F}$ is the CDF of the hypothesized distribution.
4. Plot $\hat{F}(x_{(i)})$ versus $\tilde{F}_n(x_{(i)})$ for $i = 1, 2, \ldots .n$.

The Q-Q plot is similar in spirit to the P-P plot. For the Q-Q plot, the quantiles of the empirical distribution (which are simply the order statistics) are plotted versus the quantiles from the hypothesized distribution (Figure 3-30). Let $0 \le q \le 1$, so that the qth quantile of the

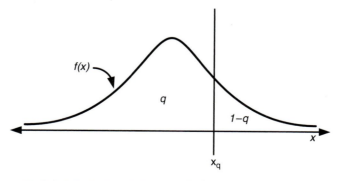

■ **FIGURE 3-30** The quantile of a distribution

distribution is denoted by x_q, and is defined by

$$q = P(X \leq x_q) = F(x_q) = \int_{-\infty}^{x_q} f(u)du$$

Thus, x_q is that value on the measurement axis such that $100q\%$ of the area under the graph of $f(x)$ lies to the left of x_q and $100(1-q)\%$ of the area lies to the right.

You have likely recognized that the "z-values" for the standard normal distribution tables are simply the quantiles of that distribution. The quantiles of a distribution are readily available as long as the inverse CDF of the distribution is available. Thus, the quantile can be defined as

$$x_q = F^{-1}(q)$$

where F^{-1} represents the inverse of the cumulative distribution function (not the reciprocal). For example, if the hypothesized distribution is N(0,1) then $1.96 = \Phi^{-1}(0.975)$ so that $x_{0.975} = 1.96$, where $\Phi(z)$ is the CDF of the standard normal distribution. When you give a probability to the inverse of the cumulative distribution function, you get back the corresponding ordinate value that is associated with the area under the curve, that is, the quantile. To make a Q-Q plot, perform the following steps:

1. Sort the data to obtain the order statistics $(x_{(1)}, x_{(2)}, \ldots x_{(n)})$.
2. Compute $q_i = \frac{i - 0.5}{n}$ for $i = 1, 2, \ldots. n$.
3. Compute $x_{q_i} = \hat{F}^{-1}(q_i)$ for $i = 1, 2, \ldots. n$, where $\hat{F}^{-1}$ is the inverse CDF of the hypothesized distribution.
4. Plot x_{q_i} versus $x_{(i)}$ for $i = 1, 2, \ldots. n$.

Thus, in order to make a P-P plot, the CDF of the hypothesized distribution must be available, and in order to make a Q-Q plot, the inverse CDF of the hypothesized distribution must be available. When the inverse CDF is not readily available, there are other methods for making Q-Q plots for many distributions. These methods are outlined in Law (2007). The following example will illustrate how to make and interpret the P-P plot and Q-Q plot for the hypothesized exponential distribution for the pharmacy service times. In Table 3-2, the CDF of the exponential distribution is given by $F(x) = 1 - e^{-\lambda x}$. To perform the Q-Q plot, the inverse CDF for the exponential distribution is needed. In brief, you need to solve for the inverse of the function, that is, $q = F(x_q)$, to get $x_q = F^{-1}(q)$:

$$q = F(x_q)$$
$$q = 1 - e^{-\lambda x_q}$$
$$1 - q = e^{-\lambda x_q}$$
$$\ln(1 - q) = -\lambda x_q$$
$$x_q = \frac{-1}{\lambda} \ln(1 - q)$$
$$x_q = \frac{-1}{\lambda} \ln(1 - q) = F^{-1}(q)$$

Before proceeding with Excel, another technical issue must be addressed. The proposed model from the Input Analyzer is really $36 + \text{EXPO}(147)$. That is, if Y is a random variable that represents the service time then $Y \sim 36 + \text{EXPO}(147)$, where 147 is the mean of the

	A	B	C	D	E	F	G	H
1					Lambda	0.006803		
2					1/Lambda	147	E[X]	
3								
4								
5	Obs#	Y	Order		q(i)	F(X(i))	X(i)	FINV(q(i))
6	81	36.84	1	0.01	0.005	0.005698	0.8	0.73684
7	84	38.5	2	0.02	0.015	0.016863	2.5	2.2217
8	54	42.93	3	0.03	0.025	0.046049	6.9	3.72172
9	24	46.23	4	0.04	0.035	0.067226	10	5.2372
10	37	46.23	5	0.05	0.045	0.067226	10	6.76846
11	85	48.3	6	0.06	0.055	0.080268	12	8.31584
12	75	52.2	7	0.07	0.065	0.104349	16	9.87969
13	51	53.46	8	0.08	0.075	0.111993	17	11.4603
14	61	55.33	9	0.09	0.085	0.123218	19	13.0582
15	56	56.11	10	0.1	0.095	0.127858	20	14.6736

B	C	E	F	G	H
		Lambda	=1/F2		
		1/Lambda	147	E[X]	
Y	Order	q(i)	F(X(i))	X(i)	FINV(q(i))
36.84	1	=(C6-0.5)/COUNT($B$6:$B$105)	=1-EXP(-$F$1*G6)	=B6-36	=-$F$2*LN(1-E6)
38.5	2	=(C7-0.5)/COUNT($B$6:$B$105)	=1-EXP(-$F$1*G7)	=B7-36	=-$F$2*LN(1-E7)
42.93	3	=(C8-0.5)/COUNT($B$6:$B$105)	=1-EXP(-$F$1*G8)	=B8-36	=-$F$2*LN(1-E8)
46.23	4	=(C9-0.5)/COUNT($B$6:$B$105)	=1-EXP(-$F$1*G9)	=B9-36	=-$F$2*LN(1-E9)
46.23	5	=(C10-0.5)/COUNT($B$6:$B$105)	=1-EXP(-$F$1*G10)	=B10-36	=-$F$2*LN(1-E10)
48.3	6	=(C11-0.5)/COUNT($B$6:$B$105)	=1-EXP(-$F$1*G11)	=B11-36	=-$F$2*LN(1-E11)
52.2	7	=(C12-0.5)/COUNT($B$6:$B$105)	=1-EXP(-$F$1*G12)	=B12-36	=-$F$2*LN(1-E12)
53.46	8	=(C13-0.5)/COUNT($B$6:$B$105)	=1-EXP(-$F$1*G13)	=B13-36	=-$F$2*LN(1-E13)
55.33	9	=(C14-0.5)/COUNT($B$6:$B$105)	=1-EXP(-$F$1*G14)	=B14-36	=-$F$2*LN(1-E14)
56.11	10	=(C15-0.5)/COUNT($B$6:$B$105)	=1-EXP(-$F$1*G15)	=B15-36	=-$F$2*LN(1-E15)

■ **FIGURE 3-31**   **Excel construction of P-P and Q-Q plots**

exponential distribution, so that $\lambda = 1/147$. Because 36 is a constant in this expression, this implies that the random variable $X = Y - 36$, has $X \sim \text{EXPO}(147)$. Thus, the model checking versus the exponential distribution can be done on X. That is, take the original data and subtract 36. The resulting data should have an EXPO(147) distribution. Figure 3-31 shows the *PharmacyInputModelingExample.xls* worksheet for developing the P-P and Q-Q plots. The P-P plot is an XY scatterplot of columns E and F starting in row 6. Cell E6 contains $(1 - 0.5)/100$ and Cell F6 contains $F(0.84) = 1 - \exp(-147*0.84)$, where F is the CDF of the exponential distribution. The Q-Q plot is an XY scatterplot of columns G and H starting in row 6. Cell G6 contains the sorted order statistics of (Y – 36) and Cell H6 contains the inverse CDF evaluated for E6. The rest of the values in the columns can be completed by using Excel's auto-fill functionality. Figure 3-31 also shows the exact cell formulas. The P-P plot and the Q-Q plot are given in Figures 3-32 and 3-33, respectively.

The Q-Q plot should appear approximately linear with intercept zero and slope 1, that is, a 45-degree line, if there is a good fit to the data. In addition, curvature at the ends implies tails that are too long or too short, and convex or concave curvature implies asymmetry, and stragglers at either end may be outliers. The P-P plot should also appear linear with intercept 0 and slope 1. As can be seen in Figures 3-32 and 3-33, both plots do not appear to show any significant departure from a straight line. In contrast, the P-P and Q-Q plots for fitting a continuous uniform distribution with minimum equal to 36 and a maximum equal to 783, that is, UNIF(36, 783) to the data are shown in Figures 3-34 and 3-35. Clearly, there is significant curvature in the plots. This indicates that the uniform distribution is inadequate, especially in regards to modeling the skewness of the data. P-P and Q-Q plots can be made in this manner in Excel if you have the CDF and inverse CDF functions. In the appendix of this chapter, some additional P-P and Q-Q plots are illustrated using the software BestFit.

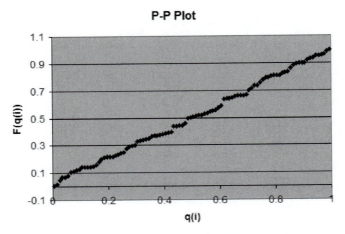

■ **FIGURE 3-32**   P-P plot for exponential fit of service times

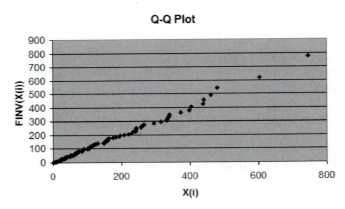

■ **FIGURE 3-33**   Q-Q plot for exponential fit of service times

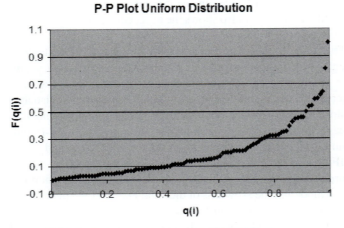

■ **FIGURE 3-34**   P-P plot of UNIF(36,783) fit

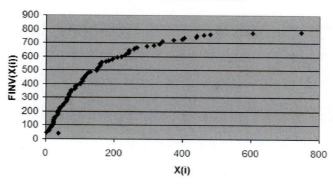

■ **FIGURE 3-35** Q-Q plot of UNIF(36, 783) fit

3.2.4 Additional Input Modeling Concepts

This section wraps up the discussion of input modeling by covering some additional topics that often come up during the modeling process.

Throughout the service time example, a continuous random variable was being modeled. But what do you do if you are modeling a discrete random variable? The basic complicating factor is that the only discrete distribution available in the Input Analyzer is the Poisson distribution and this option will only become active if the data input file only has integer values. The steps in the modeling process are essentially the same except that you cannot rely on the Input Analyzer. Software programs such as ExpertFit and BestFit have options for fitting some of the common discrete distributions. The fitting process for a discrete distribution is simplified in one way because the bins for the frequency diagram are naturally determined by the range of the random variable. For example, if you are fitting a geometric distribution, you need only tabulate the frequency of occurrence for each of the possible values of the random variable {1, 2, 3, 4, etc.}. Occasionally, you may have to group bins together to get an appropriate number of observations per bin. The fitting process for the discrete case primarily centers on a straightforward application of the chi-square goodness-of-fit test, which was outlined in this chapter, and is also covered in many introductory probability and statistics textbooks.

If you consider the data set as a finite set of values, then why can't you just reuse the data? In other words, why should you go through all the trouble of fitting a theoretical distribution when you can simply reuse the observed data, say, for example, by reading in the data from a file. There are a number of problems with this approach. The first difficulty is that the observations in the data set are a *sample*. That is, they do not necessarily cover all the possible values associated with the random variable. For example, suppose that only a sample of 10 observations was available for the service time problem. Furthermore, assume the following sample:

1	2	3	4	5	6	7	8	9	10
36.84	38.5	46.23	46.23	46.23	48.3	52.2	53.46	53.46	56.11

Clearly, this sample does not contain the high service times that were in the 100-sample case. The point is, if you resample from only these values, you will never get any values less than 36.84 or greater than 56.11. The second difficulty is that it will be more difficult to

■ **TABLE 3-8**
Simple Empirical Distribution

X	PMF	CDF
36.84	0.1	0.1
38.5	0.1	0.2
46.23	0.3	0.5
48.3	0.1	0.6
53.46	0.2	0.8
55.33	0.1	0.9
56.11	0.1	1

experimentally vary this distribution in the simulation model. If you fit a theoretical distribution to the data, you can vary the parameters of the theoretical distribution with relative ease in any experiments that you perform. Thus, it is always worthwhile to attempt to fit and use a reasonable input distribution.

But what if you cannot find a reasonable input model either because you have very limited data or because no model fits the data very well? In this situation, it is useful to try to use the data in the form of the empirical distribution. Essentially, you treat each observation in the sample as equally likely and randomly draw observations from the sample. In many situations, there are repeated observations in the sample (as above) and you can form a discrete empirical distribution over the values. If this is done for the sample of 10 data points, a discrete empirical distribution can be formed as shown in Table 3-8. In Arena™, this distribution can be modeled using the DISC() function. Again, this limits us to only the values observed in the sample.

You can also use the continuous empirical distribution, which interpolates between the distribution values. In Arena™, this is represented with the function CONT(). The discrete empirical distribution will be discussed further when randomly generation methods are presented later in this chapter. The Input Analyzer's Empirical function will fit either a discrete empirical or a continuous empirical to the data set.

What do you do if the analysis indicates that the observations are dependent or nonstationary? Either of these situations can invalidate the basic assumptions underlying the standard distribution-fitting process. First, suppose that the observations show some correlation. The first thing that you should do is to verify that the data were correctly collected. The sampling plan for the data collection effort should attempt to ensure the collection of a random sample. If the data were from an automatic collection procedure, then it is quite likely that there may be correlation in the observations. This is one of the hazards of using automatic data-collection mechanisms. You then need to decide whether modeling the correlation is important or not in the study at hand. Thus, one alternative is to simply ignore the correlation and to continue with the model-fitting process. This is not advisable for two reasons. First, the statistical tests in the model fitting process will be suspect, and second, the correlation may be an important part of the input modeling. For example, it has been shown that correlated arrivals and correlated service times in a simple queuing model can have significant effects on the values of the queue's performance measures. If you have a large enough data set, a basic approach is to form a random sample from the data set itself in order to break up the correlation. You can then proceed with fitting a distribution to the random sample; however, you should still model the dependence by trying to incorporate it into the random generation process. There are some techniques for incorporating correlation into the random variable generation process, which are discussed in Chapter 7.

■ **TABLE 3-9**
Breakpoint-Based Empirical Distribution

X	PMF	CDF
(36, 100]	0.1	0.1
(100, 200]	0.2	0.3
(200, 400]	0.3	0.6
(400, 600]	0.2	0.8
(600, ∞)	0.2	1.0

If the data show nonstationary behavior, then you can attempt to model the dependence on time using time series models or other nonstationary models. Suffice to say that these advanced techniques are beyond the scope of this text; however, the next section will discuss the modeling of a special nonstationary model, the nonhomogeneous Poisson process, which is very useful for modeling time-dependent arrival processes. For additional information on these methods, see Law (2007) and Leemis and Park (2006) and the references therein.

Finally, all of the above assumes that you have a data set on which you can perform an analysis. In many situations, you may have no data whatsoever, either because it is too costly to collect or because the system that you are modeling does not exist. In the latter case, you can look at similar systems and see how their inputs were modeled, perhaps adopting some of those input models for the current situation. In either case, you might also rely on expert opinion. In this situation, you can ask an expert in the process to describe the characteristics of a probability distribution that might model the situation. This is where the uniform and the triangular distributions can be very useful, since it is relatively easy to get an expert to indicate a minimum possible value, a maximum possible value, and even a most likely value.

Alternatively, you can ask the expert to assist in making an empirical distribution based on providing the chance that the random variable falls in various intervals. The breakpoints near the extremes are especially important to get. Table 3-9 presents a distribution for the service times based on this method.

Regardless of the amount of data you have, the key final step in the input modeling process is *sensitivity analysis*. Your ultimate goal is to use the input models to drive your larger simulation model of the system under study. You can spend significant time and energy collecting and analyzing data for an input model that has no significant effect on the output measures of interest to your study. Start out with simple input models and incorporate them into your simulation model. You can then vary the parameters and characteristics of those models in an experimental design to assess how sensitive your output is to the changes in the inputs. If you find that the output is very sensitive to particular input models, then you can plan, collect, and develop better models for those situations. The amount of sensitivity is entirely modeler dependent. Remember that in this entire process, you are in charge, not the software. The software exists only to support your decision-making process. Use the software to justify your "art."

3.3   Generating Random Numbers

In the last section, you learned how to model the input distributions for your simulation models. This modeling primarily results in the selection of an appropriate probability distribution for the situation. In Arena[TM], you will need to supply the appropriate expression to generate the necessary random variable. For example, in Chapter 1, the expression EXPO(mean) was used. This expression will generate a random variable from the exponential distribution with the given supplied mean. This section will indicate how the process of generating random variables is executed in the Arena™ Environment. The discussion will begin with how Uniform(0,1) random numbers are obtained and then the major techniques for generating random variables from the major probability distributions will be examined. While Arena™ will take care of most of this for you, you will still need to understand how this process works for the following reasons:

- The random numbers in a simulation experiment might need to be controlled in order to take advantage of them to improve decision making.
- For some situations, Arena™ lacks ready-made functions to generate the desired random variables. In these cases, you will have to do it yourself.

In addition, simulation is much broader than just using Arena™. You can perform simulation in any computer language and the informed modeler should know how the key inputs to simulations are achieved.

In simulation, a large amount of cheap (easily computed) random numbers are required. In general, consider how random numbers might be obtained:

- Dice, coins, colored balls
- Specially designed electronic equipment
- Algorithms

Clearly, in the context of computer simulation, it might be best to rely on algorithms; however, if an algorithm is used to generate random numbers, then they cannot be truly random. For this reason, the random numbers that are used in computer simulation are called *pseudo-random*.

Definition: A sequence of *pseudo-random numbers*, U_i, is a deterministic sequence of numbers in [0, 1] having the same relevant statistical properties as a sequence of truly random $U(0, 1)$ numbers (Ripley 1987).

A battery of statistical tests are performed on the pseudo-random numbers generated from algorithms in order to indicate that their properties are not significantly different from a true set of $U(0, 1)$ random numbers. The algorithms that produce pseudo-random numbers are called *random number generators*. In addition to passing a battery of tests, the random number generators need to be fast and they need to be able to reproduce a sequence of numbers if and when necessary.

Over the history of scientific computing, a wide variety of techniques/algorithms have been proposed and used for generating pseudo-random numbers. A common technique that has been used (and is still in use) in a number of simulation environments is discussed in this text. Some new types of generators that have been recently adopted in many simulation environments, especially the one used in Arena™, will also be briefly discussed. After discussing random number generators, the techniques used to convert pseudo-random numbers into random numbers from a variety of probability distributions will be presented. The approach will be practical, with just enough theory to motivate future study of this area and to allow you to understand the important implications of random number generation. A more rigorous treatment of random number and random variable generation can be found in Fishman (2006) and Devroye (1986).

3.3.1   Random Number Generators

A linear congruential generator (LCG) is a recursive algorithm for producing a sequence of pseudo-random numbers. Each new pseudo-random number from the algorithm depends on the previous pseudo-random number. Thus, a starting value called the *seed* is required. Given the value of the seed, the rest of the sequence of pseudo-random numbers can be completely determined by the algorithm. The basic definition of an LCG follows:

Definition: A LCG defines a sequence of integers, $(R_0, R_1, \ldots)$, between 0 and m-1 according to the following recursive relationship:

$$R_{i+1} = (aR_i + c) \bmod m \quad for\ i = 0, 1, 2, \ldots,$$

where

> R_0 is called the seed of the sequence
>
> a is called the constant multiplier
>
> c is called the increment
>
> m is called the modulus

with

> (m, a, c, R_0) integers and $a > 0, c \geq 0, m > 0, m > a, m > c, m > R_0, 0 \leq R_i \leq m - 1$, and a corresponding sequence of rational numbers defined by $U_i = R_i/m$.

Note that an LCG defines a sequence of integers and subsequently a sequence of real (rational) numbers that can be considered pseudo-random numbers. Remember that pseudo-random numbers can "fool" a battery of statistical tests. The choice of the seed, constant multiplier, increment, and modulus, that is, the parameters of the LCG, will determine the properties of the sequences produced by the generator. With properly chosen parameters, an LCG can be made to produce pseudo-random numbers. To make this concrete, let's look at a simple example of an LCG.

EXAMPLE 3-1   SIMPLE LCG

Consider an LCG with parameters $(m = 8, a = 5, c = 1, R_0 = 5)$, and compute the first nine R_i and U_i from the defined sequence.

First, let's review how to compute using the mod operator. The mod operator is defined as

$$z = y \bmod m \Leftrightarrow z = y - m \left\lfloor \frac{y}{m} \right\rfloor$$

where $\lfloor x \rfloor$ is the floor operator, that is, the greatest integer $\leq x$.

For example, $z = 17 \bmod 3 \Leftrightarrow z = 17 - 3 \left\lfloor \frac{17}{3} \right\rfloor = 17 - 3 \times 5 = 2$. Thus, the mod operator returns the integer remainder (including zero) when $y \geq m$ and y, when $y < m$, such as

$$z = 6 \bmod 9 \Leftrightarrow z = 6 - 9 \left\lfloor \frac{6}{9} \right\rfloor = 6 - 9 \times 0 = 6$$

Using the LCG parameters, the pseudo-random numbers are:

$$R_1 = (5R_0 + 1) \bmod 8 = 26 \bmod 8 = 2 \Rightarrow U_1 = 0.25$$
$$R_2 = (5R_1 + 1) \bmod 8 = 11 \bmod 8 = 3 \Rightarrow U_2 = 0.375$$
$$R_3 = (5R_2 + 1) \bmod 8 = 16 \bmod 8 = 0 \Rightarrow U_3 = 0.0$$
$$R_4 = (5R_3 + 1) \bmod 8 = 1 \bmod 8 = 1 \Rightarrow U_4 = 0.125$$
$$R_5 = 6 \Rightarrow U_5 = 0.75$$
$$R_6 = 7 \Rightarrow U_6 = 0.875$$
$$R_7 = 4 \Rightarrow U_7 = 0.5$$
$$R_8 = 5 \Rightarrow U_8 = 0.625$$
$$R_9 = 2 \Rightarrow U_9 = 0.25$$

Thus, the U_i are simple fractions involving $m = 8$. Certainly, this sequence does not appear very random. The U_i can only take on rational values $0, 1/m, 2/m, \ldots, (m-1)/m$, since $0 \le R_i \le m - 1$, so that if m is small there will be "gaps" on $[0, 1)$, and if m is large, then the U_i are more "dense" on $[0, 1)$. Also note that if a sequence generates the same value as a previous generated value, then the sequence will repeat or *cycle*. An important property of an LCG is that it has a long cycle, as close to m as possible. The length of the cycle is called the "period" (p) of the LCG. Ideally, the period of the LCG is equal to m. If this occurs, the LCG is said to achieve its full period. As can be seen in the example, the LCG is full period. Until recently, most computers were 32-bit machines, and thus a common value for m is $2^{31} - 1 = 2,147,483,647$, which represents the largest integer number on a 32-bit computer using 2's complement integer arithmetic. This choice of m also happens to be a prime number, which leads to special properties. A proper choice of the parameters of the LCG will allow desirable pseudo-random number properties to be obtained.

The following result, due to Hull and Dobell (1962) (see also Law 2007), indicates how to check whether an LCG will have the largest possible cycle.

Theorem: An LCG has a full period if and only if the following three conditions hold:
- The only positive integer that (exactly) divides both m and c is 1 (i.e., c and m have no common factors other than 1).
- If q is a prime number that divides m then q should divide $(a - 1)$ (i.e., $(a - 1)$ is a multiple of every prime number that divides m).
- If 4 divides m, then 4 should divide $(a - 1)$ (i.e., $(a - 1)$ is a multiple of 4 if m is a multiple of 4).

Example 3-2 illustrates the use of the LCG theorem to the LCG in Example 3-1.

Note that in the application of Condition 2 of the theorem, you need the prime numbers that divide m. This involves the prime factorization of m, which can be tedious to determine for arbitrary m. There are a number of algorithms and software implementations to perform prime factorization, many of which can be found easily via a simple Internet search.

Some simplifying conditions (Banks et al. 2005) allow for easier application of the theorem. For $m = 2^b$ (m is a power of 2) and c is not equal to 0, the longest possible period is m, and can be achieved provided that c is chosen so that the greatest common factor of c and m is 1 and $a = 4k + 1$ where k is an integer. The example LCG satisfies this situation. For $m = 2^b$ and

EXAMPLE 3-2   SIMPLE LCG CONTINUED

Consider an LCG with parameters $(m = 8, a = 5, c = 1, R_0 = 5)$. Apply the LCG theorem to check whetherthe LCG will obtain full period.

To apply the theorem, you must check whether each of the three conditions holds for the generator.

Condition 1: c and m have no common factors other than 1.

- The factors of $m = 8$ are $(1, 2, 4, 8)$, since $c = 1$ (with factor 1) condition 1 is true.

Condition 2: (a-1) is a multiple of every prime number that divides m.

- The first few prime numbers are $(1, 2, 3, 5, 7)$. The prime numbers, q, that divide $m = 8$ are $(q = 1, 2)$. Since $a = 5$ and $(a - 1) = 4$, clearly $q = 1$ divides 4 and $q = 2$ divides 4.

Condition 3: If 4 divides m, then 4 should divide $(a - 1)$.

- Since $m = 8$, clearly 4 divides m. Also, 4 divides $(a - 1) = 4$.

Given that all three conditions hold, the LCG achieves full period.

$c = 0$, the longest possible period is $m/4$, and can be achieved, provided that R_0 is odd and $a = 8k + 3$ or $a = 8k + 5$ where k is 0, 1, 2, and so on. The case of m, a prime number, and $c = 0$, defines a special case of the LCG called a "prime modulus multiplicative linear congruential generator" (PMMLCG). For this case, the longest possible period is $m - 1$, and can be achieved if the smallest integer, k, such that $a^k - 1$ is divisible by m is $m - 1$.

Because of 32-bit computers and the fact that $2^{31} - 1 = 2,147,483,647$ is prime, this choice for m with $c = 0$ has been very common. Two common values for the multiplier, a, have been

$$a = 630,360,016$$

and

$$a = 16,807$$

the latter of which was used in Arena^{TM} for a number of years. Note that for PMMLCGs the full period cannot be achieved, but with the proper selection of the multiplier, the next best period length of $m - 1$ can be obtained. In addition, for this case $R_i \in \{1, 2, \ldots, m - 1\}$ and thus $U_i \in (0, 1)$. The limitation of $U_i \in (0, 1)$ is very useful when generating random variables from various probability distributions.

When using an LCG, you must supply a starting seed as an initial value for the algorithm. This seed determines the sequence that will come out of the generator when it is called within software. Since generators cycle, you can think of the sequence as a big circular list, as indicated in Figure 3-36.

Starting with seed $R_0 = 5$, you get a sequence $\{2, 3, 0, 1, 6, 7, 4, 5\}$. Starting with seed $R_0 = 1$, you get the sequence $\{6, 7, 4, 5, 2, 3, 0, 1\}$. Note that these two sequences overlap, but that the first halves, $\{2, 3, 0, 1\}$ and $\{6, 7, 4, 5\}$, of each sequence do not overlap. If you only use four random numbers from each of these two sequences then the numbers will not overlap. The subsequence of random numbers generated from a given seed is called a *random number stream* associated with that seed. You can take the sequence produced by the random number generator and divide it up into subsequences by associating certain seeds with streams. You can call the first subsequence stream 1 and the second subsequence stream 2, and so forth. Each stream can be further divided into subsequences or substreams of nonoverlapping random numbers. In this simple example, it is easy to remember that stream 1 is defined by seed $R_0 = 5$,

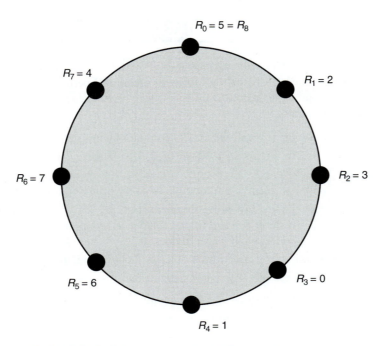

$R_0 = 5 = R_8$

$R_7 = 4$

$R_1 = 2$

$R_6 = 7$

$R_2 = 3$

$R_5 = 6$

$R_3 = 0$

$R_4 = 1$

■ **FIGURE 3-36**   Sequence for LCG example

but when m is large, the seeds will be large integer numbers, such as $R_0 = 123098345$. Rather than remember this huge integer, an assignment of stream numbers to seeds is made. The sequence can then be referenced by its stream number. Naturally, if you associate seeds with streams, you should divide the entire sequence so that the number of nonoverlapping random numbers in each stream is quite large. This ensures that as a particular stream is used, there is very little chance of continuing into the next stream. Clearly, you want m to be as large as possible and to have many streams that contain as many nonoverlapping random numbers as possible. With today's modern computers, even $m = 2^{31} - 1 = 2,147,483,647$ is not very big. For large simulations, you can easily run through all these random numbers.

Random number generators in computer simulation languages come with a default set of streams that divide the "circle" up into independent sets of random numbers. The streams are only independent if you do not use up all the random numbers in the subsequence. These streams allow the randomness associated with a simulation to be controlled. During the simulation, you can associate a specific stream with specific random processes in the model. This has the advantage of allowing you to check whether the random numbers are causing significant differences in the outputs. In addition, this allows the random numbers used across alternative simulations to be better synchronized. These issues will be discussed further in Chapter 4.

A common question for beginners using simulation languages such as Arena™ can now be answered. The question is, "If the simulation is using random numbers, why do I get the same results every time I run my program?" The corollary to this question is, "If I want to get different random results every time I run my program, how do I do it?" The answer to the first question is that the underlying random number generator is starting with the same seed every time you run your program. Thus, your program will use the same pseudo-random numbers today as it did yesterday and the day before, and so on. The answer to the corollary question is that you must tell the random number generator to use a different seed (or alternatively, a different stream) if you want different invocations of the program to produce different results.

■ **FIGURE 3-37** Simple stream demonstration

The latter is not necessarily a desirable goal. For example, when developing your simulation programs, it is desirable to have repeatable results so that you can know that your program is working correctly.

As an illustration of how random streams work, consider a simple Arena^{TM} model, *RNGExample.doe*, which writes out random numbers from the first two streams to a file. As done in the examples of Chapter 2, 10 entities are created and assigned random numbers. The numbers are written to a file and then disposed of as indicated in Figure 3-37.

Figure 3-38 shows the ASSIGN module for generating random numbers from different streams. The variable *vCount* is used to count the entities coming out of the CREATE module. Then, each entity has the attribute *myID* assigned from *vCount*. Thus, the attribute *myID* holds an identifier for each entity. In the third assignment, the function *RA(Stream Number)* is used. Unfortunately, this function is not available using the Build Expression Wizard. If you are re-creating this model, you will have to type it in directly. The RA() function will return a pseudo-random number uniformly distributed on the interval from 0 to 1 from Arena's random number generator starting from the supplied stream number. Each execution of this function returns a *different* random number continuing in the sequence defined by the seed associated with the designated stream. In this example, a variable called *vStreamNum* has been used to allow the stream numbers to be easily changed in each call to RA(). The variable *vStreamNumber* is defined in the VARIABLES module (not shown), and was initialized to the value 1. Thus, the third assignment uses RA(1) and the fourth assignment uses RA(2).

When the program is reinvoked, the underlying random number generator is reset so that all streams are associated with their original seeds. Each time the program is executed it will produce the same output as shown in Table 3-10. This output was taken from the file called *RNGOutput.txt* defined in the FILE module and written to in the READ WRITE module shown in Figure 3-39.

The answer to the corollary question should now be clear. If you want your program to produce different random outputs every time that you run it, then you must change the stream number associated with the calls to the random number generator. Actually, as you can see, when this is done, you aren't really running the same program. You have changed a key input, the stream number, and thus will get different outputs. In Figure 3-40, the third and fourth assignments were changed so that they use the same stream number.

Do you see the difference (similarity) between Tables 3-10 and 3-11? Since the second call to RA occurs in the fourth assignment, the value of myRN2 for the first entity in Table 3-11

	Type	Variable Name	Attribute Name	New Value
1	Variable	vCount	Attribute 1	vCount + 1
2	Attribute	Variable 3	myID	vCount
3	Attribute	Variable 3	myRN1	RA(vStreamNum)
4	Attribute	Variable 4	myRN2	RA(vStreamNum+1)
	Double-click here to add a new row.			

■ **FIGURE 3-38** Generating random numbers from different streams

■ **TABLE 3-10**

First 10 Random Numbers from Streams 1 and 2

myID	myRN1	myRN2
1	0.127011	0.759582
2	0.318528	0.978311
3	0.309186	0.685136
4	0.825847	0.27927
5	0.22163	0.09943
6	0.533395	0.606861
7	0.480774	0.764189
8	0.35556	0.187813
9	0.135988	0.769937
10	00.75585	0.250322

■ **TABLE 3-11**

Output for Changed Assignments

myID	myRN1	myRN2
1	0.127011	0.318528
2	0.309186	0.825847
3	0.22163	0.533395
4	0.480774	0.35556
5	0.135988	0.755852
6	0.575555	0.410064
7	0.32633	0.240378
8	0.610063	0.904181
9	0.298975	0.034154
10	00.96642	50.143495

is the same as the second entity's *myRN1* in Table 3-10. This is because every call to RA continues in the sequence defined by the designated stream, and these two cells represent the second call to RA in the two separate invocations of the program. Not very random! When you understand streams, you can completely control the randomness used in your simulation programs.

As previously mentioned, given current computing power, the previous PMMLCG that Arena^{TM} used is insufficient since it is likely that all 2 billion or so of the random numbers would be used in performing serious simulation studies. Thus, a new generation of random number generators was developed that have extremely long periods. The random number generator that Arena^{TM} has used since version 5.0 is described in L'Ecuyer et al. (2001) and is based on the combination of two multiple recursive generators, resulting in a period of approximately 3.1×10^{57}. While it is beyond the scope of this text to explore the theoretical underpinnings of this generator, it is important to note that the use of this new generator is conceptually similar to that already described. The generator allows multiple independent streams to be defined along with substreams.

The fantastic thing about this generator is the sheer size of the period. Based on their analysis, L'Ecuyer et al. (2001) state that it will be "approximately 219 years into the future before average desktop computers will have the capability to exhaust the cycle of the (generator) in a year of continuous computing." In addition to the period length, the generator has an

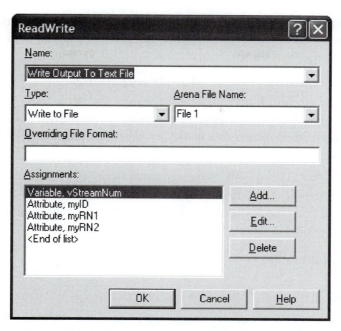

■ **FIGURE 3-39** Writing random numbers to text file

enormous number of streams, approximately 1.8×10^{19} with stream lengths of 1.7×10^{38} and substreams of length 7.6×10^{22} numbering 2.3×10^{15} per stream. Clearly, with these properties, you do not have to worry about overlapping random numbers when performing simulation experiments in Arena™. The generator was subjected to a rigorous battery of statistical tests and is known to have excellent statistical properties.

3.3.1.1 Testing Random Number Generators. This subsection closes with an overview of what is involved in testing the statistical properties of random number generators. Essentially, a random number generator is supposed to produce sequences of numbers that appear to be independent and identically distributed uniform(0,1) random variables. Thus, the lessons learned when modeling input distributions are relevant to the testing of random number generators. The hypothesis that a sample from the generator is IID uniform(0,1) must be made. Then, the hypothesis is subjected to various statistical tests. There are standard batteries of test, see for example Soto (1999), that are utilized. Typical tests examine the following:

- Distributional properties (e.g., chi-square and Kolmogorov-Smirnov test)
- Independence (e.g., correlation tests, runs tests)
- Patterns (e.g., poker test, gap test)

	Type	Variable Name	Attribute Name	New Value
1	Variable	vCount	Attribute 1	vCount + 1
2	Attribute	Variable 3	myID	vCount
3	Attribute	Variable 3	myRN1	RA(vStreamNum)
4	Attribute	Variable 4	myRN2	RA(vStreamNum)
	Double-click here to add a new row.			

■ **FIGURE 3-40** Changed assignments 3 and 4

When considering the quality of random number generators, the higher-dimensional properties of the sequence also need to be considered. For example, the serial test is a higher-dimensional chi-square test. Suppose that $U_i \sim U(0,1)$, from

$$\vec{U}_1 = (U_1, U_2, \ldots, U_d)$$
$$\vec{U}_2 = (U_1, U_2, \ldots, U_{2d})$$

The $\vec{U}_i$ should be independent and identically distributed random vectors uniformly distributed on the d-dimensional unit hypercube. To perform the serial test, take the following steps.

1. Divide [0,1] into k subintervals of equal size.
2. Generate $\vec{U}_i; i = 1, 2, \ldots n$.
3. Let $f_{j_1 j_2 \cdots j_d}$ be the number of $\vec{U}_i$'s having the first component in subinterval j_1, second component in subinterval j_2, and so on.
4. Calculate $\chi^2(d) = \dfrac{k^d}{n} \sum_{j_1=1}^{k} \sum_{j_2=1}^{k} \cdots \sum_{j_d}^{k} \left(f_{j_1 j_2 \cdots j_d} - \dfrac{n}{k^d} \right)^2$
5. Reject $H_0 : U_i \sim U(0,1)$ at α if $\chi^2(d) > \chi^2_{k^d-1, 1-\alpha}$.

The point of these types of tests is to understand how uniformly the random numbers "fill up" the multidimensional space.

A simple autocorrelation test has already been described; however, sequences that do not have other kinds of patterns are also desired. For example, the runs test attempts to test "upward" or "downward" patterns. The runs up-and-down test counts the number of runs up and down, or sometimes just total number of runs versus expected number of runs. A run is a succession of similar events preceded and followed by a different event. The length of a run is the number of events in the run. The following sequence of numbers illustrates how to compute runs up and runs down:

0.90	0.13	0.27	0.41	0.71	0.28	0.18	0.22	0.26	0.19	0.61	0.87	0.95	0.21	0.79
	−	+	+	+	−	−	−	+	−	+	+	+	−	+

In this sequence, there are 8 total runs—4 runs up and 4 runs down.

Digit patterns are also examined. The gaps test counts the number of digits that appear between repetitions of a particular digit and uses a K-S test statistic to compare with the expected number of gaps. The poker test examines for independence based on the frequency with which certain digits are repeated in a series of numbers, (e.g., pairs, three of a kind, etc.). Banks et al. (2005) discusses how to perform these tests.

Does all this testing really matter? Yes! You should always know what generator you are relying on for your simulation analysis. The development and testing of random number generators is serious business. You should stick with well-researched generators and reliable well-tested software.

3.3.2 Generating Random Variates

In simulation, pseudo-random numbers serve as the foundation for generating samples from probability input models. The value from a probability model is called a random variate. A random variate represents a generated instance of a random variable. For the pharmacy example, the expressios $36 + \text{EXPO}(147)$ represents a reasonable probability model for the data. Thus, methods for generating random variates from such expressions are required. That is, how do you generate samples from probability distributions? In generating random variates, the goal is to produce samples from a distribution, $X_i \sim F(x)$, given a source of random numbers, $U_i \sim U(0, 1)$. There are four basic strategies or methods for producing random variates:

- Inverse transform or inverse CDF method
- Convolution
- Acceptance/rejection
- Composition

The *inverse CDF* method is based on the following result. Define a random variable, Y, such that $Y = F(X)$, that is, Y is the random variable where the function F is evaluated at the random variable X where F is the CDF of X. If Y is defined in this manner, then Y is uniformly distributed on (0, 1). To show this, first note that for any probability density function, $f(x)$, the CDF, $F(x)$, has range $0 \leq F(x) \leq 1$. Thus, $Y = F(X)$ has the same range. Now, it must be shown that $P\{Y \leq y\} = y$, which is the CDF of a uniform distribution. Using the fact that F^{-1} is the inverse of F and by definition F is the CDF of X, the following can be shown:

$$
\begin{aligned}
P\{Y \leq y\} &= P\{F(X) \leq y\} \\
&= P\{F^{-1}(F(X)) \leq F^{-1}(y)\} \\
&= P\{X \leq F^{-1}(y)\} \\
&= F(F^{-1}(y)) \\
&= y
\end{aligned}
$$

This result also works in "reverse" if you start with a uniformly distributed random variable then you can get a random variable with the distribution of F. The idea is to generate $U_i \sim Uniform(0, 1)$ and then to use the inverse cumulative distribution function to transform the random number to the appropriately distributed random variate. Figure 3-41 illustrates this concept. First, generate a number, u_i, between 0 and 1 (along the U axis), and then find the corresponding x_i coordinate by using the inverse of F. For various values of u_i, the x_i will be properly "distributed" along the x-axis. The beauty of this method is that there is a one to one mapping between u_i and x_i. In other words, for every u_i there is a unique x_i because of the monotone property of the CDF.

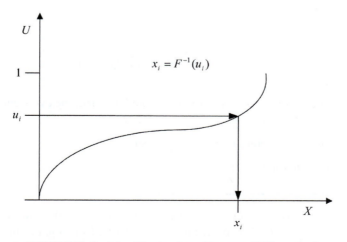

■ **FIGURE 3-41** **Illustration of inverse CDF method**

The *inverse CDF algorithm* can be stated simply as follows:

Step 1: Generate $U_i \sim Uniform(0, 1)$.

Step 2: Return $X_i = F^{-1}(U_i)$.

This algorithm depends on being able to compute the inverse cumulative distribution function. The following example illustrates the inverse CDF method for the exponential distribution.

EXAMPLE 3-3 **INVERSE CDF METHOD FOR EXPONENTIAL DISTRIBUTION**

The exponential distribution is given as

$$f(x) = \begin{cases} 0 & x < 0 \\ \lambda e^{-\lambda x} & x \geq 0 \end{cases}$$

To apply the inverse CDF method, you must first compute the CDF. For the exponential distribution (see also Table 3-2), the following is true:

$$F(x) = \int_{-\infty}^{x} f(u)du = \int_{-\infty}^{0} f(u)du + \int_{0}^{x} f(u)du = \int_{0}^{x} f(u)du = \int_{0}^{x} \lambda e^{-\lambda u}du$$

$$= -\int_{0}^{x} e^{-\lambda u}(-\lambda)du = -e^{-\lambda u}\big|_{0}^{x} = -e^{-\lambda x} - (-e^{0}) = 1 - e^{-\lambda x} \quad x \geq 0$$

Thus, the CDF of the exponential distribution is

$$F(x) = \begin{cases} 0 & x < 0 \\ 1 - e^{-\lambda x} & x \geq 0 \end{cases}$$

Now the inverse of the CDF can be computed as follows:

$$F(x) = 1 - e^{-\lambda x}$$
$$U = 1 - e^{-\lambda x}$$
$$F^{-1}(U) = \frac{-1}{\lambda}\ln(1 - U)$$

Therefore, the inverse CDF algorithm for generating exponential random variates appears as follows:

Step 1: Generate $U_i = uniform(0, 1)$.

Step 2: Set $X_i = \frac{-1}{\lambda}\ln(1 - U_i)$.

Arena's *EXPO(mean, stream number)* function uses the inverse transform method to generate exponentially distributed random variates with the supplied mean and optional stream number.

EXAMPLE 3-4   INVERSE CDF METHOD UNIFORM(a,b)

Consider generating a uniform random variable over the range from (a, b). From Table 3-2, the CDF is

$$F(x) = \frac{x-a}{b-a} \quad \text{for } a < x < b$$

Thus, applying the inverse CDF method yields

$$U = \frac{x-a}{b-a}$$
$$x - a = U(b-a)$$
$$x = a + U(b-a) = F^{-1}(U)$$

Step 1: Generate $U_i = uniform(0, 1)$.

Step 2: Set $X_i = a + U_i(b-a)$.

Example 3-4 illustrates the inverse transform method for a uniformly distributed random variable. Thus, the inverse transform method for a uniform random variable amounts to taking a U(0, 1) random variable and scaling it by $(b-a)$ and then shifting it by a, which is a simple linear transform. Arena's *UNIF*(min, max, *stream number*) function uses the inverse transform method to generate random variates from a uniform distribution over the range from min to max and optional stream number.

The inverse CDF method also works for discrete distributions. For a discrete random variable X, the cumulative distribution function is

$$F(x) = P(X \le x) = \sum_{x_i \le x} f(x_i) \quad \text{where } f(x_i) = P(X = x_i)$$

In order to apply this method, a simple search must be performed. This is illustrated in Example 3-5.

In the discrete inverse CDF algorithm, if the test $u_i \le F(x_i)$ fails, the loop moves to the next interval. If the test failed, $u_i > F(x_i)$ must be true. Thus, only the upper limit in the next interval needs to be tested. If the test passes, the algorithm stops the for/loop and immediately returns the corresponding X_i. Other more complicated and possibly more efficient methods for performing this process are discussed in Fishman (2006) and Ripley (1987). Arena's $DISC(cp_1, value_1, cp_2, value_2, \ldots, stream\ number)$ function uses the inverse transform method to generate random variates from a discrete distribution where $cp_1, value_1, cp_2, value_2$ represent the cumulative probability and the associated values. For the example, you can generate from the distribution given in the example by using DISC(0.4, 1, 0.7, 2, 0.9, 3, 1.0, 4) in Arena[TM]. Note how the last cumulative probability must be equal to 1.

Using the inverse transform method for discrete random variables, a Bernoulli random variable can easily be generated (*Bernoulli(p) algorithm*):

Step 1: Generate $u_i = uniform(0, 1)$.

Step 2: If $u \le p$, return $X = 1$; else return $X = 0$.

This is especially easy to implement in Arena[TM] because of the DISC() function. To generate a Bernoulli random variable, use *DISC(p, 1, 1.0, 0)*. To generate discrete uniform random variables, the inverse transform method yields the following *discrete uniform (a, b) algorithm*:

Step 1: Generate $u_i = uniform(0, 1)$.

Step 2: Return $X = a + \lfloor (b - a + 1)U \rfloor$.

EXAMPLE 3-5 INVERSE CDF METHOD FOR DISCRETE DISTRIBUTION

Suppose you have a random variable, X, as given in Figure 3-42.

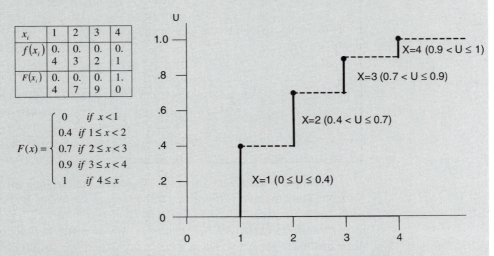

x_i	1	2	3	4
$f(x_i)$	0.4	0.3	0.2	0.1
$F(x_i)$	0.4	0.7	0.9	1.0

$$F(x) = \begin{cases} 0 & if\ x < 1 \\ 0.4 & if\ 1 \le x < 2 \\ 0.7 & if\ 2 \le x < 3 \\ 0.9 & if\ 3 \le x < 4 \\ 1 & if\ 4 \le x \end{cases}$$

X=4 (0.9 < U ≤ 1)
X=3 (0.7 < U ≤ 0.9)
X=2 (0.4 < U ≤ 0.7)
X=1 (0 ≤ U ≤ 0.4)

■ **FIGURE 3-42** **Inverse CDF method for discrete random variables**

Examining Figure 3-42 indicates that for any value of U_i in the interval $(0.4, 0.7]$, you get an X_i of 2. Thus, generating random numbers from this distribution can be accomplished by using Table 3-12.

■ **TABLE 3-12**
Discrete Random Variable Look-up Table

If U_i is in the range	Then X_i is
$0 \le U_i \le 0.4$	1
$0.4 < U_i \le 0.7$	2
$0.7 < U_i \le 0.9$	3
$0.9 < U_i \le 1.0$	4

Thus, given a U_i, pick X_i, such that $F(X_{i-1}) < U_i \le F(X_i)$. For any given value of U_i then, the generation process amounts to a table look-up of the corresponding value for X_i. This simply involves searching until the condition $F(X_{i-1}) < U_i \le F(X_i)$ is true. Since you are dealing with an increasing function (CDF), you need only check the upper limit as per the following *discrete inverse CDF* algorithm:

Step 1: Generate $u_i = uniform(0, 1)$ for $i = 1$ to n.

Step 2: If $u_i \le F(x_i)$ then return x_i.

This is easy to implement in Arena™ using any of the following expressions:

$$a + AINT((b - a + 1)RA)$$
$$AINT(UNIF(a, b) + 0.5), \text{ or}$$
$$ANINT(UNIF(a, b))$$

The inverse transform method also works for generating geometric random variables. The *geometric(p) algorithm* generates geometric random variables:

> *Step 1:* Generate $u_i = uniform(0, 1)$.
>
> *Step 2:* Return $X = 1 + \lfloor \ln(1 - U)/\ln(1 - p) \rfloor$.

In Arena™, this can be implemented with the expression:

$$1 + AINT(LN(1 - RA)/LN(1 - p))$$

The inverse transform technique is general and used when F^{-1} is closed form and easy to compute. It also has the advantage of using one $U(0, 1)$ for every X generated, which helps when applying certain techniques that are employed to improve the estimation process in simulation experiments. Because of this advantage, many simulation languages utilize the inverse transform technique even if a closed-form solution to F^{-1} does not exist by numerically inverting the function.

Many random variables are related to each other through some functional relationship. One of the most common relationships is the convolution relationship. The distribution of the sum of two or more random variables is called the "convolution." Let $Y_i \sim G(y)$ be independent and identically distributed random variables. Let $X = \sum_{i=1}^{n} Y_i$; the distribution of X is said to be the *n-fold convolution* of Y. Some common random variables that are related through the convolution operation are:

- A binomial random variable is the sum of Bernoulli random variables.
- A negative binomial random variable is the sum of geometric random variables.
- An Erlang random variable is the sum of exponential random variables.
- A Normal random variable is the sum of other normal random variables.
- A chi-squared random variable is the sum of squared normal random variables.

The basic *convolution generation algorithm* simply generates $Y_i \sim G(y)$, and then sums the generated random variables.

> *Step 1:* Generate $Y_i \sim G(y) \quad i = 1, 2, \ldots, n$
>
> *Step 2:* Return $X = \sum_{i=1}^{n} Y_i$

EXAMPLE 3-6 ERLANG CONVOLUTION EXAMPLE

Suppose X is Erlang with mean $E[X] = k\beta$. X then can be defined as the sum of k IID exponential random variables, each with mean β.

$$Y_i \sim \exp, \quad E[Y_i] = \beta$$

$$X = \sum_{i=1}^{n} Y_i = \sum_{i=1}^{n} -\beta \ln(1 - U_i) = -\beta \sum_{i=1}^{n} \ln(1 - U_i) = -\beta \ln \left(\prod_{i=1}^{n} (1 - U_i) \right)$$

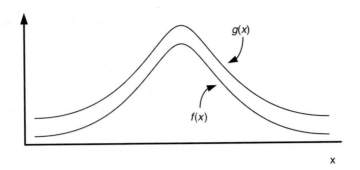

■ **FIGURE 3-43**   **Illustration of a majorizing function**

A number of the exercises for this chapter invite the reader to generate random variates via the convolution method. Because of its simplicity the convolution method is easy to implement; however, in a number of cases (in particular for large n), there are more efficient algorithms available. The previously mentioned references can provide alternatives. Arena's $ERLA(exp\,mean, k, stream\,number)$ function uses Arena's $GAMM$ ($scale$, $shape, stream\,number$) with $scale = exp\,mean$ and $shape = k$, where $exp\,mean$ is the mean of the underlying exponential distribution. Arena's GAMM()function uses the inverse transform method via a numerical inversion technique since there is no closed form inverse function in this case.

In the *acceptance-rejection method*, the PDF $f(x)$ from which a sample is desired is replaced by a "proxy" PDF, $h(x)$, that can be sampled from more easily. The following illustrates how $h(x)$ is defined such that the *selected* samples from $h(x)$ can be used directly to represent random variates from $f(x)$. The PDF $h(x)$ is based on the development of a majorizing function for $f(x)$. A majorizing function, $g(x)$, for $f(x)$, is a function such that $g(x) \geq f(x) - \infty < x < \infty$. Figure 3-43 illustrates the concept of a majorizing function for $f(x)$, which simply means a function that is bigger than $f(x)$ everywhere.

In addition to being a majorizing function for $f(x)$, $g(x)$ must have finite area. In other words,

$$c = \int_{-\infty}^{+\infty} g(x)dx < \infty$$

If $h(x)$ is defined as $h(x) = g(x)/c$, then $h(x)$ is a probability density function. The acceptance-rejection method starts by obtaining a random variate W from $h(x)$. Recall that $h(x)$ should be chosen with the stipulation that it can be easily sampled, such as via the inverse transform method. Let U be a uniform $(0, 1)$ random number. The steps of the *acceptance-rejection algorithm* are as follows:

Step 1: Generate $W \sim h(x)$.

Step 2: Generate $U \sim U(0, 1)$.

Step 3: if $U \times g(W) \leq f(W)$ accept $X = W$ and return, else reject W and return to step 1.

The validity of the procedure is based on a proof of the following statement:

$$P\{W \leq x | W = w\,is\,accepted, -\infty < w < \infty\} = \int_{-\infty}^{x} f(z)dz \quad -\infty < x < \infty$$

which is left as an exercise for the reader. The efficiency of the acceptance-rejection method is enhanced as the probability of rejection in Step 3 is reduced. This probability depends directly on the choice of the majorizing function $g(x)$. The acceptance-rejection method has a nice intuitive geometric connotation, which is best illustrated with an example.

EXAMPLE 3-7   ACCEPTANCE-REJECTION METHOD

Consider the following PDF over the range $[-1, 1]$:

$$f(x) = \begin{cases} \dfrac{3}{4}\left(1 - x^2\right) & -1 \leq x \leq 1 \\ 0 & \text{otherwise} \end{cases}$$

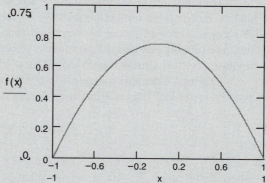

A simple method to choose a majorizing function is to set $g(x) = \max f(x)$. As can be seen from the figure of $f(x)$ the maximum $(3/4)$ occurs at 0 so that $g(x) = 3/4$. In order to proceed, you need to find $h(x) = g(x)/c$ using $g(x) = 3/4$.

$$c = \int_{-1}^{1} g(x)dx = \int_{-1}^{1} \frac{3}{4}dx = \frac{3}{2}$$

Thus,

$$h(x) = \begin{cases} \dfrac{1}{2} & -1 \leq x \leq 1 \\ 0 & \text{otherwise} \end{cases}$$

Therefore, $h(x)$ is a uniform distribution over the range $[-1, 1]$, that is, Uniform$(-1, 1)$. From Example 3-4, you know that you can generate $W \sim Uniform(a, b)$ easily using $W = a + U(b - a)$. Thus, the acceptance rejection algorithm for $f(x)$ is as follows:

Step 1: Generate $U_1 \sim U(0, 1)$.

Step 2: Set $W = -1 + 2U_1$.

Step 3: Generate $U_2 \sim U(0, 1)$.

Step 4: Set $f = \frac{3}{4}(1 - W^2)$.

Step 5: if $U_2 \times \frac{3}{4} \leq f$ accept $X = W$ and return, else reject W and return to Step 1.

In Example 3-7, Steps 1 and 2 generate a random variate, W, from $h(x)$. Note that W is in the range $[-1, 1]$ (the same as the range of X) so that Step 4 is simply finding the height associated with W in terms of f. Step 3 generates U_2. What is the range of $U_2 \times \frac{3}{4}$? The range is $[0, \frac{3}{4}]$. Note that this range corresponds to the range of possible values for f. Thus, in Step 5, a point between $[0, \frac{3}{4}]$ is being compared to the candidate point W's height along the vertical axis. If the point is under f, the W is accepted; otherwise, the W is rejected and another candidate point must be generated. In other words, if a point "under the curve" is generated, it will be accepted.

As illustrated in Example 3-7, the probability of acceptance is related to how close $g(x)$ is to $f(x)$. The ratio of the area under $f(x)$ to the area under $g(x)$ is the probability of accepting. Since the area under $f(x)$ is 1, the probability of acceptance, P_a, is

$$P_a = \frac{1}{\int_{-\infty}^{+\infty} g(x)dx} = \frac{1}{c}$$

For the example, the probability of acceptance is $P_a = 2/3$. From this, it should be clear that the more $g(x)$ is similar to the PDF $f(x)$, the better (higher) the probability of acceptance. The key to efficiently generating random variates using the acceptance-rejection method is finding a suitable majorizing function over the same range as $f(x)$. In the example, this was easy because $f(x)$ had a finite range. This is more challenging if $f(x)$ has an infinite range.

The final technique that needs to be discussed is the *composition* method. Suppose that you can write the CDF as a mixture of other CDFs,

$$F(x) = \sum_{i=1}^{n} p_i F_i(x)$$

where $p_i \geq 0$ and $\sum_{i=1}^{n} p_i = 1$, or equivalently, for the probability density function,

$$f(x) = \sum_{i=1}^{n} p_i f_i(x)$$

The idea is to first randomly select the appropriate distribution using p_i and then generate an X from $f(x)$. The composition method is used in conjunction with other methods to generate from $f(x)$. An important use of this approach is in generating random variates for situations in which the data has more than one mode.

When the distribution has a bimodal shape, there may be a logical reason. For example, suppose that someone else had collected the data and you were told to analyze it as was done in this Example 3-8. Given the difficulty in fitting the data, you become skeptical about how the data were collected and decide to go and collect some of the data yourself. During the collection process, you notice that about 60% of the time the person pays with a debit card and the other 40% of the time the person pays with a credit card. It so happens that whenever credit card payment is used, extra time is involved. The original data had both debit and credit card payments mixed together, resulting in the bimodal distribution for the overall service time. These types of distributions are often called "mixture distributions," and as indicated in the example, they can be modeled by the composition method. In this case, you probably should observe and model each type of transaction separately, and develop a distribution across the types of transactions. Remember the caveat about collecting data. It is always advisable to be very specific about how the data are to be collected so as to improve the input modeling process, and as shown here, to facilitate the random variate generation methods.

EXAMPLE 3-8   COMPOSITION METHOD

Suppose that you had collected data for the service times and the histogram appeared as in Figure 3-44. Clearly, these data have two modal points near 3.2 and 7.0. Fitting the distributions available in the Input Analyzer would appear to be fruitless in this situation.

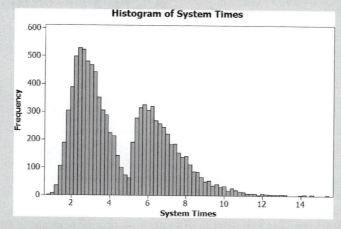

■ **FIGURE 3-44**   Bimodal data

One method applicable to this situation is to consider modeling the data as the composition of two unimodal distributions. In order to do this, you must somehow divide the data set into two separate data sets and see if a good fit for the divided data sets can be obtained. By looking at the histogram, it appears that dividing the data into values less than 5 and values greater than 5 looks reasonable. This is clearly an ad hoc and subjective method. This can be achieved simply by sorting the data and dividing it into the subsets. The spreadsheet *Compositiondata.xls* has the original and divided data. From the analysis, about 57% of the data fall below 5. An input analyzer fit to the observations that are less than 5 suggests that $0.45 + \mathrm{ERLA}(0.305, 8)$ is a reasonable model. A fit to the data greater than or equal to 5 suggests that $5 + \mathrm{GAMM}(1.07, 1.79)$ is a reasonable model. Therefore, a composition approach to generating a sample from this bimodal distribution would be as follows.

Let $\quad p_1 = 0.57, \quad X_1 \sim 0.45 + Erlang(0.305, 8), \quad p_2 = 0.43,$

$\qquad X_2 \sim 5.0 + Gamma(1.07, 1.79)$

Step 1: Generate $U \sim U(0, 1)$.

Step 2: If $U \leq p_1$, set $X = 0.45 + \mathrm{ERLA}(0.305, 8)$.

$\qquad$ Otherwise, set $X = 5.0 + \mathrm{GAMM}(1.07, 1.79)$.

Here ERLA() and GAMM() are functions that generate Erlang and Gamma random variates. This entire procedure can be easily implemented in Arena^{TM} using a one-line function:

$$\mathrm{DISC}(0.57, 0.45 + \mathrm{ERLA}(0.305, 8), 1.0, 5 + \mathrm{GAMM}(1.07, 1.79))$$

3.3.3   Modeling Randomness in Arena^{TM}

The previous sections have discussed a number of methods and the Arena^{TM} functions for generating random variates. This section presents a small example to illustrate the generation of random variates in Arena^{TM}. The example is fictitious, but has a number of features that motivate the use of Arena^{TM}. A variety of probability distributions are involved to illustrate their use.

EXAMPLE 3-9   LOTR MAKERS, INC.

Every morning the sales force at LOTR Makers, Inc. makes a number of confirmation calls to customers who have previously been visited by the sales force. They have tracked the success rate of their confirmation calls over time and have determined that the chance of success varies from day to day. They have modeled the probability of success for a given day as a beta random variable with parameters $\alpha_1 = 5$ and $\alpha_2 = 1.5$ so that the mean success rate is about 77%. They always make 100 calls every morning. Each sales call will or will not result in an order for a pair of magical rings for that day. Thus, the number of pairs of rings to produce every day is a binomial random variable, with p determined by the success rate for the day and $n = 100$ representing the total number of calls made. The sale force is large enough and the time to make the confirmation calls small enough so as to be able to complete all the calls before releasing a production run for the day. In essence, ring production does not start until all the orders have been confirmed, but the actual number of ring pairs produced every day is unknown until the sales call confirmation process is completed. The time to make the calls is negligible when compared to the overall production time.

Besides being magical, one ring is smaller than the other ring so that the smaller ring must fit snuggly inside the larger ring. The pair of rings is produced by a master ring maker and takes uniformly between 5 to 15 minutes. The rings are then scrutinized by an inspector with the time (in minutes) being distributed according to a triangular distribution with parameters (2, 4, 7) for the minimum, the mode, and the maximum. The inspection determines whether the smaller ring is too big or too small when fit inside the bigger outer ring. The inside diameter of the bigger ring, D_b, is normally distributed with a mean of 1.5 cm and a standard deviation of 0.002. The outside diameter of the smaller ring, D_s, is normally distributed with a mean of 1.49 and a standard deviation of 0.005. If $D_s > D_b$, then the smaller ring will not fit in the bigger ring; however, if $D_b - D_s > tol = 0.02\, cm$, then the rings are considered too loose.

If there are no problems with the rings, the rings are sent to a packer for custom packaging for shipment. A time study of the packaging time indicates that it is distributed according to a log-normal distribution with a mean of 7 minutes and a standard deviation of 1 minute. If the inspection shows that there is a problem with the pair of rings they are sent to a rework craftsman. The minimum time that it takes to rework the pair of rings has been determined to be 5 minutes plus some random time that is distributed according to a Weibull distribution with a scale parameter of 15 and a shape parameter of 5. After the rework is completed, the pair of rings is sent to packaging.

LOTR Makers, Inc. is interested in estimating the daily production time. In particular, management is interested in estimating the probability of overtime. Currently, the company runs two shifts of 480 minutes each. Time after the end of the second shift is considered overtime. Use 30 simulated days to investigate the situation.

Now let's proceed with the modeling of Example 3-9. We start with answering the basic model building questions.

■ What is the system? What information is known by the system?

The system is the LOTR Makers, Inc. sales calls and ring production processes. The system starts every day with the initiation of sales calls and ends when the last pair of rings produced for the day is shipped. The system "knows" the following:

Sales call success probability distribution: $p \sim BETA(5, 1.5)$

Number of calls to be made every morning: $n = 100$

Distribution of time to make the pair of rings: $UNIF(5, 15)$

Distributions associated with the big and small ring diameters: $NORM(1.5, 0.002)$ and $NORM(1.49, 0.005)$, respectively

Distribution of ring-inspection time: $TRIA(2, 4, 7)$

Distribution of packaging time: $LOGN(7, 1)$

Distribution of rework time, $5 + WEIB(15, 3)$

Length of a shift: 480 minutes

■ What are the entities? What information must be recorded for each entity?

Possible entities are the sales calls and the production job (pair of rings) for every successful sales call. Each sales call knows whether it is successful. For every pair of rings, the diameters must be known.

■ What are the resources that are used by the entities?

The sales calls do not use any resources. The production job uses a master craftsman, an inspector, and a packager. It might also use a rework craftsman.

■ What are the process flows? Write out or draw sketches of the process.

There are two processes: sales order and production.

Sales Order Process

1. Start the day.

2. Determine the likelihood of calls being successful.

3. Make the calls.

4. Determine the total number of successful calls.

5. Start the production jobs.

Production Process (for each pair of rings)

1. Make the rings (determine sizes).

2. Inspect the rings.

3. If rings do not pass inspection, perform rework.

4. Package rings and ship.

■ Develop pseudo-code for the situation.

Each of the above-described processes needs to be modeled. Note that in the problem statement there is no mention that the sales calls take any significant time. In addition, the sales order process doesn't use any resources. The calls take place before the production shifts. The major purpose of the sales order process is to determine the number of rings to produce for the daily production run. This type of situation is best modeled using a *logical entity*. In essence, the logical entity represents the entire sales order process and must determine the number of rings to produce. From the problem statement, this is a binomial random variable with $n = 100$ and $p \sim BETA(5, 1.5)$. Arena™ does not have a binomial distribution. The easiest way to accomplish this is to use the convolution property of Bernoulli random variables to generate the binomial random variable.

In the following pseudo-code for the sales order process, the logical entity is first created and then assigned the values of *p*. Then, 100 Bernoulli random variables are generated to represent the success (or not) of a sales call. The successes are summed so that the appropriate number of production jobs can be determined. In the pseudo-code, the method of summing up the number of successes is done with a "do-loop" flow of control construct. As noted in Chapter 2, Arena^TM does not have a "do-loop" construct, but this idea can be implemented in a number of ways in Arena^TM. The point is that you should make your pseudo-code something that helps you to understand what needs to be implemented. If you are comfortable with Arena^TM constructs, you can make your pseudo-code more and more "Arena-like." Arena^TM keywords have been capitalized in the pseudo-code.

Sales Order Process

```
CREATE 1 order process logical entity
ASSIGN p ~ BETA (5, 1.5),
do k = 1 to 100
      X = Bernoulli(p)
      If X = 1, add 1 to number of successes
loop
Create number of successes SEPARATE jobs and send them
to production
```

The production process pseudo-code describes what happens to a pair of rings as it moves through production.

Production Process (for each pair of rings)

```
SEIZE master ring maker
DELAY for ring making
RELEASE master ring maker
ASSIGN bigger ring inner diameter, ID = NORM(1.49, 0.005)
      Smaller ring outer diameter, OD = NORM(1.50, 0.002)
SEIZE inspector
DELAY for inspection time
RELEASE inspector
DECIDE if (ID > OD) OR (OD – ID > tol)
      SEIZE rework craftsman
      DELAY for rework
      RELEASE rework craftsman
SEIZE packager
DELAY for packaging time
RELEASE packager
DISPOSE ship rings
```

Implementing the Model in Arena^TM

Thus far, the system has been conceptualized in terms of entities, resources, and processes. This provides a good understanding of the information that must be represented in the simulation model. In addition, the logical flow of the model has been represented in pseudo-code. Now it is time to implement these ideas in Arena^TM. The following describes an Arena^TM implementation for these processes. Before proceeding, you might want to try to

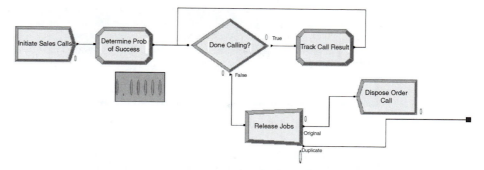

■ **FIGURE 3-45**   **Sales confirmation process in Arena**

implement the logic yourself so that you can check how you are doing against what is presented here. If not, you should try to implement the logic as you proceed through the example.

To implement the sales order process, you first need to decide how to represent the required data. In the pseudo-code, there is only one entity and no passage of time. In this situation, only 1 entity is needed to execute the associated Arena^{TM} modules. Thus, either variables or attributes can be used to implement the logic. A variable can be used because there is no danger of another entity changing the global value of the represented variables. In this model, the do-loop has been implemented using a looping DECIDE construct as discussed in Chapter 2. An overview of the model is given in Figure 3-45. If you are building the model as you follow along, you should take the time to lay down the modules shown in the figure. Each of the modules will be illustrated in what follows.

The CREATE module is very straightforward. Because the Max Arrivals field is 1 and the First Creation field is 0.0, only 1 entity will be created at time 0.0. The fields associated with the label Time between Arrivals are irrelevant in this situation since there will not be any additional arrivals generated from this CREATE module. Fill out your CREATE module as shown in Figure 3-46.

In the first ASSIGN module, the probability of success for the day is determined. The variable (*vSalesProb*) has been used to represent the probability as drawn from the BETA() distribution function (Figure 3-47). The attribute (*myCallNum*) is initialized to 1, prior to starting the "do-loop" logic. This attribute is going to play the role of "k" in the pseudo-code by counting the number of calls made.

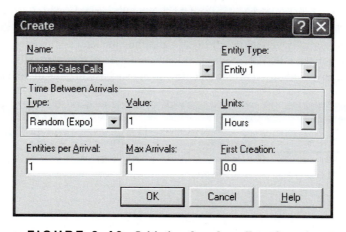

■ **FIGURE 3-46**   **Initiating the sales call confirmation process**

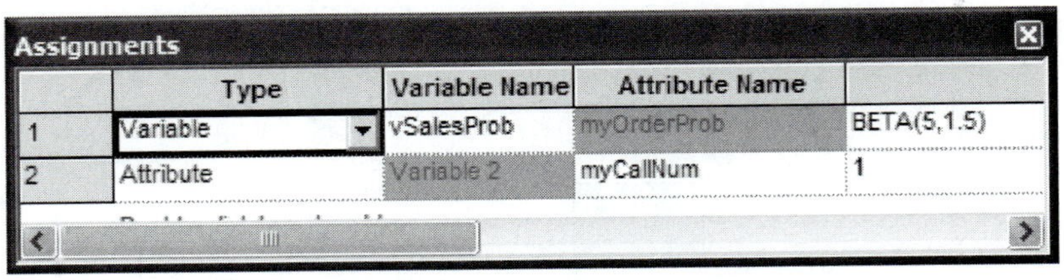

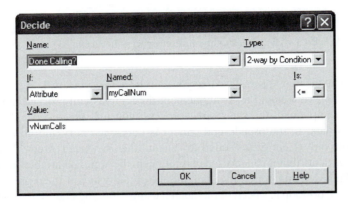

■ FIGURE 3-47 Determine the probability of success

■ FIGURE 3-48 Checking that all calls have been made

The DECIDE module uses the attribute (*myCallNum*) to check the variable *vNumCalls* (Figure 3-48). The variable (*vNumCalls*) is defined in the VARIABLE module (not shown here) and has been initialized to 100. Because the total number of calls has been represented as a variable, the desired number of calls can be easily changed in the VARIABLE module, without editing the DECIDE module.

Figure 3-49 presents the ASSIGN module in the do-loop logic. In this ASSIGN, the attribute *myCallNum* is incremented. Then, according to the pseudo-code, you must implement a Bernoulli trial. This can easily be accomplished in Arena™ by using the DISC() distribution function with only two outcomes, 1 and 0, to assign to the attribute *mySale*. You can set the probability of success for the value 1 to *vSalesProb*, which was determined via the call to the BETA function. Note that the function DISC(BETA(5, 1.5), 1, 1.0, 0) would not work in this situation. DISC(BETA(5, 1.5)) will draw a new probability from the BETA() function every time DISC() is called. Thus, each trial will have a different probability of success. This is not what is required for the problem and this is why the variable *vSaleProb* was determined outside the do-loop logic. The attribute (*myNumSales*) is incremented with the value from *mySale*, which essentially counts up all the 1's and thus the successful sales calls.

When the DECIDE module evaluates to false, all the sales calls have been made, and the attribute *myNumSales* has recorded how many successful calls there have been out of the 100

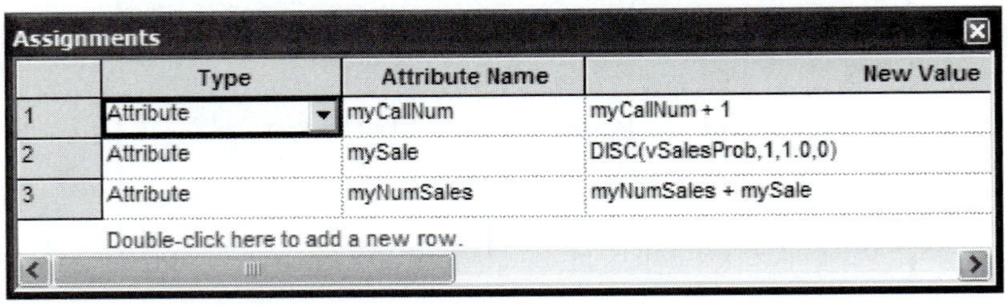

	Type	Attribute Name	New Value
1	Attribute	myCallNum	myCallNum + 1
2	Attribute	mySale	DISC(vSalesProb,1,1.0,0)
3	Attribute	myNumSales	myNumSales + mySale
	Double-click here to add a new row.		

■ FIGURE 3-49 Tracking the calls result

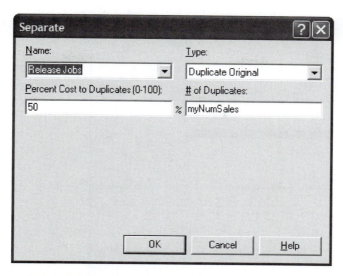

■ **FIGURE 3-50** Releasing the jobs to production

calls made. The order process entity is directed to the SEPARATE module where it creates (*myNumSales*) duplicates and sends them to the production process (Figure 3-50). The original order process logic entity is sent to a DISPOSE, which has its collect entity statistics box unchecked (not shown). By un-checking this box, no statistics will be collected for entities that exit through this DIPOSE module. Since the problem does not require anything about the statistics for the order process logic entity, this is advisable.

The production process is too long for one screen shot so this discussion is divided into three parts: the making and inspection processes, the decision concerning too big/too small, and the repair and packaging operations. Figure 3-51 presents the making and inspection processes which are implemented with PROCESS modules using the SEIZE, DELAY, RELEASE option. Figures 3-52 and 3-53 show the PROCESS modules and the appropriate delay distributions using the uniform and triangular distributions.

The ASSIGN module between the two processes is used to determine the diameters of the rings. Figure 3-54 shows that attributes (*myIDBigRing* and *myODSmallRing*) are both set using

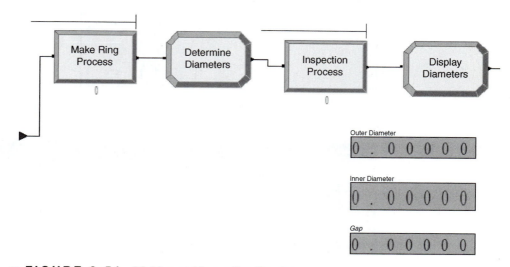

■ **FIGURE 3-51** Making and inspecting the rings

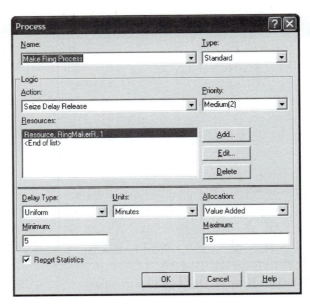

■ **FIGURE 3-52** PROCESS module for making the rings

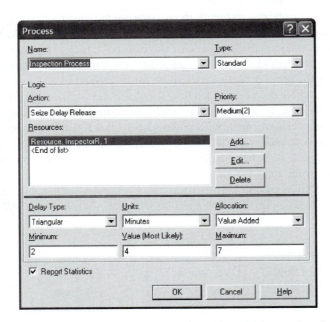

■ **FIGURE 3-53** PROCESS module for inspecting the rings

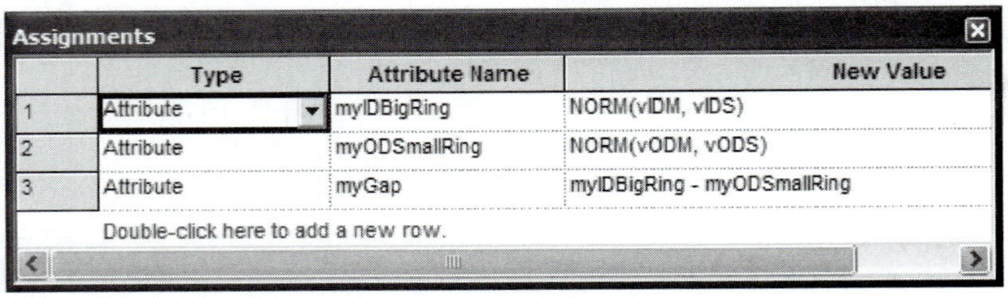

	Type	Attribute Name	New Value
1	Attribute	myIDBigRing	NORM(vIDM, vIDS)
2	Attribute	myODSmallRing	NORM(vODM, vODS)
3	Attribute	myGap	myIDBigRing - myODSmallRing
	Double-click here to add a new row.		

Assignments

■ **FIGURE 3-54** Determining the ring diameters

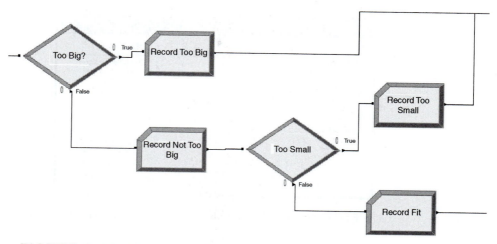

■ **FIGURE 3-55**   Checking ring diameters

the normal distribution. The parameters of the NORM() functions are defined by variables in the VARIABLE module (not shown here).

The ASSIGN module after the inspection process is used to transfer the attribute values to variables so that they can be shown in the animation. The reasons for this were described in Chapter 2. After inspection, the rings must be checked to see whether rework is necessary. An overview of this checking is given in Figure 3-55. The RECORD modules are used to collect statistics on the probability of the smaller ring being too big or the smaller ring being too small.

Figure 3-56 shows the DECIDE module for checking if the ring is too big. The 2-way by condition option was used to check if the attribute (*myODSmallRing*) is larger than the attribute (*myIDBigRing*). If the smaller ring's outer diameter is larger than the bigger ring's inner diameter, then the smaller ring will not fit inside the bigger ring and the rings will require rework.

Figure 3-57 shows the DECIDE module for checking if the smaller ring is too loose. In this DECIDE module, the two-way by condition option is used to check the expression (*myIDBigRing − myODSmallRing*) > *vTol*. If the difference between the diameters of the rings is too large (larger than the tolerance), then the rings are too loose and need rework. If the rings fit properly, then they go directly to packaging.

Figure 3-58 shows the rework and packaging processes. Again, the PROCESS module is used to represent these processes.

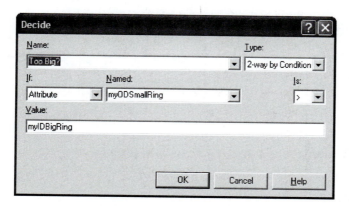

■ **FIGURE 3-56**   Checking to determine whether small ring is too big

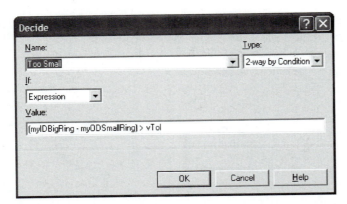

■ **FIGURE 3-57**   Checking to determine whether smaller ring is too small

Figures 3-59 and 3-60 show the rework and packaging process modules. Note that the delay type has been changed to expression so that $5 + WEIB(15, 3)$ and $LOGN(7, 1)$ can be specified for each of the respective processing times.

The problem statement requests the estimation of the probability of overtime work. The sales order process determines the number of rings to produce. The production process continues until there are no more rings to produce for that day. The number of rings to produce is a binomial random variable as determined by the sales order confirmation process. Thus, there is no clear run length for this simulation.

In Arena^{TM}, a simulation can end based on three situations:

- A scheduled run length
- A terminating condition is met
- No more events (entities) to process

Because the production for the day stops when all the rings are produced, the third situation applies for this model. The simulation will end automatically after all the rings are

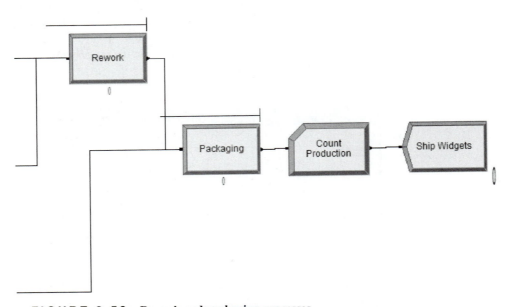

■ **FIGURE 3-58**   Rework and packaging processes

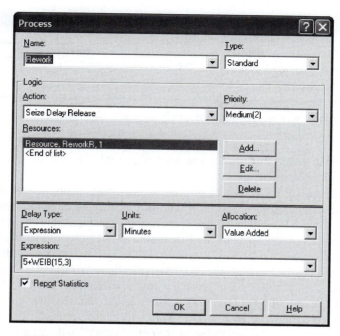

■ **FIGURE 3-59** Rework PROCESS module

produced. In essence, a day's worth of production is simulated. Because the number of rings to produce is random and it takes a random amount of time to make, inspect, rework, and package the rings, *the time that the simulation will end is a random variable*. If this time is less than 960 (the time of two shifts), there will not be any overtime. If this time is greater than 960, production will have lasted past two shifts and thus overtime will be necessary. To assess the

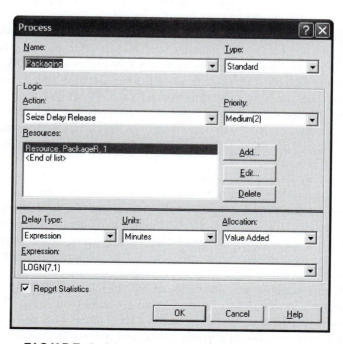

■ **FIGURE 3-60** Packaging PROCESS module

Statistic - Advanced Process				
	Name	Type	Expression	Report Label	Output File
1	TimeToMakeOrderStat	Output	TNOW	TimeToMakeOrderStat	...
2	ProbOfOT	Output	TNOW > 960	ProbOfOT	
	Double-click here to add a new row.				

■ **FIGURE 3-61** Defining OUTPUT statistics for overtime

chance that there is overtime, you need to record statistics on how often the end of the simulation is past 960.

The Arena™ special variable TNOW provides the current time of the simulation. Thus, when the simulation ends, TNOW will be the time that the simulation ended. In this case, this is the time that the last ring completed processing. To estimate the chance that there is overtime, the Advanced Process panel's Statistic module can be used. In particular, you need to define what Arena™ calls an OUTPUT statistic. An OUTPUT statistic records the final value of some system or statistical value at the end of a replication. An OUTPUT statistic can be defined by any valid Arena™ expression involving system variables, variables, statistical functions, and so on. In Figure 3-61, an OUTPUT statistic called *TimeToMakeOrderStat* has been defined, which records the value of *TNOW*. This will cause Arena™ to collect statistics (across replication) for the ending value of *TNOW*. Arena™ will record the average, minimum, maximum, and half-width across the replications for an OUTPUT statistic. You can also use the Output File field to write out the observed values to a file if necessary.

The second OUTPUT statistic in Figure 3-61 defines an OUTPUT statistic called *ProbOfOT* to represent the chance that the production lasts longer than 960 minutes. In this case, the expression is the *Boolean* value of (TNOW > 960). Arena™ takes the expression (TNOW > 960) and evaluates it. If it is true, it will evaluate to 1.0; otherwise, it will evaluate to a 0.0. This is just like an indicator variable on the desired condition. Arena™ will compute the average of the 1's and 0's, which is an estimate of the probability of the condition. Thus, you will get an estimate of the likelihood of overtime.

Figure 3-62 shows how to set up the run parameters of the simulation. You should run the simulation with a base time unit of minutes. This is important to set up here so that TNOW can now be interpreted in minutes. This ensures that the expression (TNOW > 960) makes sense in terms of the desired units. There is no replication length specified nor is there a terminating condition specified. As previously mentioned, Arena™ will end when there are no more events (entities) to process. If the model is not set up correctly (i.e., an infinite number of entities are processed), then Arena™ will never terminate. This is a logical situation that the modeler is responsible for preventing. The number of replications has been specified to 30 to represent the 30 days of production.

Running the model results in the user defined statistics for the probability of overtime and the average time to produce the orders as shown in Figure 3-63. The probability of overtime appears to be about 6%, but the 95% half-width is wide for these 30 replications. The average time to produce an order is about 780.68 minutes. While the average is under 960, there still appears to be a reasonable chance of overtime occurring. In the exercises, you are asked to explore the reasons behind the overtime and to recommend an alternative to reduce the likelihood of overtime.

In this example, you have learned how to model a small system involving random components and to translate the conceptual model into an Arena™ simulation. Using the simulation, you can then observe the statistical responses associated with the model. Chapter 4 will concentrate on how to properly collect and analyze the statistical responses from such simulations.

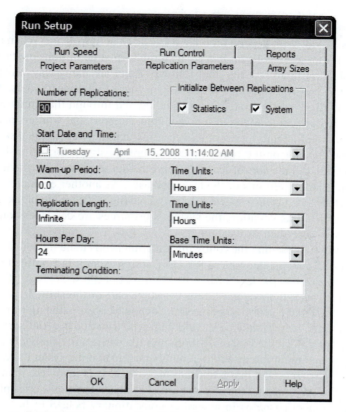

■ **FIGURE 3-62** Specifying the number of replications in run setup

Output	Average	Half Width	Minimum Average	Maximum Average
ProbOfOT	0.06666667	0.09	0.00	1.0000
TimeToMakeOrderStat	780.68	49.12	462.57	968.44

■ **FIGURE 3-63** Arena output statistics across 30 days

In addition, you have learned how to use many of the probability distributions available in Arena™ to generate random variables for the LOTR Makers, Inc. situation. There are only a couple of distributions in Arena™ that have not been covered through the previous examples (e.g., empirical continuous CONT(), Johnson JOHN(), Poisson POIS(), and nonstationary exponential NSExpo()). The following sections will cover the NSExpo() function, and an exercise will allow you to familiarize yourself with the Poisson distribution. The empirical continuous distribution works in a similar fashion as the DISC() function except that the sample values a linearly interpolated. This process is discussed in Arena's help system and a general presentation is given in Banks et al. (2005) or in Law (2007). A discussion of modeling with the Johnson distribution can also be found in Law (2007).

Two more issues remain to be addressed in input modeling: modeling nonstationary processes and correlated processes. The following section overviews methods for handling nonstationary arrival processes. Chapter 7 briefly describes a method for introducing correlation into the random generation process.

3.3.4 Nonstationary Processes

If a process does not depend on time, it is said to be *stationary*. When a process depends on time, it is said to be *nonstationary*. The more formal definitions of stationary/nonstationary are avoided in the discussion that follows. There are many ways in which you can model nonstationary (time varying) processes in your simulation models. Many of these ways depend entirely on the system being studied and through that study appropriate modeling methods will become apparent. For example, suppose that workers performing a task learn how to more efficiently perform it every time that they repeat the task. In this situation, the task time depends on the time (number of previously performed tasks). For this situation, you might use some sort of learning curve model as a basis for changing the task times as the number of repetitions of the task increases. As another example, suppose the worker's availability depends on a schedule, such as having a 30-minute break after accumulating so much time. In this situation, the system's resources are dependent on time. Modeling this situation is described in Chapter 7.

Let's suppose that the following situation needs to be modeled. Workers are assigned a shift of 8 hours and a quota to fill. Do you think that it is possible that their service times would, *on average*, be less during the latter part of the shift? Sure. They might work faster in order to meet their quota during the latter part of the shift. If you did not suspect this phenomenon and performed a time study on the workers in the morning and fit a distribution to these data, you would be developing a distribution for the service times *during the morning*. If you applied this distribution to the entire shift, then you might have a problem getting your simulation throughput numbers to match the system's real throughput. As you can see, if you suspect that a nonstationary process is involved, then it is critical that you also record the time that the observation was made. The time series plot that was discussed for input modeling helps in assessing nonstationary behavior; however, it only plots the order in which the data were collected. If you collect 25 observations of the service time every morning, you might not see any problem with nonstationary behavior. It is necessary to get observations across periods of time to see if nonstationary behavior is important. One way to better plan your observations is to randomize when you will take the observations. Work sampling methods often require this. The day is divided into intervals of time and the intervals randomly selected in which to record an activity. By comparing the behavior of the data across the intervals, you can begin to assess whether time matters in the modeling. Even if you do not divide time into logical intervals, it is good practice to record the date/time for each of the observations that you make so that it can later be placed into logical intervals. As you may now be thinking, modeling a nonstationary process will take a lot of care and more observations than a stationary process.

A very common nonstationary process that you have probably experienced is a nonstationary arrival process. In a nonstationary arrival process, the arrival process to the system will vary by time (e.g., by hour of the day). Because the arrival process is a key input to the system, the system will thus become nonstationary. For example, in the drive-through pharmacy example, the arrival of customers occurred according to a Poisson process with mean rate λ per hour. As a reminder, λ represents the mean arrival rate or the expected number of customers per unit time. Suppose that the expected arrival rate is five per hour. This does not mean that the pharmacy will get five customers every hour, but rather that on average five customers will arrive every hour. Some hours will get more than five customers and other hours will get less. The implication for the pharmacy is that they should expect about five per hour (every hour of the day) regardless of the time of day. Is this realistic? It could be, but it is more likely that the *mean number of arriving customers* will vary by time of day. For example, would you expect to get five customers per hour from 4 a.m. to 5 a.m., from 12 noon to 1 p.m., from 4 p.m. to 6 p.m.? Probably not! A system like the pharmacy will have peaks and valleys associated with the arrival of customers. Thus, the mean arrival rate of the customers may vary with the time of day. To more realistically model the arrival process to the pharmacy, you need to model a nonstationary arrival process.

Let's think about how data might be collected in order to model a nonstationary arrival process. First, as in the previous service time example, you need to think about dividing the day into logical periods of time. A pilot study may help in assessing how to divide time, or you might simply pick a convenient time division such as hours. You know that the CREATE module require a distribution that represents the time between arrivals. Ideally, you should collect the time of every arrival, T_i, throughout the day. Then, you can take the difference among consecutive arrivals, $T_i - T_{i-1}$, and fit a distribution to the interarrival times. Looking at the T_i over time may help to assess whether you need different distributions for different time periods during the day. Suppose that you do collect these data and find that three different distributions are reasonable models given the following data:

Exponential, $E[X] = 60$, for midnight to 8 a.m.
Lognormal, $E[X] = 12$, $V[X] = 4$, for 8 a.m. to 4 p.m.
Triangular, min $= 10$, mode $= 16$, max $= 20$, for 4 p.m. to midnight

One simple method for implementing this is to CREATE a logical entity that selects the appropriate distribution for the current time. In other words, schedule a logical entity to arrive at midnight so that the time between arrival distribution can be set to exponential, schedule an entity to arrive at 8 a.m. to switch to log-normal, and schedule an entity to arrive at 4 p.m. to switch to triangular. Clearly, this varies the arrival process in a time-varying manner. There is nothing wrong with this modeling approach if it works well for the situation. In taking such an approach, you need a lot of data. Not only do you need to record the time of every arrival, but you need enough arrivals in order to adequately fit distributions to the time between events for various time periods. This may or may not be feasible in practice.

Often in modeling arrival processes, you cannot readily collect the actual times of the arrivals. If you can, you should try, but sometimes you just cannot. Instead, you are limited to a count of the number of arrivals in intervals of time. For example, a computer system records the number of people arriving every hour but not their individual arrival times. This is not uncommon since it is much easier to store summary counts than it is to store all individual arrival times. Suppose that you had count data as shown in Table 3-13.

Of course, how the data are grouped into intervals will affect how the data can be modeled. Grouping is part of the modeling process. Let's accept these groups as given. Looking at the intervals, you see that more arrivals tend to occur from 12 p.m. to 2 p.m. and from 5 p.m. to

■ **TABLE 3-13**
Example Arrival Counts

Interval	Mon	Tues	Wed	Thurs	Fri	Sat	Sun
12 a.m.–6 a.m.	3	2	1	2	4	2	1
6 a.m.–9 a.m.	6	4	6	7	8	7	4
9 a.m.–12 p.m.	10	6	4	8	10	9	5
12 p.m.–2 p.m.	24	25	22	19	26	27	20
2 p.m.–5 p.m.	10	16	14	16	13	16	15
5 p.m.–8 p.m.	30	36	26	35	33	32	18
8 p.m.–12 p.m.	14	12	8	13	12	15	10

(Total arrivals)

8 p.m. Perhaps this corresponds to when people who visit the pharmacy can get off of work. Clearly, this is a nonstationary situation. What kinds of models can be used for this type of count data?

A simple and often reasonable approach to this situation is to check whether the number of arrivals in each interval occurs according to a Poisson distribution. If so, then you know that the time between arrivals in an interval is exponential. When you can divide time so that the number of events in the intervals is Poisson (with different mean rates), then you can use a nonstationary or nonhomogeneous Poisson process as the arrival process. Consider the first interval to be 12 a.m. to 6 a.m.; to check if the counts are Poisson in this interval, there are seven data points (one for every day of the week for the given period). This is not a lot of data to base a chi-square goodness-of-fit test for the Poisson distribution. So, to do a better job at modeling the situation, many more days of data are needed. The next question is: Are all the days the same? It looks like Sunday may be a little different than the rest of the days. In particular, the counts on Sunday appear to be a little less on average than the other days of the week. In this situation, you should also check whether the day of the week matters to the counts. One method of doing this is to perform a chi-square–based contingency table test, which you can find in any introductory statistics book.

Let's assume for simplicity that there is not enough evidence to suggest that the days are different and that the count data in each interval is well modeled with a Poisson distribution. In this situation, you can proceed with using a nonstationary Poisson process. From the data you can calculate the average arrival rate for each interval and the rate per hour for each interval as shown in Table 3-14. In the table, 2.14 is the average of the counts across the days for the interval 12 a.m. to 6 p.m. In the rate per hour column, the interval rate has been converted to an hourly basis by dividing the rate for the interval by the number of hours in the interval.

Figure 3-64 illustrates the time-varying behavior of the mean arrival rates over the course of a day.

The two major methods by which a nonstationary Poisson process can be generated are thinning and rate inversion. Both methods will be briefly discussed; however, the interested reader is referred to Leemis and Park (2006) or Ross (1997) for more of the theory underlying these methods. Before these methods are discussed, it is important to point out that the naïve method of using different Poisson processes for each interval and subsequently different exponential distributions to represent the interarrival times could be used to model this situation by switching the distributions as previously discussed. However, the switching method *is not technically correct* in that it will not generate a nonstationary Poisson process with the correct probabilistic underpinnings. Thus, if you are going to model a process with a nonstationary Poisson process, you should use either thinning or rate inversion to be technically correct.

■ **TABLE 3-14**
Mean Arrival Rates

Interval	Average number of arrivals for interval	Rate per hour
12 a.m.–6 a.m.	2.14 per 6 hours	0.36
6 a.m.–9 a.m.	6.0 per 3 hours	2.0
9 a.m.–12 p.m.	7.43 per 3 hours	2.48
12 p.m.–2 p.m.	23.29 per 4 hours	5.82
2 p.m.–5 p.m.	14.29 per 3 hours	4.76
5 p.m.–8 p.m.	30.0 per 3 hours	10.0
8 p.m.–12 p.m.	12.0 per 4 hours	3.0

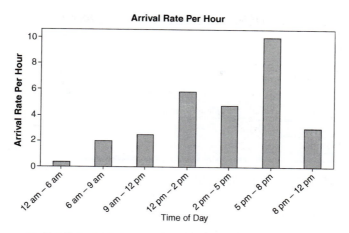

■ FIGURE 3-64 **Arrival rate per hour for intervals**

In thinning, a stationary Poisson process with a constant rate, λ^*, and arrival times, t_i^*, is first generated, and then potential arrivals, t_i^*, are rejected with probability

$$1 - \frac{\lambda(t_i^*)}{\lambda^*}$$

where $\lambda(t)$ is the mean arrival rate as a function of time. Often, λ^* is set as the maximum arrival rate over the time horizon, such as $\lambda^* = \max_{t}\{\lambda(t)\}$. In the example, λ^* would be the mean of the 5 p.m. to 8 p.m. interval, $\lambda^* = 10.0$. The following pseudo-code for implementing the thinning method creates entities at the maximum rate and thins them according to which period is currently active. An array stores each arrival rate for the intervals, and a variable is used to hold the maximum arrival rate over the time horizon. To implement the arrival rate function, the code creates a logical entity that keeps track of the current interval as a variable and that can be used to index the array of arrival rates. If the simulation lasts past the time horizon of the arrival rate function, then the variable that keeps track of the interval would need to be reset.

Pseudo-code for the thinning method follows:

```
CREATE entities with time between arrivals, expo(1/λ*)
DECIDE with chance 1−λ (period)/λ*
      If accepted, send entity to next module
      If not accepted, dispose of entity

1     CREATE time interval logical entity
2     ASSIGN period = period + 1
3     DECIDE if period < maximum number of periods
          DELAY for period length
          GOTO 2
          Else
          ASSIGN period = 0
          GOTO 2
```

One of the exercises in this chapter requires implementing the thinning algorithm in Arena™. The theoretical basis for why thinning works can be found in Ross (1997).

The second method for generating a nonstationary Poisson process is through the rate-inversion algorithm. In this method, a rate-1 Poisson process is generated, and the inverse of the

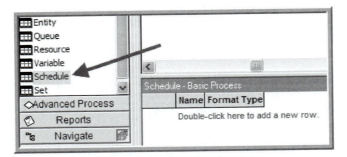

■ **FIGURE 3-65** SCHEDULE data module

mean arrival rate function is used to rescale the times of arrival to the appropriate scale. This section does not discuss the theory behind this algorithm. Instead, see Leemis and Park (2006) for further details of the theory, fitting, and the implementation of this method in simulation. This method is available for piecewise constant-rate functions in Arena™ by using an Arrivals Schedule associated with a CREATE module. Let's see how to implement the example using an Arrivals Schedule in Arena™.

The following is available in the Arena™ file *PharmacyModelNSPP.doe*. The first step in implementing a nonstationary Poisson process in Arena™ is to define the intervals and compute the rates per hour as per Table 3-14. It is extremely important to have the rates on a *per hour* basis. The SCHEDULE module is a data module on the Basic Process panel in the Arena™ Environment. Figure 3-65 illustrates where to find the SCHEDULE in the Basic Process panel. To use the SCHEDULE module, double-click on the data module area to add a new schedule. Schedules can be of type Capacity, Arrival, and Other. Right-clicking on the row and using Edit via Dialog will give you the SCHEDULE dialog window. The textboxes for the SCHEDULE dialog follow:

Name: Name of the module.

Format Type: Schedules can be defined as a collection of value, duration pairs (Duration) or they can be defined using Arena's calendar editor. The calendar editor allows for the input of a complex schedule that can be best described by a Gregorian calendar (i.e., days, months, etc.).

Type: Capacity refers to time varying schedules for resources. Arrival refers to non-stationary Poisson arrivals. Other can be used to represent other types of time delayed schedules.

Time units: This field represents how the time units for the durations will be interpreted. This only refers to the durations, not the arrival rates.

Scale Factor: This field can be used to easily change all the values in the duration value pairs.

Durations: The length of time of the interval for the associated value. Once all the durations have been executed, the schedule will repeat unless the last duration is left infinite.

Value: This field represents either the capacity to be used for the resource or the arrival rate (in entities per hour), depending on which type of schedule was specified.

The SCHEDULE module offers a number of ways in which to enter the information required for the schedule. The most straightforward method is to enter each value and duration pair using the Add button on the dialog box. You will be prompted for each pair. In this case, the arrival rate (per hour) and the duration that the arrival rate will last should be entered. The completed dialog entries are shown in Figure 3-66. Note that the arrival rates are

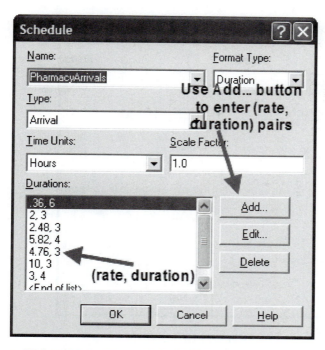

■ **FIGURE 3-66**   **SCHEDULE dialog**

given per hour. Not putting the arrival rates per hour is a common error in using this module. Even if you specify the time units of the dialog as minutes, the arrival rates must be per hour. The time units only refer to the durations. The duration information can also be entered using a spreadsheet like view. In addition, the values can be entered using the graphical schedule editor. With the values already entered, the graphical schedule editor displays the schedule over time as shown in Figure 3-67. If the durations and values have not already been entered,

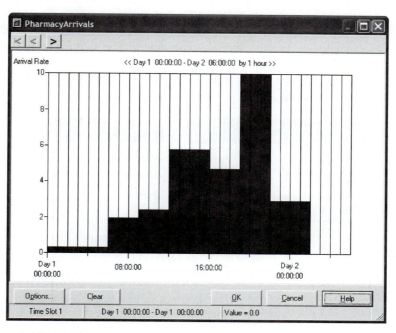

■ **FIGURE 3-67**   **Schedule graphical editor and display**

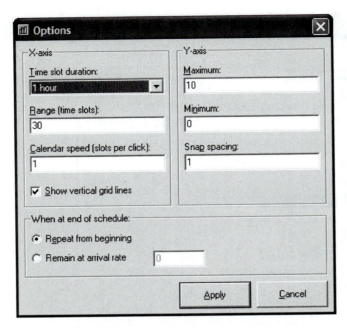

■ **FIGURE 3-68**   **Graphical editor options**

the mouse arrow can be used to "draw" the schedule over time. If you use the graphical editor, then you may need to fiddle with the options in order to facilitate the editing of the schedule (see Figure 3-68).

Once the schedule has been defined, the last step is to indicate in the CREATE module that it should use the SCHEDULE. This is shown in Figure 3-69 where the Time Between Arrivals Type has been specified as Schedule with the appropriate Schedule Name.

The schedule as in Figure 3-67 shows that the durations add up to 24 hours so that a full day's worth of arrival rates are specified. If you simulate the system for more than one day, what will happen on the second day? Because the schedule has a duration, value pair for each interval, the schedule will repeat after the last duration occurs. If the schedule is specified with an infinite last duration, the value specified will be used for the rest of the simulation and the schedule will not repeat.

As indicated in this limited example, it is relatively easy to model a nonstationary Arrival process in Arena™.

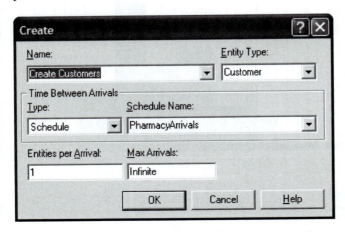

■ **FIGURE 3-69**   **CREATE module with schedule type**

3.4   Summary

This chapter covered the following important concepts used in simulation:

- Probability distributions
- Summary statistics and confidence intervals
- Hypothesis tests and their interpretation
- Fitting input models
- Generating random numbers
- Generating random variates and processes

These topics provide a solid foundation for modeling random components in your simulation models. With this chapter and the material in Chapter 2 you are now ready to model a wide variety of realistic systems using Arena^{TM}. The appendix to this chapter builds on some of these topics by further exploring other useful tools (e.g., MINITAB, BestFit, and Excel) that are helpful in simulation.

A primary use of simulation is to be able to examine and compare alternatives in order to recommend a design. Because simulations include random components, the performance measures obtained from the models should also be treated as random. Thus, the observations made on a simulation will require an appropriate statistical analysis in order make confident decisions. Chapter 4 will enable you to properly analyze the output from your simulations and help you to better plan and execute your simulation experiments.

Exercises

3.1 Provide a distribution that might best model each of the following situations:

(a) The processing time on a computer chip insertion machine

(b) The time that it takes a person to order from a fast food restaurant

(c) The time that it takes for a light-bulb to burn out

(d) The number of defects on a computer chip wafer

(e) The number of phone calls until a successful sale

3.2 The "lack of memory" property of the exponential distribution states that for all $s, t \geq 0, P\{X > s + t | X > t\} = P\{X > s\}$. For example, consider X as the time to failure for a component. Then, the lack of memory property implies that the probability that the component lasts for at least $s + t$ time units given that it has survived t time units is the same as the initial probability that it lasts for a least s time units. Prove that the exponential distribution has the lack of memory property.

3.3 Describe a simulation experiment that would allow you to test the lack of memory property empirically. Implement your simulation experiment in Arena^{TM} and test the lack of memory property empirically. Explain in your own words what "lack of memory" means.

3.4 Parts arrive to a machine center with three drill presses according to a Poisson distribution with mean λ. The arriving customers are assigned to one of the three drill presses randomly according to the respective probabilities p_1, p_2, and p_3, where $p_1 + p_2 + p_3 = 1$ and $p_i > 0$ for $i = 1, 2$, and 3. What is the distribution of the interarrival times to each drill press? Specify the parameters of the distribution.

3.5 For Problem 4, suppose that p_1, p_2, and p_3 equal to $0.25, 0.45$, and 0.3, respectively, and that λ is equal to 12 per hour and that the service time is exponentially distributed with a mean of 6 minutes. Run the simulation for 50000 hours.

(a) Using only 1 CREATE module, simulate the system using Arena^TM and estimate the average waiting time in the queue.

(b) Using 3 CREATE modules, simulate the system using Arena^TM and estimate the average waiting time in the queue.

Are the answers to (a) and (b) the same? Why? Is this generally true for other distributions?

3.6 Suppose that you wanted to develop a distribution for the processing time at two different machining centers. Suppose that you collect a sample from each of the two processes and suppose that you perform a hypothesis test to see if the mean and the variance of the two populations are the same. If the hypothesis test shows no significant difference between the mean and the variance of the two populations, can you conclude that the two distributions are the same? Explain.

3.7 Consider the following samples that represent the lifetime of two energized bunnies.

Bunny 1:	33.5	25.2	31.9	23.2	27.2
	19.3	33.5	24.7	24.0	36.1
Bunny 2:	21.5	24.6	23.2	20.9	17.5
	22.7	20.6	16.9	23.5	24.5

Suppose that you were trying to fit a distribution to the battery life of the bunnies. Can you reasonably combine these samples and use them together to develop the lifetime distribution? Base your answer on a statistical analysis.

3.8 The following data represent the interarrival times (in minutes) at a service facility.

0.96	0.65	1.09	1.40	0.24	0.74	0.51	2.25
1.09	2.31	0.99	0.18	0.73	0.07	0.59	0.19
0.28	0.46	0.32	2.20	4.81	0.50	0.09	1.20
3.13	0.07	0.17	2.70	1.71	0.35	0.52	2.74
0.64	0.41	1.75	0.28	0.25	1.54	0.95	0.14
0.05	1.81	0.56	1.48	0.10	0.99	0.64	0.52
0.22	0.13	0.02	1.68	0.71	0.81	1.47	0.37
0.80	1.16	2.04	0.06	0.27	3.85	2.04	0.06
0.01	1.80	1.30	0.62	0.01	0.36	1.25	0.55
0.25	0.62	2.54	3.62	0.77	0.63	1.87	0.07
0.53	1.39	0.41	1.50	0.15	0.93	0.33	1.18
1.22	1.75	1.96	1.87	0.27	0.35	0.86	0.72

(a) Develop a suitable histogram for the interarrival time.

(b) Estimate the mean and variance.

(c) Hypothesize a theoretical distribution for the data and give a plausible reason for the selection you make.

(d) Apply the chi-square goodness-of-fit test at a 95% confidence level.

3.9 Two different distributions may have similarly shaped functions. Thus, hypothesizing a theoretical distribution based on a histogram's shape may not clearly indicate a specific distribution. In this situation, using the characteristics of the situation being studied may be useful in hypothesizing a distribution. Identify some of the characteristics associated with the use of each of the following distributions. In other words, identify some situations that may be appropriately modeled with the provided distributions.

- Poisson
- Exponential
- Normal
- Uniform
- Gamma
- Weibull
- Beta

3.10 Consider the following sample that represents the time (in seconds) that it takes a hematology cell counter to complete a test on a blood sample.

23.79	75.51	29.89	2.47	32.37
29.72	84.69	45.66	61.46	67.23
94.96	22.68	86.99	90.84	56.49
30.45	69.64	17.09	33.87	98.04
12.46	8.42	65.57	96.72	33.56
35.25	80.75	94.62	95.83	38.07
14.89	54.80	95.37	93.76	83.64
50.95	40.47	90.58	37.95	62.42
51.95	65.45	11.17	32.58	85.89
65.36	34.27	66.53	78.64	58.24

Test the hypothesis that these data are drawn from a uniform distribution at a 95% confidence level given the following information:

(a) The interval of the distribution is between 0 and 100.

(b) The interval of the distribution is between a and b, where a and b are unknown parameters.

3.11 Consider the sample in Problem 8. Test the hypothesis that the data are drawn from the following theoretical distributions (use a 95% confidence level):

- Exponential
- Normal
- Lognormal

3.12 Consider the following frequency data on the number of orders received per day by a warehouse.

Cell number	Number of orders	Observed frequency O_i	Theoretical frequency n_i	$\dfrac{(O_i - n_i)^2}{n_i}$
1	0	10		
2	1	42		
3	2	27		
4	3	12		
5	4	6		
6	5	3		
	Totals	100		

(a) Compute the sample mean for these data.

(b) Perform a χ^2-test to test the hypothesis (use a 95% confidence level) that the data are Poisson distributed. Complete the provided table to show your calculations.

3.13 Consider the following discrete distribution of the random variable, x, whose PMF is $p(x)$.

x	0	1	2	3	4
$p(x)$	0.3	0.2	0.2	0.1	0.2
$F(x)$					

(a) Determine the CDF $F(x)$ for the random variable, x.

(b) Create a graphical summary (see Figure 3-42) of the CDF $F(x)$ with x as the x-axis and U as the y-axis.

(c) Create a look-up table (see Table 3-12) that can be used to determine a sample from the discrete distribution, $p(x)$.

3.14 Consider the following sequence of (0, 1) random numbers:

0.943	0.398	0.372	0.943	0.204	0.794
0.498	0.528	0.272	0.899	0.294	0.156
0.102	0.057	0.409	0.398	0.400	0.997

   Test to determine whether the sequence is distributed U(0, 1) using both a K-S test and a chi-square test.

3.15 Suppose that customers arrive at an ATM via a Poisson process with mean of 7 per hour. Determine the arrival time of the first 6 customers using the data given in Problem 14. Use the inverse transformation method.

3.16 The demand for parts at a repair bench per day can be described by the following discrete PMF:

Demand	0	1	2
Probability	0.3	0.2	0.5

   Generate the demand for the first 4 days using the sequence of (0, 1) random numbers in Problem 14. Use the numbers starting on row 1 reading from left to right.

3.17 Customers arrive at a service location according to a Poisson distribution with mean 10 per hour. The installation has two servers. Experience shows that 60% of the arriving customers prefer the first server. By using the (0,1) random sequence given in Problem 14, determine the arrival times and preferred server for the first nine customers. Use the numbers starting on row 1 reading from left to right.

3.18 Test the following fact by generating instances of Y: If $X_i \sim \exp(\lambda)$, then $Y = \sum_{i=1}^{r} X_i \sim erlang(r; \lambda)$. Use $r = 5$ and $\lambda = 2$ where $E[X_i] = \frac{1}{\lambda}$. Be careful to specify the correct parameters for the exponential distribution function in Arena^{TM}, such as the exponential distribution takes in the mean of the distribution as its parameter. Generate 10, 100, and 1000 instances of Y.

(a) Perform hypothesis tests to check $H_0 : \mu = \mu_0$ versus $H_1 : \mu \neq \mu_0$, where μ_0 is the true mean of Y for each of the sample sizes generated.

(b) Use the same samples to check the fit of the distribution using the Input Analyzer. Properly interpret the statistical results supplied by the Input Analyzer.

3.19 Consider the multiplicative congruential generator with (a = 13, m = 64, and seeds $X_0 = 1, 2, 3, 4$)

(a) Does this generator achieve its maximum period for these parameters? Use the theorem presented in the text to prove your answer.

(b) Generate a period's worth of uniform random variables from each of the supplied seeds.

3.20 Consider the multiplicative congruential generator with (a = 11, m = 64, and seeds $X_0 = 1, 2, 3, 4$)

(a) Does this generator achieve its maximum period for these parameters? Use the theorem presented in the text to prove your answer.

(b) Generate a period's worth of uniform random variables from each of the supplied seeds.

3.21 Analyze the following LCG: $X_i = (11X_{i-1} + 5)(\mod(16), \quad X_0 = 1$

(a) What is the maximum possible period length for this generator? Does this generator achieve the maximum possible period length? Justify your answer using the theorem presented in the text.

(b) Generate two pseudo-random uniform numbers for this generator.

3.22 Generate 100 uniform (0, 1) numbers using Arena™.

- Use a CREATE module to create 100 entities.
- Use an ASSIGN module and the expression builder to assign the randomly generated number to a variable.
- Use a WRITE module to write the value of the variable to a text file.
- Use a DISPOSE module to dispose of the entities.

(a) Use the K-S test with alpha = 0.05 to test whether the hypothesis that the numbers are uniformly distributed on the [0, 1] interval can be rejected. Implement the test using Excel.

(b) Use a chi-square goodness-of-fit test with 10 intervals and alpha = 0.05 to test whether the hypothesis that the numbers are uniformly distributed on the [0,1] interval can be rejected. Implement this test using Excel.

(c) If you have access to a suitable statistical package, perform a runs-above-the-mean test to check the randomness of the numbers. Use the nonparametric runs test functionality of your statistical software to perform this test.

(d) What is your conclusion concerning the suitability of the Arena's random number generator?

3.23 Repeat the previous problem using Excel's RAND() function. Search on the Internet to find documentation on Excel's random number generator. Briefly discuss what you learn about its implementation.

3.24 Consider the following triangular distribution:

$$F(x) = \begin{cases} 0 & x < a \\ \dfrac{(x-a)^2}{(b-a)(c-a)} & a \le x \le c \\ 1 - \dfrac{(b-x)^2}{(b-a)(b-c)} & c < x \le b \\ 1 & b < x \end{cases}$$

Implement a VBA function with the following signature to generate random variates from the triangular distribution. Use the Rnd() function of VBA as your source of uniform random variates.

```
Public Function Triangular(min As Double, mode As Double, max As Double)
As Double
End Function
```

Use Arena's Input Analyzer to check your work. For a test case, let a = 2, b = 6, c = 3, generate 1000 samples.

3.25 The times to failure for an automated production process have been found to be randomly distributed according to a Rayleigh distribution:

$$f(x) = \begin{cases} 2\beta^{-2}x\exp(-(x/\beta)^2) & x > 0 \\ 0 & \text{otherwise} \end{cases}$$

(a) Derive an algorithm for generating random variables from this distribution.

(b) Use your algorithm to generate five values using the first five uniform numbers given in Problem 14 with $\beta = 2$.

3.26 Consider the LOTR Makers, Inc. example and suppose that the number of calls to be made for a given day is not known a priori. Suppose that the sales force will always make 100 calls, but occasionally they have the opportunity to make a few extra calls. Let N be the number of extra calls made each day.

(a) Suppose N ∼ geometric(mean = 5). Modify the LOTR Makers, Inc. simulation to accommodate this change. What is the effect on the probability of overtime?

(b) Suppose that N ∼ Poisson(mean = 5). Modify the LOTR Makers, Inc. simulation to accommodate this change. What is the effect on the probability of overtime?

3.27 Consider the LOTR Makers, Inc. example, and suppose that the calling time is no longer negligible. Suppose that there are 10 salespeople and that the 100 calls for the day go to a queue to be pulled from by whomever of the 10 sales people is available to make the call. The call takes approximately 10 minutes on average according to an exponential distribution. If the call's order is confirmed, the order to produce the rings is *immediately* released to the production line. Implement this new release logic by modifying the LOTR Maker, Inc.'s model. Estimate the effect on the probability of overtime? Discuss.

3.28 Lead-time demand may occur in an inventory system when the lead time is other than instantaneous. The lead time is the time from the placement of an order until the order is received. The lead time is a random variable. During the lead time, demand also occurs at random. Lead-time demand is thus a random variable defined as the sum of the demands during the lead time, or $LTD = \sum_{i=1}^{T}D_i$, where i is the time period of the lead time and T is the lead time. The distribution of lead-time demand is determined by simulating many cycles of lead time and the demands that occur during the lead time to get many realizations of the random variable LTD. Note that LTD is the *convolution* of a random number of random demands. Suppose that the daily demand for an item is given by the following PMF.

Daily demand (items)	4	5	6	7	8
Probability	0.10	0.30	0.35	0.10	0.15

(a) Assume that the lead time is distributed according to a geometric distribution with a mean of 5 days. Use Arena™ to simulate 1000 instances of LTD, and save the samples to a file.

(b) Report the statistics for the 1000 observations using Arena™.

(c) Estimate the chance that LTD is greater than or equal to 10. Report a 95% confidence interval on your estimate.

(d) Use your favorite statistical program to develop a frequency diagram for LTD.

3.29 Using the random numbers from Problem 14, generate 1 random number from the negative binomial distribution with parameters $(r = 4, p = 0.4)$ using a) the convolution method, and b) Bernoulli trials method.

3.30 If $Z \sim N(0, 1)$, then $Y = \sum_{i=1}^{k} Z_i^2 \sim \chi_k^2$, where χ_k^2 is a chi-squared random variable with k degrees of freedom. Use Arena™ to generate 1000 χ_5^2 random variables using the convolution method. Test the goodness of your implementation with the Input Analyzer. Hint: The chi-squared random variable is a special case of the gamma distribution.

3.31 Prove that the acceptance-rejection method for continuous random variables is valid by showing that for any x, $P\{X \leq x\} = \int_{-\infty}^{x} f(y)dy$. Hint: Let E be the event that the acceptance occurs and use conditional probability.

3.32 This problem is based on Cheng (1977); see also Ahrens and Dieter (1972). Consider the gamma distribution

$$f(x) = \beta^{-\alpha} x^{\alpha-1} \frac{e^{-x/\beta}}{\Gamma(\alpha)}$$

where $x > 0$, and $\alpha > 0$ is the shape parameter and $\beta > 0$ is the scale parameter. In the case where α is a positive integer, the distribution reduces to the Erlang distribution and $\alpha = 1$ produces the negative exponential distribution. Acceptance-rejection techniques can be applied to the cases of $0 < \alpha < 1$ and $\alpha > 1$. For the case of $0 < \alpha < 1$, see Ahrens and Dieter (1972). For the case of $\alpha > 1$, Cheng (1977) proposed the following majorizing function:

$$g(x) = \left[\frac{4\alpha^\alpha e^{-\alpha}}{a\Gamma(\alpha)} \right] h(x)$$

where $a = \sqrt{(2a - 1)}, b = \alpha^a$, and $h(x)$ is the resulting probability distribution function when converting $g(x)$ to density function

$$h(x) = ab \frac{x^{a-1}}{(b + x^a)^2} \qquad for\ x > 0$$

(a) Develop an inverse transform algorithm for generating from $h(x)$.

(b) Using the data given in Problem 14, determine the arrival time of the first customer, given that the interarrival time is gamma distributed with parameters $\alpha = 2$ and $\beta = 10$, via the acceptance/rejection method.

3.33 This procedure derives from Box and Muller (1958). Let U_1 and U_2 be two independent uniform (0,1) random variables and defined as

$$X_1 = \cos(2\pi U_2)\sqrt{-2\ln(U_1)}$$

$$X_2 = \sin(2\pi U_2)\sqrt{-2\ln(U_1)}$$

It can then be shown that X_1 and X_2 will be independent, standard normal random variables, that is, N(0, 1). Use Arena™ to implement the Box and Muller algorithm for generating normal random variables. Generate 1000 $N(\mu = 2, \sigma = 0.75)$ random variables via this method, and use the Input Analyzer to test your work.

3.34 Consider the nonstationary arrival process example of Section 3.3.4 involving the pharmacy model. Run the model under the same conditions as those given in Chapter 1 and compare the results to those from Chapter 1. Chapter 1 presents steady-state results for the pharmacy model. Do these results apply to this situation? Why or why not?

APPENDIX 1

Using MINITAB and BestFit during Input Modeling

This appendix illustrates the use of MINITAB and BestFit during the input modeling process. MINITAB is a full featured statistical analysis program, see www.minitab.com. With MINITAB you can easily perform the following statistical procedures for analyzing the input data: summary statistics, histogram, time series plot, autocorrelation plot, scatterplot, P-P plots, and empirical CDF versus theoretical CDF. These can be obtained via the following menu sequences:

> Summary Statistics: Stat > Basic Statistics > Graphical Summary
>
> Histogram: Graph > Histogram
>
> Time Series plot: Graph > Time Series Plot
>
> Autocorrelation plot: Stat > Time Series > Autocorrelation
>
> Scatterplot: Stat > Time Series > Lag (to create another column) then use Graph > Scatterplot to make the plot
>
> P-P plots: Graph > Probability Plot > Single (then pick your distribution)
>
> Empirical CDF versus theoretical CDF: Graph > Empirical CDF > Single (then pick your distribution)

Figures A1-1, A1-2, and A1-3 illustrate some of the output from using these MINITAB commands. The P-P plot and Empirical CDF functions will fit the parameters of various distributions.

Summary for Service Times

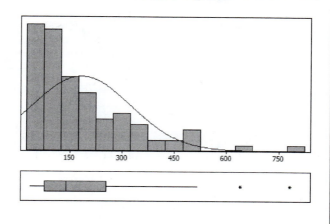

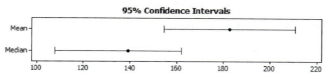

Anderson-Darling Normality Test	
A-Squared	4.86
P-Value <	0.005
Mean	182.78
StDev	141.61
Variance	20053.73
Skewness	1.66396
Kurtosis	3.19897
N	100
Minimum	36.84
1st Quartile	77.39
Median	139.11
3rd Quartile	254.15
Maximum	782.22
95% Confidence Interval for Mean	
154.68	210.88
95% Confidence Interval for Median	
107.77	161.99
95% Confidence Interval for StDev	
124.34	164.51

■ **FIGURE A1-1**   **MINITAB summary statistics**

Note that in Figure A1-2, MINITAB has placed a confidence interval band about the autocorrelation estimates. The plot with the band suggests that there is no significant correlation present in these data. Figure A1-3 shows the P-P plot for the shifted exponential distribution. The plot is very linear, which indicates that the shifted exponential is a reasonably good model for these data. Based on the plot, there appears to be one outlier in the tail of the data.

The BestFit program from Palisades is designed to facilitate all aspects of the distribution fitting process: (1) defining the data, (2) specifying the distribution, (3) fitting the parameters, and (4) interpreting the results. BestFit's extensive help system provides a

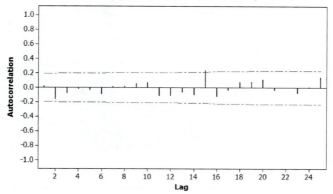

■ **FIGURE A1-2**   **MINITAB autocorrelation plot**

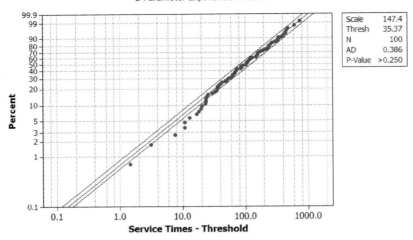

■ **F I G U R E   A 1 - 3**   **MINTAB P-P plot**

detailed tutorial on using BestFit for the distribution fitting process. Figures A1-4 and A1-5 illustrate the BestFit program Fitting menu options and the specify distributions dialog. Once the data distributions have been specified, the user simply selects Run Fit from the Fitting menu. BestFit will fit each of the distributions specified, compute P-P plots and Q-Q plots, and rank the distributions. BestFit allows the user to select the ranking method. For continuous data, you can choose to rank the distributions via the chi-square statistic, the Anderson-Darling statistic, or the Kolmogorov-Smirnov statistic. For discrete data, only the chi-square test statistic is used for determining the rank.

After the fitting process has completed, BestFit displays the results as shown in Figure A1-6. For each of the distributions, a histogram, difference, P-P plot, and Q-Q plot is computed. In addition, the summary statistics for the fit are displayed.

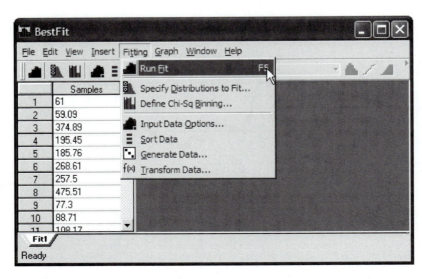

■ **F I G U R E   A 1 - 4**   **BestFit by palisades**

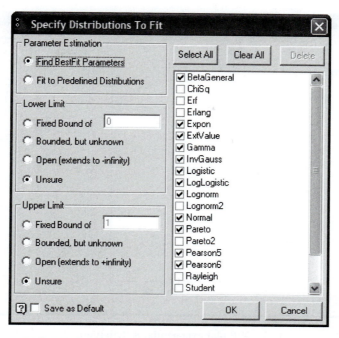

■ **FIGURE A1-5**   **Specifying the distribution**

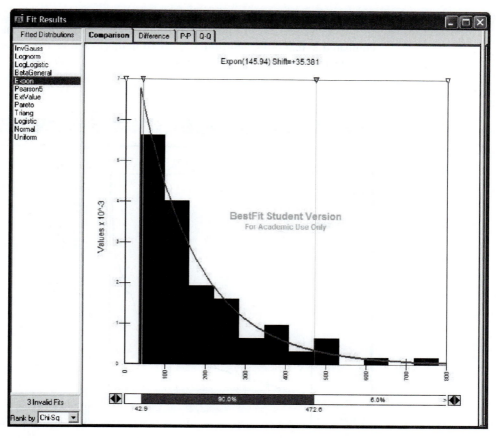

■ **FIGURE A1-6**   **BestFit results for exponential fit to service times**

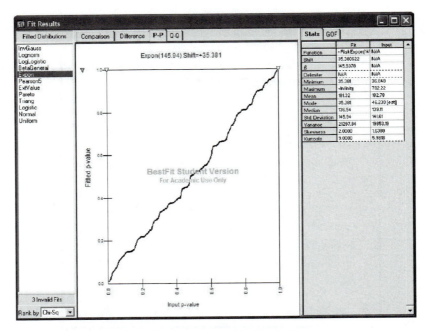

■ **FIGURE A1-7** P-P plot exponential distribution

As can be seen in Figures A1-7 and A1-8, the P-P plot of the exponential has been compared to that of the uniform. Clearly from this example, the uniform distribution is not a good fit for these data. BestFit makes this process extremely easy.

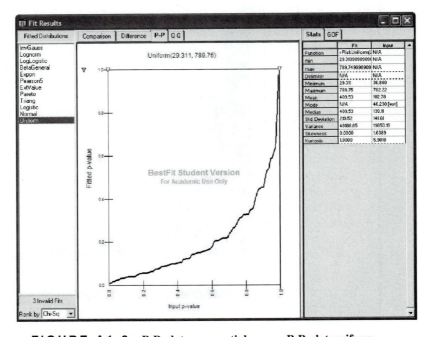

■ **FIGURE A1-8** P-P plot exponential versus P-P plot uniform

APPENDIX 2

Basic Spreadsheet Simulation Concepts

This appendix discusses how to implement a few of the random number generation techniques mentioned in this chapter in Excel. The purpose is to give you another look at how these concepts can be used. The immediacy of Excel is useful in better understanding some of these concepts. In addition, Excel can be an extremely useful tool for performing non-discrete event simulation. This introduction should give you enough information to begin to appreciate the more advanced tools like @Risk for performing simulations in a spreadsheet environment.

The Excel workbook, *SimulatingRandomnessInExcel.xls*, contains the examples that are discussed in this appendix. The RAND() function in Excel will generate a random number uniformly distributed on the interval from (0,1). From this random number, many other random numbers can be generated. The RAND() function is an "active" function. That is, each time the worksheet is calculated a new value from RAND() will be returned. For example, a worksheet will be recalculated if a cell is evaluated. As you will see, this can be inconvenient when trying to work with large spreadsheets that simulate many random variables. Excel also has the worksheet function RANDBETWEEN(a,b), which will generate a random integer between [a,b]. This function is also an "active" function. To turn off Excel's automatic calculation, you can use the Tools > Options > Calculation as shown in A2-1. When you turn off automatic calculation, you can use F9 to cause the spreadsheet to recalculate.

If you are following along to reproduce the spreadsheets shown in this appendix, keep in mind that your random numbers may be different than that shown in the screenshots. This is because there is no way to "set the seed" for the RAND() and the RANDBETWEEN() functions. Using these two functions, a great deal of useful work can be performed (Figures A2-2 and A2-3).

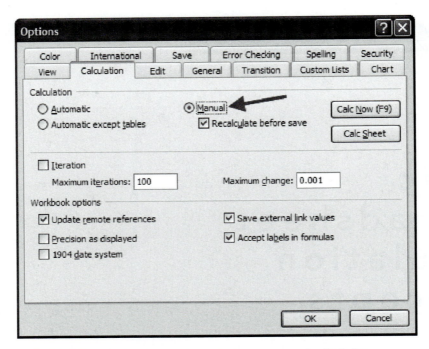

■ **FIGURE A2-1**   Setting the worksheet to manual calculation

To get a uniform (a, b) number, use the cell formula $= a + (b - a)*RAND()$; $(b - a)*RAND()$ gets a number between 0 and $(b - a)$. Then by adding "a," the number is translated to the appropriate spot on the x-axis. If the RANDBETWEEN() function was not available, you could use

$$= a + INT((b - a + 1) * RAND())$$

	B5	▼	f_x =RAND()	
	A	B	C	D
1		**min=**	2	
2		**max=**	10	
3				
4		**U(0,1)**	**DU(min,max)**	**U(min,max)**
5	1	0.478231133	5	9.229413517
6	2	0.786127707	3	9.153368806
7	3	0.158092971	2	8.131687769
8	4	0.605374938	8	4.690445215
9	5	0.138187052	10	8.877661769
10	6	0.909570015	5	3.748755936
11	7	0.700547878	4	7.361288708
12	8	0.539418158	2	2.451534615
13	9	0.543234406	4	5.705660412
14	10	0.508808182	5	5.739570912

■ **FIGURE A2-2**   Use of RAND() function

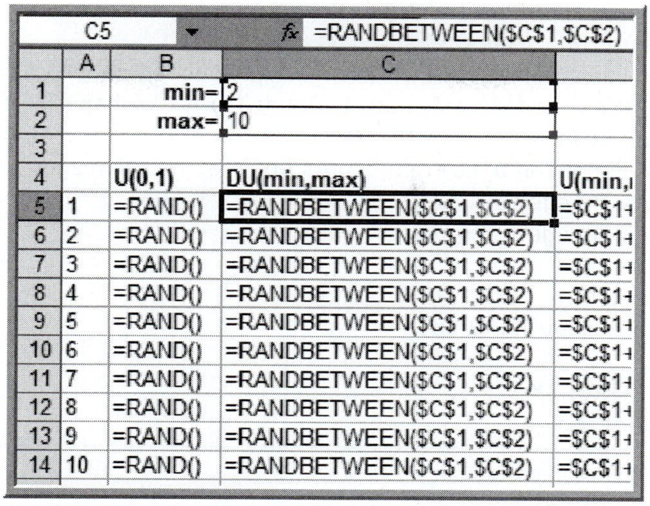

	C5	▼	*fx* =RANDBETWEEN($C$1,$C$2)

	A	B	C	
1		min=	2	
2		max=	10	
3				
4		U(0,1)	DU(min,max)	U(min,!
5	1	=RAND()	=RANDBETWEEN($C$1,$C$2)	=$C$1+
6	2	=RAND()	=RANDBETWEEN($C$1,$C$2)	=$C$1+
7	3	=RAND()	=RANDBETWEEN($C$1,$C$2)	=$C$1+
8	4	=RAND()	=RANDBETWEEN($C$1,$C$2)	=$C$1+
9	5	=RAND()	=RANDBETWEEN($C$1,$C$2)	=$C$1+
10	6	=RAND()	=RANDBETWEEN($C$1,$C$2)	=$C$1+
11	7	=RAND()	=RANDBETWEEN($C$1,$C$2)	=$C$1+
12	8	=RAND()	=RANDBETWEEN($C$1,$C$2)	=$C$1+
13	9	=RAND()	=RANDBETWEEN($C$1,$C$2)	=$C$1+
14	10	=RAND()	=RANDBETWEEN($C$1,$C$2)	=$C$1+

■ **FIGURE A2-3** Use of RAND() and RANDBETWEEN()

to generate discrete uniform random variables. A Bernoulli (p) random variable can be implemented by using Excel's *IF()* function in combination with the RAND() function.

$$= IF(RAND() <= p, 1, 0)$$

To implement binomial random variables, you can easily sum the cells containing the Bernoulli trials. The geometric distribution can be implemented with the cell formula:

$$= 1 + INT(LN(1 - RAND())/LN(1 - p))$$

where *P* is the probability of success. A number of continuous distributions are easy to implement in Excel due to the fact that Excel provides the inverse CDF function for the distribution. In these formulas, you simply replace the argument requesting a probability with the RAND() function. For example, to simulate a Normal random variable, you use the cell formula:

$$= NORMINV(RAND(), mean, std\ dev)$$

where mean and std dev are the mean and the standard deviation for the desired normal variates. Figure A2-4 illustrates the various random variables that are easy to implement in Excel. This comes from the sheet labeled *Distributions* in the example spreadsheet.

If spreadsheet cell formulas are inadequate for your work, you can always implement random variate generation algorithms in an Excel VBA function. The following functions use the VBA function Rnd() to generate a U(0,1) random number and then implement the algorithms discussed in the text. These functions can be found in the VBA module named Random in the example spreadsheet via Tools > Macro > Visual Basic Editor.

```
Public Function bernoulliRV(p As Double) As Integer
If (Rnd() < (1 − p)) Then
bernoulliRV = 0
Else
bernoulliRV = 1
End If
End Function
Public Function expo(mean As Double) As Double
```

```
expo = −mean * Log(1 − Rnd())
End Function
Public Function weibullRV(scaleB As Double, shapeA As Double) As Double
weibullRV = scaleB * (−Log(1 − Rnd())) ^ (1/shapeA)
End Function
Public Function binomialRV(p As Double, n As Integer) As Integer
Dim X As Integer
Dim i As Integer
X = 0
For i = 1 To n
X = X + bernoulliRV (p)
Next i
binomialRV = X
End Function
```

	A	B	C	D	E
1					Formula
2	Uniform(0,1)			0.845009617	=RAND()
3		min	max		
4	Discrete Uniform(min,max)	3	10	8	=RANDBETWEEN(B5,C5)
5		min	max		
6	Uniform(min,max)	4	20	18.66724518	=B7+(C7-B7)*RAND()
7			p		
8	Bernoulli(p)		0.7	1	=IF(RAND()<=C9,1,0)
9			p		
10	Geometric(p)		0.3	2	=1+INT(LN(1-RAND())/LN(1-C11))
11			mean		
12	Exponential(mean)		5	4.359013352	=-C13*LN(1-RAND())
13		scale	shape		
14	Weibull(scale, shape)	5	3	3.975632185	=B15*POWER(-LN(1-RAND()),1/C15)
15		alpha1	alpha2		
16	Beta(alpha1, alpha2)	5	1.5	0.807041407	=BETAINV(RAND(),B17,C17)
17			DF		
18	ChiSquare		5	14.12091438	=CHIINV(RAND(),C19)
19		scale	shape		
20	Gamma(scale,shape)	2	10	19.06188666	=GAMMAINV(RAND(),C21,B21)
21		mean	std dev		
22		5	3		
23	Lognormal(mean, std dev)	1.455695563	0.554513	5.012218975	=LOGINV(RAND(),B24,C24)
24		mean	std dev		
25	Normal(mean, std dev)	10	2	10.50135199	=NORMINV(RAND(),B25,C25)

■ **FIGURE A2-4**   Generating from distributions in Excel

A general discrete distribution can also be implemented as a VBA function, but fortunately the VLOOKUP cell function in Excel provides an easy way to implement the inverse transform method. Let's take a look at an example and implement it with the VLOOKUP function. Let X be a discrete random variable, with $X \in \{1, 2, 3, 4\}$ and the PMF and cumulative distribution function as given in Table A2-1. This is the same

■ **TABLE A2-1**

Example PMF for Excel

X	PMF	CDF
1	0.4	0.4
2	0.3	0.7
3	0.2	0.9
4	0.1	1.0

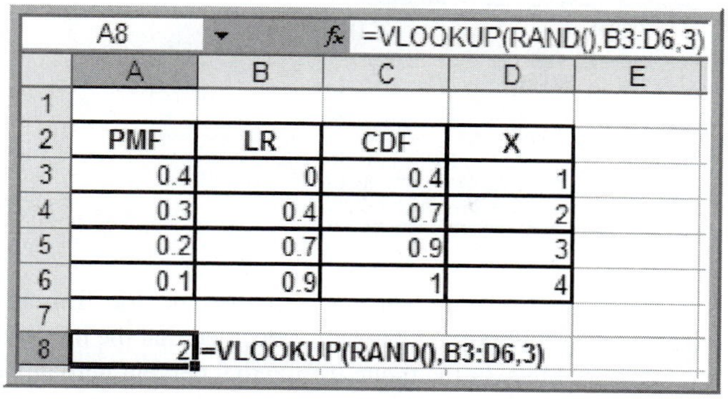

A8 =VLOOKUP(RAND(),B3:D6,3)

	A	B	C	D	E
1					
2	PMF	LR	CDF	X	
3	0.4	0	0.4	1	
4	0.3	0.4	0.7	2	
5	0.2	0.7	0.9	3	
6	0.1	0.9	1	4	
7					
8	2	=VLOOKUP(RAND(),B3:D6,3)			

■ **FIGURE A2-5** **Setting up the look-up table**

distribution as that used in 3-5. The key to using the VLOOKUP function in Excel is to make a look-up table similar to Table 3-12 in Example 3-5.

To create the table for VLOOKUP, you should set up your spreadsheet as indicated in A2-5.

The PMF column contains the probability mass function values

The LR column contains the lower limit on the range for the value. For example, if the random number U falls between (0, 0.4) then the X is set to 1 where the CDF column specifies the upper range.

The X column should contain the possible values for the random variable.

Now you can use the following:

$$= VLOOKUP(RAND()),\ Cell\,Range,\ 3)$$

where Cell Range is the range for the LR-X columns. In the example, Cell Range is B3:D6. RAND() generates a number between 0 and 1. This number is "looked up" in the table's first column and the corresponding value from the third column is returned. This works because of the way in which VLOOKUP does its search. This general scheme will work for any discrete random variable. In fact, the X column does not require numbers. It could contain text and the VLOOKUP function will randomly select the value with the appropriate probabilities. The following example illustrates how to perform a simple simulation in Excel using the concepts discussed in this appendix.

EXAMPLE A2-1 SIMULATION IN EXCEL

A firm is trying to decide whether it should purchase a new scale to check a package filling line in the plant. The scale would allow for better control over the filling operation and result in less overfilling. It is known for certain that the scale costs $800 initially. The annual cost has been estimated to be normally distributed with a mean of $150 and a standard deviation of $10. The extra savings associated with better control of the filling process has been estimated to be normally distributed with a mean of $300 and a standard deviation of $50. The salvage value has been estimated to be uniformly distributed between $90 and $100. The useful life of the scale varies according to the amount of usage of the scale. The manufacturing has estimated that the useful can vary between 4 to 7 years with the chances given in the following table.

Life	PMF	CDF
4	0.3	0.3
5	0.4	0.7
6	0.1	0.8
7	0.2	1.0

The interest rate has been varying recently and the firm is unsure of the rate for performing the analysis. To be safe, they have decided that the interest rate should be modeled as a beta random variable over the range from 6 to 9 percent with alpha = 5.0 and beta = 1.5. Given all the uncertain elements in the situation, management has decided to perform a simulation analysis in order to assess the expected present value of the decision and the chance that the decision has a negative return.

The *SimulatingRandomnessInExcel.xls* spreadsheet with sheet labeled *PV Example* contains the solution to this problem in Excel. This problem is an economic analysis problem and can be solved using present value calculations. The Excel function *PV* will compute the present value of a future value that occurs at the end of a period as follows:

$$P = F\left[\frac{1}{(1+i)^n}\right] = F\,(P/F, i, n) = PV(i, n, , -F)$$

The Excel function PV will compute the present value of a series of future values that are assumed to occur at the end of the periods as follows:

$$P = A\left[\frac{(1+i)^n - 1}{i(1+i)^n}\right] = A(P/A, i, n) = PV(i, n, -A)$$

However, in computing the present value of the salvage value, annual cost, and annual savings you should see that the inputs to the above functions will be random variables in this problem. That is, i, n, F, A must be determined from the given probability distributions. The basic formulation is as follows:

Let S be a random variable that represents the salvage value.
Let C be a random variable that represents the annual cost.
Let Y be a random variable that represents the annual savings.
Let N be a random variable that represents the useful life.
Let I be a random variable that represents the interest rate.
Present value of cost: $P_1 = C(P/A, I, N)$.
Present value of savings: $P_2 = Y(P/A, I, N)$.
Present value of salvage value: $P_3 = S(P/F, I, N)$.
Net present value: $P = -1000 - P_1 + P_2 + P_3$.

Using the PV() function in Excel and randomly generated values for the salvage value, annual cost, annual savings, useful life, and the interest rate, you can simulate values of the net present value. The following discusses how to set up the spreadsheet to accomplish the simulation.

A good spreadsheet simulation should be separated into three parts: the inputs, the simulation, and the outputs. Each of these in turn will be discussed. A2-6 shows the input area for the simulation. The distribution's parameters have been organized into a common area. This allows for easily changing the parameters to rerun different simulations.

	A	B	C	D	E
1	Inputs				
2	Initial Cost	-$800.00			
3	Annual Cost				
4	mean	-$100.00			
5	std dev	$10.00			
6	nnual Savings				
7	mean	$300.00			
8	std dev	$50.00			
9	Salvage Value				
10	min	$90.00			
11	max	$100.00			
12	Useful Life	PMF		LR	CDF	Life
13		0.30	0	0.3	4
14		0.40	0.3	0.7	5
15		0.10	0.7	0.8	6
16		0.20	0.8	1	7
17	Interest Rate				
18	alpha	5.00			
19	beta	1.50			
20	min	6.00%			
21	max	9.00%			

■ **FIGURE A2-6** Inputs for the Excel simulation

The simulation area of the spreadsheet consists of two parts. The first part implements the formulas necessary to compute the simulated values. In this example, the simulation is tied to the input area and uses the NORMINV function to simulate the annual cost and annual savings. The salvage value is simulated by cell 29, where a U(0, 1) is shifted and scaled over the desired range. The useful life is simulated using the VLOOKUP function. Finally, the interest rate is simulated using the BETAINV function. Then the PV() function of Excel is used to compute the desired present values. Figure A2-7 illustrates the cell formulas necessary to implement the simulation.

Pressing F9 will cause a new set of random numbers to be drawn for the RAND() function and thus a new net present value. You could keep on pressing F9 in order to get a sample of the net present values; however, there is an easier way to "trick" Excel into repeating (replicating) the simulation a given number of times. You can do this by defining a data table in Excel that involves our simulated net present value. To define the data table, do the following:

25	Simulation	
26	Initial Cost	=B2
27	Annual Cost	=NORMINV(RAND(),B4,B5)
28	Annual Savings	=NORMINV(RAND(),B7,B8)
29	Salvage Value	=B10+(B11-B10)*RAND()
30	Useful Life	=VLOOKUP(RAND(),C13:E16,3)
31	Interest Rate	=BETAINV(RAND(),B18,B19,B20,B21)
32	PV of Salvage	=PV(B31,B30,,-B29)
33	PV of Annual Cost	=PV(B31,B30,-B27)
34	PV of Annual Savings	=PV(B31,B30,-B28)
35	Net Present Value	=B26+B33+B34+B32

■ **FIGURE A2-7** First part of the simulation area of Excel example

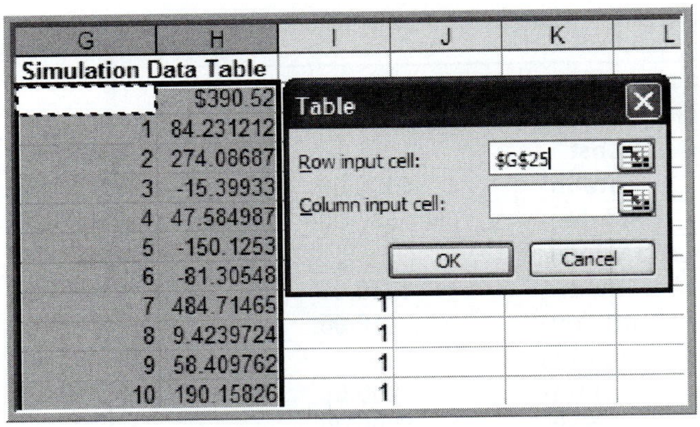

■ **FIGURE A2-8** Defining the data table

1. In a column (e.g., G), list your replication numbers (e.g., 1 to 30).
2. In a cell (e.g., H25), the upper rightmost cell, above the start of the replication numbers), type in your simulation equation. In this case, make it equal to B35 (the cell for the net present value).
3. Select the range of the data table. In the example, this is G25:H55.
4. Use the Data > Table menu and enter ANY BLANK CELL as the column input cell (e.g., G25). A row input cell is not needed. Excel will repeat the calculation for the data table for the number of replications that you specified. Figure A2-8 illustrates this process. With your data table range selected, the Table pop-up dialog requests either a row input cell or a column input cell. Use an empty cell as the row input cell. After entering the cell, the data table will be filled with repeated values of the net present value as per column H in the figure. Press F9 to refresh the simulation.

In this simulation, you want an estimate of the mean expected net present value. In addition, an estimate of the probability of a positive net present value is desired. To estimate whether the present value is positive, simply define an indicator variable and use an IF() statement in Excel to implement it. Let X_i be a 1 if NPV > 0, else let X_i be a 0. You can average these zeros and ones to estimate the probability. This is illustrated in Figure A2-9.

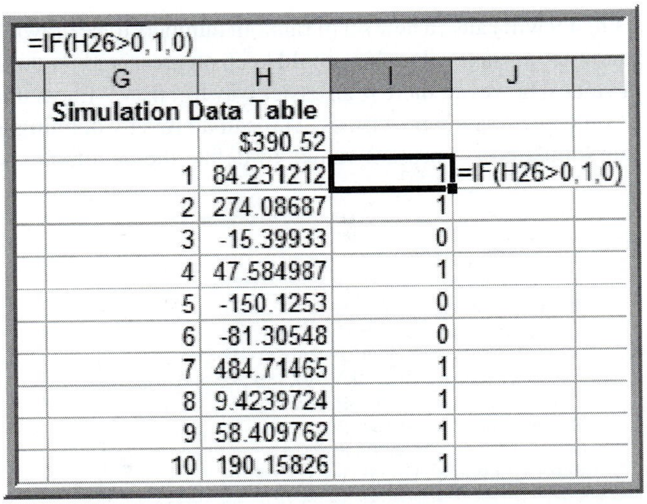

■ **FIGURE A2-9** Indicator variable for positive NPV

G	H	I
Output	NPV	P(NPV>0)
Sample Average	91.57328	0.6
Standard Error	51.34126	0.090972
Median	55.36079	1
Standard Deviation	281.2077	0.498273
Sample Variance	79077.76	0.248276
Range	1154.26	1
Minimum	-463.832	0
Maximum	690.4279	1
Sum	2747.198	18
Count	30	30
Alpha	0.05	0.05
T-Value(Alpha/2,n-1)	2.04523	2.04523
CI Half-Width	105.0047	0.186058
UL-Confidence Interval	196.5779	0.786058
LL-Confidence Interval	-13.4314	0.413942

■ **FIGURE A2-10**   Summary statistics output area

The last step of the simulation is to summarize the output. Figure A2-10 shows the results for the simulation in the output area. For these particular input values, the simulation estimates that the expected net present value will be positive ($51.127); however, there is only a 0.6 chance that net present value will have a positive value. There is a lot of variability in these results as indicated by the size of the half-widths for the sample averages. Figure A2-11 shows the formulas for implementing these statistical calculations from the simulated data table.

G	H	I
Output	NPV	P(NPV>0)
Sample Average	91.57328	0.6
Standard Error	51.34126	0.090972
Median	55.36079	1
Standard Deviation	281.2077	0.498273
Sample Variance	79077.76	0.248276
Range	1154.26	1
Minimum	-463.832	0
Maximum	690.4279	1
Sum	2747.198	18
Count	30	30
Alpha	0.05	0.05
T-Value(Alpha/2,n-1)	2.04523	2.04523
CI Half-Width	105.0047	0.186058
UL-Confidence Interval	196.5779	0.786058
LL-Confidence Interval	-13.4314	0.413942

■ **FIGURE A2-11**   Summary statistics cell formulas

Based on this simple example, you should have a basic understanding of how simulation can be performed in Excel. You have also learned how to generate from a variety of probability distributions in Excel. These concepts should give you a better understanding of how to treat randomness in simulation. There are a number of sophisticated software packages that facilitate spreadsheet simulation (e.g., @Risk) that you should explore if you are interested in furthering your understanding of simulation in Excel.

CHAPTER 4

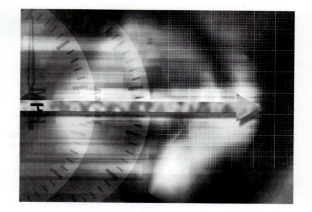

Analyzing Simulation Output

LEARNING OBJECTIVES

After completing this chapter, you should be able to:

- Recognize the various types of statistics used in and produced by simulation models
- Analyze finite-horizon simulations via the method of replications
- Analyze infinite-horizon simulations via the batch means and replication-deletion methods
- Compare simulation alternatives and make valid decisions based on the statistical output of a simulation
- Model systems with shared resources and apply the Arena™ constructs involved in using resource sets

Chapter 3 explored the issue of randomness in simulation model inputs. Because the inputs to the simulation are random, the outputs from the simulation are also random. This concept is illustrated in Figure 4-1. You can think of a simulation model as a function that maps inputs to outputs. This chapter presents the statistical analysis of simulation model output.

In addition, a number of issues that are related to the proper execution of simulation experiments are presented. For example, the simulation outputs are dependent on the input random variables, input parameters, and initial conditions of the model. "Initial conditions" refer to the starting conditions for the model, that is, whether the system starts empty and idle. The effect of initial conditions on steady-state simulations will be discussed in this chapter.

Input parameters are related to the controllable and uncontrollable factors associated with the system. For a simulation model, *all* input parameters are controllable; however, in the

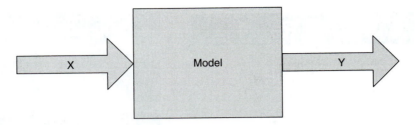

■ **FIGURE 4-1**   **Random input implies random output**

system being modeled you typically have control over only a limited set of parameters. Thus, in simulation you have the unique ability to control the random inputs into your model. This chapter will discuss how to take advantage of controlling the random inputs.

Input parameters can be further classified as decision variables, that is, the parameters of interest that you want to change in order to test model configurations for decision making. The structure of the model itself may be considered a decision variable when you are trying to optimize the performance of the system. When you change the input parameters for the simulation model and then execute the simulation, you are simulating a different design alternative. This chapter describes how to analyze the output from a single design alternative as well as the results of multiple design alternatives. To begin the discussion, an understanding of the types of statistics that may be produced by a simulation experiment is necessary.

4.1   Types of Statistical Variables

A simulation experiment occurs when the modeler sets the input parameters to the model and executes the simulation. This causes events to occur and the simulation model to evolve over time. During the execution of the simulation, the behavior of the system is observed and various statistical quantities computed. When the simulation reaches its termination point, the statistical quantities are summarized in the form of output reports. A simulation experiment may be for a single replication of the model or may have multiple replications. A *replication* is the generation of one sample path, representing the evolution of the system from its initial conditions to its ending conditions. If you have multiple replications within an experiment, each replication represents a different sample path, starting from the same initial conditions and being driven by the same input parameter settings. Because the randomness within the simulation can be controlled, the underlying random numbers used within each replication of the simulation can be made to be independent. Thus, as the name implies, each replication is an independently generated "repeat" of the simulation. Figure 4-2 illustrates the concept of replications being repeated (independent) sample paths.

Within a single sample path (replication), the statistical behavior of the model can be observed. The statistical quantities collected within a replication are called *within replication statistics*. The statistical quantities collected across the replications are called *across replication statistics*. The statistical properties of within and across replication statistics are inherently different and require different methods of analysis. Of the two, within replication statistics are the more challenging from a statistical standpoint.

For within replication statistical collection, there are two primary types of statistical data: *observation based* and *time based*. Observation-based data represent a sequence of equally weighted data values that do not persist over time. This type of data is associated with the duration or interval of time that an object is in a particular state. As such it is observed by marking the time that the object enters the state and the time that the object exits the state. Time-based data represent a sequence of values that persist over some specified amount of time with

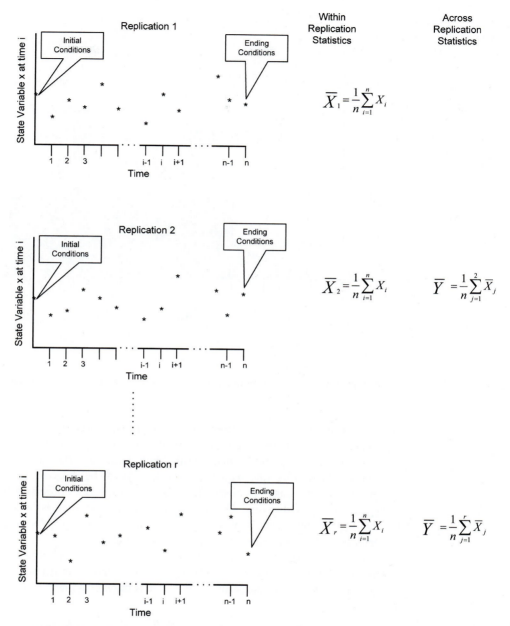

■ **FIGURE 4-2**   The concept of replicated sample paths

that value being weighted by the amount of time over which the value persists. This type of data is often associated with state variables in the model.

Figure 4-3 illustrates a single sample path for the number of customers in a queue over a period of time. From this sample path, events and subsequent statistical quantities can be observed.

1. Let $A_i; i = 1 \ldots n$ represent the time that the ith customer enters the queue.

2. Let $D_i; i = 1 \ldots n$ represent the time that the ith customer exits the queue.

3. Let $W_i = D_i - A_i; i = 1 \ldots n$ represent the time that the ith customer spends in the queue.

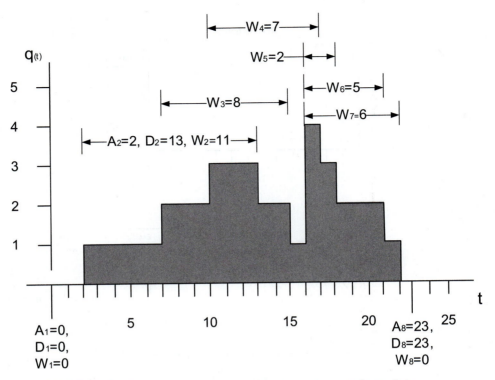

■ **FIGURE 4-3** Sample path for observational and time-persistent data

Thus, $W_i; i = 1 \ldots n$ represents the sequence of wait times for the queue, each of which can be individually observed. These are observational data because the customer enters a state (the queued state) at time A_i and exits the state at time D_i. When the customer exits the queue at time D_i, the waiting time in queue, $W_i = D_i - A_i$, can be observed. W_i is only observable at the instant D_i. This makes W_i observation-based data and, once observed, the value never changes again with respect to time. Observational data are most often associated with an entity that is moving through states implied by the simulation model. An observation becomes available each time the entity enters and subsequently exits the state.

With observational data, it is natural to compute the sample average as a measure of the data's central tendency. Assume that you can observer n customers entering and exiting the queue. Thus, the average waiting time across the n customers is given by

$$\overline{W}(n) = \frac{\sum_{i=1}^{n} W_i}{n}$$

Many other statistical quantities, such as the minimum, maximum, and sample variance, can also be computed from these observations. Unfortunately, within replication observational data are often (if not always) correlated with respect to time. In other words, within replication observational data, such as $W_i; i = 1 \ldots n$, are not statistically independent. In fact, they are likely to also not be identically distributed. Both of these issues will be discussed when the analysis of infinite-horizon or steady-state simulation models is presented.

The other type of statistical variable encountered within a replication is derived from time-based data. Let $q(t); t_0 \leq t \leq t_n$ be the number of customers in the queue at time t. Note that $q(t) \in \{0, 1, 2, \ldots\}$. As illustrated in Figure 4-3, $q(t)$ is a function of t (a step function in this particular case). That is, for a given (realized) sample path, $q(t)$ is a function that returns the number of customers in the queue at time t. The mean value theorem of calculus for integrals states that given a function, f, continuous on an interval [a, b], there exists a constant, c, such that

$$\int_a^b f(x)dx = f(c)(b-a)$$

The value $f(c)$ is called the "mean" value of the function. A similar function can be defined for $q(t)$. In simulation, this function is called the "time average":

$$\overline{L}_q(n) = \frac{\int_{t_0}^{t_n} q(t)dt}{t_n - t_0}$$

This function represents the average with respect to time of the given state variable. This type of statistical variable is called "time based" because $q(t)$ is a function of time.

In the particular case where $q(t)$ represents the number of customers in the queue, $q(t)$ will take on constant values during intervals of time corresponding to when the queue has a certain number of customers. Let $q(t) = q_k$ for $t_{k-1} \le t \le t_k$ and define $v_k = t_k - t_{k-1}$; the time average can now be rewritten as follows:

$$\overline{L}_q(n) = \frac{\int_{t_0}^{t_n} q(t)dt}{t_n - t_0} = \sum_{k=1}^{n} \frac{q_k(t_k - t_{k-1})}{t_n - t_0} = \frac{\sum_{k=1}^{n} q_k v_k}{t_n - t_0} = \frac{\sum_{k=1}^{n} q_k v_k}{\sum_{k=1}^{n} v_k}$$

Note that $q_k(t_k - t_{k-1})$ is the area under $q(t)$ over the interval $t_{k-1} \le t \le t_k$, and

$$t_n - t_0 = \sum_{k=1}^{n} v_k = (t_1 - t_0) + (t_2 - t_1) + \cdots (t_{n-1} - t_{n-2}) + (t_n - t_{n-1})$$

is the total time period in which the variable is observed. Thus, the time average is simply the area under the curve divided by the amount of time over which the curve is observed. From this equation, it should be noted that each value of q_k is weighted by the length of time that the variable has the value. This is why the time average is often called the time-weighted average. If $v_k = 1$, then the time average is the same as the sample average.

With time-based data, you often want to estimate the percentage of time that the variable takes on a particular value. Let T_i denote the *total* time during $t_0 \le t \le t_n$ that the queue had $q(t) = i$ customers. To compute T_i, you sum all the rectangles corresponding to $q(t) = i$ in the sample path. In this example; because $q(t) \in \{0, 1, 2, \ldots\}$, there are an infinite number of possible values for $q(t)$ however, in a finite sample path you can only observe a finite number of the possible values. The ratio of T_i to $T = t_n - t_0$ can be used to estimate the percentage of time that the queue had i customers. That is, define $\hat{p}_i = T_i/T$ as an estimate of the proportion of time that the queue had i customers during the interval of observation.

EXAMPLE 4-1   QUEUE EXAMPLE STATISTICS

Consider Figure 4-3, which shows the variation in queue length over a simulated period of 25 time units. Since the queue length is a time-based variable, the time-average queue length can be computed as:

$$\overline{L}_q = \frac{0(2-0) + 1(7-2) + 2(10-7) + 3(13-10) + 2(15-13) + 1(16-15)}{25}$$

$$+ \frac{4(17-16) + 3(18-17) + 2(21-18) + 1(22-21) + 0(25-22)}{25}$$

$$\overline{L}_q = \frac{39}{25} = 1.56$$

To estimate the percentage of time that the queue had $\{0, 1, 2, 3, 4\}$ customers, the values of $v_k = t_k - t_{k-1}$ need to be summed for whenever $q(t) \in \{0, 1, 2, 3, 4\}$. This results in the following:

$$\hat{p}_0 = \frac{T_0}{T} = \frac{(2-0) + (25-22)}{25} = \frac{2+3}{25} = \frac{5}{25} = 0.2$$

$$\hat{p}_1 = \frac{T_1}{T} = \frac{(7-2) + (16-15) + (22-21)}{25} = \frac{5+1+1}{25} = \frac{7}{25} = 0.28$$

$$\hat{p}_2 = \frac{T_2}{T} = \frac{3+2+3}{25} = \frac{8}{25} = 0.28$$

$$\hat{p}_3 = \frac{T_3}{T} = \frac{3+1}{25} = \frac{4}{25} = 0.16$$

$$\hat{p}_4 = \frac{T_4}{T} = \frac{1}{25} = \frac{1}{25} = 0.04$$

Note that the sum of the $\hat{p}_i$ adds to one. To compute the average waiting time in the queue, use the supplied values for each waiting time.

$$\overline{W}(8) = \frac{\sum_{i=1}^{n} W_i}{n} = \frac{0 + 11 + 8 + 7 + 2 + 5 + 6 + 0}{8} = \frac{39}{8} = 4.875$$

Note that there were two customers, one at time 1.0 and another at time 23.0 who had waiting times of zero. The state graph did not move up or down at those times. Each unit increment in the queue length is equivalent to a new customer entering (and staying in) the queue. On the other hand, each unit decrement in the queue length signifies a departure of a customer from the queue. If you assume a first-in, first-out (FIFO) queue discipline, the waiting times of the six customers that entered the queue (and had to wait) are shown in the figure.

4.2   Types of Simulation with Respect to Output Analysis

When modeling a system, specific measurement goals for the simulation outputs are often required. The goals, coupled with how the system operates, will determine how you execute and analyze the simulation experiments. In planning the experimental analysis, it is useful to think of simulations as consisting of two main categories related to the period of time over which a decision needs to be made:

Finite horizon: In a finite-horizon simulation, a well define ending time or ending condition can be specified, which clearly demarks the end of the simulation. Finite horizon simulations are often called *terminating* simulations, since there are clear terminating conditions.

Infinite horizon: In an infinite-horizon simulation, there is no well-defined ending time or condition. The planning period is over the life of the system, which from a conceptual standpoint lasts forever. Infinite-horizon simulations are often called *steady-state* simulations because one is often interested in the long-term or steady-state behavior of the system.

For a finite-horizon simulation, an event or condition associated with the system is present that indicates the end of each simulation replication. This event can be specified in advance or its time of occurrence can be a random variable. If the event is specified in advance, it is often because you do not want information past that point in time (e.g., a 3-month planning horizon). It might be a random variable in the case of the system stopping when a condition is met. For example, an ending condition may be specified to stop the simulation when there are no entities left to process. This was the case in the LOTR, Inc. example of Chapter 3. Finite horizon simulations are very common since most planning processes are finite. A few example systems involving a finite horizon include:

- *Bank:* Bank doors open at 9 a.m. and close at 5 p.m.
- *Military battle:* Simulate until force strength reaches a critical value.
- *Filling a customer order:* Suppose that a new contract is accepted to produce 100 products, you might simulate the production of the 100 products to determine cost, delivery time, and so on. The LOTR, Inc. system of Chapter 3 is an example of this situation.

For a finite-horizon simulation, each replication represents a sample path of the model for one instance of the finite horizon. The length of the replication corresponds to the finite horizon of interest. For example, in modeling a bank that opens at 9 a.m. and closes at 5 p.m., the length of the replication is 8 hours.

In contrast to a finite-horizon simulation, an infinite-horizon simulation has no natural ending point. Of course, when you actually simulate an infinite-horizon situation, a finite replication length must be specified. Hopefully, the replication length will be long enough to satisfy the goal of observing long run performance. Examples of infinite-horizon simulations include:

- A factory in which you are interested in measuring the steady-state throughput
- A hospital emergency room that is open 24 hours a day, 7 days a week
- A telecommunications system that is always operational

Infinite-horizon simulations are often tied to systems that operate continuously and for which the long-run or steady-state behavior needs to be estimated.

Because infinite-horizon simulations often model situations where the system is always operational, they often involve the modeling of nonstationary processes. In such situations, care must be taken in defining what is meant by long-run or steady-state behavior. For example, in an emergency room that is open 24 hours a day, 365 days per year, the arrival pattern in such a system probably depends on time. Thus, the output associated with the system is also nonstationary. The concept of steady-state implies that the system has been running so long that the system's behavior (in the form of performance measures) no longer depends on time; however, in the case of the emergency room, since the inputs depend on time, so do the outputs. In such cases, it is often possible to find a period of time or cycle over which the nonstationary behavior repeats. For example, the pattern of arrivals at the emergency room may depend on the day of the week, such that every Monday has the same characteristics, every Tuesday has the same characteristics, and so on for each day of the week. Thus, on a weekly basis the nonstationary behavior repeats. You can then define your performance measure of interest based on the appropriate nonstationary cycle of the system. For example, you can define Y as the expected waiting time of patients *per week*. This random variable may have performance that can be described as long term. In others, the long-run weekly performance of the system may be stationary. This type of simulation has been termed "steady-state cyclical parameter estimation" in Law (2007).

Of the two types of simulations, finite-horizon simulations are easier to analyze. Luckily, finite-horizon simulation is the more typical type of simulation found in practice. In fact, when you think that you are faced with an infinite-horizon simulation, you should very carefully evaluate the goals of your study to see if they can just as well be met with a finite planning horizon. The analysis of both of these types of simulations will be discussed in this chapter through Arena™ examples.

4.3 Analysis of Finite-Horizon Simulations

This section illustrates how observation-based and time-based statistics are collected within a replication, and how statistics are collected across replications in Arena™. Finite horizon simulations can be analyzed by traditional statistical methodologies that assume a random sample—that is, independent and identically distributed random variables. A simulation experiment is the collection of experimental design points (specific input parameter values) over which the behavior of the model is observed. For a particular design point, you may want to repeat the execution of the simulation multiple times to form a sample at that design point. To get a random sample, you execute the simulation starting from the same initial conditions and ensure that the random numbers used in each replication are independent. Each replication must also be terminated by the same conditions. It is very important to understand that independence is achieved *across* replications, which means that the replications are independent. The data *within a replication* may or may not be independent.

The method of *independent replications* is used to analyze finite-horizon simulations. Suppose that R replications of a simulation are available where each replication is terminated by some event E and begun with the same initial conditions. Let Y_{ri} be the ith observation on replication r for $i = 1, 2, \ldots, n_r$ and $r = 1, 2, \ldots, R$, and define the sample average for each replication to be

$$\bar{Y}_r = \frac{\sum_{i=1}^{n_r} Y_{ri}}{n_r} \quad for\ r = 1, 2, \ldots, R$$

If the data are time based, then

$$\bar{Y}_r = \frac{\int_0^{T_e} Y_r(t)dt}{T_e}$$

where T_e is the ending time of the simulation. Then, $\bar{Y}_r$ is the sample average of the *within replication* statistics. It is a random variable that can be observed at the end of each replication; therefore, $\bar{Y}_r$ for $r = 1, 2, \ldots, R$ forms a random sample. Thus, typical statistical analysis of the random sample can be readily performed.

To make this concrete, suppose that you are examining a bank that opens with no customers at 9 a.m. and closes its doors at 5 p.m. to prevent further customers from entering. Let $W_{ri}; i = 1 \ldots n_r$ represent the sequence of waiting times for the customers that entered the bank between 9 a.m. and 5 p.m. on day (replication), r, where n_r is the number of customers who were served between 9 a.m. and 5 p.m. on day r. For simplicity, ignore the customers who entered before 5 p.m. but who were not served until after 5 p.m. Let $N_r(t)$ be the number of customers in the system at time t for day (replication) r. Suppose that you are interested in the mean daily customer waiting time and the mean number of customers in the bank on any given day from 9 a.m. to 5 p.m.; that is, you are interested in $E[W_r]$ and $E[N_r]$ for any given day r. At the end of each replication, the following can be computed:

$$\bar{W}_r = \frac{\sum_{i=1}^{n_r} W_{ri}}{n_r} \qquad \bar{N}_r = \frac{\int_0^8 N_r(t)dt}{8}$$

At the end of all replications, random samples $(\overline{W}_1, \overline{W}_2, \ldots, \overline{W}_R)$ and $(\overline{N}_1, \overline{N}_2, \ldots, \overline{N}_R)$ are available from which sample averages, standard deviations, confidence intervals, and so on can be computed. Both of these statistics are based on observations of within replication data.

Both $\overline{W}_r$ and $\overline{N}_r$ are averages of many observations within the replication. Sometimes, there may only be one observation based on the entire replication. For example, suppose that you are interested in the probability that someone is still in the bank when the doors close at 5 p.m.; that is, you are interested in $\theta = P\{N(t = 5\,pm) > 0\}$. In order to estimate this probability, an indicator variable can be defined in the simulation and observed each time the condition is met or not. For this situation, an indicator variable, I_r, for each replication can be defined as follows:

$$I_r = \begin{cases} 1 & \text{if } N(t = 5\,pm) > 0 \\ 0 & \text{if } N(t = 5\,pm) = 0 \end{cases}$$

Therefore, at the end of the replication, the simulation must tabulate whether there are customers in the bank and record the value of this indicator variable. Since this happens only once per replication, a random sample of the $(I_1, I_2, \ldots, I_R)$ will be available after all replications have been executed.

Since the analysis of the system will be based on a random sample, the key design criteria for the experiment will be the required number of replications. In other words, you need to determine the sample size. Because confidence intervals may form the basis for decision making, you can use the confidence interval half-width in determining the sample size. For example, in estimating $E[W_r]$ for the bank example, you might want to be 95% confident that you have estimated the true waiting time within ± 2 minutes. A sample size that will ensure this requirement can be easily determined. There are three related methods that are commonly used for determining the sample size for this situation: an iterative method based on the t-distribution, an approximate method based on the normal distribution, and the half-width ratio method.

4.3.1 Determining the Number of Replications

The following assumes that a random sample, $(X_1, X_2, \ldots X_n)$, of size n needs to be formed (n is the number replications of the simulation). Assume that the X_i are independent and identically distributed random variables from a normal distribution with $\theta = E[X_i]$ and variance $Var[X_i] = \sigma^2$. From confidence interval theory, it is known that a $100 \times (1 - \alpha)$ percent confidence interval for θ is given by

$$\bar{x} \pm t_{\alpha/2, n-1} \frac{s}{\sqrt{n}}$$

where $t_{\alpha/2, n-1}$ is the upper $100(\alpha/2)$ percentage point of the t-distribution with $n - 1$ degrees of freedom. The quantity

$$h = t_{\alpha/2, n-1} \frac{s}{\sqrt{n}}$$

is called the half-width of the confidence interval. You can place a bound, E, on the half-width by picking a sample size that satisfies

$$h = t_{\alpha/2, n-1} \frac{s}{\sqrt{n}} \leq E$$

$$n \geq \left(\frac{t_{\alpha/2, n-1} s}{E} \right)^2$$

	B9	▼	fx	=B8-B4		
	A	B	C	D	E	
1						
2	alpha	0.05	Sample size determination			
3	S	66.083				
4	bound	20	Run goal seek to find the values			
5	n	44.40159	of n such that the half-width is			
6	alpha/2	0.025	less than the specified desired			
7	t-alpha	2.016692	bound.			
8	half-width	20.00001				
9	difference	5.56E-06	1) Specify alpha			
10			2) Specify S (standard deviation)			
11			3) Specify bound			
12			4) Specify initial n			
13			5) Run Goal Seek			
14			a) Tools> Goal Seek			
15			b) Set cell b9 to Value: 0			
16			by changing cell b5			
17						

■ **FIGURE 4-4** Determining sample size in Excel

Unfortunately, $t_{\alpha/2,n-1}$ depends on n, and thus this is an iterative equation. That is, you must try different values of n until the condition is satisfied. Figure 4-4 is a snapshot from the spreadsheet *SampleSizeDetermination.xls* that accompanies this chapter. In the spreadsheet, Excel's Goal Seek functionality is used to iteratively search for the required value of n.

Alternatively, the required sample size can be approximated using the normal distribution via

$$n \geq \left(\frac{z_{\alpha/2}S}{E}\right)^2$$

This equation generally works well for large n, say $n > 50$. Both of these methods require an initial value for the standard deviation. In order to use these methods, you should make an initial pilot set of replications (e.g., five replications) in order to get an initial estimate of the standard deviation. Given a value for s, you can then set a desired bound and use the formulas. The bound is problem and performance measure dependent and is under your subjective control. You must determine what bound is reasonable for your given situation. One thing to remember is that the bound is squared in the denominator for evaluating n. Thus, very small values of E can result in very large sample sizes.

When multiple replications are executed in Arena™, Arena™ will automatically compute a 95% confidence interval for your performance measures. Arena™ does not report the standard deviation, but instead directly reports the half-width value for a confidence interval. If you make a pilot run of n_0 replications, you can use the half-width reported from

Arena[TM] to determine how many replications you need to have to be close to a desired half-width bound. This method is called the half-width ratio method. Let h_0 be the initial value for the half-width from the pilot run of n_0 replications.

$$h_0 = t_{\alpha/2, n_0-1} \frac{s_0}{\sqrt{n_0}}$$

Solving for n_0 yields,

$$n_0 = t_{\alpha/2, n_0-1}^2 \frac{s_0^2}{h_0^2}$$

Now, consider making another set of n replications,

$$n = t_{\alpha/2, n-1}^2 \frac{s^2}{h^2}$$

Taking the ratio of n_0 to n, and assuming that $t_{\alpha/2, n-1}$ is approximately equal to $t_{\alpha/2, n_0-1}$, and s^2 is approximately equal to s_0^2, yields

$$n \cong n_0 \frac{h_0^2}{h^2}$$

Figure 4-5 illustrates the spreadsheet in *SampleSizeDetermination.xls* that facilitates this calculation. Each of these methods assumes that the data are independent and identically distributed from a normal distribution. In addition, the methods also assume that the pilot replications are representative of the population under study. When the data are not normally distributed, you must rely on the central limit theorem to form approximate confidence intervals. The assumption of normality is typically justified when across replication statistics are based on within replication averages. This is due to the fact that the central limit theorem can typically be applied.

In the case of the indicator variable, I_r, which was suggested for use in estimating the probability that there are customers in the bank after 5 p.m., the observed data are clearly not normally distributed. In this case, since you are estimating a proportion, you can use the sample size determination techniques for estimating proportions.

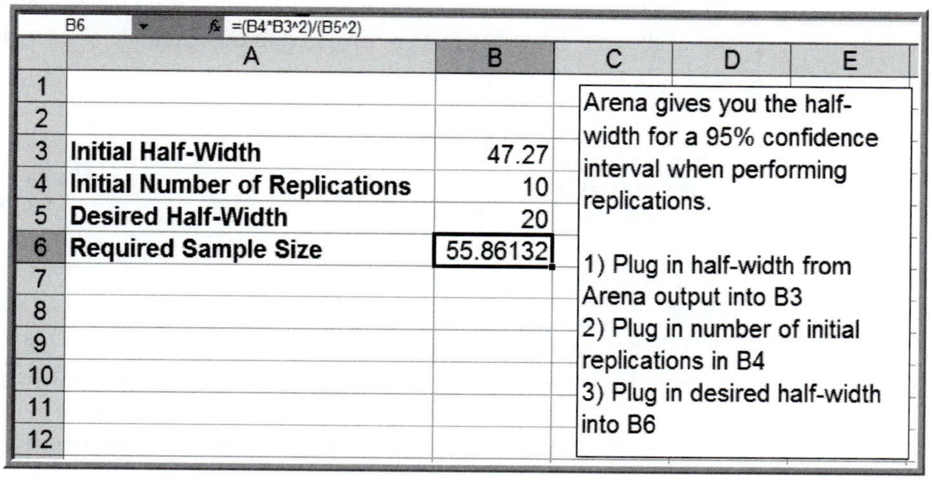

	B6	▼	f_x =(B4*B3^2)/(B5^2)			
	A		**B**	**C**	**D**	**E**
1				Arena gives you the half-width for a 95% confidence interval when performing replications.		
2						
3	**Initial Half-Width**		47.27			
4	**Initial Number of Replications**		10			
5	**Desired Half-Width**		20			
6	**Required Sample Size**		55.86132	1) Plug in half-width from Arena output into B3		
7						
8				2) Plug in number of initial replications in B4		
9						
10				3) Plug in desired half-width into B6		
11						
12						

■ **FIGURE 4-5** Half-width ratio method in Excel

In particular, a $100 \times (1 - \alpha)\%$ large sample confidence interval for a proportion, p, has the following form:

$$\hat{p} \pm z_{\alpha/2} \sqrt{\frac{\hat{p}(1 - \hat{p})}{n}}$$

where $\hat{p}$ is the estimate for p. From this, you can determine the sample size via the following equation:

$$n = \left(\frac{z_{\alpha/2}}{E}\right)\hat{p}(1 - \hat{p})$$

Again, a pilot run is necessary for obtaining an initial estimate, $\hat{p}$, for use in determining the sample size. If no pilot run is available, $\hat{p} = 0.5$ is often assumed. If you have more than one performance measure of interest, you can use these sample size techniques for each of your performance measures and then use the maximum sample size required across the performance measures.

4.3.2 Finite-Horizon Example in Arena™

This example re-examines the LOTR Ring Maker, Inc. model of Chapter 3 in order to form confidence intervals on the following additional quantities:

- The average number of pairs of rings in both the ring-making process and the ring-inspection process.
- The average time that it takes for a pair of rings to go through both the ring-making process and the ring-inspection process. In addition, a 95% confidence interval for the mean time to complete these processes to within ±20 minutes is desired.

Arena™ can automatically report the average number of pairs of rings in the ring-making process, the average number of pairs of rings in the inspection process, and the average time spent in each process separately. However, for this example, confidence intervals on the combined total are required, which Arena™ will not automatically collect. The following illustrates how to add logic to collect the quantities of interest. The completed example is available in the file *LOTRCh4Example.doe*. If you are following along with the text, you can make a copy of the file *LOTRCh3Example.doe* and complete the steps outlined here. Open the file *LOTRCh3Example.doe* and double-click on the Ring Processing submodel. The start and end of the two processes that are involved in this example are denoted in Figure 4-6.

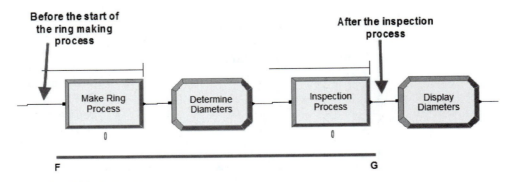

■ FIGURE 4-6 Identifying the processes

The pair of rings flows through each of these modules, and the number of pairs of rings that are between the denoted points and how long that each pair of rings spends in the identified part of the flow process must be captured.

Let A_i denote the time that the ith pair of rings arrives to the Make Ring Process, let D_i represent the time that the ith pair of rings departs from the Inspection Process, and let $N(t)$ be the number of pairs of rings between the points demarked as F and G in Figure 4-6. The time spent in both processes for the ith pair of rings is $D_i - A_i$. Clearly, $D_i - A_i$ is an observation-based quantity, and $N(t)$ is a time-based quantity. Within Arena^{TM}, you can use a variable to keep track of $N(t)$. Each time a pair of rings passes point F, the variable should be incremented and each time a pair of rings passes point G, the variable should be decremented. Thus, the variable will keep track of the current number of entities (rings) in both processes. In the Arena^{TM} model, this variable is denoted as *vNumInMakeAndInspect*. You should use the VARIABLE data module to define this variable. Because the arrival time to the Make Ring Process is related to each entity, an attribute will be used to remember when the entity enters into the ring-making process. In the Arena^{TM} model, this attribute is denoted by *myArriveToMakeRingTime*.

There are two methods by which you can have time-based statistics collected on the variable *vNumInMakeAndInspect*. The first method can be accessed through the VARIABLE module, and the second method involves using the STATISTICS module on the Advanced Process panel. Both methods will be illustrated. For the first method, select the VARIABLE data module and use the spreadsheet view to enter the name of the variable. Make sure that the Report Statistics option is checked as shown in the top part of Figure 4-7. By checking the Report Statistics option, you are telling Arena^{TM} to collect and report time-based statistics on this variable during the simulation. This option is only available if the variable is a scalar (i.e., neither a 1D or 2D array variable).

The second option involves developing a user-defined, time-based STATISTIC on the variable. You should experiment with both ways to prove to yourself that they are the same. Go to the Advanced Process panel and select the STATISTIC data module. Double-click on the spreadsheet add row area to add another row and then right-click on the row. This will give the context menu to Edit Via Dialog. You should fill out the dialog box as shown in the bottom part of Figure 4-7.

Use the name *AvgNumInMakeAndInspect* for the statistic, specify its type as time-persistent (time-based), and indicate that the expression to be used for collecting the statistics is simply the variable name that is being using to track the number of entities in the Make and Inspect processes. Arena^{TM} will create what is called a DSTAT variable to collect the time-based statistics for the given expression. The advantage of using the STATISTIC module over the Report Statistics option in the VARIABLE module is that the STATISTIC module allows you to provide a different name for the reports and to save the data to a file. When the quantity to be collected is more complicated than a simple variable, then the STATISTIC module's expression field can be used to define the appropriate expression. Thus, to collect time-based statistics on a quantity of interest in the simulation, you should use the time-persistent option in the STATISTIC module. Now, let us examine how to collect observation-based statistics.

To collect observation-based statistics, you need to place RECORD modules at appropriate locations in the model. In the case of collecting the quantity, $D_i - A_i$, you can first use an ASSIGN module to assign the value of A_i to *myArriveToMakeRingTime* and then use a RECORD module to collect the difference. As previously mentioned, in Arena^{TM} the special variable TNOW will return the current simulation time. In this situation, you can use TNOW to get the values of A_i and D_i when the entity passes the appropriate points in the model. When the entity passes the point F in Figure 4-6 make the assignment:

$$myArriveToMakeRingTime = TNOW$$

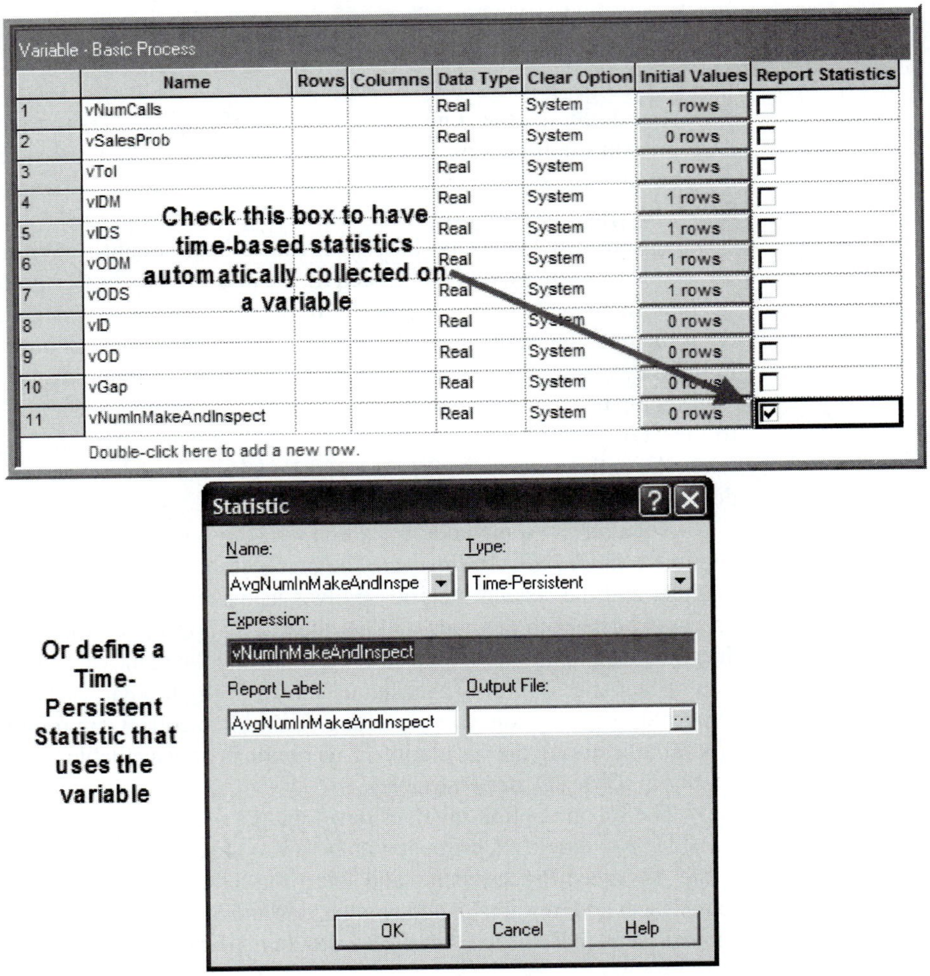

■ FIGURE 4-7   Two ways to collect statistics on a time-based variable

This is the value of A_i for the entity. Then, when the entity reaches point G of Figure 4-6, record the expression:

$$\text{TNOW} - myArriveToMakeRingTime$$

This expression represents the elapsed *interval* of time that the entity took to "move" from point F to point G of Figure 4-6. Since this type of recording is so common Arena™ provides a special option for the RECORD module to facilitate this operation. As shown in Figure 4-8, delete the appropriate connectors, place the ASSIGN module, fill out the ASSIGN module, and reconnect the modules.

You should now delete the appropriate connectors and place a RECORD module as shown in Figure 4-9. The RECORD module contains five different record types:

Count: When Count is chosen, Arena™ defines a COUNTER variable. This variable will be incremented or decremented by the value specified.

Entity Statistics: This option will cause entity statistics to be recorded. Entity statistics will be discussed in Chapter 7.

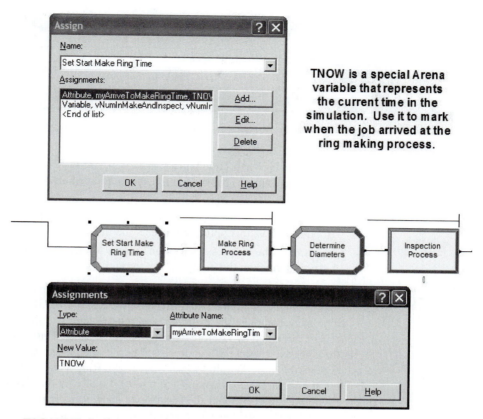

TNOW is a special Arena variable that represents the current time in the simulation. Use it to mark when the job arrived at the ring making process.

■ **FIGURE 4-8**   Capturing the start of ring making

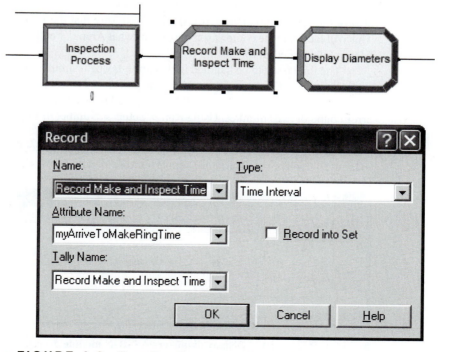

■ **FIGURE 4-9**   Recording the make and inspect time

Assignments

	Type	Variable Name	Attribute Name	New Value
1	Attribute	Variable 1	myArriveToMakeRingTi	TNOW
2	Variable	vNumInMakeAndInspect	Attribute 2	vNumInMakeAndInspect +1

Double-click here to add a new row.

■ **FIGURE 4-10** **Set start make ring time ASSIGN module**

Time Interval: When Time Interval is chosen, Arena™ will take the difference between the current time, TNOW, and the specified attribute.

Time Between: When Time Between is chosen, Arena™ will take the difference between the current time, TNOW, and the last time an entity passed through the RECORD module. Thus, this option represents the time between arrivals to the module.

Expression: The Expression option allows statistics to be collected on a general expression when the entity passes through the module.

In this situation, you can use either the Time Interval option as shown in Figure 4-9 or you can select the expression option and use TNOW – *myArriveToMakeRingTime* as the expression.

In order to complete the example, you must ensure that the value of *vNumInMakeAndInspect* is incremented by 1 before the rings enter the Make Ring Process and decremented by 1 after the rings exit the Inspection process. The already placed ASSIGN modules can be used to accomplish this task. Within the Set Start Make Ring Time ASSIGN module, you should increment the variable as shown in Figure 4-10. Finally, in the Display Diameters ASSIGN module, you should decrement the variable as shown in Figure 4-11. This ensures that the variable, *vNumInMakeAndInspect*, correctly tracks how many rings are currently in the area of the process marked off between points *F* and *G* of Figure 4-6.

As mentioned at the beginning of the example, Arena™ will automatically collect statistical information about individual processes. Figure 4-12 shows the Statistical Collection area of the Run Setup dialog. You should make sure that the Process Check Box is checked. This will facilitate the comparison of some of what Arena™ collects automatically to what has been implemented in this example.

The final issue to be handled in this finite-horizon simulation is to specify the number of replications. First, a small pilot run will be performed to get initial half-widths, and then the required number of replications will be computed to ensure a 95% confidence interval with an error bound of ±20 minutes. Using Run > Setup > Number of Replications, specify *10* as the number of replications and run the model.

The results for the combined process time based on the direct RECORD module are given in Figure 4-13. The results for the average number of pairs of rings in both processes are given in

Assignments

	Type	Variable Name	New Value
1	Variable	vNumInMakeAndInspect	vNumInMakeAndInspect - 1
2	Variable	vID	myIDBigRing
3	Variable	vOD	myODSmallRing
4	Variable	vGap	myGap

Double-click here to add a new row.

■ **FIGURE 4-11** **Display diameters ASSIGN module decrementing the number in Make and Inspect**

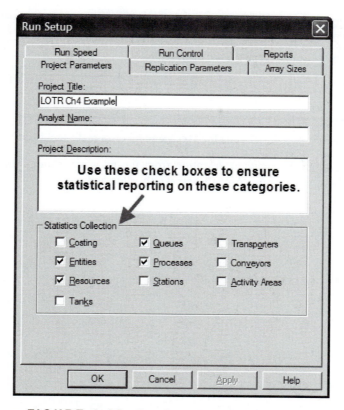

■ **FIGURE 4-12** **Run Setup Statistical Collection**

Figure 4-14. On average, it takes about 365.79 minutes for all the rings to get through these two processes. This is occurring because all the rings are released at once to the Make Ring Process. If you examine Figure 4-15, you can see that Arena^{TM} will collect the total time spent in each of the individual processes. Thus, based on this output you can add up the process times from Arena's process report. However, this does not automatically give a confidence interval on the total time. Based on the pilot run, the number of replications necessary to meet the confidence interval half-width target can be determined.

Using $n \cong n_0 \left(h_0^2 / h^2 \right)$ with $n_0 = 10$, $h_0 = 47.27$, and $h = 20$, approximately $n = 56$ replications are needed to meet the criteria. If you wanted to use the iterative method, you must

Tally		
Expression	Average	Half Width
ProbTooBig	0.03133511	0.01
ProbTooSmall	0.03571710	0.01
Interval	Average	Half Width
Record Make and Inspect Time	365.79	47.27

Tree panel:
- Queue
- Resource
- User Specified
  - Tally
  - Counter
  - Time Persistent
  - Output

■ **FIGURE 4-13** **Result for RECORD module**

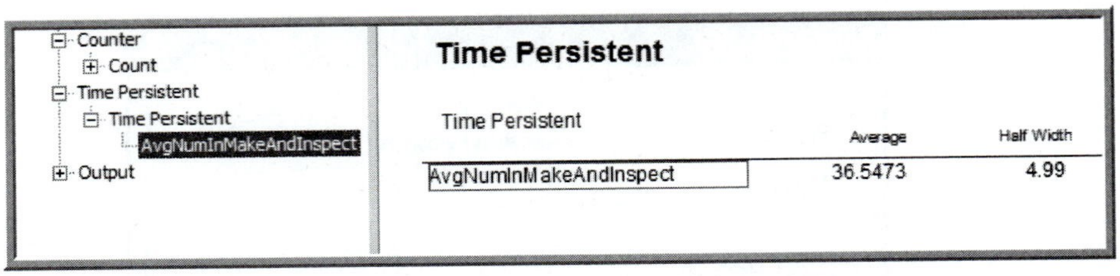

Time Persistent		
Time Persistent	Average	Half Width
AvgNumInMakeAndInspect	36.5473	4.99

(Tree view: Counter → Count; Time Persistent → Time Persistent → AvgNumInMakeAndInspect; Output)

■ **FIGURE 4-14**   Results for average number in make and inspect

first determine the standard deviation from the pilot replications. In the case of multiple replications, you can use Arena's half-width value and

$$h_0 = t_{\alpha/2, n_0-1} \frac{s_0}{\sqrt{n_0}}$$

$$s_0 = \frac{h_0 \sqrt{n_0}}{t_{\alpha/2, n_0-1}}$$

to compute s_0. For the example, this yields

$$s_0 = \frac{h_0 \sqrt{n_0}}{t_{\alpha/2, n_0-1}}$$

$$s_0 = \frac{47.27\sqrt{10}}{t_{0.25,9}} = \frac{47.27\sqrt{10}}{2.262} = 66.083$$

Then, iteratively solving,

$$n \geq \left(\frac{t_{\alpha/2, n-1} 66.083}{20} \right)^2$$

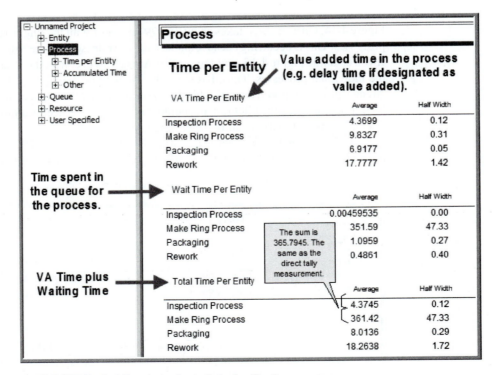

■ **FIGURE 4-15**   Arena's statistical collection process

Interval	n = 45	Average	Half Width
Record Make and Inspect Time		400.83	20.50
Interval	n = 56	Average	Half Width
Record Make and Inspect Time		388.80	20.34

■ **FIGURE 4-16**   **Results from running $n = 45$ and 56 replications**

yields a recommended sample size of $n = 44.4 \approx 45$. The half-width ratio method tends to be more conservative in recommending the required sample size (number of replications).

The results in Figure 4-16 indicate that the half-width criterion is almost met by both sample sizes. Note that the make and inspection time is highly variable. The methods to determine the half-width assume that the standard deviation, s_0, in the pilot runs will be similar to the standard deviation observed in the final set of replications. However, when the full set of replications is run, the actual standard deviation may be different than that which occurred in the pilot run. Thus, the half-width criterion might not be met in your final set of replications. If the assumptions are reasonably met, there will be a high likelihood that the desired half-width will be very close to the desired criteria, as shown in this example.

4.3.3   Sequential Sampling for Finite-Horizon Simulations in Arena™

The methods discussed for determining the sample size are based on pre-determining a *fixed* sample size and then making the replications. If the half-width equation is considered as an iterative function of n,

$$h(n) = t_{\alpha/2, n-1} \frac{s(n)}{\sqrt{n}} \leq E$$

Then it becomes apparent that additional replications of the simulation can be executed until the desired half-width bound is met. This is called sequential sampling, and in this case the sample size of the experiment is not known in advance. The brute force method for implementing this approach would be to run and rerun the simulation each time increasing the number of replications until the criterion is met. By defining an OUTPUT statistic and using the *ORUNHALF(Output ID)* function, you can have Arena™ stop when the criterion is met. The pseudo-code for the Arena™ code is as follows:

```
CREATE a logical entity at time 0.0
DECIDE if number of replications <=2
     DISPOSE
Else DECIDE if half-width <= bound
     ASSIGN maximum number of replications = current number of replications
ELSE
     DISPOSE
```

In order to implement this code, you must first define an OUTPUT statistic. This is done in Figure 4-17, where the Arena™ function *TAVG(Tally ID)* has been used to get the end of the replication average for the time spent in both the make and inspection processes. This ensures

	Name	Type	Expression
	Statistic - Advanced Process		
1	TimeToMakeOrderStat	Output	TNOW
2	ProbOfOT	Output	TNOW > 960
3	AvgNumInMakeAndInspect	Time-Persistent	vNumInMakeAndInspect
4	MakeAndInspectTimeAvgAcrossRep	Output	TAVG(Record Make and Inspect Time)
	Double-click here to add a new row.		

■ **FIGURE 4-17   OUTPUT statistic for record make and inspect time**

that you can use the *ORUNHALF(Output ID)* function. The *ORUNHALF(Output ID)* function returns the current half-width for the specified OUTPUT statistic based on all the replications that have been fully completed. This function can be used to check against the half-width bound.

Figure 4-18 illustrates the logic for implementing sequential sampling in Arena™. The CREATE module creates a single entity at time 0.0 for each replication. That entity then proceeds to the Check Num Reps DECIDE module. The special Arena™ variable, NREP, holds the *current* replication number. When NREP equals 1, no replications have yet been completed. When NREP equals 2, the first replication has been completed and the second replication is in progress. When NREP equals 3, two replications have been completed. At least two completed replications are needed to form a confidence interval (since the formula for the standard deviation has $n - 1$ in its denominator). Since at least two replications are needed to start the checking, the entity can be disposed if the number of replications is less than or equal to 2. Once the simulation is past two replications, the entity will then begin checking the half-width.

The logic shown in Figure 4-19 indicates that the second DECIDE module uses the Arena™ function *ORUNHALF(Output ID)* to check the current half-width versus the error bound, which was 20 minutes in this case. If the half-width is less than or equal to the bound, the entity goes through the ASSIGN module to trigger the stopping of the replications. If the half-width is larger than the bound, the entity is disposed.

Arena™ has a special variable called MREP that represents the maximum number of replications to be run for the simulation. When you fill out the Run > Setup dialog and specify the number of replications, MREP is set to the value specified. At the beginning of each replication NREP is checked against MREP. If NREP equals MREP, that replication will be the final replication run for the experiment. Therefore, if MREP is set equal to NREP in the check, the next replication will be the last replication. This actually causes one more replication to be executed than needed, but this cannot be avoided because of how ARENA functions. The only way to prevent this is to use some VBA code. Figure 4-20 shows the ASSIGN module

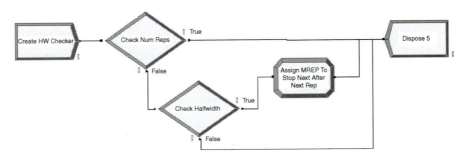

■ **FIGURE 4-18   Logic for sequentially checking half-width**

■ FIGURE 4-19   DECIDE logic for sequential sampling

■ FIGURE 4-20   Logic to stop the replication using MREP

for setting MREP equal to NREP. Note that the assignment type is Other because MREP is a special variable rather than a standard variable in Arena™ (as defined in the VARIABLE module).

This has been implemented in the file *LOTRCh4ExampleSequential.doe*. Before running the model, you should set the number of replications (MREP) in the Run > Setup > Replication Parameters dialog to some arbitrarily large integer. If you execute the model, you will get the result indicated in Figure 4-21.

In the sequential sampling experiment there were 51 replications. This is actually less than the recommended 56 replications for the fixed half-width method in Figure 4-16, and the half-width met the criteria. The reason why the half-width in Figure 4-16 is larger than the half-width in Figure 4-21 is that replications 52 to 56 must have had significantly more variability. This is perfectly possible, and emphasizes the fact that in the sequential sampling method, the number of replications is actually a random variable. If you were to use different streams and rerun the sequential sampling experiment, the number of replications completed may be different each time.

4.3.4   Using the Arena-Generated Access File

Each time the simulation experiment is executed Arena™ writes the statistical information collected during the run to a Microsoft Access database that has the same name as the Arena™ model file. This database is what is used by the Crystal Reports report generator in Arena™. You can also access this database and extract the statistical information yourself. This is useful if you need to post-process the statistical information in a statistical package. This section demonstrates how to extract information from this database. The intention is to make you aware of the database and to give you enough information so that you can use it for simple data extraction. There is detailed information concerning the Reports Database in the Arena™ Help system (search under "The Report Database").

■ FIGURE 4-21   Output for sequential sampling

When you create and run an Arena[TM] model—for example, *yourmodel.doe*—Arena[TM] will create a corresponding Microsoft Access[TM] file called *yourmodel.mdb*. Each time the Arena[TM] model is run, Arena[TM] writes the statistical information from the simulation to the database. If you change the input parameters to your model (without doing anything else), and then rerun the model, that model's results are written to the database. Any previous data will be overwritten. There are two basic ways to have Arena[TM] save information from multiple simulation runs. The first is to simply rename the database file to a different name before rerunning the simulation. Arena[TM] will automatically create a new database file for the model. This approach creates a new database for each simulation execution.

The second approach is to use the Project Parameters panel in Arena's Run > Setup menu dialog, as illustrated in Figure 4-12. The Project Title field identifies a particular project. This field is used by the database to identify a given project's results. If this name is not changed when the simulation is rerun then the results are overwritten for the named project. Thus, by changing this name before each experiment, the database will contain the information for each experiment accessible by this project name. Now let's take a look at the type of information stored in the database.

The Reports database consists of a number of tables and queries. A table is a set of records with each row in the table having a unique identifier called its primary key and a set of fields (columns) that hold information about the current record. A query is a database instruction for extracting particular information from a table or a set of related tables. The Arena[TM] Help system describes in detail the structure of the database. For accessing statistical information, key tables include:

Definition: Holds the name of the statistical item from within the Arena[TM] model, its type (e.g., DSTAT, TALLY, FREQ, etc.) and information about its collection and reporting.

Statistic: The summary within replication results for a given statistic on a given replication

Count: The ending value for a RECORD module designated as Count for a given replication

Frequency: The ending values for tabulated frequency statistics by replication

Output: The value of a defined OUTPUT statistic for each replication

The Definition table with the Statistic table can be used to access the replication information. Figure 4-22 illustrates the fields in the Definition table. The important fields in the Definition table are ID, Name, and DefinitionTypeID. The ID is the primary key of the table, Name represents the name to appear on the report that was assigned in Arena[TM] for the statistic, DefinitionTypeID indicates the type of statistical element. For the purposes of this section, the statistic types of interest are DSTAT (time based), TALLY (observation based), COUNTER (RECORD with count option), OUTPUT (captured at end of replication), and FREQUENCY (tabulates percentage of time spent in defined categories).

Field Name	Data Type	Description
ID	AutoNumber	
ReportID	Number	Report that this line belongs to
Name	Text	Name to be used to identify this report line
Format	Text	Format specifier in 'C' or FORTRAN format, refer to documentation for limitations in database processing
ReportLineDefinitionID	Number	
ArgumentIndex	Number	Order of the argument in the parameter list
Expression	Text	Expression used to develop the value
Type	Text	Data type for report line, typically SMINT, SMREAL, STR
RunOutputID	Number	The run that is associated with this row
OutputFileID	Number	
SourceDataTypeID	Number	Data type (like VACost or NVATime)
SourceCategoryID	Number	Generic type (like Entity, Resource, or Queue), or that it is a particular template and module (like "Basic-Create" or "Advanced-Process")
SourceProcessID	Number	Identify the module
DefinitionTypeID	Number	Type of statistical element represeted ie:DSTAT, CSTAT, TALLY ...... foreign key to Definitiontypes
Limit	Number	

■ **FIGURE 4-22** Definition table field design

Field Name	Data Type	Description
ID	AutoNumber	
ReplicationID	Number	
DefinitionID	Number	
MinObs	Number	The minimum observation value within this scenario/replicaiton.
MaxObs	Number	The maximum observation value within this scenario/replication.
AvgObs	Number	The average observation value for this scenario/replicaiton.
HalfWidth	Number	The .95 half width for this scenario/ replication.
LastValue	Number	The final variable value for this scenario/replicaiton.
NumObs	Number	The number of observations in this stat
StdDev	Number	The standard deviation of observations

Statistic : Table

■ **FIGURE 4-23**　Statistic table field design

The Statistic table holds the statistical values for the within replication statistics as shown in Figure 4-23. Note that the Statistic table has a field called DefinitionID. This field is a database foreign key to the Definition table. With a foreign key field, you can relate the two tables together and get the name and type of statistic.

To explore these tables, you will work with the database called *CH4DB-Stat-Example.mdb*. This file was produced by running the *LOTRCh4Example.doe* with 56 replications and renaming the created database file. In Microsoft Access, open the database and then open the table called Definition. Select the desired row (e.g., row 8), and use Insert > Subdatasheet to get the Insert Subdatasheet dialog as shown in Figure 4-24. Then, select the Statistic table and press the OK button as shown in the figure.

Figure 4-25 shows the result of inserting the linked data sheet and selecting the + symbol for the statistic named Record Make and Inspect Time. To sort the subdatasheet by increasing replication number, right-click on the ReplicationID field. From this table, you can readily assess the replication statistics. For example, the AvgObs column refers to the ending average

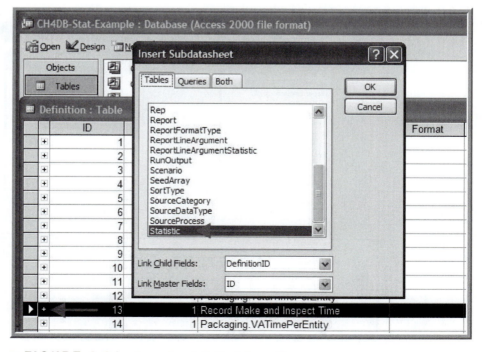

■ **FIGURE 4-24**　Accessing statistics information

ID	ReportID	Name		Format	ReportLineDefin	ArgumentIndex
12	1	Packaging.TotalTimePerEntity				0
13	1	Record Make and Inspect Time				0

ID	ReplicationID	MinObs	MaxObs	AvgObs	HalfWidth	LastValue
13	1	11.4505300186	705.241399798	363.804363517	2E+20	705.241399798
56	2	18.2185229538	700.783930168	346.058804617	2E+20	700.783930168
99	3	15.5303417880	496.832758865	259.099768822	2E+20	496.832758865
142	4	18.1115237371	659.372679112	328.811820096	2E+20	659.372679112
185	5	11.6530644408	879.533296880	449.363748219	2E+20	879.533296880
228	6	12.5886395300	708.877836415	365.714157645	2E+20	708.877836415
271	7	16.4254930053	664.036713358	339.121842804	2E+20	664.036713358
314	8	19.0659549956	614.273960189	316.357093763	2E+20	614.273960189
357	9	12.8852738538	941.455156957	486.854846717	2E+20	941.455156957
400	10	15.9273709035	806.236078717	402.755368380	2E+20	806.236078717

Record: 9 of 56

■ **FIGURE 4-25**   Replication statistics for make and inspect time

across all the observations recorded during each replication. From this you can easily cut and paste any required information into Excel or a statistical package such as MINITAB.

For those familiar with writing queries, this same information can be extracted by writing the query as shown in Figure 4-26. To summarize the statistics across the replications, you can write a group by query as shown in Figure 4-27.

In the original LOTR Ring Maker, Inc. model, an OUTPUT statistic was defined to collect the time that the simulation ended. The OUTPUT statistic values collected at the end of each replication are saved in the Output table as shown in Figure 4-28.

To see the collected values for a specific defined statistic, you can again use the Definition table. Open up the Definition table and scroll down to the row 48, which is the *Time-ToMakeOrderStat* statistic. Select the row and using the Insert > Subdatasheet dialog, insert the Output statistic table as the subsheet. You should see the subsheet as shown in Figure 4-29. These same basic techniques can be used to examine information on the COUNTERS and

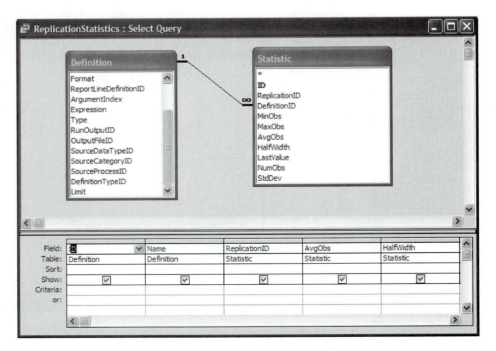

■ **FIGURE 4-26**   Query to access replication statistics

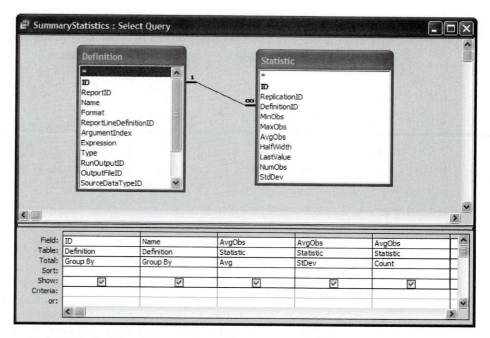

■ FIGURE 4-27   Query to summarize across replications

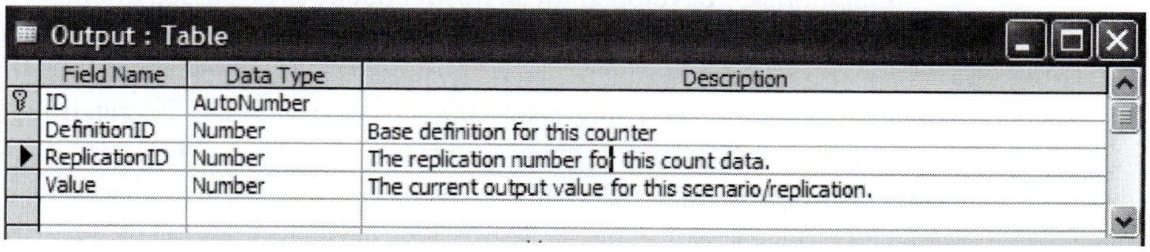

■ FIGURE 4-28   Output table field design

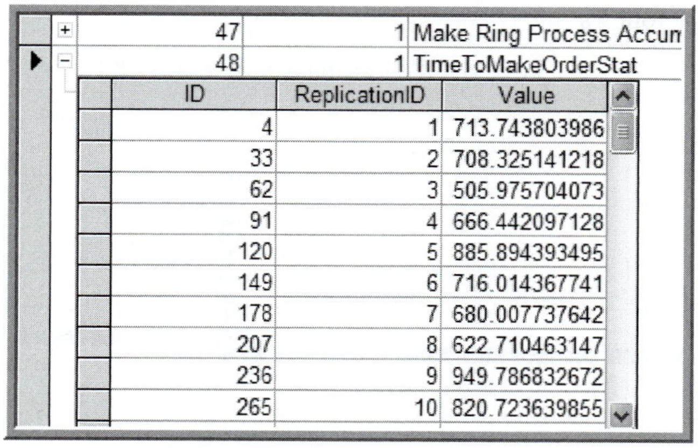

■ FIGURE 4-29   Expanded datasheet view for OUTPUT statistics

FREQUENCY statistics that have been defined in the Arena™ model. If you name each run with a different Project Name, then you can do queries across projects. Thus, after you have completed your simulation analysis you do not need to rerun your model, you can simply access the statistics that were collected from the Reports database. An experienced Microsoft Access user can then create custom queries, reports, charts, and so on for simulation results.

4.4   Analysis of Infinite-Horizon Simulations

This section discusses how to plan and analyze infinite-horizon simulations. When analyzing infinite-horizon simulations, the primary difficulty is the nature of *within replication* data. In the finite-horizon case, the statistical analysis is based on three basic requirements:

- Observations are independent.
- Observations are sampled from identical distributions.
- Observations are drawn from a normal distribution (or enough observations are present to invoke the central limit theorem).

These requirements were met by performing independent replications of the simulation to generate a random sample. In a direct sense, the outputs within a replication do not satisfy any of these requirements; however, certain procedures can be imposed on the manner in which the observations are gathered to ensure that these statistical assumptions are not grossly violated. The following will first explain why within replication data typically violate these assumptions and then will provide some methods for mitigating the violations in the context of infinite-horizon simulations.

To illustrate the challenges related to infinite-horizon simulations, a simple spreadsheet simulation was developed for an M/M/1 queue as discussed in Chapter 1. The final spreadsheet is given in the spreadsheet file *MM1-QueueingSimulation.xls* that accompanies this chapter. The immediate feedback from a spreadsheet model should facilitate understanding the concepts. Consider a single-server queuing system as illustrated Figure 4-30.

For a single-server queuing system, there is an equation that allows the computation of the waiting times of each of the customers based on knowledge of the arrival and service times. Let $X_1, X_2, \ldots$ represent the successive service times and $Y_1, Y_2, \ldots$ represent the successive interarrival times for each of the customers who visit the queue. Let $E[Y_i] = 1/\lambda$ be the mean of the interarrival times so that λ is the mean arrival rate. Let $E[X_i] = 1/\mu$ be the mean of the service times so that μ is the mean service rate. Let W_i be the waiting time in the queue for the ith customer—that is, the time between when the customer arrives until they enter service. Lindley's equation (see Gross and Harris 1998) relates the waiting time to the arrivals and services as follows:

$$W_{i+1} = \max(0, W_i + X_i - Y_i)$$

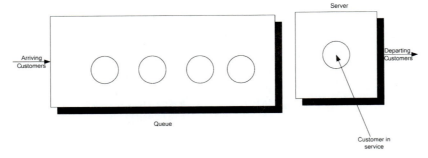

■ **FIGURE 4-30**   **Single-server queuing system**

	A	B	C	D	E	F	G	H
20								
21		Customer	Waiting	Service	Interarrival			
22		Number	time	time	time		Cumulative	Cumulative
23		n	W(n)	X(n)	Y(n)	W(n)+X(n)-Y(n)	Sum	Avg
24		0	0	0.165049697	3.205865303	-3.040815606		
25		1	0	0.52814479	1.34872615	-0.82058136	0	0
26		2	0	0.338501251	2.146635912	-1.808134661	0	0
27		3	0	1.435505306	1.442529663	-0.007024357	0	0
28		4	0	0.436372252	0.184805878	0.251566375	0	0
29		5	0.251566375	0.600203813	0.576593493	0.275176695	0.251566375	0.050313275
30		6	0.275176695	1.499580523	3.497237695	-1.722480477	0.52674307	0.087790512
31		7	0	2.087126617	0.225136044	1.861990574	0.52674307	0.07524901
32		8	1.861990574	0.940140227	4.372192785	-1.570061985	2.388733644	0.298591705
33		9	0	0.55141614	0.710258739	-0.158842598	2.388733644	0.265414849
34		10	0	0.489774047	0.0205816	0.469192446	2.388733644	0.238873364

■ **FIGURE 4-31** Spreadsheet for Lindley's equation

The relationship says that the time that the $(i + 1)st$ customer must wait is the time the ith waited, plus the ith customer's service time, X_i (because that customer is in front of the ith customer), less the time between arrivals of the ith and $(i + 1)st$ customers, Y_i. If $W_i + X_i - Y_i$ is less than zero, then the $(i + 1)st$ customer arrived after the ith finished service, and thus the waiting time for the $(i + 1)st$ customer is zero, because his service starts immediately. Suppose that $X_i \sim \exp(E[X_i] = 0.7)$ and $Y_i \sim \exp(E[Y_i] = 1.0)$. This is an M/M/1 queue with $\lambda = 1$ and $\mu = 10/7$. Thus, using the equations given in Chapter 1 yields:

$$\rho = 0.7 \quad L_q = \frac{0.7 \times 0.7}{1 - 0.7} = 1.6\overline{33} \quad W_q = \frac{L_q}{\lambda} = 1.6\overline{33} \text{ minutes}$$

The spreadsheet model involves generating X_i and Y_i and implementing Lindley's equation in the rows of the spreadsheet. Figure 4-31 shows sample output from the simulation. The simulation was initialized with $W_0 = 0$ and random draws for X_0 and Y_0. The generated values for X_i and Y_i are based on the inverse transform technique for exponential random variables with the cell formulas given in cells D24 and E24, as shown in Figure 4-32.

Figure 4-32 shows that cell F25 is based on the value of cell F24. This implements the recursive nature of Lindley's formula. Column G holds the cumulative sum

$$\sum_{i=1}^{n} W_i \quad \text{for } n = 1, 2, \dots$$

	A	B	C	D	E	F	G	H
20								
21		Customer	Waiting	Service	Interarrival			
22		Number	time	time	time		Cumulative	Cumulative
23		n	W(n)	X(n)	Y(n)	W(n)+X(n)-Y(n)	Sum	Avg
24		0	0	=-$B$4*LN(1-RAND())	=-$B$5*LN(1-RAND())	=C24+D24-E24		
25		1	=MAX(0,F24)	=-$B$4*LN(1-RAND())	=-$B$5*LN(1-RAND())	=C25+D25-E25	=C25	=G25/B25
26		2	=MAX(0,F25)	=-$B$4*LN(1-RAND())	=-$B$5*LN(1-RAND())	=C26+D26-E26	=C26+G25	=G26/B26
27		3	=MAX(0,F26)	=-$B$4*LN(1-RAND())	=-$B$5*LN(1-RAND())	=C27+D27-E27	=C27+G26	=G27/B27
28		4	=MAX(0,F27)	=-$B$4*LN(1-RAND())	=-$B$5*LN(1-RAND())	=C28+D28-E28	=C28+G27	=G28/B28
29		5	=MAX(0,F28)	=-$B$4*LN(1-RAND())	=-$B$5*LN(1-RAND())	=C29+D29-E29	=C29+G28	=G29/B29
30		6	=MAX(0,F29)	=-$B$4*LN(1-RAND())	=-$B$5*LN(1-RAND())	=C30+D30-E30	=C30+G29	=G30/B30
31		7	=MAX(0,F30)	=-$B$4*LN(1-RAND())	=-$B$5*LN(1-RAND())	=C31+D31-E31	=C31+G30	=G31/B31
32		8	=MAX(0,F31)	=-$B$4*LN(1-RAND())	=-$B$5*LN(1-RAND())	=C32+D32-E32	=C32+G31	=G32/B32
33		9	=MAX(0,F32)	=-$B$4*LN(1-RAND())	=-$B$5*LN(1-RAND())	=C33+D33-E33	=C33+G32	=G33/B33
34		10	=MAX(0,F33)	=-$B$4*LN(1-RAND())	=-$B$5*LN(1-RAND())	=C34+D34-E34	=C34+G33	=G34/B34

■ **FIGURE 4-32** Spreadsheet formulas for queuing simulation

	A	B
1		
2	Simple Queueing Simulation	
3		
4	Mean Service Time	0.7
5	Mean Interarrival time	1
6	Waiting time in Queue	1.633333333
7		
8	Sample Average	1.1872560
9	Sample Variance	3.4886921
10	StdDev	1.86780
11	Count	1000
12	StdError	0.0590652
13	conf level	0.95000
14	degrees of freedom	999.00000
15	alpha for mean CI	0.05000
16	t-value	1.962341416
17	half-width	0.11590599
18	CI Lower Limit for mean	1.07135003
19	CI Upper Limit for mean	1.30316201

■ **FIGURE 4-33**   Results across 1000 customers

and column H holds the cumulative average

$$\frac{\sum_{i=1}^{n} W_i}{n} \quad \text{for } n = 1, 2, \ldots$$

Thus, cell H26 is the average of the first two customers, cell H27 is the average of the first three customers, and so forth. Thus, the final cell of column H represents the average of all of the customer waiting times.

Figure 4-33 shows the results of the simulation across 1000 customers. The analytical results indicate that the true long-run expected waiting time in the queue is 1.633 minutes. The average over the 1000 customers in the simulation is 1.187 minutes. Figure 4-33 indicates that the sample average is significantly lower than the true expected value. Figure 4-34 presents the

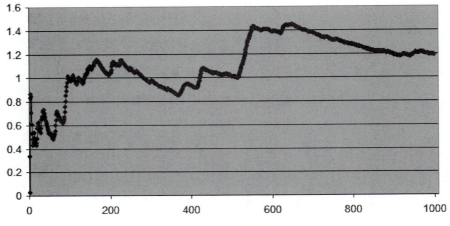

Cumulative Average Waiting Time In Q

■ **FIGURE 4-34**   Cumulative average waiting time of 1000 customers

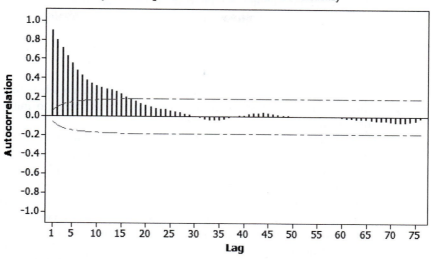

■ **FIGURE 4-35** Autocorrelation plot for waiting times

cumulative average plot of the first 1000 customers. As seen in the plot, the cumulative average starts out low and then eventually trends toward 1.2 minutes.

The first issue to consider with these data is independence. To do this, you should analyze the 1000 observations in terms of the autocorrelation. From Figure 4-35, it is readily apparent that the data has strong positive correlation using the methods discussed in Chapter 3. The lag-1 correlation for these observations is estimated to be about 0.9. Figure 4-36 clearly indicates the strong first-order linear dependence between W_i and W_{i-1}. This positive dependence implies that if the previous customer waited a long time the next customer is likely to wait a long time. If the previous customer had short wait, then the next customer is likely to have a short wait. This

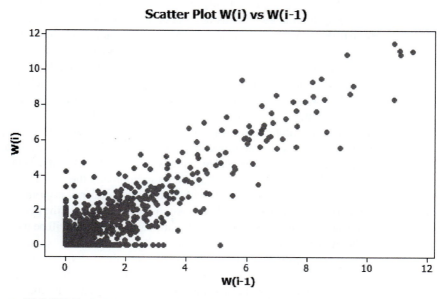

■ **FIGURE 4-36** Scatterplot of W_{i-1} versus W_i

makes sense with respect to how a queue operates. This correlation has serious implications when developing confidence intervals on the mean customer waiting time because the usual estimator for the sample variance,

$$S^2(n) = \frac{\sum_{i=1}^{n}\left(X_i - \overline{X}\right)^2}{n-1}$$

is a biased estimator for the true population variance when there is correlation in the observations. This issue will be re-examined when ways to mitigate these problems are discussed.

The second issue that needs to be discussed is that of the nonstationary behavior of the data. As discussed in Chapter 3, nonstationary data indicates some dependence on time. More generally, nonstationary implies that the $W_1, W_2, W_3, \ldots, W_n$ are not obtained from identical distributions. Why should the distribution of W_1 not be the same as the distribution of W_{1000}? The first customer is likely to enter the queue with no previous customers present and thus it is very likely that the first customer will experience little or no wait (the way W_0 was initialize in this example allows a chance of waiting for the first customer). However, the 1000th customer may face an entirely different situation. Between the first and the 1000th customer there might likely be a line formed. In fact from the M/M/1 formula, it is known that the "steady-state" expected number in the queue is 1.633. Clearly, the conditions that the first customer faces are different than the 1000th customer. Thus, the distributions of their waiting times are likely to be different. Recall the definition of covariance stationary from Chapter 3. A time series, $(X_1, X_2, \ldots X_n)$, is said to be *covariance stationary* if the following conditions hold:

- The mean exists and $\theta = E[X_i]$, for $i = 1, 2, \cdots, n$
- The variance exists and $Var[X_i] = \sigma^2 > 0$, for $i = 1, 2, \cdots, n$
- The lag-k autocorrelation, $\rho_k = cor(X_i, X_{i+k})$, is not a function of i; that is, the correlation between any two points in the series does not depend on where the points are in the series, it depends only on the distance between them in the series.

In the case of the customer waiting times, you should conclude from the discussion that it is very likely that $\theta \neq E[W_i]$ and $Var[W_i] \neq \sigma^2$ for *each* $i = 1, 2, \cdots, n$ for the time series.

Do you think that is it likely that the distributions of W_{9999} and W_{10000} will be similar? The argument, that the 9999th customer is on average likely to experience similar conditions as the 10,000th customer, sure seems reasonable. Figure 4-37 shows 10 different replications of the cumulative average for a 10,000-customer simulation. This was developed using the sheet labeled *10000-Customers*. From the figure, you can see that the cumulative average plots can vary significantly over the 10000 customers with the average tracking above the true expected value, below the true expected value, and possibly toward the true expected value. Each of these plots was generated by pressing the F9 function key to have the spreadsheet recalculate with new random numbers and then capturing the observations to another sheet for plotting. Essentially, this is like running a new replication. You are encouraged to try generating multiple sample paths using the provided spreadsheet. For the case of 10,000 customers, you should notice that the cumulative average starts to approach the expected value of the steady-state mean waiting time in the queue with increasing number of customers. This is the law of large numbers in action. It appears that it takes a period of time for the performance measure to *warm up* toward the true mean. Determining the warm-up period will be the basic way to mitigate the problem of nonidentical distributions.

From this discussion, you should conclude that the second basic statistical assumption of identically distributed data is not valid for within replication data. From this, you can also conclude that it is very likely that the data are not normally distributed. In fact, for the M/M/1 it can be shown that the steady-state distribution for the waiting time in the queue is not a normal

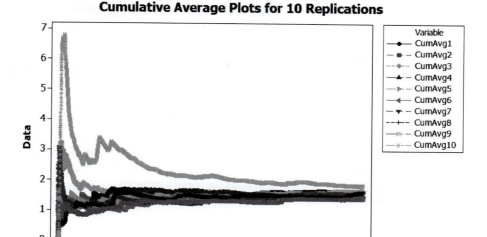

■ FIGURE 4-37   Multiple sample paths of queuing simulation

distribution. Thus, all three of the basic statistical assumptions are violated for the within replication data of this example. This problem needs to be addressed in order to properly analyze infinite-horizon simulations.

There are two basic methods for performing infinite-horizon simulations. The first is to perform multiple replications. This approach addresses independence and normality in a similar fashion as the finite-horizon case, but special procedures will be needed to address the nonstationary aspects of the data. The second basic approach is to work with one very long replication. Both of these methods depend on first addressing the problem of the nonstationary aspects of the data. The next section looks at ways to mitigate the nonstationary aspect of within-replication data for infinite-horizon simulations.

4.4.1   Assessing the Effect of Initial Conditions

Consider the output stochastic process $\{X_i\}$ of the simulation. Let $F_i(x \mid I)$ be the conditional cumulative distribution function of $\{X_i\}$, where I represents the initial conditions used to start the simulation at time 0. If $F_i(x \mid I) \to F(x)$ when $i \to \infty$, for all initial conditions I, then $F(x)$ is called the steady-state distribution of the output process (Law 2007). In infinite-horizon simulations, estimating parameters of the steady-state distribution, $F(x)$, such as the steady-state mean, θ, is often the key objective. The fundamental difficulty associated with estimating steady-state performance is that unless the system is initialized using the steady-state distribution (which is not known), there is no way to directly observe the steady-state distribution.

It is true that if the steady-state distribution exists and you run the simulation long enough, the estimators will tend to converge to the desired quantities. Thus, in infinite-horizon simulations, you must decide on how long to run the simulations and how to handle the effect of the *initial conditions* on the estimates of performance. The initial conditions of a simulation represent the state of the system when the simulation is started. For example, in simulating the pharmacy system, the simulation was started with no customers in service or in the line. This is referred to as "empty and idle." The initial conditions of the simulation affect the rate of convergence of estimators of steady-state performance.

Because the distributions $F_i(x \mid I)$ at the start of the replication tend to depend more heavily on the initial conditions, estimators of steady-state performance such as the sample average, $\overline{X} = (1/n)\sum_{i=1}^{n} X_i$, will tend to be *biased*. A point estimator, $\hat{\theta}$, is an *unbiased* estimator of the parameter of interest, θ, if $E[\hat{\theta}] = \theta$. That is, if the expected value of the sampling distribution is equal to the parameter of interest, then the estimator is said to be unbiased. If the estimator is biased, then the difference, $E[\hat{\theta}] - \theta$, is called the bias of the estimator, $\hat{\theta}$. Note that any individual difference between the true parameter, θ, and a particular observation, X_i, is called error, $\varepsilon_i = X_i - \theta$. If the expected value of the errors is not zero, then there is bias. A particular observation is not biased. Bias is a property of the estimator. Bias is analogous to being consistently off target when shooting at a bull's-eye. It is as if the sights on your gun are crooked. In order to estimate the bias of an estimator, you must have multiple observations of the estimator. Suppose that you are estimating the mean waiting time in the queue as per the previous example and that the estimator is based on the first 20 customers. That is, the estimator is

$$\overline{W}_r = \frac{\sum_{i=1}^{20} W_{ir}}{20}$$

and there are $r = 1, 2, \ldots 10$ replications. Table 4-1 shows the sample average waiting time for the first 20 customers for 10 different replications. In the table, B_r is an estimate of the bias for the rth replication, where $W_q = 1.6\overline{33}$. Upon averaging across the replications, it can be seen that $\overline{B} = -0.9536$, which indicates that the estimator based only on the first 20 customers has significant negative bias; that is, on average it is less than the target value.

This is the so called *initialization bias problem* in steady-state simulation. Unless the initial conditions of the simulation can be generated according to $F(x)$, which is not known, you must focus on methods that detect and/or mitigate the presence of initialization bias. One strategy for initialization bias mitigation is to find an index, d, for the output process, $\{X_i\}$, so that $\{X_i; i = d + 1, \ldots\}$ will have substantially similar distributional properties as the steady-state distribution, $F(x)$. This is called the simulation warm-up problem, where d is called the *warm-up point*, and $\{i = 1, \ldots, d\}$ is called the *warm-up period* for the simulation. Then the estimators of steady-state performance are based only on $\{X_i; i = d + 1, \ldots\}$. For example, when estimating the steady-state mean waiting time for each replication r, the estimator can be computed as

$$\overline{W}_r = \frac{\sum_{i=d+1}^{n} W_{ir}}{n - d}$$

■ **TABLE 4-1**
Ten Replications of 20 Customers

r	$\overline{W}_r$	$B_r = \overline{W}_r - W_q$
1	0.194114	−1.43922
2	0.514809	−1.11852
3	1.127332	−0.506
4	0.390004	−1.24333
5	1.05056	−0.58277
6	1.604883	−0.02845
7	0.445822	−1.18751
8	0.610001	−1.02333
9	0.52462	−1.10871
10	0.335311	−1.29802
	$\overline{\overline{W}} = 0.6797$	$\overline{B} = -0.9536$

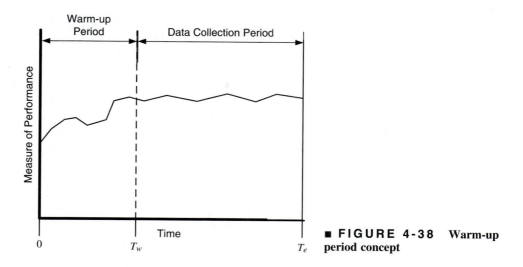

■ **FIGURE 4-38** **Warm-up period concept**

For time-based performance measures, such as the average number in queue, a time T_w can be determined past which the data collection process can begin. Estimators of time-based performance such as the sample average are computed as

$$\overline{Y}_r = \frac{\int_{T_w}^{T_e} Y_r(t)dt}{T_e - T_w}$$

Figure 4-38 shows the concept of a warm-up period for a simulation replication. When you perform a simulation in Arena^{TM}, you can easily specify a time-based warm-up period using the Run > Setup > Replication Parameters panel. In fact, even for observation-based data, it will be more convenient to specify the warm-up period in terms of time. A given value of T_w implies a particular value of d and vice versa. Specifying a warm-up period in Arena^{TM} causes Arena^{TM} to schedule a warm-up event for time T_w. At that time, all the accumulated statistical counters are cleared so that the net effect is that statistics are only collected over the period from T_w to T_e. The problem then becomes one of finding an appropriate warm-up period.

Before proceeding with how to assess the length of the warm-up period, the concept of steady state needs to be further examined. This subtle concept is often misunderstood or misrepresented. Often you will hear the phrase, "The system has reached steady state." The correct interpretation of this phrase is that the distribution of the desired performance measure has reached a point where it is sufficiently similar to the desired steady-state distribution. Steady state is a concept involving the performance measures generated by the system as time goes to infinity. However, sometimes this phrase is interpreted incorrectly to mean that the system *itself* has reached steady state. Let me state emphatically that the system *never* reaches steady state. If the system itself reached steady state, then by implication it would never change with respect to time. It should be clear that the system continues to evolve with respect to time; otherwise, it would be a very boring system! Thus, it is incorrect to indicate that the system has reached steady state. Because of this, do not use the phrase, "The system has reached steady state."

Understanding this subtle issue raises an interesting implication concerning the notion of deleting data to remove the initialization bias. Suppose that the state of the system at the end of the warm-up period, T_w, is exactly the same as at $T = 0$. For example, it is certainly possible that at time T_w for a particular replication that the system was empty and idle. Since the state of the system at T_w is the same as that of the initial conditions, there will be no effect of deleting the warm-up period for this replication. In fact there will be a negative effect, in the sense that data will have been thrown away for no reason. Deletion methods are predicated on the likelihood that the state of the system seen at T_w is more representative of steady-state conditions. At the end of the warm-up period, *any of the possible* states of the system may be observed. Some states will be more likely than others. If multiple replications are made, then at T_w each replication may experience a different set of conditions at T_w. Let $I^r_{T_w}$ be the "initial" conditions (state) at time T_w on replication r. By setting a warm-up period and performing multiple replications, you are in essence sampling from the distribution governing the state of the system at time T_w. If T_w is long enough, then on average across the replications, you will start collecting data in states that are "representative" of steady-state conditions.

Many methods and rules have been proposed to determine the warm-up period. The interested reader is referred to Wilson and Pritsker (1978), Lada et al. (2003), Litton and Harmonsoky (2002), White et al. (2000), Cash et al. (1992), and Rossetti and Delaney (1995) for an overview of such methods. This discussion will concentrate on the visual method proposed in Welch (1983b). The basic idea behind Welch's graphical procedure is simple:

1. Make R replications. Typically, $R \geq 5$ is recommended.
2. Let Y_{ri} be the ith observation within replication r for $i = 1, 2, \cdots, n_r$ and $r = 1, 2, \cdots, R$.
3. Compute the averages across the replications for each $i = 1, 2, \cdots, n$.

$$\overline{Y}_{\bullet i} = \frac{\sum_{r=1}^{R} Y_{ri}}{R}$$

4. Plot $\overline{Y}_{\bullet i}$ for each $i = 1, 2, \cdots, n$.
5. Apply smoothing techniques to $\overline{Y}_{\bullet i}$, $i = 1, 2, \cdots, n$.
6. Visually assess where the plots start to converge.

Let's apply Welch's procedure to the replications generated from the Lindley equation simulation. Using the 10 replications stored on sheet *10Replications*, compute the average across each replication for each customer. In Figure 4-39, cell B2 represents the average across the 10 replications for the first customer. Column D represents the cumulative average associated with column B.

Figure 4-40 is the plot of the cumulative average (column D) superimposed on the averages across replications (column B). The cumulative average is one method of "smoothing" the data. From the plot, you can infer that after about customer 3000, the cumulative average has "settled down." Thus, from this analysis, you might infer that $d = 3000$.

When you perform an infinite-horizon simulation by specifying a warm-up period and make multiple replications, you are using the method of *replication-deletion*. If the method of

	B2	▼	f_x =AVERAGE(E2:N2)						
	A	B	C	D	E	F	G	H	I
1		Average	Sum	CumAvg	Rep 1	Rep 2	Rep 3	Rep 4	Rep 5
2	1	0.548433	0.548433	0.548432719	0	0.337306	0	1.79165	0.716249
3	2	0.359163	0.907596	0.453798017	0.213726	0.427596	0	1.892545	0
4	3	0.793544	1.70114	0.567046583	0.428974	1.498441	0	2.921952	1.373886
5	4	0.679799	2.380939	0.595234777	0.141178	2.265391	0	1.559509	1.523948
6	5	0.772271	3.15321	0.630642059	0.255458	0.839788	0	1.930919	1.569935
7	6	0.598509	3.75172	0.625286585	0.022953	0.909185	0	1.677104	1.460226
8	7	0.760241	4.51196	0.644565755	0	0.917162	0	0.152849	1.979963
9	8	0.879366	5.391326	0.673915781	0	0.350159	0	0	3.596641
10	9	1.169247	6.560573	0.728952544	0.161749	0.457508	0	0	2.767113
11	10	0.897993	7.458566	0.745856589	0.351444	1.45561		0	1.213456

■ **FIGURE 4-39** Computing the averages for Welch plot

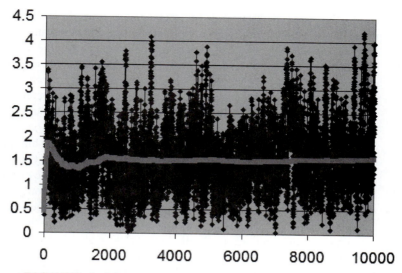

■ **FIGURE 4-40**   Welch plot with superimposed cumulative average line

replication-deletion with $d = 3000$ is used for the current example, a slight reduction in the bias can be achieved as indicated in Table 4-2.

While not definitive for this simple example, the results suggest that deleting the warm-up period helps to reduce initialization bias. This model's warm-up period will be further analyzed in the next section using the tools available in Arena™.

In performing the method of replication-deletion, a fundamental trade-off occurs. Because observations are deleted, the variability of the estimator will tend to increase while the bias will tend to decrease. This is a trade-off between a reduction in bias and an increase in variance. That is, accuracy is being traded off against precision when deleting the warm-up period. In addition to this trade-off, data from each replication is also being thrown away. This takes computational time that could be expended more effectively on collecting usable data. Another disadvantage of performing replication-deletion is that the techniques for assessing the warm-up period

■ **TABLE 4-2**
Replication-Deletion Results, $d = 3000$

r	$\overline{W}_r(d = 0)$	$\overline{W}_r(d = 3000)$	$B_r(d = 0)$	$B_r(d = 3000)$
1	1.594843	1.592421	−0.03849	−0.04091
2	1.452237	1.447396	−0.1811	−0.18594
3	1.657355	1.768249	0.024022	0.134915
4	1.503747	1.443251	−0.12959	−0.19008
5	1.606765	1.731306	−0.02657	0.097973
6	1.464981	1.559769	−0.16835	−0.07356
7	1.621275	1.75917	−0.01206	0.125837
8	1.600563	1.67868	−0.03277	0.045347
9	1.400995	1.450852	−0.23234	−0.18248
10	1.833414	1.604855	0.20008	−0.02848
	$\overline{\overline{W}} = 1.573617$	$\overline{\overline{W}} = 1.603595$	$\overline{B} = -0.05972$	$\overline{B} = -0.02974$
	$s = 0.1248$	$s = 0.1286$	$s = 0.1248$	$s = 0.1286$
95% LL	1.4843	1.5116	−0.149023	−0.121704
95% UL	1.6629	1.6959	−0.029590	0.062228

(especially graphical) may require significant data storage. The Welch plotting procedure requires the saving of data points for post-processing after the simulation run. In addition, significant time by the analyst may be required to perform the technique and the technique is subjective.

When a simulation has many performance measures, you may have to perform a warm-up period analysis for every performance measure. This is particularly important, since in general, the performance measures of the same model may converge toward steady-state conditions at different rates. In this case, the length of the warm-up period must be sufficiently long to cover all the performance measures.

Finally, replication-deletion may simply compound the bias problem if the warm-up period is insufficient relative to the length of the simulation. If you have not specified a long enough warm-up period, you are potentially compounding the problem for R replications. Despite all these disadvantages, replication-deletion is commonly used in practice because of the simplicity of the analysis after the warm-up period has been determined. Once you are satisfied that you have a good warm-up period, the analysis of the results is the same as that of finite-horizon simulations. Replication-deletion also facilitates the use of experimental design techniques that rely on replicating design points. In addition, replication-deletion facilitates the use of such tools as Arena's Process Analyzer and OptQuest. The Process Analyzer will be discussed later in this chapter and OptQuest in subsequent chapters. The next section illustrates how to perform the method of replication-deletion on this simple M/M/1 model in Arena^{TM}.

4.4.2   Performing the Method of Replication-Deletion in Arena^{TM}

The first step in performing the method of replication-deletion is to determine the length of the warm-up period. This example illustrates how to:

- Save the values from observation and time-based data to files in Arena^{TM} for post processing in the Output Analyzer
- Assess the warm-up period using the Output Analyzer
- Make Welch plots based on Arena^{TM} data using Excel and Microsoft Access
- Set up and run the multiple replications
- Interpret the results

The file *CH4-MM1-ReplicationDeletion.doe* contains the M/M/1 model with the time between arrivals $\exp(E[Y_i] = 1.0)$ and the service times $\exp(E[X_i] = 0.7)$ set in the CREATE and DELAY modules respectively. Figure 4-41 shows the overall flow of the model. Since the waiting times in the queue need to be captured, an ASSIGN module has been used to mark the time that the customer arrives to the queue. Once the customer exits the SEIZE module they have been removed from the queue to start service. At that point a RECORD module is used to capture the time interval representing the queuing time.

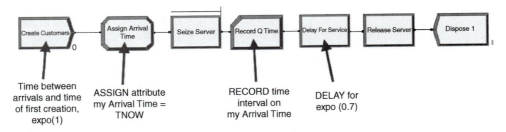

■ **FIGURE 4-41**   **M/M/1 Arena model**

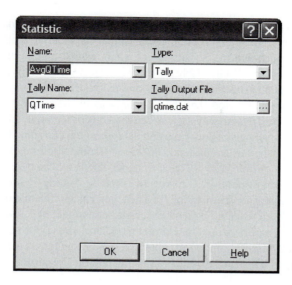

■ **FIGURE 4-42** Capturing observation-based data to a file

Using the STATISTIC data module, the observations collected within the replications can be captured to files for post processing by the Output Analyzer. This example will collect the queue time of each customer and the number of customers in the queue over time. Figure 4-42 shows how to add a statistic using the STATISTIC data module that will capture observation-based data (e.g., queue time) to a file. The Type of the statistic is Tally and the Tally Name corresponds to the name specified in the RECORD module used to capture the observations. When a name for the output file is specified, Arena™ will store every observation and the time of the observation to a file with the extension (*.dat*). These files are not human-readable, but can be processed by the Output Analyzer.

Figure 4-43 shows how to create a statistic to capture the number in queue over time to a file. The type of the statistic is indicated as Time-Persistent, and an expression must be given that represents the value of the quantity to be observed through time. In this case, the expression builder has been used to select the current number of entities in the Seize Server.Queue (using the NQ(queue ID) function). When the Output File is specified, the value of the expression and the time of the observation are written to a (*.dat*) file.

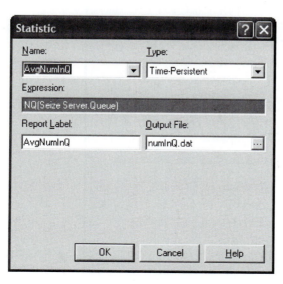

■ **FIGURE 4-43** Capturing time-based data to a file

When performing a warm-up period analysis, the first decision to make is the length of each replication. In general, there is very little guidance that can be offered other than to try different run lengths and check for the sensitivity of your results. Within the context of queuing simulations, the work by Whitt (1989) offers some ideas on specifying the run length, but these results are difficult to translate to general simulations. Since the purpose here is to determine the length of the warm-up period, then the run length should be bigger than what you suspect the warm-up period to be. In this analysis, it is better to be conservative. You should make the run length as long as possible given your time and data storage constraints. Banks et al. (2005) offer the rule of thumb that the run length should be at least 10 times the amount of data deleted. That is, $n \geq 10d$, or in terms of time, $T_e \geq 10T_w$. Of course, this is a "catch-22" situation because you need to specify n, or equivalently, T_e, in order to assess T_w. Setting T_e very large is recommended when doing a preliminary assessment of T_w. Then, you can use the rule of thumb of 10 times the amount of data deleted when doing a more serious assessment of T_w (e.g., using Welch plots).

A preliminary assessment of the current model has already been performed based on the previously described Excel simulation. That assessment suggested a deletion point of at least $d = 3000$ customers. This can be used as a starting point in the current effort. Now, T_w needs to be determined based on d. The value of d represents the customer number for the end of the warm-up period. To get T_w, you need to answer the question: How long (on average) will it take for the simulation to generate d observations. In this model, the mean number of arrivals is 1 customer per minute. Thus, the initial T_w is

$$3000 \ customers \times \frac{minute}{1 \ customer} = 3000 \ minutes$$

and therefore, the initial T_e should be 30,000 minutes. Specify 30,000 minutes for the replication length and 10 replications on the Run > Setup > Replication Parameters as shown in Figure 4-44.

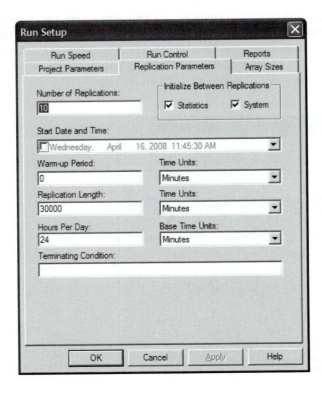

■ **FIGURE 4-44**   Replication parameters tab

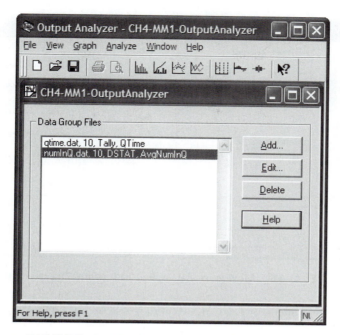

■ **FIGURE 4-45**   Output Analyzer data group

The replication length is the time when the simulation will end. The Base Time Units can be specified for how Arena^{TM} will report statistics on the default reports and represents the units for TNOW. The Time Units for the Warm-up Period and the Replication Length can also be set. The Warm-up period is the time where the statistics will be cleared. Thus, data will be reported over net (Replication Length – Warm-up Period) time units. Running the simulation will generate two files *numInQ.dat* and *qtime.dat* in the current working directory for the Arena^{TM} model.

4.4.3   Looking for the Warm-up Period in the Output Analyzer

Open up the Output Analyzer and create a new data group. You should then add the two files *numInQ.dat* and *qtime.dat* to the data group.

As previously discussed, the Welch plot is a reasonable approach to assessing the warm-up period. Unfortunately, the Output Analyzer does not automatically perform a Welch plot analysis. The best that can be done with the Output Analyzer is to look at each replication individually using the Moving Average command from the Plot menu as shown in Figure 4-46. In this command, the Output Analyzer allows data to be smoothed for the replication selected by the moving average, exponential, or cumulative options. The Moving option allows a moving average smoothing technique to be applied to the data. The exponential option allows for the application of an exponential moving-average technique. The cumulative option superimposes the cumulative average on the plot of the data. Figure 4-47 shows the cumulative average plot for the first replication for the waiting times in the queue. From this plot, it appears that the warm-up period is at least 4000 time units (minutes); however, this is just one replication. You can make a plot for each replication and try to "eyeball" the warm-up time on each plot. This approach can also be applied to time-based data after *filtering* the data into batches.

Time-based data are saved in an Arena^{TM} file such that the time of the observation and the value of the state variable at the time of change are recorded. Thus, the observations are not equally spaced in time. In order to apply the moving average command, you need to cut the data into discrete equally spaced intervals of time as illustrated in Figure 4-48. Suppose that you

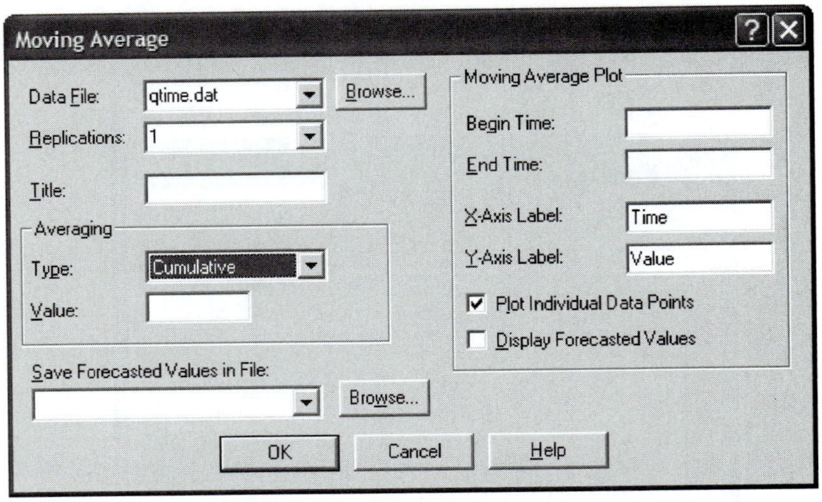

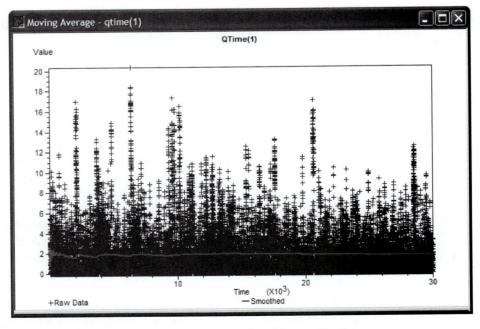

■ **FIGURE 4-46**   Moving average option with cumulative average selected

divide T_e into k intervals of size Δt, so that $T_e = k \times \Delta t$. The time average over the jth interval is given by:

$$\overline{Y}_{rj} = \frac{\int_{(j-1)\Delta t}^{j\Delta t} Y_r(t)dt}{\Delta t}$$

Thus, the overall time average can be computed from the time average associated with each interval as shown below:

$$\overline{Y}_r = \frac{\int_0^{T_e} Y_r(t)dt}{T_e} = \frac{\int_0^{T_e} Y_r(t)dt}{k\Delta t} = \frac{\sum_{j=1}^k \int_{(j-1)\Delta t}^{j\Delta t} Y_r(t)dt}{k\Delta t} = \frac{\sum_{j=1}^k \overline{Y}_{rj}}{k}$$

■ **FIGURE 4-47**   Cumulative average plot of first replication

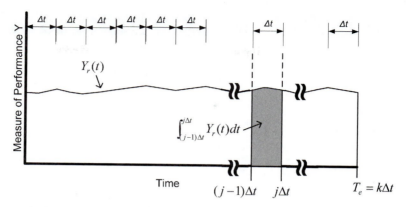

■ **FIGURE 4-48**   Time-based data

Each of the $\overline{Y}_{rj}$ are computed over intervals of time that are equally spaced and can be treated as if they are observation-based data.

The computation of the $\overline{Y}_{rj}$ for time-based data can be achieved in the Output Analyzer by using the Batch/Truncate Obs'ns option in the Analyze menu. Figure 4-49 illustrates the Batch/Truncate dialog. Using this dialog, you can select the replication you want to batch, the type of batching (time based or observation based), and the size of the batch (either in time units or number of observations). Since the number in queue data is time based, time-based batches are selected, and the batch size is specified in terms of time. In this case, the observations are being batched based on a time interval of 10 minutes.

This produces a file *numInQfilter.flt*, which contains the $\overline{Y}_{rj}$ as observations. This file can then be added to the Output Analyzer's data group and processed using the Moving Average command from the Plot menu as previously illustrated. The resulting plot is shown in Figure 4-50.

You can make a plot for each replication and try to "eyeball" the warm-up time on each plot; however, as previously discussed, a Welch plot may be more effective in helping this process. To make a Welch plot, you need to process the data in Excel or Microsoft Access in order to compute the averages across the replications.

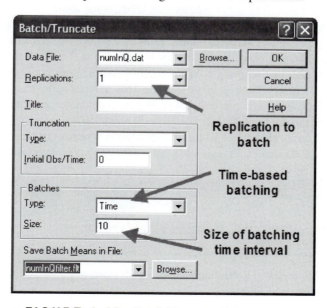

■ **FIGURE 4-49**   Batch/Truncate dialog

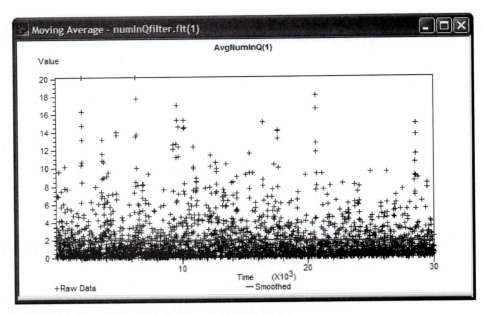

■ **FIGURE 4-50**   Plot of filtered time-based data

In order to use Excel or Access to process the recorded data from the simulation, you must direct Arena™ to produce a text file rather than a *dat* file. There are two changes that need to be made in Arena™ to facilitate this process. The first is to tell Arena™ to write output files as text using the Run Setup > Run Control > Advanced dialog as shown in Figure 4-52. The second is to change the file extension to "csv" in the STATISTICS module as shown in Figure 4-51. This will facilitate Excel recognizing the file as a comma-separate-variable (CSV) file. Thus, when you double-click on the file in Windows Explorer, the file will be opened in Excel. There is one caveat with respect to using Excel. Excel is limited to 65,536 rows of data.[1] If you have n observations per replication and r replications, then $n \times r + (r + 7)$ needs to be less than 65,536. You should use Access to process the data if you have a large amount of data. The analysis using both Excel and Access will be demonstrated.

Figure 4-53 illustrates the structure of the resulting text file that is produced by Arena™ when using the Write Statistics Output Files as Text option. Each time a new observation is recorded (either time based or observation based), Arena™ records the time of the observation and the value of the observation. When multiple replications are executed, Arena™ uses the number (-1) in the time field to indicate that a replication has ended. Using these file characteristics, software can be written to post-process the *csv* files in any fashion that is required for the analysis.

To illustrate the use of Excel, the simulation was run for three replications with a run length of 10,000 minutes. This should produce about 30,000 observations of the waiting time in

	Name	Type	Tally	Tally Output File	Expression	Report Label	Output File
1	AvgQTime	Tally	QTime	qtime.csv		AvgQTime	
2	AvgNumInQ	Time-Persistent	Tally 3		NQ(Seize Server.Queue)	AvgNumInQ	numInQ.csv
	Double-click here to add a new row.						

Statistic - Advanced Process

■ **FIGURE 4-51**   Renaming files with CSV extension

[1]Excel 2007 permits more than 1 million rows.

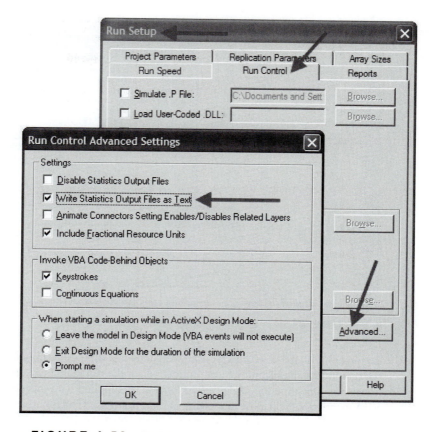

■ **FIGURE 4-52** Writing statistics output files as text

	A	B	C	
1	Project:	Unnamed Project		
2	User:		#Replications	
3	Data item:	QTime		
4	Run date:	7/24/2006		
5	Options:	YDT	3	
6				
7	Time	Observation		
8	0.346152	0		
9	0.791373	0		
10	0.889847	0.03830269		
11	2.888587	1.91214956		
12	4.178476	2.41026417		
13	6.350378	2.66263546		
9985	9998.654	1.52816907		
9986	9998.997	1.5158467		
9987	9999.233	1.46073417	End of	
9988	-1	0	replication	
9989	0.107442	0	indicator	

■ **FIGURE 4-53** Resulting CSV file as shown in Excel

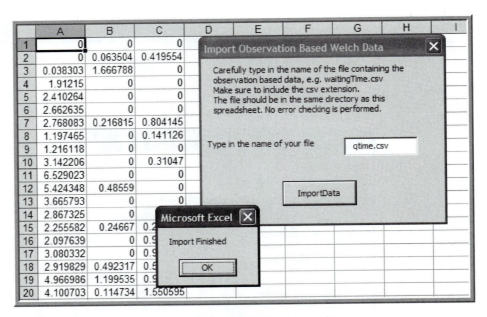

■ **FIGURE 4-54**   Importing observation data into Excel

the queue. With these settings, you are well under the limitations imposed by Excel, but you can see that increasing the number of replications or the run length can easily exceed Excel's row limitation. Open the file *MakeWelchPlotNew.xls* supplied with this chapter. This file contains VBA macros that will allow importing observation-based and time-based data values generated from Arena™. Using Tools > Macro > Macros, open the defined macros for the workbook and select the macro called *ShowObservationBasedWelchForm*. This will bring up an Excel VBA form for importing the data. You should then type in the name of the file containing the observations. The macro does no error checking so make sure that you type in the name of the file correctly (with the *csv* extension) and that the file is located in the same directory as the workbook. Once you press the Import button, the observations from each replication will be imported into columns starting at column A of a worksheet named WelchData as shown in Figure 4-54. The other macro, *ShowTimeBasedWelchForm*, will allow you to import time-based data into Excel. When using this form, you must provide a batching interval so that the time-based data can be batched into equal intervals. Then the analysis is the same except that when you identify a deletion point, it will be based on the number of intervals, from which you can then compute the time of the warm-up period by multiplying the number of intervals times the length of the interval.

   After the import is completed, you should scroll down through the values. You will see that the number of observations per replication is not the same. When you build the Welch plot, you will want to include only those points where you have values for each replication. In this particular example, the last row containing values for each replication is row 9980. To build the Welch plot, you need to compute the average across replications for each observation. It is also useful to compute the cumulative average for plotting it on top of the averages. The spreadsheet *MakeWelchPlotNew.xls* has the completed Welch plot on the worksheet labeled WelchPlot. To make this plot, the imported observations were used. Then, columns were created to hold the observation index, the average across the replications, the running sum of the across replication averages, and the cumulative sum of the across replication averages as shown in Figure 4-55. Then, an Excel scatterplot of the data was made using the observation index as the x-axis series and the average and cumulative average for the other series of the chart. The plot indicates that after about number 4000, there is a decrease in the cumulative average and that after about

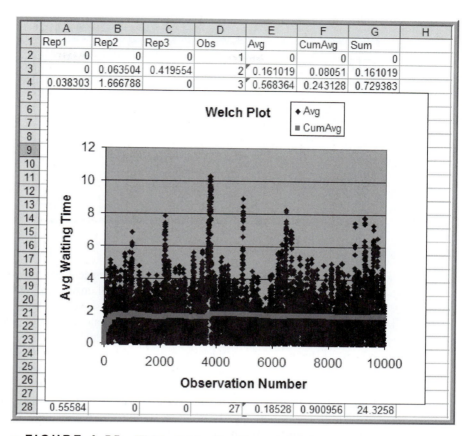

	A	B	C	D	E	F	G	H
1	Rep1	Rep2	Rep3	Obs	Avg	CumAvg	Sum	
2	0	0	0	1	0	0	0	
3	0	0.063504	0.419554	2	0.161019	0.08051	0.161019	
4	0.038303	1.666788	0	3	0.568364	0.243128	0.729383	
28	0.55584	0	0	27	0.18528	0.900956	24.3258	

■ **FIGURE 4-55**   **Welch plot for Excel imported data**

8000, there does not appear to be much change. For these observations, additional analysis can be performed. In addition, if more replications are available, you will more likely be able to discern a pattern. VBA code was also developed for Microsoft Access to allow the processing of larger data sets. The database file *MakeWelchPlot.mdb* that accompanies this chapter can be used for analysis of larger data sets.

The processing of the data in Microsoft Access is very similar to that done in Excel except that queries can be used to compute many of the quantities. To illustrate the use of Access, the simulation was re-executed with 30 replications with a replication length of 30,000 minutes. This will generate very large files for *numInQ.csv* and *qtime.csv* (at least 22 MB of data). For the sake of convenience, these files were renamed *numInQ30R.csv* and *qtime30R.csv*. Open the database file *MakeWelchPlot.mdb* (answer No when asked to block expressions, and Yes to the unblocked expression warning). Within Access, navigate to the Forms and open the form named *frmImportObsBased*. This form will walk you through the steps to import the data. Since the files are so large, you should expect a significant delay as the observations are read in and processed. The import process will create a table called Observations, which contains the time of the observation, the observation, the observation number, and the replication associated with the observation as shown in Figure 4-56.

Selecting the MakeWelchData button will cause the creation of a table with the averages across the replications for each observation as shown in Figure 4-57. You will then be offered the opportunity to mark the records in the WelchData table into batches. This process assigns each observation to a batch to facilitate plotting in Access. This also computes the cumulative average column. You now have everything that you need to make Welch plots. This data can be

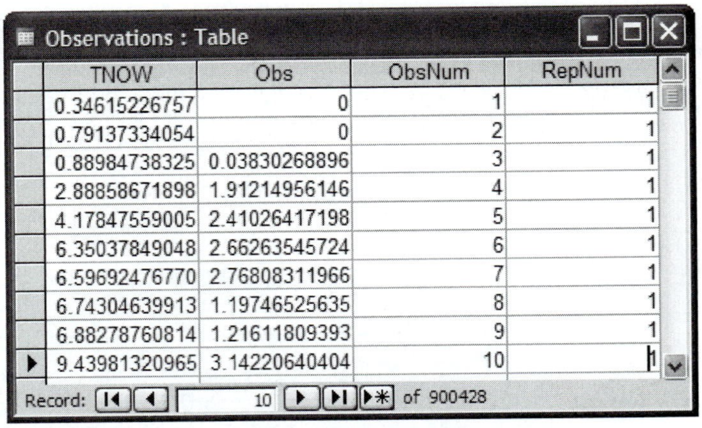

Observations : Table

TNOW	Obs	ObsNum	RepNum
0.34615226757	0	1	1
0.79137334054	0	2	1
0.88984738325	0.03830268896	3	1
2.88858671898	1.91214956146	4	1
4.17847559005	2.41026417198	5	1
6.35037849048	2.66263545724	6	1
6.59692476770	2.76808311966	7	1
6.74304639913	1.19746525635	8	1
6.88278760814	1.21611809393	9	1
9.43981320965	3.14220640404	10	1

Record: 10 of 900428

■ **FIGURE 4-56 Observations table after import**

exported to other plotting packages or plots can be made using Access's charting capabilities. Unfortunately, at the time of this writing, the size of charts in Access is limited to 4000 data points. To collapse the data further by averaging across batches, you can use the Make Batched Welch Data Table button, which takes the assigned batches and computes the average for each batch producing *BatchWelchDataTable* as shown in Figure 4-58. After the entire import process has been completed, the importing form should resemble the one shown in Figure 4-59.

The Report section of the database has a number of premade reports that will show charts of the data as illustrated in Figure 4-60. Unfortunately, the chart size limitation of Access will truncate the plot after only 4000 observations. The figure indicates that the observations are less variable because they have been averaged over 30 replications. It is also clear that is the observations are centered around 1.6, and that at least 2000 data points are needed for the warm-up period.

Figure 4-61 illustrates a plot of the data from the WelchData table using MINITAB. This graph includes all the observations. For illustrative purposes, the analytical answer of 1.63 has been included on the graph. From this graph, it is clear that after 6000 data points, the Welch plot has settled down. Based on all analyses, at least 6000 data points should be used for the

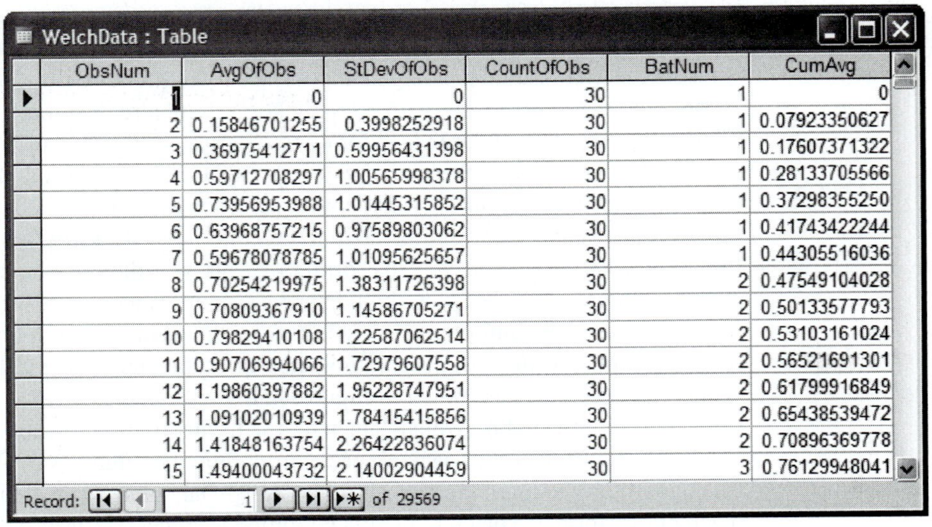

WelchData : Table

ObsNum	AvgOfObs	StDevOfObs	CountOfObs	BatNum	CumAvg
1	0	0	30	1	0
2	0.15846701255	0.3998252918	30	1	0.07923350627
3	0.36975412711	0.59956431398	30	1	0.17607371322
4	0.59712708297	1.00565998378	30	1	0.28133705566
5	0.73956953988	1.01445315852	30	1	0.37298355250
6	0.63968757215	0.97589803062	30	1	0.41743422244
7	0.59678078785	1.01095625657	30	1	0.44305516036
8	0.70254219975	1.38311726398	30	2	0.47549104028
9	0.70809367910	1.14586705271	30	2	0.50133577793
10	0.79829410108	1.22587062514	30	2	0.53103161024
11	0.90706994066	1.72979607558	30	2	0.56521691301
12	1.19860397882	1.95228747951	30	2	0.61799916849
13	1.09102010939	1.78415415856	30	2	0.65438539472
14	1.41848163754	2.26422836074	30	2	0.70896369778
15	1.49400043732	2.14002904459	30	3	0.76129948041

Record: 1 of 29569

■ **FIGURE 4-57 WelchData table**

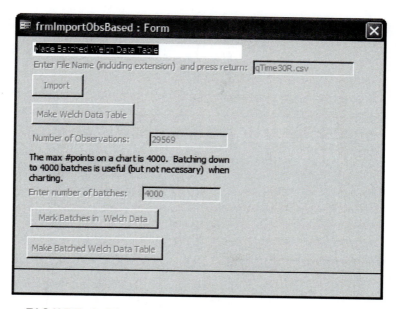

■ **FIGURE 4-58**   Importing observation-based data form

warm-up period. To repeat the process on another data file, you must either delete or rename the tables (Observation, WelchData, BatchWelchDataTable). To reassign observations in WelchData to different batches (i.e., to re-batch the data), you can use the macro called *MarkBatchNumbersInWelchDataTable*. This macro calls a VBA function that requires the desired number of observations per batch.

Once you have performed the warm-up analysis, you still need to use your simulation model to estimate system performance. Because the process of analyzing the warm-up period involves saving the data you could use the already saved data to estimate system performance after truncating the initial portion of the data from the data sets. Arena's Output Analyzer

BatNum	AvgOfAvgOfObs	CumAvg
1	0.44305516036	0.44305516036
2	0.97487223519	0.70896369778
3	1.48210633283	0.96667790946
4	1.28369342375	1.04593178803
5	1.56587888426	1.14992120728
6	1.31879948855	1.17806758749
7	1.34084573780	1.20132160896
8	1.80004622952	1.27616218653
9	1.45954992211	1.2965386016
10	1.35987011345	1.30287175278
11	1.65848389806	1.33520012963
12	1.51605364164	1.35027125563
13	1.64511306143	1.37295139453
14	1.60635840576	1.38962332391
15	1.73317539592	1.41252679538

Record: I◄ ◄ 1 ► ►I ►* of 4225

■ **FIGURE 4-59**   Batch averages and cumulative average

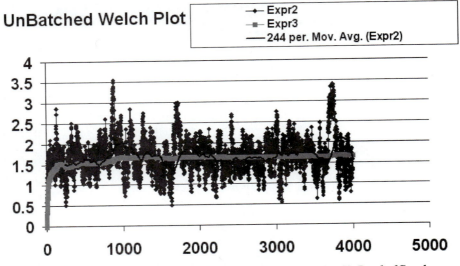

■ **FIGURE 4-60** Result of running report *CumulativeAvgUnBatchedGraph*

facilitates this via the Batch/Truncate dialog, which allows you to truncate the data. If the analysis was done in Excel or Access, it is simply a matter of deleting the portion of the observations contained in the warm-up period. If rerunning the simulation is relatively inexpensive, then you can simply set the warm-up period in the Run Setup > Replication Parameters dialog and execute the model. Following the rule of thumb that the length of the run should be at least 10 times the warm-up period, the simulation was rerun with the settings given

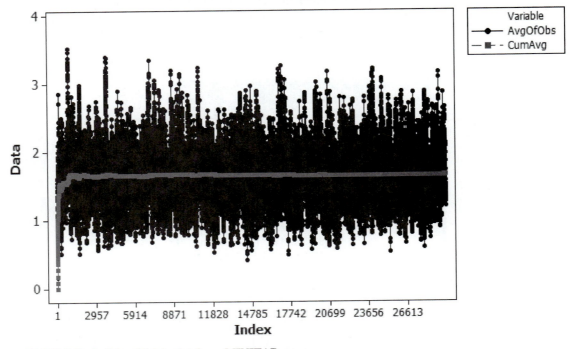

■ **FIGURE 4-61** Welch plot from MINITAB

Waiting Time	Average	Half Width
Seize Server.Queue	1.6354	0.02

Other

Number Waiting	Average	Half Width
Seize Server.Queue	1.6355	0.02

■ **FIGURE 4-62**   **Results based on 30 replications**

in Figure 4-63 (30 replications, 6000-minute warm-up period, 60,000-minute replication length). The results shown in Figure 4-62 indicate the absence of significant bias with these replication settings. The true waiting time in the queue is $1.6\overline{33}$, and it is clear that the 95% confidence interval produced by Arena^{TM} contains this value.

The process described here for determining the warm-up period for steady-state simulation is tedious and time consuming. Research into automating this process is still an active area of investigation. The recent work by Robinson (2005) and Rossetti et al. (2005) holds some promise in this regard; however, the need to integrate these methods into computer simulation software remains. Even though determining the warm-up period is tedious, some consideration of the warm-up period should be done for infinite-horizon simulations.

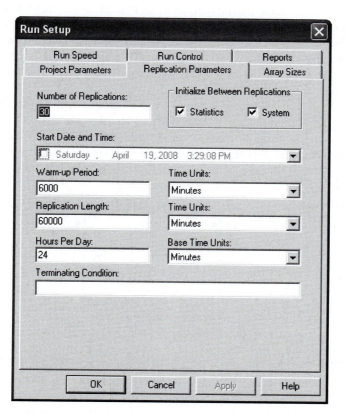

■ **FIGURE 4-63**   **Replication-deletion settings for the example**

Once the warm-up period has been found, you can set the Warm-up Period field in the Run Setup > Replication Parameters dialog to the time of the warm-up. Then you can use the method of replication-deletion to perform your simulation experiments (Figure 4-63). Thus, all the discussion previously presented regarding analysis of finite-horizon simulations can be applied. When determining the number of replications, you can apply the fixed sample size procedure after performing a pilot run. If the analysis indicates that you need to make more runs to meet your confidence interval half-width, you have two alternatives: (1) increase the number of replications, or (2) keep the same number of replications but increase the length of each replication. If n_0 was the initial number of replications and n is the number of replications recommended by the sample size determination procedure, then you can instead set T_e equal to $(n/n_0)T_e$ and run n_0 replications. Thus, you will still have approximately the same amount of data collected over your replications, but the longer run length may reduce the effect of initialization bias.

As previously mentioned, the method of replication-deletion causes each replication to "throw away" the initial portion of the run. As an alternative, you can make one long run and delete the initial portion only once. When analyzing an infinite-horizon simulation based on a single long replication, a method is needed to address the correlation present in the within replication data. The batch means method is often used in this case and has been automated in Arena^{TM}. The next section discusses the statistical basis for the batch means method and addresses some of the practical issues of using it in Arena^{TM}.

4.4.4   Batch Means Method

In the batch means method, only one simulation run is executed. After deleting the warm-up period, the remainder of the run is divided into k batches, with each batch average representing a single observation as illustrated in Figure 4-64.

The advantages of the batch means method are that it entails a long simulation run, thus dampening the effect of the initial conditions. The disadvantage is that the within replication data are correlated and unless properly formed the batches may also exhibit a strong degree of correlation.

The following presentation assumes that a warm-up analysis has already been performed and that the observations that have been collected occur after the warm-up period. For simplicity, the presentation assumes observation-based data. The discussion also applies to time-based data that has been cut into discrete equally spaced intervals of time as described in

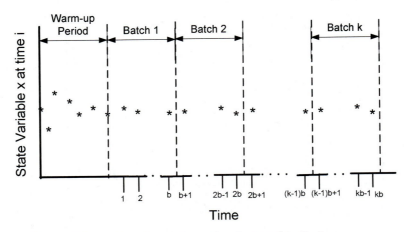

■ **FIGURE 4-64**   **Illustration of batch means method**

Section 4.4.3. Therefore, assume that a series of observations, $(X_1, X_2, X_3, \ldots, X_n)$, is available from within the single long replication after the warm-up period. As shown earlier at the beginning of Section 4.4, within replication data can be highly correlated. In that section, it was mentioned that standard confidence intervals based on

$$S^2(n) = \frac{\sum_{i=1}^{n}(X_i - \overline{X})^2}{n-1}$$

are not appropriate for this type of data. Suppose you were to ignore the correlation, what would be the harm? In essence, a confidence interval implies a certain level of confidence in the decisions based on the confidence interval. When you use $S^2(n)$ as defined above, you will not achieve the desired level of confidence because $S^2(n)$ is a biased estimator for the variance of $\overline{X}$ when the data are correlated. Under the assumption that the data are covariance stationary, an assessment of the harm in ignoring the correlation can be made. For a series that is covariance stationary, one can show that

$$Var(\overline{X}) = \frac{\gamma_0}{n}\left[1 + 2\sum_{k=1}^{n-1}\left(1 - \frac{k}{n}\right)\rho_k\right]$$

where $\gamma_0 = Var(X_i)$, $\gamma_k = Cov(X_i, X_{i+k})$, and $\rho_k = \gamma_k/\gamma_0$ for $k = 1, 2, \ldots, n-1$. When the data are correlated, S^2/n is a biased estimator of $Var(\overline{X})$. To show this, you need to compute the expected value of S^2/n

$$E\left[\frac{S^2}{n}\right] = \frac{\gamma_0}{n}\left[1 - \frac{2R}{n-1}\right] \text{ where } R = \sum_{k=1}^{n-1}\left(1 - \frac{k}{n}\right)\rho_k$$

Bias is defined as the difference between the expected value of the estimator and the quantity being estimated. In this case, the bias can be computed with some algebra as

$$Bias = E\left[\frac{S^2}{n}\right] - Var(\overline{X}) = \frac{-2\gamma_0 R}{n-1}$$

Since $\gamma_0 > 0$ and $n > 1$, the sign of the bias depends on the quantity R and thus on the correlation. There are three cases to consider: zero correlation, negative correlation, and positive correlation. Since $-1 \leq \rho_k \leq 1$, examining the limiting values for the correlation will determine the range of the bias. One can easily show that for negative correlation, $-1 \leq \rho_k \leq 0$, the bias will be positive $(0 \leq Bias \leq \gamma_0)$ and for positive correlation, $0 \leq \rho_k \leq 1$, the bias will be negative, $(-\gamma_0 \leq Bias \leq 0)$. Thus, the bias is negative if the correlation is positive, and the bias is positive if the correlation is negative. In the case of positive correlation, S^2/n underestimates the $Var(\overline{X})$. Thus, using S^2/n to form confidence intervals will make the confidence intervals too short. You will have unjustified confidence in the point estimate in this case. The true confidence will not be the desired $1 - \alpha$. Decisions based on positively correlated data will have a higher than planned risk of making an error based on the confidence interval. In the case of negatively correlated data, S^2/n overestimates the $Var(\overline{X})$. A confidence interval based on S^2/n will be too wide and the true quality of the estimate will be better than indicated. The true confidence coefficient will not be the desired $1 - \alpha$, but rather greater than $1 - \alpha$. Of the two cases, the positively correlated case is the more severe in terms of its effect on the decision-making process; however, both are problems. Thus, the naïve use of S^2/n for dependent data is highly unwarranted. If you want to build confidence intervals on $\overline{X}$, you need to find an unbiased estimator of the $Var(\overline{X})$.

The batch means method provides a way to develop (at least approximately) an unbiased estimator for $Var(\overline{X})$. Assuming that you have a series of data points, the batch means method divides the data into subsequences of contiguous batches,

$$\underbrace{X_1, X_2, \ldots, X_b}_{batch\ 1} \underbrace{X_{b+1}, X_{b+2}, \ldots, X_{2b}}_{batch\ 2} \cdots \underbrace{X_{(j-1)b+1}, X_{(j-1)b+2}, \ldots, X_{jb}}_{batch\ j} \cdots \underbrace{X_{(k-1)b+1}, X_{(k-1)b+2}, \ldots, X_{kb}}_{batch\ k}$$

and computes the sample average of the batches. Let k be the number of batches, each consisting of b observations, so that $k = \lfloor n/b \rfloor$. If b is not a divisor of n, then the last $(n - kb)$ data points will not be used. Define $\overline{X}_j(b) = \frac{1}{b}\sum_{i=1}^{b} X_{(j-1)b+i}$ as the jth batch mean for $j = 1, 2, \ldots, k$.

Each of these batch means are treated like observations in the batch means series. For example, if the batch means are relabeled as $Y_j = \overline{X}_j(b)$, the batching process simply produces another series of data, $(Y_1, Y_2, Y_3, \ldots, Y_k)$, which may be more like a random sample. To form a $(1 - \alpha)\%$ confidence interval, you simply treat this new series like a random sample and compute approximate confidence intervals using the sample average and sample variance of the batch means series, as follows:

$$\overline{Y}(k) = \frac{1}{k}\sum_{j=1}^{k} Y_j \quad S_b^2(k) = \frac{\sum_{j=1}^{k}(Y_j - \overline{Y})^2}{k - 1} \quad \overline{Y}(k) \pm t_{\alpha/2, k-1}\frac{S_b(k)}{\sqrt{k}}$$

Since the original $X's$ are covariance stationary, it follows that the resulting batch means are also covariance stationary. One can show (see Alexopoulos and Seila 1998), that the correlation in the batch means reduces as both the size of the batches, b, and the number of data points, n, increases. In addition, one can show that $S_b^2(k)/k$ approximates $Var(\overline{X})$ with error that reduces as both b and n increase toward infinity.

The basic difficulty with the batch means method is determining the batch size or alternatively the number of batches. Larger batch sizes are good for independence but reduce the number of batches, resulting in higher variance for the estimator. Schmeiser (1982) performed an analysis suggesting that there is little benefit if the number of batches is larger than 30, and recommended that the number of batches remain in the range from 10 to 30. However, when trying to access whether the batches are independent, it is better to have a large number of batches (>100) so that tests on the lag-k correlation have better statistical properties.

Various procedures have been developed that will automatically batch the data as the observations are collected; see, for example, Fishman and Yarberry (1997), Steiger and Wilson (2002), and Banks et al. (2005). Arena^{TM} has its own batching algorithm. Arena's batching algorithm is described in Kelton et al. (2006, 311). See also Fishman (2001, 254) for an analysis of the effectiveness of Arena's algorithm. The discussion here is based on the description in Kelton et al. (2006). When Arena^{TM} has recorded a sufficient amount of data, it begins by forming $k = 20$ batches. As more observations are collected, additional batches are formed until $k = 40$ batches are collected. When 40 batches are formed, Arena^{TM} collapses the number of batches back to 20 by averaging each pair of batches. This has the net effect of doubling the batch size. This process is repeated as more observations are collected, thereby ensuring that the number of batches is between 20 and 39. Arena^{TM} begins the formation of batches when it has at least 320 observations of observation-based data. For time-based data, Arena^{TM} requires that there were at least five time units during which the time-based variable changed 320 times. If there are not enough observations in a run, then Arena^{TM} reports "Insufficient" for the half-width value on the output reports. In addition, Arena^{TM} also tests to see if the lag-1 correlation is significant by testing the hypothesis that the batch means are uncorrelated using the following test statistic (see Alexopoulos and Seila 1998):

$$C = \sqrt{\frac{k^2 - 1}{k - 2}} \left[\hat{\rho}_1 + \frac{[Y_1 - \overline{Y}]^2 + [Y_k - \overline{Y}]^2}{2\sum_{j=1}^{k}(Y_j - \overline{Y})^2} \right]$$

$$\hat{\rho}_1 = \frac{\sum_{j=1}^{k-1}(Y_j - \overline{Y})(Y_{j+1} - \overline{Y})}{\sum_{j=1}^{k}(Y_j - \overline{Y})^2}$$

The hypothesis is rejected if $C > z_\alpha$ for a given confidence level, α. If the batch means do not pass the test, Arena^{TM} reports "Correlated" for the half-width on the statistical reports.

4.4.5 Performing the Method of Batch Means in Arena^{TM}

Performing the batch means method in Arena^{TM} is relatively straightforward. The following assumes that a warm-up period analysis has already been performed. Since batches are formed during the simulation run and the confidence intervals are based on the batches, the primary concern will be to determine the run length that will ensure a desired half-width on the confidence intervals. A fixed sampling-based method and a sequential sampling method will be illustrated.

The analysis performed to determine the warm-up period should give you some information concerning how long to make this single run and how long to set its warm-up period. Assume that a warm-up analysis has been performed using n_0 replications of length T_e, and that the analysis has indicated a warm-up period of length T_w. As previously discussed, the method of replication-deletion spreads the risk of choosing a "bad" initial condition across multiple replications. The batch means method relies on only a single replication. If you were satisfied with the warm-up period analysis based on n_0 replications and you were going to perform replication-deletion, then you are willing to throw away the observations contained in at least $n_0 \times T_w$ time units and you are willing to use the data collected over $n_0 \times (T_e - T_w)$ time units. Therefore, the warm-up period for the single replication can be set at $n_0 \times T_w$ and the run length can be set at $n_0 \times T_e$. For example, suppose your warm-up analysis was based on the initial results shown in Section 4.4.3, that is, $n_0 = 10, T_e = 30000, T_w = 3000$. Thus, your starting run length would be $n_0 \times T_e = 10 \times 30,000 = 300,000$ and the warm-up period will be $n_0 \times T_w = 30000$. For these settings, the results shown in Figure 4-65 are very close to the results for the replication-deletion example.

Waiting Time	Average	Half Width
Seize Server.Queue	1.6225	0.061752921
Other		
Number Waiting	Average	Half Width
Seize Server.Queue	1.6241	0.064281088

■ **FIGURE 4-65** Initial batch means results

Statistic - Advanced Process

	Name	Type	Expression	Report Label
1	NBforQT	Output	TNUMBAT(QTime)	NBforQT
2	BSforQT	Output	TBATSIZ(QTime)	BSforQT
	Double-click here to add a new row.			

■ **FIGURE 4-66** **OUTPUT statistics to get number of batches and batch size**

Suppose now you want to ensure that the half-widths from a single replication are less than a given error bound. The half-widths reported by Arena™ for a single replication are based on the *batch means*. You can get an approximate idea of how much to increase the length of the replication by using Arena's functions TNUMBAT(Tally ID) and TBATSIZ(Tally ID) for observation-based statistics, or DNUMBAT(DSTAT ID) and DBATSIZ(DSTAT ID) in conjunction with the half-width, sample-size determination formula:

$$n \cong n_0 \frac{h_0^2}{h^2}$$

In this case, you interpret n and n_0 as the number of batches. OUTPUT statistics can be added to the model to observe the number of batches for the waiting time in queue and for the size of each batch, as shown in Figure 4-66. The resulting values for the number of batches formed for the waiting times and the size of the batches are given in Figure 4-67. Using this information in the half-width based sample size formula with $n_0 = 32$, $h_0 = 0.06$, and $h = 0.02$, yields

$$n \cong n_0 \frac{h_0^2}{h^2} = 32 \times \frac{(0.06)^2}{(0.02)^2} = 288 \text{ batches.}$$

Since each batch in the run had 8192 observations, this yields the need for 288 × 8192 = 2,359,296 additional observations for the waiting time in the queue. Since, in this model, customers arrive at a mean rate of 1 per minute, this requires about 2,359,296 additional time units of simulation. Because of the warm-up period, you therefore need to set T_e equal to (2,359,296 + 30,000 = 2389296). Rerunning the simulation yields the results shown in Figure 4-68. The results show that the half-width meet the desired criteria. This approach is approximate since the characteristics of the batches in the final run may be different than those used in the initial runs.

Rather than trying to fix the amount of sampling, you might instead try to use a sequential sampling technique that is based on the half-width computed by Arena™ during the simulation run. This is easy to do by supplying the appropriate expression in the Terminating Condition field on the Run Setup > Replication Parameters dialog. Figure 4-69 illustrates that you can use a Boolean expression in the Terminating Condition field. In this case, the THALF(Tally ID) function is used to specify that the simulation should terminate when the half-width criterion is

Output	Value
BSforQT	8192.00
NBforQT	32.0000

■ **FIGURE 4-67** **Results for number of batches and batch size**

Waiting Time	Average	Half Width
Seize Server.Queue	1.6403	0.017292365
Other		
Number Waiting	Average	Half Width
Seize Server.Queue	1.6422	0.018253294

■ **FIGURE 4-68**   **Batch means results for fixed sample size**

met. Arena's batching mechanism computes the value of THALF(Tally ID) after sufficient observations have been observed. This expression can be expanded to include other performance measures in a compound Boolean statement.

The results of running the simulation based on the sequential method are given in Figure 4-70. In this case, the simulation run ended at approximately time 1,928,385. This is lower than the time specified for the fixed sampling procedure (but the difference is not excessive).

Once the warm-up period has been analyzed, performing infinite-horizon simulations using the batch means method in Arena™ is relatively straightforward. A disadvantage of this method is that it will be more difficult to use the statistical methods available in Arena's Process Analyzer or in OptQuest because they assume a replication-deletion approach. If you are faced with an infinite-horizon simulation, then you can use either the replication-deletion approach or

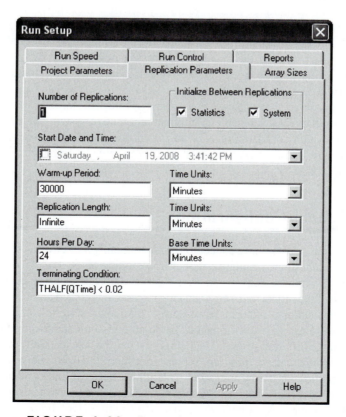

■ **FIGURE 4-69**   **Sequential sampling using terminating condition**

Waiting Time	Average	Half Width
Seize Server.Queue	1.6366	0.0199291 23

Other

Number Waiting	Average	Half Width
Seize Server.Queue	1.6384	0.021182914

■ **FIGURE 4-70**   **Results for infinite-horizon sequential sampling method**

the batch means method readily from within Arena^TM. In either case, you should investigate for problems related to initialization bias. If you use the replication-deletion approach, you should play it safe when specifying the warm-up period. Making the warm-up period longer than you think it should be is better than replicating a poor choice. When performing an infinite-horizon simulation based on one long run, you should make sure that your run length is long enough. A long run length can help to "wash out" the effects of initial condition bias.

Ideally, in the situation where you have to make many simulation experiments using different parameter settings of the same model, you should perform a warm-up analysis for each design configuration. In practice, this is not readily feasible when there are a large number of experiments. In this situation, you should use your common sense to pick the design configurations (and performance measures) that you feel will most likely suffer from initialization bias. If you can determine long enough warm-up periods for these configurations, the other configurations should be relatively safe from the problem by using the longest warm-up period found.

A number of other techniques have been developed for the analysis of infinite-horizon simulations, including the standardized time series method, the regenerative method, and spectral methods. An overview of these methods and others can be found in Alexopoulos and Seila (1998) and Law (2007).

Thus far you have learned how to analyze the data from one design configuration. A key use of simulation is to be able to compare alternative system configurations and to assist in choosing which configurations are best according to the decision criteria. The next section presents how to use simulation and the tools available in Arena^TM to compare various system configurations.

4.5   Comparing System Configurations

The previous sections have concentrated on estimating the performance of a system through the execution of a single simulation model. The running of the model requires the specification of the input variables (e.g., mean time between arrivals, service distribution, etc.), and the structure of the model (e.g., FIFO queue, process flow, etc.). The specification of a set of inputs (variables and/or structure) represents a particular system configuration, which is then simulated to estimate performance. To be able to simulate design configurations, you may have to build different Arena^TM models or you may be able to use the same model supplied with different values of the program inputs. In either situation, you now have different design configurations that can be compared. This allows the performance of the system to be estimated under a wide variety of controlled conditions. It is this ability to easily perform these "what-if" simulations that make simulation such a useful analysis tool. Figure 4-71 represents the notion of using different inputs to get different outputs.

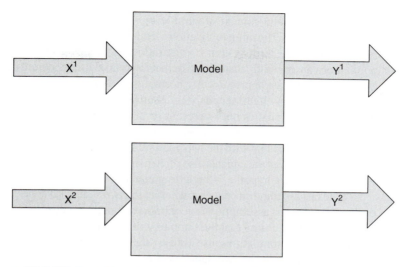

■ **FIGURE 4-71**   Multiple inputs on models represent different system configurations

Naturally, when you have different design configurations, you would like to know which configurations are better than others. Since the simulations are driven by random variables, the outputs from each configuration (e.g., Y^1, Y^2) are also random variables. The estimate of the performance of each system must be analyzed using statistical methods to ensure that the differences in performance are not simply due to sampling error. In other words, you want to be confident that one system is statistically better (or worse) than the other system.

4.5.1   Comparing Two Systems

The techniques for comparing two systems via simulation are essentially the same as that found in books that cover the statistical analysis of two samples (e.g., Montgomery and Runger 2006). This section begins with a review of these methods and then discusses them in the context of Arena^{TM}. Assume that samples from two different populations (system configurations) are available:

$$X_{11}, X_{12}, \ldots, X_{1n_1} \text{ a sample of size } n_1 \text{ from system configuration 1}$$

$$X_{21}, X_{22}, \ldots, X_{2n_2} \text{ a sample of size } n_2 \text{ from system configuration 2}$$

These samples represent a particular performance measure of the system that will be used in a decision regarding which system configuration is preferred. For example, the performance measure may be the average system throughput per day, and you want to pick the design configuration that has highest throughput. Assume that each system configuration has an unknown population mean for the performance measures of interest, $E[X_1] = \theta_1$ and $E[X_2] = \theta_2$. Thus, the problem is to determine, with some statistical confidence, whether $\theta_1 < \theta_2$ or alternatively $\theta_1 > \theta_2$. Since the system configurations *are* different, an analysis of the situation of whether $\theta_1 = \theta_2$ is of less relevance in this context. Define $\theta = \theta_1 - \theta_2$ as the mean difference in performance between the two systems. Clearly, if you can determine whether $\theta > 0$ or $\theta < 0$, you can determine whether $\theta_1 < \theta_2$ or $\theta_1 > \theta_2$. Thus, it is sufficient to concentrate on the difference in performance between the two systems.

Given samples from two different populations, there are a number of ways in which the analysis can proceed based on different assumptions concerning the samples. The first common assumption is that the observations in each sample for each configuration form a random sample. That is, the samples represent independent and identically distributed random variables. Within the context of simulation, this can be easily achieved for a given system configuration by performing replications. Thus, $X_{11}, X_{12}, \ldots, X_{1n_1}$ are the observations from n_1 replications of the first system configuration. A second common assumption is that both populations are normally distributed or that the central limit theorem can be used so that sample averages are at least approximately normal.

To proceed with further analysis, assumptions concerning the population variances must be made. Many statistics books present results for the case of the population variance being known. In general, this is not the case in simulation contexts, so the assumption here will be that the variances associated with the two populations are unknown. The assumption on whether the population variances are equal is also often made. Rather than directly making that assumption, it is better to test a hypothesis regarding equality of population variances.

The last assumption concerns whether the two samples can be considered independent of each other. This last assumption is very important in the context of simulation. Unless you take specific actions to ensure that the samples will be independent, they will, in fact, be dependent because of how simulations use (reuse) the same random number streams. The possible dependence between the two samples is not necessarily a bad thing. In fact, under certain circumstance it can be a good thing.

The following section first presents the methods for analyzing the case of unknown variance with independent samples. Then, it focuses on the case of dependence between the samples. Finally, how to use Arena™ to do the work of the analysis will be illustrated.

4.5.1.1 Analyzing Two Independent Samples.

Although the variances are unknown, the unknown variances are either equal or not equal. In the situation where the variances are equal, the observations can be pooled when developing an estimate for the variance. In fact, rather than just assuming equal or not equal variances, you can (and should) use an F-test to test for the equality of variance. The F-test can be found in most elementary probability and statistics books (see Montgomery and Runger 2006). The decision regarding whether $\theta_1 < \theta_2$ can be addressed by forming confidence intervals on $\theta = \theta_1 - \theta_2$. Let $\overline{X}_1, \overline{X}_2, S_1^2,$ and S_2^2 be the sample averages and sample variances based on the two samples ($k = 1, 2$):

$$\overline{X}_k = \frac{1}{n_k} \sum_{j=1}^{n_k} X_{kj}$$

$$S_k^2 = \frac{1}{n_k - 1} \sum_{j=1}^{n_k} (X_{kj} - \overline{X}_k)^2$$

An estimate of $\theta = \theta_1 - \theta_2$ is desired, which can be achieved with $\hat{D} = \overline{X}_1 - \overline{X}_2$. To form confidence intervals on $\hat{D} = \overline{X}_1 - \overline{X}_2$, an estimator for the variance of $\hat{D} = \overline{X}_1 - \overline{X}_2$ is required. Because the samples are independent, the computation of the variance of the difference is

$$Var(\hat{D}) = Var(\overline{X}_1 - \overline{X}_2) = \frac{\sigma_1^2}{n_1} + \frac{\sigma_2^2}{n_2}$$

where σ_1^2 and σ_2^2 are the unknown population variances. Under the assumption of equal variance, this can be written as

$$Var(\hat{D}) = Var(\overline{X}_1 - \overline{X}_2) = \frac{\sigma_1^2}{n_1} + \frac{\sigma_2^2}{n_2} = \sigma^2 \left(\frac{1}{n_1} + \frac{1}{n_2} \right)$$

where σ^2 is the common unknown variance. A pooled estimator of σ^2 can be defined as

$$S_p^2 = \frac{(n_1 - 1)S_1^2 + (n_2 - 1)S_2^2}{n_1 + n_2 - 2}$$

Thus, a $(1 - \alpha)\%$ confidence interval on $\theta = \theta_1 - \theta_2$ is

$$\hat{D} \pm t_{\alpha/2, \nu} s_p \sqrt{\frac{1}{n_1} + \frac{1}{n_2}} \quad \text{where } \nu = n_1 + n_2 - 2$$

For the case of unequal variances, an approximate $(1 - \alpha)\%$ confidence interval on $\theta = \theta_1 - \theta_2$ is given by

$$\hat{D} \pm t_{\alpha/2, \nu} \sqrt{\frac{S_1^2}{n_1} + \frac{S_2^2}{n_2}} \quad \text{where } \nu = \left\lfloor \frac{\left(\frac{S_1^2}{n_1} + \frac{S_2^2}{n_2}\right)^2}{\frac{\left(S_1^2/n_1\right)^2}{n_1 + 1} + \frac{\left(S_2^2/n_2\right)^2}{n_2 + 1}} - 2 \right\rfloor$$

Let $[l, u]$ be the resulting confidence interval where l and u represent the lower and upper limits of the interval, via the construction $l < u$. Thus, if $u < 0$, you can conclude with $(1 - \alpha)\%$ confidence that $\theta = \theta_1 - \theta_2 < 0$ (i.e., that $\theta_1 < \theta_2$). If $l > 0$, you can conclude with $(1 - \alpha)\%$ that $\theta = \theta_1 - \theta_2 > 0$ (i.e., that $\theta_1 > \theta_2$). If $[l, u]$ contains 0, then no conclusion can be made at the given sample sizes about which system is better. This does not indicate that the system performance is the same for the two systems. You *know* that the systems are different. Thus, their performance will be different. This only indicates that you have not taken enough samples to detect the true difference. If sampling is relatively cheap, then you may want to take additional samples in order to discern an ordering between the systems.

The confidence interval can assist in making decisions regarding relative performance of the systems from a *statistically significant* standpoint. However, if you make a conclusion about the ordering of the system, it still may not be practically significant. That is, the difference in the system performance is statistically significant but the actual difference is of no practical use. For example, suppose you compare two systems in terms of throughput with resulting output $\overline{X}_1 = 5.2$ and $\overline{X}_2 = 5.0$, and the difference is statistically significant. While the difference of 0.2 may be statistically significant, you might not be able to achieve this in the actual system. After all, you are making a decision based on a *model of the system* not on the real system. If the costs of the two systems are appreciably different, you should prefer the cheaper of the two systems since there is no practical difference between the two systems. The fidelity of the difference is dependent on your modeling assumptions. Other modeling assumptions may overshadow such a small difference.

The notion of practical significance is model and performance measure dependent. One way to characterize the notion of practical significance is to conceptualize a zone of performance for which you are indifferent between the two systems. Figure 4-72 illustrates the concept of an indifference zone around the difference between the two systems. If the difference between the two systems falls in this zone, you are indifferent between the two systems (i.e., there is no practical difference).

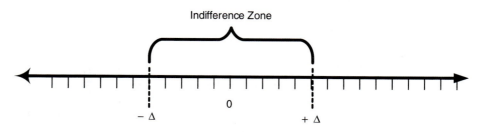

■ **FIGURE 4-72** **Indifference zone**

Using the indifference zone to model the notion of practical significance, if $u < -\Delta$, you can conclude with confidence that $\theta_1 < \theta_2$, and if $l > \Delta$, you can conclude with confidence that $\theta_1 > \theta_2$. If l falls in the indifference zone and u does not (or vice versa), then there is not enough evidence to confidently reach a conclusion. If $[l, u]$ is totally contained in the indifference zone, then you can conclude with confidence that there is no practical difference between the two systems.

4.5.1.2 Analyzing Two Dependent Samples.

In this situation, continue to assume that the observations in a sample are independent and identically distributed random variables; however, the samples are not independent. That is, assume that the $(X_{11}, X_{12}, \ldots, X_{1n_1})$ and $(X_{21}, X_{22}, \ldots, X_{2n_2})$ from the two systems are dependent. For simplicity, suppose configuration 2 is a simple parameter change in the model. First, run the model to produce $(X_{11}, X_{12}, \ldots, X_{1n_1})$ for configuration 1. Then, change the parameter and re-execute the model to produce $(X_{21}, X_{22}, \ldots, X_{2n_2})$ for configuration 2. Assuming that you did nothing with respect to the random number streams, the second configuration used the same random numbers that the first configuration used. Thus, the generated responses will be correlated (dependent). In this situation, it is convenient to assume that each system is run for the same number of replications, that is, $n_1 = n_2 = n$. Since each replication for the two systems uses the same random number streams, the correlation between $(X_{1,j}, X_{2,j})$ will not be zero; however, each pair will still be independent *across* the replications. The basic approach to analyzing this situation is to compute the difference for each pair:

$$D_j = X_{1,j} - X_{2,j} \quad \text{for } j = 1, 2, \ldots, n$$

The $(D_1, D_2, \ldots, D_n)$ will form a random sample, which can be analyzed via traditional methods. Thus, a $(1 - \alpha)\%$ confidence interval on $\theta = \theta_1 - \theta_2$ is

$$\overline{D} \pm t_{\alpha/2, n-1} \frac{S_D}{\sqrt{n}} \quad \text{where } \overline{D} = \frac{1}{n} \sum_{j=1}^{n} D_j \quad \text{and } S_D^2 = \frac{1}{n-1} \sum_{j=1}^{n} \left(D_j - \overline{D}\right)^2$$

The interpretation of the resulting confidence interval $[l, u]$ is the same as in the independent sample approach. Of the two approaches (independent versus dependent) samples, the latter is much more prevalent in simulation contexts. The approach is called the method of *common random numbers* and is a natural by product of how most simulation languages handle their assignment of random number streams.

To understand why this method is the preferred method for comparing two systems, you need to understand the method's effect on the variance of the estimator. In the case of independent samples, the estimator of performance was $\hat{D} = \overline{X}_1 - \overline{X}_2$. Since

$$\overline{D} = \frac{1}{n} \sum_{j=1}^{n} D_j = \frac{1}{n} \sum_{j=1}^{n} \left(X_{1,j} - X_{2,j}\right) = \frac{1}{n} \sum_{j=1}^{n} X_{1,j} - \frac{1}{n} \sum_{j=1}^{n} X_{2,j} = \overline{X}_1 - \overline{X}_2 = \hat{D}$$

the two estimators are the same, when $n_1 = n_2 = n$; however, their variances are not the same. Under the assumption of independence, computing the variance of the estimator yields

$$V_{IND} = Var(\overline{X}_1 - \overline{X}_2) = \frac{\sigma_1^2}{n} + \frac{\sigma_2^2}{n}$$

Under the assumption that the samples are not independent, the variance of the estimator is

$$V_{CRN} = Var(\overline{X}_1 - \overline{X}_2) = \frac{\sigma_1^2}{n} + \frac{\sigma_2^2}{n} - 2 \operatorname{cov}\left(\overline{X}_1, \overline{X}_2\right)$$

If you define $\rho = corr\left(\overline{X}_1, \overline{X}_2\right)$, the variance for the common random number situation is $V_{CRN} = V_{IND} - 2\sigma_1\sigma_2\rho$. Therefore, whenever there is positive correlation in the pairs resulting in $\rho > 0$, $V_{CRN} < V_{IND}$. If the variance of the estimator in the case of common random numbers is smaller than the variance of the estimator under independent sampling, then a *variance reduction* has been achieved. The method of common random numbers is called a variance reduction technique. If the variance reduction results in a confidence interval for θ that is tighter than the independent case, the use of common random numbers should be preferred. The variance reduction needs to be big enough to overcome any loss in the number of degrees of freedom caused by the pairing. When the number of replications is relatively large ($n > 30$) this will generally be the case since the student-t value does not vary appreciatively for large degrees of freedom. Note that the method of common random numbers might backfire and cause a variance increase if there is negative correlation between the pairs. An overview of the conditions under which common random numbers may work is given in Law (2007).

This notion of pairing the outputs from each replication for the two system configurations makes common sense. When trying to discern a difference, you want the two systems to experience the same randomness so that you can more readily infer that any difference in performance is due to the inherent difference between the systems and not caused by the random numbers. In experimental design settings, this is called blocking on a factor. For example, if you wanted to perform an experiment to determine whether a change in a work method was better than the old method, you should use the same worker to execute both methods. If instead you had different workers execute the methods, you could not be certain that the resulting difference was due to the workers or to the proposed change in the method. In this context, the worker is the factor that should be "blocked." In the simulation context, the random numbers are being "blocked" when using common random numbers.

4.5.1.3 Example of CRN in Arena™. Example 4.2 explores how independent sampling and common random numbers in Arena™ can be implemented. This example returns to the LOTR Makers system. This example also illustrates the use of resources sets in Arena™ modeling.

EXAMPLE 4-2 | **LOTR MAKERS, INC. CRN EXAMPLE**

The company is interested in reducing the amount of overtime. They have noticed that a bottleneck forms at the ring-making station early in the production release and that the rework station is not sufficiently utilized. Thus, they would like to test the sharing of the rework worker between the two stations. The rework worker should be assigned to both the rework station and the ring-making station. If a pair of rings arrives to the ring-making station, the master ring maker or the rework worker can make the rings. If both ring makers are available, the master ring maker should be preferred. If a pair of rings needs rework, the rework worker should give priority to the rework. The rework worker is not as skilled as the master ring maker and the time to make the rings varies depending on the maker. The master ring maker still takes UNIF(5,15) minutes to make rings; however, the shared rework worker has a much more variable process time. The rework worker can make rings in 30 minutes according to an exponential distribution. In addition, the rework worker must walk from station to station to perform the tasks and needs a bit more time to change from one task to the next. It is estimated that it will take the rework work an extra 10 minutes per task. Thus, the rework worker's ring-making time is on average 10 + EXPO(30) minutes, and the rework worker's time to process rework is now 15 + WEIB(15, 3) minutes.

In addition to the sharing of the rework craftsman, management has noted that the release of the jobs at the beginning of the day is not well represented by the previous model. In fact, after the jobs are released the jobs go through an additional step before reaching the ring-making station. After being released all the paperwork and raw materials for the job are found and must travel to the ring station for the making of the rings. The process takes 12 minutes on average according to an exponential distribution for each pair of rings. There are always sufficient workers available to move the rings to the ring station at the beginning of the day. LOTR Makers Inc. would like an analysis of the time to complete the orders for each of the following systems:

Configuration 1 The system with ring preparation/travel delay with no sharing of the rework craftsman.

Configuration 2 The system with ring preparation/travel delay and the sharing of the rework craftsman.

Figure 4-73 illustrates the two system configurations in the form of activity diagrams. Configuration two illustrates that the two resources (master craftsman and rework craftsman) are shared at the make ring activity by placing the two resources in a larger oval. This oval represents the fact that these two resources are in a set. Note how the SEIZE and RELEASE arrows from the make ring activity go to the boundary of the oval. This indicates that the make ring activity pulls resources from this set of resources.

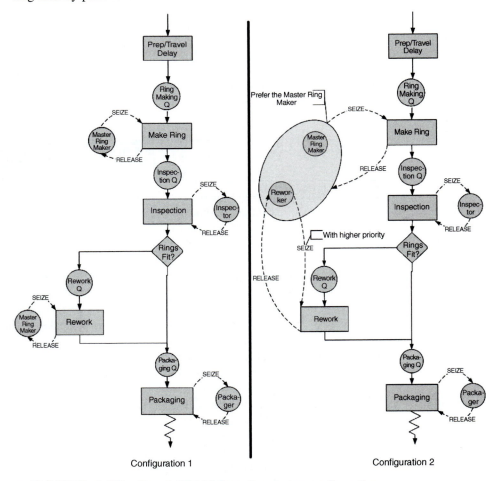

■ FIGURE 4-73   Two LOTR Makers, Inc. system configurations

■ **TABLE 4-3**

Arena's Set Concept as a List

RingMakers

Index	Member
1	RingMakerR
2	ReworkR

The rework activity still uses the rework craftsman. In particular, the SEIZE and RELEASE arrows go directly to the rework craftsman. The SEIZE arrows have been augmented to indicate that the master craftsman is to be preferred and that the rework activity has higher priority for the rework craftsman. In both activity diagrams the preparation/travel time has been represented with an activity that does not use any resources. This can be modeled in Arena™ with a DELAY module. The implementation of configuration 1 in Arena™ poses no additional modeling challenges; however, for configuration 2, the resource sharing must be modeled in Arena™.

The diagram indicates that the master craftsman and the rework craftsman are in the same set. Arena™ has the capability of defining sets to hold various object types (e.g., resources, queues, etc.). Thus, Arena's set construct can be used to model the sharing of the resources. Arena's set construct is simply a named list of objects of the same type. Thus, a resource set is a list of resources. Each resource listed in the set is called a member of the set. Unlike the mathematical set concept, Arena's sets are *ordered* lists. Each member of the set is associated with an index representing its order in the list. The first member has an index of 1, the second has an index of 2, and so forth. If you know the index, you can look up which member of the set is associated with the index. In this sense, an Arena™ set is like an array of objects found in other programming languages. Table 4-3 illustrates Arena's resource set concept to hold the ring makers.

A set named *RingMakers* can be defined with the two previously defined resources as members of the set. In this case, the resource representing the master craftsman (RingMakerR) is placed first in the set and the resource representing the rework craftsman (ReworkR) is placed second in the set. The name of the set can be used to return the associated object:

RingMakers(1) will return the resource *RingMakerR*.

RingMakers(2) will return the resource *ReworkR*.

There are three useful functions for working with sets:

MEMBER(Set ID, Index): The MEMBER function returns the construct number of a particular set member. Set ID identifies the set to be examined and index is the index into the set. Using the name of the set with the index number is functionally equivalent to the MEMBER function.

MEMIDX(Set ID, Member ID): MEMIDX returns the index value of a construct in the specified Set ID. Member ID is the name of the construct.

NUMMEM(Set ID): NUMMEM returns the number of members in the specified Set ID.

The ordering of the members in the set may be important to the modeling because of how Arena™ defines rules for selecting members from the set. When an entity attempts to seize a member of the resource set, a resource selection rule may be invoked. The rule will be invoked if there is more than one resource idle at the time that the entity attempts to seize a member of the resource set. Arena™ has seven default rules and two ways to specify user-defined rules.

CYC: Selects the first available resource beginning with the successor of the last resource selected. This has the effect of cycling through the resources. For example, if there are 5 members in the set {1,2,3,4,5} and 3 was the last resource selected, then 4 will be the next resource selected if it is available.

POR: Selects the first resource for which the required resource units are available. Each member of the set is checked in the order listed. The order of the list specifies the preferred order of selection.

LNB: Selects the resource that has the largest number of resource units busy; ties are broken using the POR rule.

LRC: Selects the resource that has the largest remaining resource capacity; ties are broken using the POR rule.

SNB: Selects the resource that has the smallest number of resource units busy; ties are broken using the POR rule.

SRC: Select the resource that has the smallest remaining resource capacity; ties are broken using the POR rule.

RAN: Selects randomly from the resources of the set for which the required resource units are available.

ER(User Rule): Selects the resource based on rule User Rule defined in the experiment frame.

UR(User Rule): Selects the URth resource where UR is computed in a user-coded rule function.

Since the master ring maker should be preferred if both are available, the *RingMakerR* resource should be listed first in the set and the POR resource selection rule should be used. The only other modeling issue that must be handled is the fact that the rework worker should show priority for rework jobs. From the activity diagram, you can see that there will be two SEIZE modules attempting to grab the rework worker. If the rework worker becomes idle, which SEIZE should have preference? According to the problem, the SEIZE related to the rework activity should be given preference. In Arena^{TM}, you can specify a priority level associated with the SEIZE to handle this case. A lower number results in the SEIZE having a higher priority.

You are now ready to implement this situation in Arena^{TM}. Starting with the file name, *LOTRCh4Example.doe*, the necessary model changes can be made. Open the submodel named *Ring* Processing, and insert a DELAY module at the beginning of the process as shown in Figure 4-74. The DELAY module can be found on the Advanced Process panel.

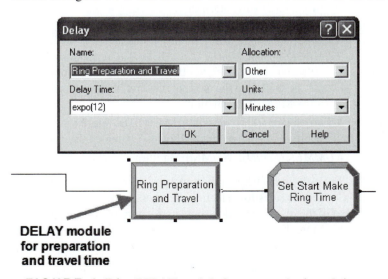

DELAY module for preparation and travel time

■ FIGURE 4-74   DELAY module for preparation/travel time

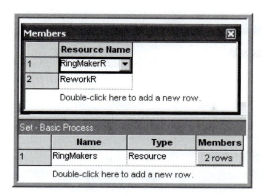

■ **FIGURE 4-75**   **Adding members to set**

Specify an expo(12) distribution for the delay time. After making this change, you should save the model under the name, *LOTRCh4Config1.doe*, to represent the first system configuration. You will now edit this model to create the second system configuration.

The first step will be to define the resource set. This can be done using the SET module on the Basic Process panel. Within the SET module, double-click on a row to start a new set. Then, you should name the set RingMakers and indicate that the type of set is Resource as shown in Figure 4-75. Now, you can click on the Members area to add rows for each member. Do this as shown in Figure 4-75 and make sure to place *RingMakerR* and *ReworkR* as the first and second members in the set. Since the resources had already been defined it is a simple matter of using the drop down textbox to select the proper resources. If you define a set of resources before defining the resources using the RESOURCE module, Arena™ will automatically create the listed resources in the RESOURCE module. You will still have to edit the RESOURCE module.

After defining and adding the resources to the set, you should save your model as *LOTRCh4Config2.doe*. You are now ready to specify how to use the sets in model. Open the PROCESS module named, Make Ring Process, in order to edit the previously defined resource specification for the SEIZE, DELAY, RELEASE logic. In Figure 4-76, the resource type has been specified as Set. Then, you should select the RingMakers set using the Preferred Order resource selection rule. You should close the PROCESS module and save your model.

Now, you must handle the fact that the processing time to make a ring depends on which resource is selected. In Figure 4-76, there is a text field labeled Save Attribute. This attribute will hold the index number of the resource that was selected by the SEIZE. In the current situation, this attribute can be used to have the rings (entity) remember which resource was selected. This index will have the value 1 or 2 according to whichever member of the *RingMakers* set was selected. This index can then be used to determine the appropriate processing time. Since there are only two ring makers, an arrayed EXPRESSION can be defined (see Figure 4-77) to represent the different processing times. The first expression represents the processing time for the master ring maker and the second expression represents the ring-making time of the rework worker. The Save Attribute index can be used to select the appropriate processing time distribution in this arrayed expression. After defining your expressions as shown in Figure 4-77, open your Make Ring Process module and edit it according to Figure 4-78. Note how an attribute has been used to remember the index and then that attribute is used to select the appropriate processing time from the EXPRESSION array.

The next required model change is to ensure that the rework craftsman gives priority to rework jobs. This can be done by editing the PROCESS module for the rework process as shown in Figure 4-79. In addition, you need to add an additional 10 minutes for the rework

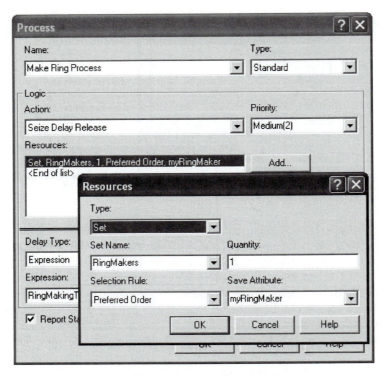

■ **FIGURE 4-76** Using a resource set in PROCESS module

worker's time to perform the rework due to the job sharing (Figure 4-80). You are now almost ready to run the models.

The final change to the model will enable the time to produce the rings to be captured to a file for analysis in Arena's Output Analyzer. Go to the STATISTICS module and add a file name (*prodTimeC2.dat*) to the OUTPUT statistic for the time to make the order statistics. Arena™ will write the time to make the orders to the file so that the analysis tools in the Output Analyzer can be used. You should then open the model file for the first configuration and add an output file (e.g., *prodTimeC1.dat*) to capture the statistics for the first configuration. Next, run each model for 30 replications.

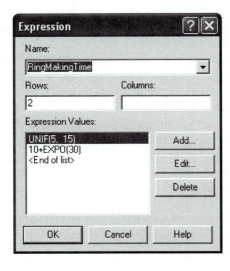

■ **FIGURE 4-77** Expressions for ring-making time by type of worker

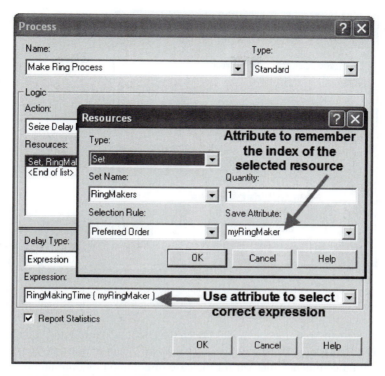

■ FIGURE 4-78   Handling processing time by resource selected

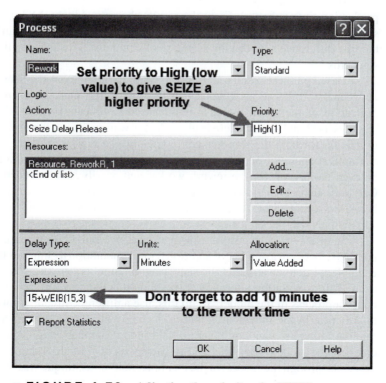

■ FIGURE 4-79   Adjusting the priority of a SEIZE

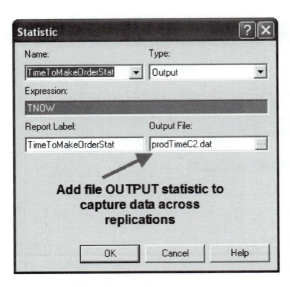

■ **FIGURE 4-80** Capturing across replication results

After running the models, open the Output Analyzer and add the two generated files to a new data group. Use the Analyze > Compare Means option of the Output Analyzer to develop a paired confidence interval for the case of common random numbers. Figure 4-81 illustrates how to set up the Compare Means options. Note that configuration 1 is associated with data file A and configuration 2 is associated with data file B. Arena^{TM} computes the difference between A and B ($\theta = \theta_1 - \theta_2$). Thus, if $l > 0$, you can conclude that configuration 1 has the higher mean time to produce the rings. If this is the case, configuration 2 would be preferred (shorter production time is better). From the results shown in Figure 4-82, you can clearly see that system configuration 2 has the smaller production time. In fact, you can be 95% confident that the true difference between the systems is 92 minutes. This is a practical difference by most standards.

Based on these results, LOTR Makers, Inc. should consider sharing the rework worker between the work stations. If you check the other performance measures, you will see that the utilization of the rework worker is increased significantly (near 97%) in configuration 2 (Figures 4-83 and 4-84). There is also a larger waiting line at the rework station. Such a high utilization for both the ring makers (especially the rework worker) is a bit worrisome. For example, suppose the quality of the work suffered in the new configuration. Then, there would be even more work for the rework worker and possibly a bottleneck at the rework station. Certainly such a high utilization is troublesome from a human factors standpoint, since worker breaks have not even been modeled! These and other trade-offs can be examined using simulation.

In order to try to ensure a stronger variance reduction when using common random numbers, there are a number of additional implementation techniques that can be applied. For example, to help ensure that the same random numbers are used for the same processes in each of the simulation alternatives you should dedicate a different stream number to each random process in the model. To do this, use a different stream number in each probability distribution in the model. In addition, to help with the synchronization of the use of the random numbers, you can generate the random numbers that each entity will need upon entering the model. Then each entity carries its own random numbers and uses them as it proceeds through the model. In the example, neither technique was done in order to simplify the exposition.

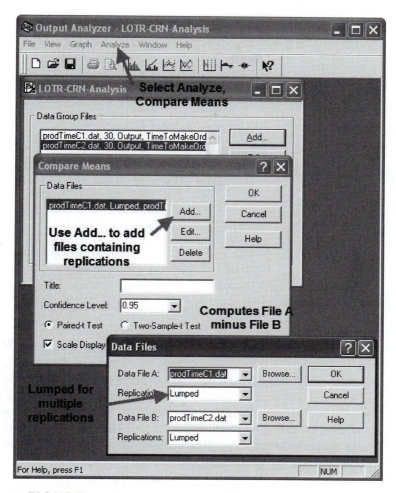

■ **FIGURE 4-81**   **Setting up paired difference analysis**

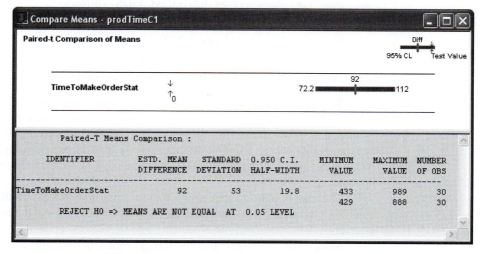

■ **FIGURE 4-82**   **Results for comparison of production times**

Instantaneous Utilization	Average	Half Width
InspectorR	0.4282	0.01
PackageR	0.6883	0.01
ReworkR	0.1106	0.02
RingMakerR	0.9785	0.00

■ **FIGURE 4-83**   **Configuration 1 resource utilization**

Instantaneous Utilization	Average	Half Width
InspectorR	0.4832	0.01
PackageR	0.7780	0.02
ReworkR	0.9795	0.00
RingMakerR	0.9221	0.02

■ **FIGURE 4-84**   **Configuration 2 resource utilization**

4.5.1.4 Implementing Independent Sampling in Arena^{TM}. This section outlines how to perform the independent sampling case in Arena^{TM}. The completed model files are available as *LOTRCh4Config1IND.doe* and *LOTRCh4Config2IND.doe*. The basic approach to implementing independent sampling in Arena^{TM} is to utilize different random number streams for each configuration. There are a number of different ways to implement the models so that independent sampling can be achieved. The simplest method is to use a variable to represent the stream number in each of the distributions in the model. In *LOTRCh4Config1IND.doe*, every distribution was changed as follows:

Sales probability distribution: BETA(5,1.5, vStream)

Success of sale distribution: DISC(vSalesProb,1,1.0,0, vStream)

Preparation and travel time distribution: EXPO(12, vStream)

Master ring make processing time: UNIF(5,15, vStream)

Inner ring diameter: NORM(vIDM, vIDS, vStream)

Outer ring diameter: NORM(vODM, vODS, vStream)

Inspection time distribution: TRIA(2,4,7, vStream)

Rework time distribution: 5+WEIB(15,3, vStream)

Packaging time distribution: LOGN(7,1, vStream)

The same procedure was used for *LOTRCh4Config2IND.doe*. Then the variable vStream was set to different stream numbers so that each model uses different random number streams. Both models were executed first using vStream equal to 1 for configuration 1 and vStream equal to 2 for configuration 2. The Output Analyzer can again be used to compare the results using the Analyze > Compare Means > Two sample *t*-test option. Figure 4-85 presents the results from the analysis.

The results indicate that configuration 2 has a shorter production time. Note that the confidence interval for the independent analysis is wider than in the case of common random numbers.

Since comparing two systems through independent samples takes additional work in preparing the model, one may wonder when it should be applied. If the results from one of the

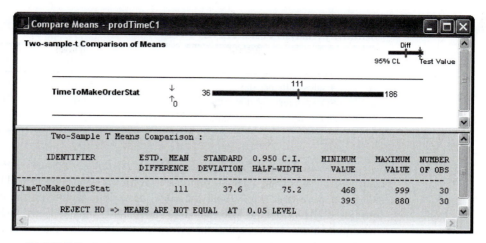

■ **FIGURE 4-85** Independent sample analysis from Output Analyzer

alternatives are already available, but the individual replication values are unavailable to perform the pairing, then the independent sample approach might be used. For example, this might be the case if you were comparing your model with already published results. In addition, suppose the model takes a very long time to execute and only the summary statistics are available for one of the system configurations. You might want to ensure that the running of the second alternative is independent of the first so that you do not have to re-execute the first alternative. Finally, even though the situation of comparing two simulation runs has been the primary focus of this section, you might have the situation of comparing the results of the simulation to the performance of the *actual* system. In this case, you can use the data gathered on the actual system and use the two sample independent analysis to compare the results. This is useful in the context of validating the results of the model against the actual system.

This section presented the results for how to compare two alternatives; however, in many situations, you might have many more than two alternatives to analyze. For a small number of alternatives, you might take the approach of making all pairwise comparisons to develop an ordering. This approach has limitations, which will be discussed in the next section along with how to handle the multiple comparison of alternatives in Arena[TM].

4.5.2 Analyzing Multiple Systems

The analysis of multiple systems stems from a number of objectives. First, you may want to perform a *sensitivity analysis* on the simulation model. In a sensitivity analysis, you want to measure the effect of small changes to key input parameters to the simulation. For example, you might have assumed a particular arrival rate of customers to the system and want to examine what would happen if the arrival rate decreased or increased by 10%. Sensitivity analysis can help you to understand how robust the model is to variations in assumptions. When you perform a sensitivity analysis, there may be a number of factors that need to be examined in combination with other factors. This may result in a large number of experiments to run. This section discusses how to use Arena's Process Analyzer to analyze multiple experiments. Besides performing a sensitivity analysis, you may want to compare the performance of multiple alternatives in order to choose the best alternative. This type of analysis is performed using multiple comparison statistical techniques. This section will also illustrate how to perform a multiple comparison with the best analysis using Arena's Process Analyzer.

■ **TABLE 4-4**
Sensitivity Analysis Factors/Levels

	Factor	Base level	% Change	Low level	High level
A	# sales calls per day	100	±10%	90	110
B	SD of outer ring diameter	0.005	±2%	0.0049	0.0051
C	SD of inner ring diameter	0.002	±2%	0.00196	0.00204

4.5.2.1 Sensitivity Analysis Using Arena's Process Analyzer. For illustrative purposes, a sensitivity analysis on the LOTR Makers, Inc. system will be performed. Suppose that you were concerned about the effect of various factors on the operation of the newly proposed system and that you wanted to understand what happens if these factors vary from the assumed values. In particular, the effects of the following factors in combination with each other are of interest:

Number of sales calls made each day: What if the number of calls made each day was 10% higher or lower than its current value?

Standard deviation (SD) of inner ring diameter: What if the SD were reduced or increased by 2% compared to its current value?

SD of outer ring diameter: What if the SD were reduced or increased by 2% compared to its current value?

The resulting factors and their levels are given in Table 4-4.

Arena's Process Analyzer allows the setting up and the batch running of multiple experiments. With the Process Analyzer, you can control certain input parameters (variables, resource capacities, replication parameters) and define specific response variables (COUNTERS, DSTATS [time-based statistics], TALLY [observation-based statistics], OUTPUT statistics) for a given simulation model. An important aspect of using the Process Analyzer is to plan the development of the model so that you can have access to the items that you want to control. In the current example, the factors that need to vary have already been specified as variables; however, the Process Analyzer has a four-decimal-place limit on control values. Thus, for specifying the standard deviations for the rings, a little creativity is required. Since the actual levels are simply a percentage of the base level, you can specify the percent change as the level of the factor and multiply accordingly in the model. For example, for factor (C), the standard deviation of the inner ring, you can specify 0.98 as the low value since $0.98 \times 0.002 = 0.00196$. Table 4-5 shows the resulting experimental scenarios in terms of the multiplying factor. Within the model, you need to ensure that the variables are multiplied. For example, define three new variables vNCF, vIDF, and vODF to represent the multiplying factors and multiply the original variables wherever they are used:

In Done Calling? DECIDE module: vNumCalls*vNCF

In Determine Diameters ASSIGN module: NORM(vIDM, vIDS*vIDF, vStream)

In Determine Diameters ASSIGN module: NORM(vODM, vODS*vODF, vStream)

Another approach to achieving this would be to define an EXPRESSION and use the expression in the model. For example, you can define an expression *eNumCalls = vNumCalls*vNCF* and use eNumCalls in the model. The advantage of this is that you do not have to hunt throughout the modules for all the changes. The definitions can be easily located in the EXPRESSION module.

The Process Analyzer is a separate program that is installed when the Arena™ Environment is installed. Since it is a separate program, it has its own help system, which

■ TABLE 4-5
Combinations of Factors and Levels

Scenario	vNCF (A)	vIDF (B)	vODF (C)
1	0.90	0.98	0.98
2	0.90	0.98	1.02
3	0.90	1.02	0.98
4	0.90	1.02	1.02
5	1.1	0.98	0.98
6	1.1	0.98	1.02
7	1.1	1.02	0.98
8	1.1	1.02	1.02

you can access after starting the program. You can access the Process Analyzer through your Start Menu > or through the Tools menu in the Arena™ Environment. The first thing to do after starting the Process Analyzer is to start a new PAN file via the File > New menu. You should then see a screen similar to Figure 4-86. In the Process Analyzer, you define the scenarios that you want to be executed. A scenario consists of an Arena™ model in the form of a (.p) file, a set of controls, and a set of responses. To generate a (.p) file, you can use Run > Check Model. If the check is successful, this will create a (.p) file with the same name as your model and located in the same directory as the model. Each scenario refers to one (.p) file, but the set of scenarios can

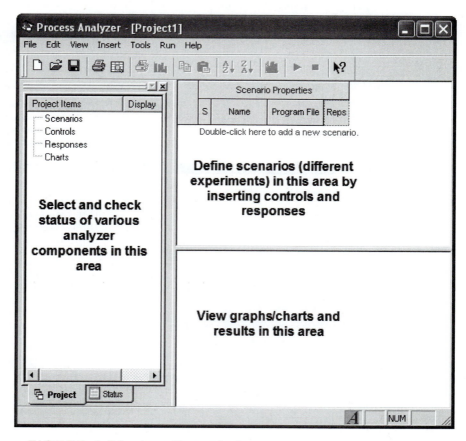

■ FIGURE 4-86 Arena Process Analyzer

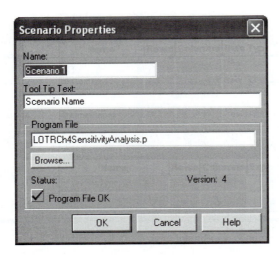

■ FIGURE 4-87 Adding a new scenario

consist of scenarios that have different (.p) files. Thus, you can easily specify models with different structure as part of the set of scenarios.

To start a scenario, double-click on the add new scenario row in the scenario properties area. You can type in the name for the scenario and the name of the (.p) file (or use the file browser) to specify the scenario as shown in Figure 4-87. Then you can use the Insert menu to insert controls and response variables for the scenario. The insert control dialog is shown in Figure 4-88. Using this, define a control for Num Reps (number of replications), vStream

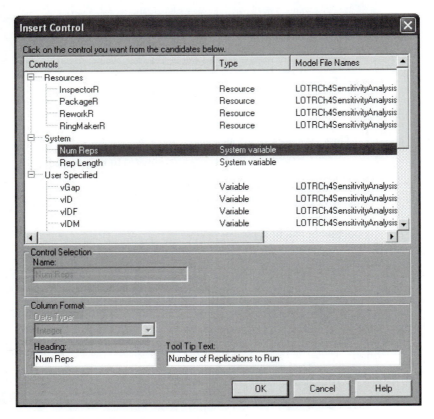

■ FIGURE 4-88 Inserting controls

	S	Name	Program File	Reps	Num Reps	vStream	vNCF	vIDF	vODF	Rework.Queue.WaitingTime	ThroughPut	PairOfRings.TotalTime
		Scenario Properties				Controls					Responses	
1		Scenario 1	0 : LOTRCh4	0	30	1	0.9000	0.9800	0.9800	---	---	---
2		Scenario 2	0 : LOTRCh4	0	30	2	0.9000	0.9800	1.0200	---	---	---
3		Scenario 3	0 : LOTRCh4	0	30	3	0.9000	1.0200	0.9800	---	---	---
4		Scenario 4	0 : LOTRCh4	0	30	4	0.9000	1.0200	1.0200	---	---	---
5		Scenario 5	0 : LOTRCh4	0	30	5	1.1000	0.9800	0.9800	---	---	---
6		Scenario 6	0 : LOTRCh4	0	30	6	1.1000	0.9800	1.0200	---	---	---
7		Scenario 7	0 : LOTRCh4	0	30	7	1.1000	1.0200	0.9800	---	---	---
8		Scenario 8	0 : LOTRCh4	0	30	8	1.1000	1.0200	1.0200	---	---	---

■ **FIGURE 4-89**   **Experimental setup controls/responses**

(random number stream), vNCF (number of sales calls factor), vIDF (standard deviation of inner ring factor), and vODF (standard deviation of outer ring factor). For each of the controls, be sure to specify the proper data type (e.g., vStream is an integer, vIDF is a real number).

After specifying the first scenario, select the row, right-click, and choose Duplicate Scenario(s) until you have defined eight scenarios (Figures 4-87 and 4-88). You should then complete the scenario definitions as shown in Figure 4-89). To insert a response, use the Insert menu and select the appropriate responses. In this analysis, you will focus on the average time it takes a pair of rings to complete production (*PairOfRings.TotalTime*), the throughput per day (*Throughput*), and the average time spent waiting at the rework station (*Rework.Queue. WaitingTime*).

Since a different stream number has been specified for each of the scenarios, all will be independent. Within the context of experimental design, this is useful because traditional experimental design techniques (such as response surface analysis) depend on the assumption that the experimental design points are independent. If you use the same stream number for each of the scenarios, then you will be using common random numbers. From the standpoint of analysis via traditional experimental design techniques, this poses an extra complication. The analysis of experimental designs with common random numbers is beyond the scope of this text. The interested reader should refer to Kleijnen (1988, 1998).

To run all the experiments consecutively, select the scenarios that you want to run and use the Run menu or the VCR-like run button. Each scenario will run the number of specified replications and the results for the responses will be tabulated in the response area as shown in Figure 4-90. After the simulations have been completed, you can add more responses and the results will be shown.

The purpose in performing a sensitivity analysis is twofold: (1) to see if small changes in the factors result in significant changes in the responses, and (2) to check if the expected

	S	Name	Program File	Reps	Num Reps	vStream	vNCF	vIDF	vODF	Rework.Queue.WaitingTime	ThroughPut	PairOfRings.TotalTime
		Scenario Properties				Controls					Responses	
1		Scenario 1	4 : LOTRCh4	30	30	1	0.9000	0.9800	0.9800	27.864	70	314.814
2		Scenario 2	4 : LOTRCh4	30	30	2	0.9000	0.9800	1.0200	29.309	70	314.696
3		Scenario 3	4 : LOTRCh4	30	30	3	0.9000	1.0200	0.9800	36.677	69	311.969
4		Scenario 4	4 : LOTRCh4	30	30	4	0.9000	1.0200	1.0200	36.813	70	322.541
5		Scenario 5	4 : LOTRCh4	30	30	5	1.1000	0.9800	0.9800	35.630	83	370.346
6		Scenario 6	4 : LOTRCh4	30	30	6	1.1000	0.9800	1.0200	33.003	82	366.924
7		Scenario 7	4 : LOTRCh4	30	30	7	1.1000	1.0200	0.9800	31.104	90	400.028
8		Scenario 8	4 : LOTRCh4	30	30	8	1.1000	1.0200	1.0200	36.775	79	359.321

■ **FIGURE 4-90**   **Results after running the scenarios**

direction of change in the response is achieved. Intuitively, if there is low variability in the ring diameters, then you should expect less rework and thus less queuing at the rework station. In addition, if there is more variability in the ring diameters then you might expect more queuing at the rework station. From the responses for Scenarios 1 and 4, this basic intuition is confirmed. The results in Figure 4-90 indicate that Scenario 4 has a slightly higher rework waiting time and that Scenario 1 has the lowest rework waiting time. Further analysis can be performed to examine the statistical significance of these differences.

In general, you should examine the other responses to validate that your simulation is performing as expected for small changes in the levels of the factors. If the simulation does not perform as expected, then you should investigate the reasons. A more formal analysis involving experimental design and analysis techniques may be warranted to ensure statistical confidence in your analysis. Again, in a simulation context, you know that there should be differences in the responses, so that standard ANOVA tests of differences in means are not really meaningful. In other words, they simply tell you whether you took enough samples to detect a difference in the means. Instead, you should be looking at the magnitude and direction of the responses. A full discussion of the techniques of experimental design is beyond the scope of this chapter. For a more detailed presentation of the use of experimental design in simulation, see Law (2007) and Kleijnen (1998).

Selecting a particular response cell will cause additional statistical information concerning the response to appear in the status bar at bottom left of the Process Analyzer main window. In addition, if you place your cursor in a particular response cell, you can build charts associated with the individual replications associated with that response. Place your cursor in the cell as shown in Figure 4-90 and right-click to insert a chart. The Chart Wizard will start and walk you through chart building. In this case, you will make a simple bar chart of the rework waiting times for each of the 30 replications of scenario 1. In the chart wizard, select "Compare the replication values of a response for a single scenario" and choose the Column chart type. Proceed through the chart wizard by selecting Next (and then Finish) making sure that the waiting time is your performance measure. You should see a chart similar to that shown in Figure 4-91.

If you right-click on the chart options pop-up menu, you can gain access to the data and properties of the chart. From this dialog you can copy the data associated with the chart. This gives you an easy mechanism for cutting and pasting the data into another application for additional analysis.

To create a chart across the scenarios, select the column associated with the desired response and right-click. Then, select Insert Chart from the pop-up menu. This will bring up

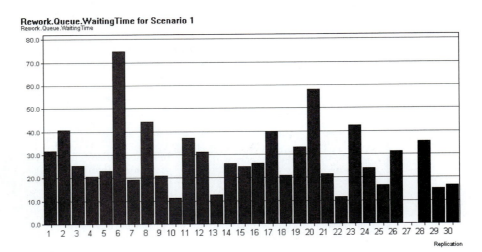

Rework.Queue.WaitingTime for Scenario 1

■ **FIGURE 4-91** Individual scenario chart for rework waiting time

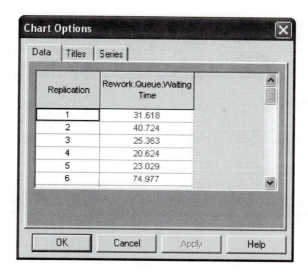

■ **FIGURE 4-92** Chart options

the chart wizard with the "Compare the average values of a response across scenarios" option selected. In this example, you will make a box-whiskers chart of the waiting times. Follow the wizard through the process until you have created a chart as shown in Figure 4-93.

There are varying definitions of what constitutes a box-whiskers plot. In the Process Analyzer, the box-whiskers plot shows whiskers (narrow thin lines) to the minimum and maximum of the data and a box (dark solid rectangle) that encapsulates the interquartile range for the data (3rd Quartile–1st Quartile). Fifty percent of the observations are in the box. This plot can give you an easy way to compare the distribution of the responses across the scenarios. Right-clicking on the chart give you access to the data used to build the chart, and in particular it includes the 95% half-widths for confidence intervals on the responses. Table 4-6 shows the data copied from the box-whiskers chart.

This section illustrated how to set up experiments to test the sensitivity of various factors in your models by using the Process Analyzer. After you are satisfied that your simulation model is working as expected, you often want to use the model to help to pick the best design out of a set of alternatives. The case of two alternatives has already been

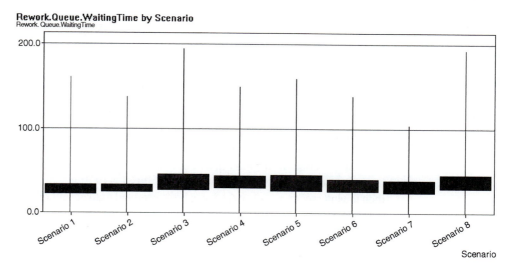

■ **FIGURE 4-93** Box-whiskers chart across scenarios

■ **TABLE 4-6**
Data from Box-Whiskers Chart

Scenario	Min	Max	Low	Hi	95% CI
Scenario 1	0	160.8	22.31	33.42	5.553
Scenario 2	0	137.2	24.93	33.69	4.382
Scenario 3	0	194.1	27.47	45.88	9.205
Scenario 4	0	149.8	29.69	43.93	7.12
Scenario 5	0	159.2	26.58	44.68	9.052
Scenario 6	0	138.3	26.11	39.9	6.895
Scenario 7	0	103.5	24	38.2	7.101
Scenario 8	0	193.1	28.82	44.73	7.956

discussed; however, when there are more than two alternatives, more sophisticated statistical techniques are required in order to ensure that a certain level of confidence in the decision making process is maintained. The next section overviews why more sophisticated techniques are needed. In addition, the section illustrates how Arena^{TM} facilitates the analysis using the Process Analyzer.

4.5.2.2 Multiple Comparisons with the Best in Arena^{TM}.

Suppose that you are interested in analyzing k systems based on performance measures $\theta_i\ i = 1, 2, \ldots, k$. The goals may be to compare each of the k systems with a base case (or existing system), to develop an order among the systems, or to select the best system. In any case, assume that some decision will be made based on the analysis and that the risk associated with making a bad decision needs to be controlled. In order to perform an analysis, each θ_i must be estimated, which results in sampling error for each individual estimate of θ_i. The decision will be based on all the estimates of θ_i (for every system) thereby compounding the risk associated with an overall decision. To see this more formally, the "curse" of the Bonferroni inequality needs to be understood.

The Bonferroni inequality states that given a set of (not necessarily independent) events, E_i, which occur with probability $1 - \alpha_i$ for $i = 1, 2, \ldots, k$, then a lower boundary on the probability of the intersection of the events is given by

$$P\{\cap_{i=1}^{k} E_i\} \geq 1 - \sum_{i=1}^{k} \alpha_i$$

In words, the Bonferroni inequality states that the probability of all the events occurring is at least one minus the sum of the probability of the individual events occurring. This inequality can be applied to confidence intervals, which are probability statements concerning the chance that the true parameter falls in intervals formed by the procedure.

Suppose that you have c confidence intervals each with confidence $1 - \alpha_i$. The ith confidence interval is a statement S_i that the confidence interval procedure will result in an interval that contains the parameter being estimated. The confidence interval procedure forms intervals such that S_i will be true with probability $1 - \alpha_i$. If you define events, $E_i = \{S_i\ is\ true\}$, then the intersection of the events can be interpreted as the event representing all the statements being true.

$$P\{all\ S_i\ true\} = P\{\cap_{i=1}^{k} E_i\} \geq 1 - \sum_{i=1}^{k} \alpha_i = 1 - \alpha_E$$

where $\alpha_E = \sum_{i=1}^{k} \alpha_i$. The value α_E is called the overall error probability. This statement can be restated in terms of its complement event as

$$P\{\text{one or more } S_i \text{ are false}\} \leq \alpha_E$$

This gives an upper bound on the probability of a false conclusion based on the confidence intervals.

This inequality can be applied to the use of confidence intervals when comparing multiple systems. For example, suppose that you have, $c = 10$, 90% confidence intervals to interpret. Thus, $\alpha_i = 0.10$, so that

$$\alpha_E = \sum_{i=1}^{10} \alpha_i = \sum_{i=1}^{10} (0.1) = 1.0$$

Consequently, $P\{\text{all } S_i \text{ true}\} \geq 0$ or $P\{\text{one or more } S_i \text{ are false}\} \leq 1$. In words, this implies that the chance that all of the confidence interval procedures result in confidence intervals that cover the true parameter is greater than zero and less than 1.

Think of it this way: If your boss asked you how confident you were in your decision, you would have to say that your confidence is somewhere between zero and one. This would not be very reassuring to your boss (or for your job!). To combat the "curse of Bonferroni," you can adjust your confidence levels in the individual confidence statements in order to obtain a desired overall risk. For example, suppose that you wanted an overall confidence of 95% on making a correct decision based on the confidence intervals. That is, you desire $\alpha_E = 0.05$. You can prespecify the α_i for each individual confidence interval to whatever values you want provided that you get an overall error probability of $\alpha_E = 0.05$. The simplest approach is to assume $\alpha_i = \alpha$. The question then becomes: What should α be to get $\alpha_E = 0.05$? Assuming that you have c confidence intervals, this yields

$$\alpha_E = \sum_{i=1}^{c} \alpha_i = \sum_{i=1}^{c} \alpha = c\alpha$$

So, you should set $\alpha = \alpha_E / c$. For the case of $\alpha_E = 0.05$ and $c = 10$, this implies that $\alpha = 0.005$. What does this do to the width of the individual confidence intervals? Since the $\alpha_i = \alpha$ have gotten smaller, the confidence coefficient (e.g., z value or t value) used in confidence intervals will be larger, resulting in a wider confidence interval. Thus, you must trade off your overall decision error against wider (less precise) individual confidence intervals.

Because the Bonferroni inequality does not assume independent events, it can be used for the case of comparing multiple systems when common random numbers are used. In the case of independent sampling for the systems, you can do better than simply bounding the error. For the case of comparing k systems based on independent sampling, the overall confidence is

$$P\{\text{all } S_i \text{ true}\} = \prod_{i=1}^{c} (1 - \alpha_i)$$

If you are comparing k systems where one of the systems is the standard (e.g., base case, existing system, etc.), you can reduce the number of confidence intervals by analyzing the difference between the other systems and the standard. That is, assuming that system $k = 1$ is the base case, you can form confidence intervals on $\theta_1 - \theta_i$ for $i = 2, 3, \ldots, k$. Since there are $k - 1$ differences, there are $c = k - 1$ confidence intervals to compare.

If you are interested in developing an ordering between all the systems, then one approach is to make all the pairwise comparisons between the systems. That is,

	S	Name	Program File	Reps	Num Reps	vStream	vNCF	vIDF	vODF	Rework.Queue.WaitingTime	ThroughPut	PairOfRings.TotalTime
1		Scenario 1	4 : LOTRCh4	30	30	1	0.9000	0.9800	0.9800	27.864	70	314.814
2		Scenario 2	4 : LOTRCh4	30	30	2	0.9000	0.9800	1.0200	29.309	70	314.696
3		Scenario 3	4 : LOTRCh4	30	30	3	0.9000	1.0200	0.9800	36.677	69	311.969
4		Scenario 4	4 : LOTRCh4	30	30	4	0.9000	1.0200	1.0200	36.813	70	322.541
5		Scenario 5	4 : LOTRCh4	30	30	5	1.1000	0.9800	0.9800	35.630	83	370.346
6		Scenario 6	4 : LOTRCh4	30	30	6	1.1000	0.9800	1.0200	33.003	82	366.924
7		Scenario 7	4 : LOTRCh4	30	30	7	1.1000	1.0200	0.9800	31.104	90	400.028
8		Scenario 8	4 : LOTRCh4	30	30	8	1.1000	1.0200	1.0200	36.775	79	359.321

■ **FIGURE 4-94**   Results for MCB analysis using CRN

construct confidence intervals on $\theta_j - \theta_i$ for $i \neq j$. The number of confidence intervals in this situation is

$$c = \binom{k}{2} = \frac{k(k-1)}{2}$$

The trade-off between overall error probability and the width of the individual confidence intervals will become severe in this case for most practical situations.

Because of this, a number of techniques have been developed to allow the selection of the best system (or the ranking of the systems) and still guarantee an overall prespecified confidence in the decision. The Process Analyzer uses a method based on multiple comparison procedures as described in Goldsman and Nelson (1998) and the references therein. See also Law (2007) for how these methods relate to other ranking and selection methods.

Using the scenarios that were already defined in the sensitivity analysis section, the following example illustrates how you can use Arena^{TM} to select the best system with an overall confidence of 95%. The procedure built into the Process Analyzer can handle common random numbers. The previously described scenarios were set up and re-executed, as shown in Figure 4-94. Note in the figure that the stream number for each scenario was set to the same value, thereby applying common random numbers. The PAN file for this analysis is called *LOTR-MCB.pan* and can be found in the supporting files for this chapter.

Suppose you want to pick the best scenario in terms of the average time that a pair of rings spends in the system. Furthermore, suppose that you are indifferent between the systems if they are within 5 minutes of each other. Thus, the goal is to pick the system that has the smallest time with 95% confidences.

To perform this analysis, right-click on the *PairOfRings.TotalTime* response column and choose insert chart. Make sure that you have selected "Compare the average values of a response across scenarios" and select a suitable chart in the first wizard step. This example uses a Hi-Lo chart, which displays the confidence intervals for each response as well has the minimum and maximum value of each response. The comparison procedure is available with the other charts as well. On the second wizard step, you can pick the response that you want (*PairOfRings.TotalTime*) and choose *next*. On the third wizard step, you can adjust your titles for your chart. When you get to the fourth wizard step, you have the option of identifying the best scenario. Figure 4-95 illustrates the settings for the current situation. Select the "identify the best scenarios option" with the smaller is better option, and specify an indifference amount (Error tolerance) as 5 minutes. Using the Show Best Scenarios button, you can see the best scenarios listed. Clicking Finish causes the chart to be created and the best scenarios to be identified in red (Scenarios 1, 2, 3, and 4) as shown in Figure 4-96.

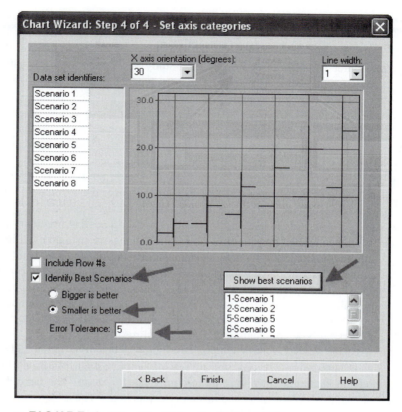

■ **FIGURE 4-95**   **Identifying the best scenario**

As indicated in Figure 4-96, Arena™ has recommended four possible scenarios as the best. This means that you can be 95% confident that any of these four scenarios is the best based on the 5-minute error tolerance. This analysis has narrowed down the set of scenarios, but has not recommended a specific scenario because of the variability and closeness of the responses. To further narrow the set of possible best scenarios down, you can run additional replications of the identified scenarios and adjust your error tolerance. Thus, with the Process Analyzer you can easily screen systems out of further consideration and adjust your statistical analysis in order to meet the objectives of your simulation study.

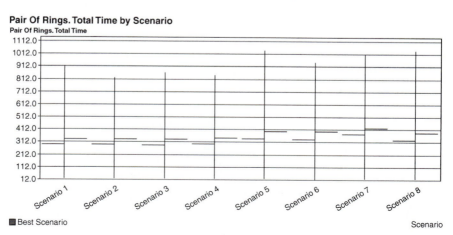

■ **FIGURE 4-96**   **Possible best scenarios**

4.6   Summary

This chapter described many of the statistical aspects of simulation that you will typically encounter in performing a simulation study. An important aspect of performing a correct simulation analysis is to understand the type of data associated with your performance measures (time-based versus observation-based) and how to collect/analyze such data. In your modeling, you will be faced with specifying the time horizon of your simulation. Most situations involve finite horizons, which are fortunately easy to analyze via the method of replications. This allows a random sample to be formed across replications and to analyze the simulation output via traditional statistical techniques.

In the case of infinite-horizon simulations, things are more complicated. You must first analyze the effect of any warm-up period on the performance measures and decide whether you should use the replication-deletion or batch means method. Regardless of the situation, Arena™ has tools to handle the situation or facilitates the necessary analysis.

Since you often want to use simulation to make a recommendation concerning a design configuration, an analysis across system configurations must be carefully planned. Arena™ facilitates the analysis of two or more systems. The Process Analyzer makes the setting up and execution of experimental designs easier and facilitates the ranking and selection of design configurations. When performing your analysis, you should consider how and when to use the method of common random numbers, and you should consider the impact of common random numbers on how you analyze the simulation results.

Now that you have a solid understanding of how to program and model in Arena™ and how to analyze your results, you are ready to explore the application of Arena™ to additional modeling situations involving more complicated systems. The next chapter concentrates on systems involving queuing and inventory components. These systems form the building blocks for modeling more complicated systems in manufacturing, transportation, and service industries.

Exercises

4.1 In a four-server facility, the number of busy servers changes with time as shown in Figure 4-97. Compute the following measures of performance over a period of 25 time units.

(a) Average number of busy servers

(b) Busy time per server

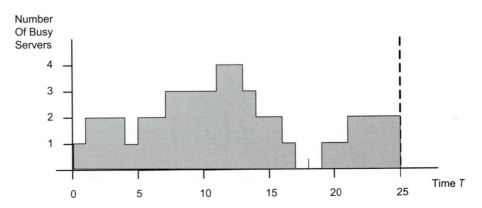

■ **FIGURE 4-97**   **Plot of number of busy servers with time**

(c) Idle time per server

(d) The utilization of the servers

4.2 The usage of a resource with a capacity of three can be summarized graphically as shown in Figure 4-98. Compute the following:

(a) Average utilization

(b) Percentage of idleness

(c) Idle time per server

(d) Busy time per server

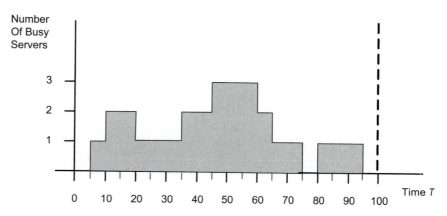

■ **FIGURE 4-98** Usage of resource with 3 capacity

4.3 Figure 4-99 summarizes the changes in queue length with time. Compute the average queue length.

(a) Average queue length

(b) Busy time per server

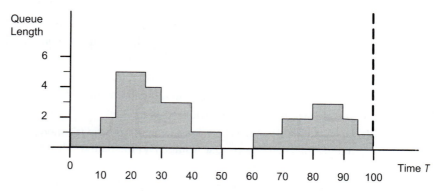

■ **FIGURE 4-99** Changes in queue length

4.4 Figure 4.100 shows the changes in queue length for a single-server model over a run length of 35 time units. The first 5 time units are estimated to represent the warm-up period. The remaining 30 time units are divided equally among 5 batches. The mean and variance of queue length are of interest. For each batch, compute the time average batch mean of the queue length. Use your results to estimate the mean and variance of the queue length.

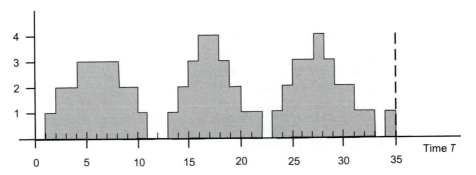

■ **FIGURE 4-100** Changes in queue length for single-server model

4.5 Using the supplied data set, draw the sample path for this state variable. Compute the time average over the supplied time range.

n	Time state variable changed	Value of state variable immediately after the change
1	0	0
2	1.5	2
3	6	1
4	10	1
5	15	1
6	18	2
7	20	2
8	25	3
9	30	2
10	34	1
11	39	0
12	42	1
13	45	0
14	48	1
15	50	1

4.6 Compute the required sample size necessary to ensure a 95% confidence interval with a half-width of no larger than 30 minutes for the total time to produce the rings for the LOTR System of Section 4.3.2. Use the following three methods:

(a) Sample size based on the normal distribution

(b) Sample size based on the t-distribution approximation

(c) Sample size using the half-width ratio method

(d) Discuss the differences between the methods. Run the model for the specified number of replications and report your results. Did the half-width from your replications meet the target value? Discuss.

4.7 Using the Lindley equation spreadsheet simulation, supplied with this Chapter perform the following:

(a) Develop a 95% confidence interval for your estimate of the mean waiting time based on the data from one replication of 1000 customers. Discuss why this is inappropriate. How does your simulation estimate compare to the theoretical value?

(b) How does your running average track the theoretical value? What would happen if you increased the number of customers?

(c) Construct a Welch plot using five replications of the 1000 customers. Determine a warm-up point for this simulation. Do you think that 1000 customers are enough?

(d) Make an autocorrelation plot of your 1000 customer wait times using your favorite statistical analysis package. What are the assumptions for forming the confidence interval in part (a). Are these observations independent and identically distributed? What is the implication of your answer for your confidence interval in part (a)?

(e) Use your warm-up period from part (c) and generate an addition 1000 customers after the warm-up point. Use the batch means method to batch the 1000 observations into 40 batches of size 25. Make an autocorrelation plot of the 40 batch means. Compute a 95% confidence interval for the mean waiting time using the 40 batches.

(f) Use the replication-deletion method to develop a 95% confidence interval for the mean waiting time. Use your warm-up period from part (c). Compare the result with that of (a) and (e) and discuss.

4.8 Create a spreadsheet simulation to simulate n observations from a $N(\mu, \sigma^2)$ random variable.

(a) Use your simulation to generate two independent samples of size $n_1 = 20$ and $n_2 = 30$ from normal distributions having $\mu_1 = 2, \sigma_1^2 = 0.64$ $\mu_2 = 2.2, \sigma_2^2 = 0.64$. Assume that you don't know the true means and variances. Use the method of independent samples to examine whether $\mu_2 > \mu_1$. Show all your work.

(b) Use your simulation to generate two independent samples of size $n_1 = 20$ and $n_2 = 30$ from normal distributions having $\mu_1 = 2, \sigma_1^2 = 0.64$ $\mu_2 = 2.2, \sigma_2^2 = 0.36$. Assume that you don't know the true means and variances. Use the method of independent samples to examine whether $\mu_2 > \mu_1$. Show all your work.

(c) Use your simulation to generate two independent samples of size $n_1 = 30$ and $n_2 = 30$ from normal distributions having $\mu_1 = 2, \sigma_1^2 = 0.64$ $\mu_2 = 2.2, \sigma_2^2 = 0.36$. Assume that you don't know the true means and variances. Use the paired t method to examine whether $\mu_2 > \mu_1$.

(d) Repeat part (c), but instead generate two dependent samples. Use the paired t method to examine whether $\mu_2 > \mu_1$. What is the effect of using common random numbers? Show all your work.

4.9 Recall problem 2.10 in Chapter 2. Determine the number of days (replications) to run so that you can be 95% confident that you are within 2 minutes of the true mean time in the system for damaged orders. Rerun the model and summarize your statistics in a table.

(a) What would happen if the arrival rate were to double (i.e., the inter-arrival time mean were 5 minutes instead of 10 minutes)? In this case, if you could place another person anywhere in the system to help out with one of the five tasks, where would you place the person? Justify your answer with simulation results. Note: If you do not have the professional version of Arena^{TM} available, then work the problem with a mean inter-arrival time of 7.5 minutes.

(b) Using the original arrival rate, report confidence interval statistics on the number of suits left in the system at the end of the 12 hour shift. In other words, how many suits on average are left for the skeleton crew to finish up?

4.10 YBox video game players arrive at a two-person station for testing. The inspection time per YBox set is EXPO(10) minutes. On the average, 82% of the sets pass inspection. The remaining 18% are routed to an adjustment station with a single operator. Adjustment time per YBox is UNIF(7,14) minutes. After adjustments are made, the units are routed back to the inspection station to be retested. Build a simulation model of this system using ARENA. Use a replication length of 30,000 minutes.

(a) Perform a warm-up analysis of the total time a set spends in the system and estimate the system time to within 2 minutes with 95% confidence.

(b) Collect statistics to estimate the average number of times a given job is adjusted.

(c) Suppose that any one job is not allowed more than two adjustments, after which time the job must be discarded. Modify your simulation model and estimate the number of discarded jobs.

4.11 Cars arrive every EXPO(18) minutes at a car-wash facility that also offers vacuum cleaning. It takes EXPO(12) minutes to wash and EXPO(15) minutes to vacuum clean. When a car arrives, it can go to either wash or vacuum cleaning, depending on which queue is shorter. After the first activity (wash or clean), the car must go to the remaining activity (clean or wash). Assuming infinite queue sizes, determine the average time that a car spends in washing and the average time that a car spends in cleaning, as well as the time it spends in the system using a simulation model built using ARENA. Include a warm-up analysis of the system time and estimate the system time to within ± 2 minutes with 95% confidence.

4.12 Jobs arrive in batches of 10 items each. The interarrival time is EXPO(2) hours. The machine shop contains two milling machines and one drill press. About 30% of the items require drilling before being processed on the milling machine. Drilling time per item is UNIF(10,15) minutes. The milling time is EXPO(15) minutes for items that do not require drilling, and UNIF(15,20) minutes for items that do. Assume that the shop has two 8-hour shifts each day and that you are only interested in the *first* shift's performance. Any jobs left over at the end of the first shift are left to be processed by the second shift. Estimate the average number of jobs left for the second shift to complete at the end of the first shift to within plus or minus five jobs with 95% confidence. What is your replication length? Number of replications? Determine the utilization of the drill press and the milling machines as well as the average time an item spends in the system.

4.13 A repair and inspection facility consists of two stations, a repair station with two technicians, and an inspection station with one inspector. Each repair technician works at a rate of three items per hour, while the inspector can inspect eight items per hour each exponentially distributed. Approximately 10% of all items fail inspection and are sent back to the repair station (this percentage holds even for items that have been repaired two to three times). If an item fails inspection three times, it is scrapped. When an item is scrapped, the item is sent to a disassembly station to recover the usable parts. At the disassembly station, the items wait in a queue until a technician is available. The disassembly time is distributed according to a log-normal distribution with a mean of 20 minutes and a standard deviation of 10 minutes. Assume that items arrive according to a Poisson arrival process with a rate of four per hour. The weekly performance of the system is the key objective of this simulation analysis. Assume that the system starts empty and idle on Monday mornings and runs continuously for two 8-hour shifts per day for 5 days. Any jobs not completed by the end of second shift are carried over to the first shift of the next day. Any jobs left over at the end of the week are handled by a separate weekend staff that is not of concern to the current study. Estimate the following:

■ The average system time of items that pass inspection on the first attempt. Measure this quantity such that you are 95% confident to within 3 minutes.

■ The average number of jobs completed per week.

(a) Sketch an activity diagram for this situation.

(b) Assume that there are two technicians at the repair station, one inspector at the inspection station, and one technician at the disassembly station. Develop an Arena^TM model for this situation.

(c) Assume that there are two technicians at the repair station and one inspector at the inspection station. The disassembly station is also staffed by the two technicians who are assigned to the repair station. Develop an Arena^{TM} simulation model for this situation.

4.14 A small manufacturing system produces three types of parts. There is a 30% chance of getting a Type 1 part, a 50% chance of getting a Type 2 part, and a 20% chance of getting a Type 3 part. The parts arrive from an upstream process such that the time between arrivals is exponentially distributed with a mean of 3 minutes. All parts that enter the system must go through a preparation station where there are two preparation workers. The preparation time is exponentially distributed with means 3, 5, and 7, for part types 1, 2, and 3, respectively.

There is space for only 6 parts in the preparation queue. Any parts that that arrive to the system when there are 6 or more parts in the preparation queue cannot enter the system. These parts are shunted to a recirculating conveyor, which takes 10 minutes to recirculate the parts before they can try again to enter the preparation queue. Hint: Model the recirculating conveyor as a simple deterministic delay.

After preparation, the parts are processed on two different production lines. A production line is dedicated to type 1 parts and a production line is dedicated to type 2 and 3 parts. Part types 2 and 3 are built one at a time on their line by 1 of 4 operators assigned to the build station. The time to build a part type 2 or 3 part is triangularly distributed with a (min = 5, mode = 10, max = 15) minutes. After the parts are built, they leave the system.

Part type 1 has a more complicated process because of some special tooling that is required during the build process. In addition, the build process is separated into two different operations. Before starting operation 1, the part must have 1 of 10 special tooling fixtures. It takes between 1 and 2 minutes uniformly distributed to place the part in the tooling fixture. An automated machine places the part in the tooling so that the operator at operation 1 does not have to handle the part. There is a single operator at the first operation, which takes 3 minutes on average exponentially distributed. The part remains in the tooling fixture after the first operation and proceeds to the second operation. There is one operator at the second operation, which takes between 3 and 6 minutes uniformly distributed. After the second operation is complete, the part must be removed from the tooling fixture. An automated machine removes the part from the tooling so that the operator at operation two does not have to handle the part. It takes between 20 and 30 seconds uniformly distributed to remove the part from the tooling fixture. After the part is built, it leaves the system.

In this problem, the steady-state performance of this system is required in order to identify potential long-term bottlenecks in this process. For this analysis, collect statistics on the following quantities:

- Queue statistics for all stations. Utilization statistics for all resources.
- The system time of parts by part type. The system time should not include the time spent on the recirculating conveyor.
- The average number of parts on the conveyor.
- Perform a warm-up analysis on the system time of a part regardless of type.

4.15 Reconsider exercise 4.14. A process change is being recommended for the build station for part types 2 and 3. In particular, a machine change will cause the processing time to be log-normal distributed with a mean of 10 minutes and a standard deviation of 2 minutes. Use Arena's Output Analyzer to compare the system time of the old configuration and the new configuration based on 30 replications of length 1000 hours with a warm-up of 200 hours. Which configuration would you recommend?

4.16 Reconsider exercise 4.15. Suppose that you are interested in checking the sensitivity of a number of configurations and possibly picking the best configuration based on the part system time. The factors of interest are given in the following table.

Factor	Levels
Preparation space	6 or 8
Number of tooling fixtures	10 or 15
Parts 2 and 3 processing distribution	TRIA(5, 10, 15) or LOGN(10, 2)

(a) Use Arena's Process Analyzer to examine these system configurations in an experimental design.

(b) Based on system time, recommend a configuration that has the shortest system time with 95% confidence on your overall decision.

Base your analysis on 30 replications of length 1000 hours and a warm-up period of 200 hours.

CHAPTER 5

Modeling Queuing and Inventory Systems

LEARNING OBJECTIVES

After completing this chapter, you should be able to:

- Define and explain the key performance measures in queuing and inventory systems
- Identify and apply standard queuing models
- Simulate variations of standard queuing models
- Simulate networks of queues using the STATION, ROUTE, and SEQUENCE modules
- Know and understand the capabilities of OptQuest
- Simulate classic inventory models
- Use the HOLD and SIGNAL modules
- Simulate multiechelon inventory systems

This chapter examines how to model queuing and inventory systems. As illustrated in previous chapters, many real-life situations involve the possible waiting of entities (e.g., customers, parts, etc.) for resources (e.g., bank tellers, machines, etc.). Systems that involve waiting lines are called queuing systems. In addition to queuing systems, situations involving inventory will also be discussed. In the classic sense, inventory is a build-up of items for future allocation to customers. Queuing systems and inventory systems are closely related because inventory can be considered a resource in a system. Just like a customer arriving at a pharmacy to request service from the pharmacy, customers can arrive at a store to request units of an item. These requests are called demands, and if the item is unavailable then the customer might wait until a sufficient quantity of the item becomes available. This chapter introduces both analytical (formula-based) and simulation-based approaches to modeling the performance of these systems.

Once the performance of the system is modeled, the design of the system to meet operational requirements becomes an important issue. For example, in the simple pharmacy example, you might want to determine the number of spaces that should be available so that arriving customers can have a high chance of entering the line. In the case of inventory systems, you might want to design the system so that the quantity of inventory on hand is sufficient to meet customer demand with a high chance that the customer will be able to get the item requested on the first request. In both of these situations, having more of the resource (pharmacist or inventory) available at any time will assist in meeting the design criteria; however, an increase in a resource typically comes at some cost. Thus, design questions in queuing and inventory systems involve a fundamental trade-off between customer service and the cost of providing that service.

To begin the analysis of these systems, a brief analytical treatment of the key modeling issues is presented. In both cases (queuing and inventory), analytical results are available only for simplified situations; however, the analytical treatment will serve two purposes. First, it will provide an introduction to the key modeling issues, and second, it can provide approximate models for more complicated situations. After understanding some of the basic models, the chapter examines situations for which simulation is the most appropriate mechanism for estimating performance. Along the way, some of the design issues relevant to these systems will be illustrated.

5.1   Single-Line Queuing Stations

Chapter 1 presented the pharmacy model and analyzed it with a single server, single queue queuing system called the M/M/1. This section shows how the formulas for the M/M/1 model in Chapter 1 were derived and discusses the key notation and assumptions of analytical models for systems with a single queue. In addition, you will also learn how to simulate variations of these models in Arena^{TM}.

In a queuing system, customers compete for resources by moving through processes. The competition for resources causes waiting lines (queues) to form and delays to occur in the customer's process. In these systems, the arrivals and/or service processes are often stochastic. Queuing theory is a branch of mathematical analysis of systems that involves waiting lines in order to predict (and control) their behavior over time. The basic models in queuing theory involve a single line that is served by a set of servers. Figure 5-1 illustrates the major components of a queuing system with a single queue feeding into a set of servers.

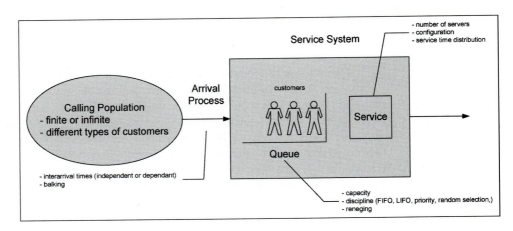

■ **FIGURE 5-1**   **Single-queue system figure**

In queuing theory, the term "customer" is used as a generic term to describe the entities that flow and receive service. A "resource" is a generic term used to describe the components of the system that are required by a customer as the customer moves through the system. The individual units of the resource are often called "servers." In Figure 5-1, potential customers can be described as coming from a calling population. The term "calling population" comes from the historical use of queuing models in the analysis of phone calls to telephone trunk lines. Customers in the calling population may arrive at the system according to an arrival process. In the finite population case, the arrival rate that the system experiences will quite naturally decrease as customers arrive since fewer customers are available to arrive if they are in the system. In the infinite calling population case, the rate of arrivals to the system does not depend on how many customers have already arrived. In other words, there are so many potential customers in the population that the arrival rate to the system is not affected by the current number of customers in the system. Besides characterizing the arrival process by the rate of arrivals, it is useful to think in terms of the interarrival times, and in particular the interarrival time distribution. In general, the calling population may also have different types of customers that arrive at different rates. In the analytical analysis presented here, there will only be one type of customer.

The queue is that portion of the system that holds waiting customers. The two main characteristics for the queue are its size (or capacity) and its discipline. If the queue has a finite capacity, this indicates that there is only enough space in the queue for a certain number of customers to be waiting at any given time. The queue discipline refers to the rule that will be used to decide the order of the customers in the queue. A first-come, first-served (FCFS) queue discipline orders the queue by the order of arrival, with the most recent arrival always joining the end of the queue. A last-in, first-out (LIFO) queue discipline has the most recent arrival joining the beginning of the queue. A LIFO queue discipline acts like a "dish" stack. The first dish is at the bottom of the stack, and the last dish added to the stack is at the top of the stack. Thus, when a dish is needed, the next dish to be used is the last one added to the stack. This type of discipline often appears in manufacturing settings when the items are placed in bins, with newly arriving items being placed on top of items that have previously arrived. Other disciplines include random and priority. You can think of a random discipline modeling the situation of a server randomly picking the next part to work on (e.g., reaches into a shallow bin and picks the next part). A priority discipline allows the customers to be ordered in the queue by a specified priority or characteristic. For example, the waiting items may be arranged by due date for a customer order.

The resource is that portion of the system that holds customers that are receiving service. Each customer arriving at the system may require a particular number of units of the resource. The resource component of this system can have one or more servers. In the analytical treatment, each customer will require only one server and the service time will be governed by a probability distribution called the "service time distribution." In addition, for the analysis the service time distribution will be the same for all customers. Thus, the servers are all identical in how they operate and there is no reason to distinguish between them. Queuing systems that contain servers that operate in this fashion are often referred to as parallel server systems. When a customer arrives to the system, the customer will either be placed in the queue or placed in service. After waiting in line, the customer must select one of the (identical) servers to receive service. The following analytical analysis assumes that the only way for the customer to depart the system is to receive service.

Specification of how the major components of the system operate provides the basic system configuration. To help in classifying and identifying the appropriate modeling situations, Kendall's notation (Kendall 1953) can be used. The basic format is *arrival process/ service process/number of servers/system capacity/size of the calling population/queue discipline*. For example, the notation M/M/1 specifies that the arrival process is Markovian (M) (exponential time between arrivals), the service process is Markovian (M) (exponentially distributed service times, and that there is 1 server. When the system capacity is not specified, it is assumed to be infinite. Thus, in this case, there is a single queue that can hold any arriving customer. The calling population is also assumed to be infinite if not explicitly specified. Unless otherwise noted, the queue discipline is assumed to be FCFS. Traditionally, the first letter(s) of the appropriate distribution is used to denote the arrival and service processes. Thus, the case LN/D/2 represents a queue with 2 servers having log-normal (LN) distributed time between arrivals and deterministic (D) service times. Unless otherwise specified, it is typically assumed that the arrival process is a renewal process; in other words, that the times between arrivals are independent. In addition, the service times are often assumed to be independent random variables. Finally, the standard models assume that the arrival process is independent of the service process and vice versa. To denote any distribution, the letter G for general (any) distribution is used. The notation (GI) is often used to indicate a general distribution in which the random variables are independent. Thus, the G/G/5 represents a queue with an arrival process having any general distribution, a general distribution for the service times, and 5 servers (see Figure 5-2).

A number of quantities form the basis for measuring the performance of queuing systems:

The time that a customer spends waiting in the queue: T_q

The time that a customer spends in the system (queue time plus service time): T

The number of customers in the queue at time t: $N_q(t)$

The number of customers that are in the system at time t: $N(t)$

The number of customer in service at time t: $N_b(t)$

The previous quantities will be random variables when the queuing system has stochastic elements. Assuming that the system is work conserving—that is, the customers do not exit without receiving all required service and that a customer uses one and only one server to receive service—the following is true:

$$N(t) = N_q(t) + N_b(t)$$

This indicates that the number of customers in the system must be equal to the number of customers in queue plus the number of customers in service. Since the number of servers is known, the number of busy servers can be determined from the number of customers in the system. Under the assumption that each customer requires 1 server (1 unit of the resource), then $N_b(t)$ is also the current number of busy servers. For example, if the number of servers is 3 and the number of customers in the system is 2, then there must be 2 customers in service (2 servers that are busy). Therefore, knowledge of $N(t)$ is sufficient to describe the state of the system. Because $N(t) = N_q(t) + N_b(t)$ is true, the following relationship between expected values is also true:

$$E[N(t)] = E[N_q(t)] + E[N_b(t)]$$

If ST is the service time of an arbitrary customer, then it should be clear that

$$E[T] = E[T_q] + E[ST]$$

In other words, the expected system time is equal to the expected waiting time in the queue plus the expected time spent in service.

Chapter 4 presented time-based data and computed time averages. For queuing systems, you can show that relationships exist between such quantities as the expected number in the queue and the expected waiting time in the queue. Figure 5-3 illustrates the sample path for the number of customers in the queue over a period of time.

Let $A_i; i = 1 \ldots n$ represent the time that the ith customer enters the queue, $D_i; i = 1 \ldots n$ represents the time that the ith customer exits the queue, and $T_{q_i} = D_i - A_i; i = 1 \ldots n$

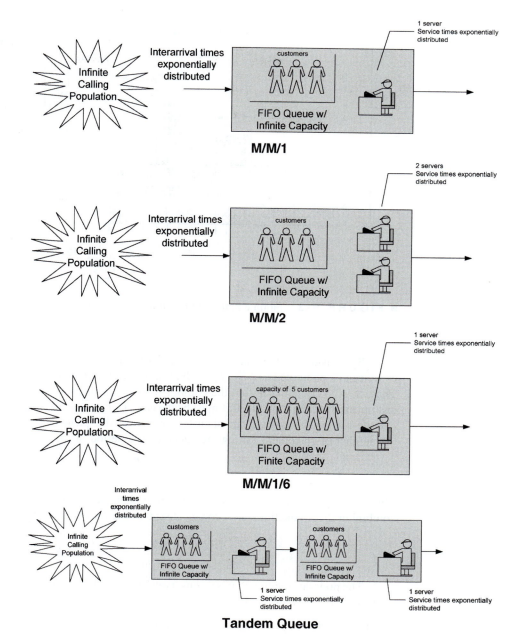

■ **FIGURE 5-2**   Illustrations of common queuing situations

represents the time that the ith customer spends in the queue. Recall that the average time spent in the queue was

$$\overline{T}_q = \frac{\sum_{i=1}^n T_{q_i}}{n} = \frac{0 + 11 + 8 + 7 + 2 + 5 + 6 + 0}{8} = \frac{39}{8} = 4.875$$

The average number of customers in the queue was

$$\overline{L}_q = \frac{\int_{t_0}^{t_n} q(t)dt}{t_n - t_0} = \frac{39}{25} = 1.56$$

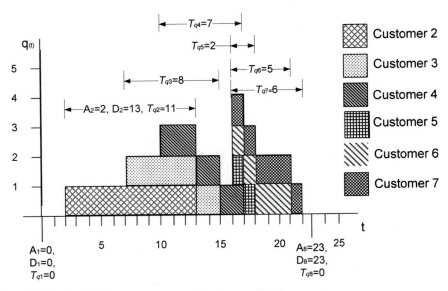

■ **FIGURE 5-3**   Sample path for the number in a queue

By considering the waiting time lengths in the figure, the area under the sample path curve can be computed as $\sum_{i=1}^{n} T_{q_i} = 39$, but by definition the area should also be $\int_{t_0}^{t_n} q(t)dt$. Thus, it is no coincidence that the computed value for the numerators in $\overline{T}_q$ and $\overline{L}_q$ for the example is 39. Operationally, this must be the case. Define $\overline{R}$ as the average rate that customers exit the queue. The average rate of customers exiting the queue can be estimated by counting the number of customers exiting the queue over a period of time. That is,

$$\overline{R} = \frac{n}{t_n - t_0}$$

This quantity is often called the average throughput rate. For the example, $\overline{R} = 8/25$. By combining these equations, it becomes clear that the following relationship holds:

$$\overline{L}_q = \frac{\int_{t_0}^{t_n} q(t)dt}{t_n - t_0} = \frac{n}{t_n - t_0} \times \frac{\sum_{i=1}^{n} T_{q_i}}{n} = \overline{R} \times \overline{T}_q$$

The previous relationship is a conservation law and can also be applied to other portions of the queuing system as well. In words, the relationship states that the

Average number in the queue = average throughput rate × average waiting time in the queue

When the service portion of the system is considered, then the relationship can be translated as

Average number in service = average throughput rate × average time in service

When the entire queuing system is considered, the relationship yields

Average number in the system = average throughput rate × average time in the system

These relationships hold operationally for these statistical quantities as well as for the expected values of the random variables that underlie the stochastic processes. This relationship is called Little's formula after the queuing theorist who first formalized the technical conditions of its applicability to the stochastic processes in queues of this nature. See Little (1961) and Glynn and Whitt (1989) for more on these relationships. In particular, Little's formula states a relationship between the steady-state expected values for these processes.

To develop these formulas, define N, N_q, N_b as random variables that represent the number of customers in the system, in the queue, and in service at an arbitrary point in time in steady state. Also, let λ be the expected arrival rate so that $1/\lambda$ is the mean of the interarrival time distribution and let $\mu = 1/E[ST]$ so that $E[ST] = 1/\mu$ is the mean of the service time distribution. The expected values of the quantities of interest can be defined as:

$$L \equiv E[N] \quad L_q \equiv E[N_q] \quad B \equiv E[N_b]$$
$$W \equiv E[T] \quad W_q \equiv E[T_q]$$

Thus, it should be clear that $L = L_q + B$ and $W = W_q + E[ST]$. In steady state, the mean arrival rate to the system should also be equal to the mean throughput rate. Thus, based on Little's relationship the following are true:

$$L = \lambda W \quad L_q = \lambda W_q \quad B = \lambda E[ST] = \frac{\lambda}{\mu}$$

To gain an intuitive understanding of Little's formulas in this situation, consider that in steady state, the mean rate that customers exit the queue must also be equal to the mean rate that customers enter the queue. Suppose that you are a customer departing the queue and you "look back" to see how many customers are left in the queue. This quantity should be L_q on average. If it took you an average W_q to get through the queue, how many customers would have arrived on average during this time? If the customers arrive at rate λ, then $\lambda \times W_q$ is the number of customers (on average) that would have arrived during your time in the queue, but these are the customers that you would see (on average) when "looking back." Thus, $L_q = \lambda W_q$.

Note that λ and μ must be given, and therefore B is known. The quantity B represents the expected number of customers in service in steady state, but since a customer uses only one server while in service, B also represents the expected number of busy servers in steady state. If there are c identical servers in the resource, then the quantity B/c represents the fraction of the servers that are busy. This quantity can be interpreted as the utilization of the resource as a whole or the average utilization of a server, since they are all identical. This quantity is defined as

$$\rho = \frac{B}{c} = \frac{\lambda}{c\mu}$$

The quantity $c\mu$ represents the maximum rate at which the system can perform work on average. Because of this, $c\mu$ can be interpreted as the mean capacity of the system. One of the technical conditions required for Little's formula to be applicable is that $\rho < 1$. That is, the mean arrival rate to the system must be less than mean capacity of the system. This also implies that the utilization of the resource must be less than 100%.

The queuing system can also be characterized in terms of the *offered load*. The offered load is a dimensionless quantity that gives the average amount of work offered per time unit to the c servers. The offered load is defined as $r = \lambda/\mu$. Note that this can be interpreted as each customer arriving with $1/\mu$ average units of work to be performed. The steady-state conditions thus indicate that $r < c$. In other words, the arriving amount of work to the queue cannot exceed the number of servers. These conditions make sense for steady-state results to be applicable, since if the mean arrival rate of work was greater than the mean capacity of the system, the waiting line would continue to grow over time.

5.1.1 Deriving Formulas for Markovian Single-Queue Systems

Note that with $L = \lambda W$ and the other relationships, all of the major performance measures for the queue can be computed if a formula for the expected number of customers in the system, L, is available. In order to derive a formula for L, the arrival and service processes must be specified. This section shows that for the case of exponential time between arrivals and exponential service times, the necessary formulas can be readily derived. It is useful to work through the basic derivations in order to better understand the interpretation of the various performance measures, the implications of the assumptions, and the concept of steady state.

Let's consider a simple example. Suppose you want to model an old-style telephone booth, which can hold only one person while the person uses the phone. Also assume that any people that arrive while the booth is in use immediately leave. In other words, nobody waits to use the booth. For this system, it is important to understand the behavior of the stochastic process $\{N(t); t \geq 0\}$, where $N(t)$ represents the number of people that are in the phone booth at any time t. Clearly, the possible values of $N(t)$ are 0 and 1, that is, $N(t) \in \{0, 1\}$. Developing formulas for the probability that there are 0 or 1 customers in the booth at any time t, that is, $P_i(t) = P\{N(t) = i\}$, will be the key to modeling this situation. Let λ be the mean arrival rate of customers to the booth, and let $E[ST] = 1/\mu$ be the expected length of a telephone call. For example, if the mean time between arrivals is 12 minutes, then $\lambda = 5/hr$, and if the mean length of a call is 10 minutes, then $\mu = 6/hr$. The following reasonable assumptions will be made:

1. The probability of a customer arriving in a small interval of time, Δt, is roughly proportional to the length of the interval, with the proportionality constant equal to the mean rate of arrival.

2. The probability of a customer completing an ongoing phone call during a small interval of time is roughly proportional to the length of the interval, and the proportionality constant is equal to the mean service rate.

3. The probability of more than one arrival in an arbitrarily small interval, Δt, is negligible. In other words, Δt can be made small enough so that only one arrival can occur in the interval.

4. The probability of more than one service completion in an arbitrarily small interval, Δt, is negligible. In other words, Δt can be made small enough so that only one service can occur in the interval.

Let $P_0(t)$ and $P_1(t)$ represent the probability that there is 0 or 1 customer using the booth respectively. Suppose that you observe the system at time t, and you want to derive the probability for 0 customers in the system at some future time, $t + \Delta t$. Thus, you want $P_0(t + \Delta t)$. For there to be 0 customers in the booth at time $t + \Delta t$, two possible situations could occur. First, there could have been no customers in the system at time t and no arrivals during the interval Δt, or there could have been one customer in the system at time t and the customer completed service during Δt. Thus, the following relationship should hold:

$$\{N(t + \Delta t) = 0\} = \{\{N(t) = 0\} \cap \{\text{no arrivals during } \Delta t\}\} \cup$$
$$\{\{N(t) = 1\} \cap \{\text{service completed during } \Delta t\}\}$$

It follows that

$$P_0(t + \Delta t) = P_0(t)P\{\text{no arrivals during } \Delta t\} + P_1(t)P\{\text{service completed during } \Delta t\}$$

In addition, at time $t + \Delta t$, there might be a customer using the booth: $P_1(t + \Delta t)$. For there to be one customer in the booth at time $t + \Delta t$, there are two possible situations that could occur. First, there could have been no customers in the system at time t and one arrival during the interval Δt, or there could have been one customer in the system at time t and the customer did not complete service during Δt. Thus, the following holds:

$$P_1(t + \Delta t) = P_0(t)P\{1 \text{ arrival during } \Delta t\} + P_1(t)P\{\text{service not completed during } \Delta t\}$$

Because of assumptions 1 and 2, the following probability statements can be used:

$$P\{1 \text{ arrival during } \Delta t\} \cong \lambda \Delta t$$
$$P\{\text{no arrivals during } \Delta t\} \cong 1 - \lambda \Delta t$$
$$P\{\text{service during } \Delta t\} \cong \mu \Delta t$$
$$P\{\text{no service during } \Delta t\} \cong 1 - \mu \Delta t$$

This results in the following:

$$P_0(t + \Delta t) = P_0(t)[1 - \lambda \Delta t] + P_1(t)[\mu \Delta t] \tag{1}$$

$$P_1(t + \Delta t) = P_0(t)[\lambda \Delta t] + P_1(t)[1 - \mu \Delta t] \tag{2}$$

Collecting the terms of the equations, rearranging, dividing by Δt, and taking the limit as Δt goes to zero, yields the following set of differential equations:

$$\frac{dP_0(t)}{dt} = \lim_{\Delta t \to 0} \frac{P_0(t + \Delta t) - P_0(t)}{\Delta t} = -\lambda P_0(t) + \mu P_1(t) \tag{3}$$

$$\frac{dP_1(t)}{dt} = \lim_{\Delta t \to 0} \frac{P_1(t + \Delta t) - P_1(t)}{\Delta t} = \lambda P_0(t) - \mu P_1(t) \tag{4}$$

It is also true that $P_0(t) + P_1(t) = 1$. Assuming $P_0(0) = 1$ and $P_1(0) = 0$ as the initial conditions, the solutions to these differential equations are

$$P_0(t) = \left(\frac{\mu}{\lambda + \mu}\right) + \left(\frac{\lambda}{\lambda + \mu}\right) e^{-(\lambda + \mu)t} \tag{5}$$

$$P_1(t) = \left(\frac{\lambda}{\lambda + \mu}\right) - \left(\frac{\lambda}{\lambda + \mu}\right) e^{-(\lambda + \mu)t} \tag{6}$$

These equations represent the probability of having either 0 or 1 customer in the booth at any time. If the limit as t goes to infinity is considered, the *steady state probabilities*, P_0 and P_1, can be determined:

$$P_0 = \lim_{t \to \infty} P_0(t) = \frac{\mu}{\lambda + \mu} \tag{7}$$

$$P_1 = \lim_{t \to \infty} P_1(t) = \frac{\lambda}{\lambda + \mu} \tag{8}$$

These probabilities can be interpreted as the chance that an arbitrary customer finds the booth empty or busy after an infinitely long period of time has elapsed. If only the steady-state probabilities are desired, there is an easier method to perform the derivation, both from a conceptual and a mathematical standpoint. The four previous assumptions ensure that the arrival and service processes will be Markovian. In other words, the time between arrivals of the customer is exponentially distributed and the service times are exponentially distributed. In addition, the concept of steady state can be used. Consider the differential equations given by equations 3 and 4. These equations govern the rate of change of the *probabilities* over time. Consider the analogy of water to probability and think of a dam or container that holds an amount of water. The rate of change of the level of water in the container can be thought of as

Rate of change of level = rate into container − rate out of the container

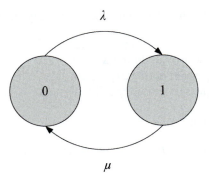

λ

μ

■ **FIGURE 5-4**   **Two-state transition diagram**

In steady state, the level of the water should not change; thus, the rate into the container must equal the rate out of the container. Using this analogy,

$$\frac{dP_i(t)}{dt} = rate\ in - rate\ out$$

and for steady state, *rate in = rate out* with the probability "flowing" between the states. Figure 5-4 illustrates this concept via a state-transition diagram. If N represents the steady-state number of customers in the system (booth), the two possible states that the system can be in are 0 and 1. The rate of transition from state 0 to state 1 is the rate that an arrival occurs, and the rate of transition from state 1 to state 0 is the service rate.

The rate of transition into state 0 can be thought of as the rate that probability "flows" from state 1 to state 0 times the chance of being in state 1, that is, μP_1. The rate of transition out of state 0 can be thought of as the rate from state 0 to state 1 times the chance of being in state 0, that is, λP_0. Using these ideas yields

state	rate in	=	rate out
0	μP_1	=	λP_0
1	λP_0	=	μP_1

Note that these are identical equations, but with the fact that $P_0 + P_1 = 1$, the equations can be easily solved to yield the same results as equations 7 and 8. Sets of equations derived in this manner are called steady-state equations.

Now more general situations can be examined. Consider a general queuing system with a given number of servers. An arrival at the system represents a "birth" and a departure from the system represents a "death." Figure 5-5 illustrates a general state-transition diagram for a system with births and deaths. Let N be the number of customers in the system in steady state, and define $P_n = P\{N = n\} = \lim_{t \to \infty} P\{N(t) = n\}$ as the steady-state probability that there are n customers in the system. Let λ_n be the mean arrival rate of customers entering the system when there are n customers in the system, $\lambda_n \geq 0$. Let μ_n be the mean service rate for the overall

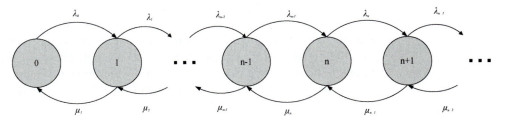

■ **FIGURE 5-5**   **General state-transition diagram**

system when there are n customers in the system. This is the rate at which customers depart when there are n customers in the system. In this situation, the number of customers may be infinite, that is, $N \in \{0, 1, 2, \ldots\}$.

The steady-state equations for this situation are as follows:

$$
\begin{array}{cccc}
state & rate\ in & = & rate\ out \\
0 & \mu_1 P_1 & = & \lambda_0 P_0 \\
1 & \lambda_0 P_0 + \mu_2 P_2 & = & \mu_1 P_1 + \lambda_1 P_1 \\
2 & \lambda_1 P_1 + \mu_3 P_3 & = & \mu_2 P_2 + \lambda_2 P_2 \\
\vdots & \vdots & \vdots & \vdots \\
n & \lambda_{n-1} P_{n-1} + \mu_{n+1} P_{n+1} & = & \mu_n P_n + \lambda_n P_n
\end{array}
$$

These equations can be solved recursively starting with state 0, which yields

$$
P_1 = \frac{\lambda_0}{\mu_1} P_0
$$

$$
P_2 = \frac{\lambda_1 \lambda_0}{\mu_2 \mu_1} P_0
$$

$$
\vdots
$$

$$
P_n = \frac{\lambda_{n-1} \lambda_{n-2} \cdots \lambda_0}{\mu_n \mu_{n-1} \cdots \mu_1} P_0 = \prod_{j=0}^{n-1} \left(\frac{\lambda_j}{\mu_{j+1}} \right) P_0 \quad \text{for } n = 1, 2, 3, \ldots
$$

Provided that $\sum_{n=0}^{\infty} P_n = 1$, P_0 can be computed as

$$
P_0 = \left[\sum_{n=0}^{\infty} \prod_{j=0}^{n-1} \left(\frac{\lambda_j}{\mu_{j+1}} \right) \right]^{-1}
$$

Therefore, for any given set of λ_n and μ_n, P_n can be computed. The P_n represent a probability distribution, and therefore the expected value of this distribution can be computed. What is the expected value for the P_n distribution? The expected number of customers in the system in steady state. The expected number of customers in the system and the queue are given by

$$
L_s = \sum_{n=0}^{\infty} n P_n
$$

$$
L_q = \sum_{n=c}^{\infty} (n - c) P_n
$$

where c is the number of servers. One more formula is needed before Little's formula can be applied with other known relationships. For certain systems, such as finite-size systems, not all arriving customers will enter the queue. Little's formula is true for the customers that are able to enter the system. Thus, the effective arrival rate, or mean rate of arrivals entering the system, must be defined. This is given by computing the expected arrival rate across the states:

$$
\lambda_e = \sum_{n=0}^{\infty} \lambda_n P_n
$$

for infinite system size and

$$
\lambda_e = \sum_{n=0}^{k-1} \lambda_n P_n
$$

for system size k, since $\lambda_n = 0$ for $n \geq k$ (because nobody can enter when the system is full). All these relationships yield

$$L_s = \lambda_e W_s \quad L_q = \lambda_e W_q \quad B = \frac{\lambda_e}{\mu} \quad \rho = \frac{\lambda_e}{c\mu} \quad L_s = L_q + B \quad W_s = W_q + \frac{1}{\mu}$$

Tables 5-1 and 5-2 present the results of applying the general solution for P_n to various queuing system configurations. Table 5-3 presents specific results for the M/M/c queuing system for $c = 1, 2, 3$. The results for the single server, general and deterministic service distribution and Poisson arrivals are given in Table 5-4. These results make possible the evaluation of various queuing situations.

This section has only "scratched the surface" of queuing theory. A vast amount of literature is available on queuing theory and its applications. See, for instance, Gross and Harris (1998), Cooper (1990), and Kleinrock (1975). A number of free on-line resources are also available on the topic (see http://web2.uwindsor.ca/math/hlynka/qonline.html).

5.1.2 Examples and Applications of Queuing Analysis

The formulas in the previous section certainly appear to be intimidating and they can be tedious to apply. Fortunately, readily available spreadsheet software can be used to perform the calculations. On-line resources for queuing-analysis spreadsheet software follow:

- Excel-based queuing-theory software that accompanies Gross and Harris (1998), at www.geocities.com/qtsplus
- Excel-based add-in that accompanies Jensen and Bard (2003), at www.me.utexas.edu/~jensen/ORMM/

Thus, the most important part of performing a queuing analysis is to identify the most appropriate queuing model for a given situation. This section provides a number of examples and discusses the differences among systems so that you can better apply results of previous sections. The solutions to these types of problems involve the following steps:

1. Identify the arrival and service processes.
2. Identify the size of the arriving population and the size of the system.
3. Specify the appropriate queuing model and its input parameters.
4. Identify the desired performance measures.
5. Compute the required performance measures.

Example 1. Customers arrive at a one window drive through pharmacy according to a Poisson distribution with a mean of 10 per hour. The service time per customer is exponential with a mean of 5 minutes. There are 3 spaces in front of the window, including that for the car being served. Other arriving cars can wait outside these 3 spaces. The pharmacy is interested in answering the following questions:

(a) What is the probability that an arriving customer can enter one of the 3 spaces in front of the window?
(b) What is the probability that an arriving customer will have to wait outside the 3 spaces?
(c) What is the probability that an arriving customer has to wait?
(d) How long is an arriving customer expected to wait before starting service?
(e) How many car spaces should be provided in front of the window so that an arriving customer has a 40% chance of being able to wait in one of the provided spaces?

■ **TABLE 5-1**
Results for P_0 and P_n for Various Queuing Systems

Notation	Parameters	P_0	P_n
M/M/1	$\lambda_n = \lambda; \mu_n = \mu;$ $c = 1; \rho = \dfrac{\lambda}{c\mu} = r$	$P_o = 1 - r$	$P_n = P_o r^n$
M/M/c	$\lambda_n = \lambda;$ $\mu_n = \begin{cases} n\mu & 0 \le n < c \\ c\mu & n \ge c \end{cases};$ $\rho = \lambda/c\mu; r = \lambda/\mu$	$P_0 = \left[\displaystyle\sum_{n=0}^{c-1}\frac{r^n}{n!} + \frac{r^c}{c!(1-\rho)}\right]^{-1}$	$P_n = \begin{cases} \dfrac{r^n}{n!}P_0 & 1 \le n < c \\[2mm] \dfrac{r^n}{c!\,c^{n-c}}P_0 & n \ge c \end{cases}$
M/M/1/k	$\lambda_n = \begin{cases} \lambda & n < k \\ 0 & n \ge k \end{cases};$ $\mu_n = \begin{cases} \mu & 0 \le n \le k \\ 0 & n > k \end{cases}$ $c = 1; \rho = \dfrac{\lambda}{c\mu} = r;$ $\lambda_e = \lambda(1 - P_k)$	$P_0 = \begin{cases} \dfrac{1-r}{1-r^{k+1}} & r \ne 1 \\[2mm] \dfrac{1}{k+1} & r = 1 \end{cases}$	$P_n = \begin{cases} P_0 r^n & r \ne 1 \\[2mm] \dfrac{1}{k+1} & r = 1 \end{cases}$
M/M/c/k	$\lambda_n = \begin{cases} \lambda & n < k \\ 0 & n \ge k \end{cases};$ $\mu_n = \begin{cases} n\mu & 0 \le n < c \\ c\mu & c \le n \le k \end{cases}$ $\rho = \lambda/c\mu; r = \lambda/\mu;$ $\lambda_e = \lambda(1 - P_k)$	$P_0 \begin{cases} \left[\displaystyle\sum_{n=0}^{c-1}\frac{r^n}{n!} + \frac{r^c}{c!}\frac{1-\rho^{k-c+1}}{1-\rho}\right]^{-1} & \rho \ne 1 \\[4mm] \left[\displaystyle\sum_{n=0}^{c-1}\frac{r^n}{n!} + \frac{r^c}{c!}(k-c+1)\right]^{-1} & \rho = 1 \end{cases}$	$P_n = \begin{cases} \dfrac{r^n}{n!}P_0 & 1 \le n < c \\[2mm] \dfrac{r^n}{c!\,c^{n-c}}P_0 & c \le n \le k \end{cases}$
M/M/c/c M/G/c/c	$\lambda_n = \begin{cases} \lambda & n < c \\ 0 & n \ge c \end{cases};$ $\mu_n = \begin{cases} n\mu & 0 \le n \le c \\ 0 & n > c \end{cases}$ $\rho = \lambda/c\mu; r = \lambda/\mu;$ $\lambda_e = \lambda(1 - P_c)$	$P_0 = \left[\displaystyle\sum_{n=0}^{c}\frac{r^n}{n!}\right]^{-1}$	$P_n = \dfrac{r^n}{n!}P_0 \quad 0 \le n \le c$
M/M/1/k/k	$\lambda_n = \begin{cases} (k-n)\lambda & 0 \le n < k \\ 0 & n \ge k \end{cases}$ $\mu_n = \begin{cases} \mu & 0 \le n \le k \\ 0 & n > k \end{cases}$ $r = \lambda/\mu; \lambda_e = \lambda(k-L)$	$P_0 = \left[\displaystyle\sum_{n=0}^{k}\prod_{j=0}^{n-1}\left(\frac{\lambda_j}{\mu_{j+1}}\right)\right]^{-1}$	$P_n = \dbinom{k}{n}n!\,r^n P_0$ $0 \le n \le k$
M/M/c/k/k	$\lambda_n = \begin{cases} (k-n)\lambda & 0 \le n < k \\ 0 & n \ge k \end{cases}$ $\mu_n = \begin{cases} n\mu & 0 \le n < c \\ c\mu & n \ge c \end{cases}$ $r = \lambda/\mu; \lambda_e = \lambda(k-L)$	$P_0 = \left[\displaystyle\sum_{n=0}^{k}\prod_{j=0}^{n-1}\left(\frac{\lambda_j}{\mu_{j+1}}\right)\right]^{-1}$	$P_n = \begin{cases} \dbinom{k}{n}r^n P_0 & 1 \le n < c \\[2mm] \dbinom{k}{n}\dfrac{n!}{c^{n-c}c!}r^n P_0 & c \le n \le k_0 \end{cases}$

■ **TABLE 5-2**
Results for L_q for Various Queuing Systems

Notation	L_q
M/M/1	$L_q = \dfrac{r^2}{1-r}$
M/M/c	$L_q = \left(\dfrac{r^c \rho}{c!(1-\rho)^2}\right) P_0$
M/M/1/k	$L_q = \begin{cases} \dfrac{\rho}{1-\rho} - \dfrac{\rho(k\rho^k + 1)}{1-\rho^{k+1}} & \rho \neq 1 \\[2ex] \dfrac{k(k-1)}{2(k+1)} & \rho = 1 \end{cases}$
M/M/c/k	$L_q = \begin{cases} \dfrac{P_0 r^c \rho}{c!(1-\rho)^2}\left[1 - \rho^{k-c} - (k-c)\rho^{k-c}(1-\rho)\right] & \rho < 1 \\[2ex] \dfrac{r^c(k-c)(k-c+1)}{2c!}P_0 & \rho = 1 \end{cases}$
M/M/c/c M/G/c/c	$L_q = 0$
M/M/1/k/k	$L_q = k - \left(\dfrac{\lambda + \mu}{\lambda}\right)(1 - P_0)$
M/M/c/k/k	$L_q = \displaystyle\sum_{n=c}^{k}(n-c)P_n$

■ **TABLE 5-3**
Results for M/M/c $\rho = \lambda/c\mu$

c	P_0	L_q
1	$1 - \rho$	$\dfrac{\rho^2}{1-\rho}$
2	$\dfrac{1-\rho}{1+\rho}$	$\dfrac{2\rho^3}{1-\rho^2}$
3	$\dfrac{2(1-\rho)}{2 + 4\rho + 3\rho^2}$	$\dfrac{9\rho^4}{2 + 2\rho - \rho^2 - 3\rho^3}$

■ **TABLE 5-4**
Results for M/G/1 and M/D/1

	Parameters	L_q
M/G/1	$E[ST] = \dfrac{1}{\mu}; Var[ST] = \sigma^2; r = \lambda/\mu$	$L_q = \dfrac{\lambda^2\sigma^2 + r^2}{2(1-r)}$
M/D/1	$E[ST] = \dfrac{1}{\mu}; Var[ST] = 0; r = \lambda/\mu$	$L_q = \dfrac{r^2}{2(1-r)}$

Example 1 Solution. The customers arrive according to a Poisson process, which implies that the time between arrivals is exponentially distributed. Thus, the arrival process is Markovian (M). The stated service process is exponential. Thus, the service process is Markovian (M). There is only 1 window and customers wait in front of this window to receive service. Thus, the number of servers is $c = 1$. The problem states that customers arriving when the 3 spaces are filled, wait for service outside the 3 spaces. Thus, there does not appear to be a restriction on the size of the waiting line, making this an infinite-size system. The arrival rate is specified for the group of customers and there is no information given concerning the total population of customers. Thus, it appears that an infinite population of customers can be assumed. Consequently, this is an M/M/1 queuing situation with an arrival rate of $\lambda = 10/hr$ and a service rate of $\mu = 12/hr$. Note that the input parameters have been converted to a common unit of measure (customers/hour). Let's consider each question in turn:

(a) Probability statements of this form are related to the underlying state variable for the system. In this case, let N represent the number of customers in the system. To find the probability that an arriving customer can enter one of the 3 spaces in front of the window, consider the question, "When can a customer enter one of the 3 spaces?" A customer can enter one of the three spaces when there are $N = 0$, 1, or 2 customers in the system. This is not $N = 0$, 1, 2, or 3, because if there are 3 customers in the system, then the 3rd space is taken. Therefore, $P\{N \leq 2\}$ needs to be computed.

(b) An arriving customer will have to wait outside the 3 spaces, when there are more than 2 (3 or more) customers in the system. Thus, $P\{N > 2\}$ needs to be computed. This is the complement event for (a).

(c) An arriving customer has to wait when there are 1 or more customers already at the pharmacy, or $P\{N \geq 1\} = 1 - P\{N < 1\} = 1 - P_0$.

(d) The waiting time that does not include service is the queuing time. Thus, W_q needs to be computed.

(e) This is a design question for which the probability of waiting in one of the provided spaces is used to determine the number of spaces to provide. Suppose that there are m spaces. An arriving customer can waiting in one of the spaces if there are $m - 1$ or less customers in the system. Thus, m needs to be chosen such that $P\{N \leq m - 1\} = 0.4$.

The equations for the M/M/1 can be readily applied. From Tables 5-1 and 5-2, the following formulas can be applied:

$$\lambda_n = \lambda \qquad\qquad \mu_n = \mu \qquad \rho = \frac{\lambda}{\mu} = r < 1 \qquad \lambda_{eff} = \lambda$$

$$P_o = 1 - \frac{\lambda}{\mu} = 1 - r \qquad P_n = P_o r^n \qquad L_s = \frac{r}{1-r} \qquad\qquad L_q = \frac{r^2}{1-r}$$

$$P\{N \geq n\} = \sum_{j=n}^{\infty} P_0 r^j = (1-r)\sum_{j=n}^{\infty} r^{j-n} = (1-r)\frac{r^n}{(1-r)} = r^n$$

$$P\{N \leq n\} = 1 - P\{N > n\} = 1 - P\{N \geq n+1\} = 1 - r^{n+1}$$

For this problem, utilization can be computed as $\rho = \lambda/\mu = (10/12) = 5/6$, which yields

$$P\{N \leq 2\} = 1 - r^3 \cong 0.42$$
$$P\{N > 2\} = 1 - P\{N \leq 2\} \cong 0.58$$
$$P\{N \geq 1\} = 1 - P\{N < 1\} = 1 - P_0 = 1 - (1-r) = \rho = 5/6$$

This is also the probability that the pharmacist will be busy.

$$W_q = L_q/\lambda = r/(\mu(1-r)) \cong 0.417 \, hour = 25.02 \, \text{min}$$
$$P\{N \leq m-1\} = \gamma \quad 1 - r^m = \gamma \quad r^m = 1 - \gamma$$
$$m = \frac{\ln(1-\gamma)}{\ln r} = \frac{\ln(1-0.4)}{\ln(5/6)} = 2.8 \approx 3 \, spaces$$

(Rounding up guarantees $P\{N \leq m-1\} \geq 0.4$.)

Example 2. Student union copy center managers are considering the installation of self-service copiers. They predict that arrivals will be governed by a Poisson distribution with a rate of 30 per hour and that the time spent copying is exponentially distributed with a mean of 1.75 minutes. They would like the probability of 4 or more people in the copy center to be less than 5%. How many copiers should they install?

Example 2 Solution. The Poisson arrivals and exponential service times make this situation an M/M/c where c is the number of copiers to install and $\lambda = 0.5$ and $\mu = 1/1.75$ per minute. To meet the design criteria, $P\{N \geq 4\} = 1 - P\{N \leq 3\}$ needs to be computed for systems with $c = 1, 2, \ldots$ until this probability is less than 5%. This can be readily achieved with the provided formulas or by using the aforementioned software. In the following, the QTSPlus software (found at http://www.geocities.com/qtsplus/) was used (Figures 5-6 and 5-7). By changing the number of servers, $c = 3$ meets the probability requirement as shown in Table 5-5.

Often in the case of systems with multiple servers such as the M/M/c, you want to determine the best value of c, as in Example 2. Another common design situation is to determine the value of c such that there is an acceptable probability that an arriving customer will have to wait. For the case of Poisson arrivals, this is the same as the steady-state probability that there are more than c customers in the system. For the M/M/c model, this is called the Erlang delay probability:

$$P_w = P\{N \geq c\} = \sum_{n=c}^{\infty} P_n = \frac{\frac{r^c}{c!}}{\frac{r^c}{c!} + (1-\rho)\sum_{j=0}^{c-1}\frac{r^j}{j!}}$$

Even though this is a relatively easy formula to use (especially in view of available spreadsheet software), an interesting and useful approximation has been developed called the *square root staffing rule*. The derivation of the square root staffing rule is given in Tijms (2003). In the following, the usefulness of the rule is discussed. The square root staffing rule states that

M/M/c: POISSON ARRIVALS TO MULTIPLE EXPONENTIAL SERVERS

Input Parameters:

Arrival rate (λ)	0.5
Mean service time ($1/\mu$)	1.75
Number of servers in the system (c)	3

Plot Parameters:

Maximum size for probability chart	15
Total time horizon for probability plotting	2.

Results:

Mean interarrival time ($1/\lambda$)	2.
Service rate (μ)	0.571429
Average # arrivals in mean service time (r)	0.875
Server utilization (ρ)	29.17%
Fraction of time all servers are idle (p_0)	0.414003
Mean number of customers in the system (L)	0.901871
Mean number of customers in the queue (Lq)	0.026871
Mean wait time (W)	1.803743
Mean wait time in the queue (Wq)	0.053743

■ **FIGURE 5-6**   QTS M/M/c spreadsheet

	A	B	C	D
1	**Customer Size Distribution**			
2	n	prob(n)	CDF(n)	1-CDF(n)
3	0	0.414003	0.414003	0.585997
4	1	0.362253	0.776256	0.223744
5	2	0.158486	0.934741	0.065259
6	3	0.046225	0.980966	0.019034
7	4	0.013482	0.994448	0.005552
8	5	0.003932	0.998381	0.001619
9	6	0.001147	0.999528	0.000472
10	7	0.000335	0.999862	0.000138
11	8	0.000098	0.999960	0.000040
12	9	0.000028	0.999988	0.000012
13	10	0.000008	0.999997	0.000003
14	11	0.000002	0.999999	0.000001
15	12	0.000001	1.000000	0.000000
16	13	0.000000	1.000000	0.000000
17	14	0.000000	1.000000	0.000000
18	15	0.000000	1.000000	0.000000

■ **FIGURE 5-7**   QTS results for Example 2 for c = 3

■ **TABLE 5-5**

Results for Example 2, c = 1, 2, 3, and 4

c	$P\{N \geq 4\} = 1 - P\{N \leq 3\}$
1	0.586182
2	0.050972
3	0.019034
4	0.013022

	A	B	C	D	E
1	**Square Root Staffing Rule**				
2					
3	Offered Load = r	0.875			
4	Probability of delay criteria = alpha	0.1			
5	Initial Factor = gamma	1.420189435			
6	LHS of factor equation	9.000052301	$\frac{\gamma\Phi(\gamma)}{\varphi(\gamma)}$ $\frac{1-\alpha}{\alpha}$		
7	RHS of factor equation	9			
8	Root finding equation	-5.23013E-05			
9	Approximate number of servers needed	2.203465572	$c^* \cong r + \gamma_\alpha \sqrt{r}$		
10	Number of servers rounded up	3			
11					
12	Goal Seek ☒	1) Enter the offered load and delay criteria			
13		2) Enter an initial value for gamma. The value 1 will			
14	Set cell: B8 ▦	always work.			
15	To value: 0	3) Run goal seek as shown (setting B8 to zero by			
16	By changing cell: $B$5 ▦	changing B5)			
17					
18	OK Cancel	Once the factor gamma has been found for the given			
19		alpha, you can vary the offered load and get the			
20		number of servers needed for different values of the			
21		offered load.			
22					

■ **FIGURE 5-8** Spreadsheet for square root staffing rule

the least number of servers c^* required to meet the criteria $P_w \leq \alpha$ is given by $c^* \cong r + \gamma_\alpha \sqrt{r}$, where the factor γ_α is the solution to the following equation:

$$\frac{\gamma\Phi(\gamma)}{\varphi(\gamma)} = \frac{1-\alpha}{\alpha}$$

The functions $\Phi(\cdot)$ and $\varphi(\cdot)$ are the cumulative distribution function (CDF) and the probability density function (PDF) of a standard normal random variable.

Suppose that you want to determine the lowest number of servers such that the probability of an arriving customer waiting is less than or equal to 0.10. Thus, for $\lambda = 0.5$ and $\mu = 1/1.75$ per minute, you have that $r = 0.875$. Figure 5-8 illustrates the use of the *SquareRootStaffingRule.xls* spreadsheet that accompanies this chapter. The spreadsheet uses Excel's built-in Goal-Seek solver to solve for γ_α. For this problem, the result indicates that about 3 servers are needed to ensure that the wait probability will be less than 10%. The square root staffing rule has been shown to be quite a robust approximation, and can be useful in many design settings involving staffing.

Example 3. A single machine is connected to a conveyor system. The conveyor causes parts to arrive at the machine at a rate of 1 part per minute according to a Poisson distribution. There is a finite buffer of size 5 in front of the machine. The machine's processing time is considered to be exponentially distributed with a mean rate of 1.2 parts per minute. Any parts that arrive on the conveyor when the buffer is full are carried to other machines that are not part of this analysis. What are the expected system time and the expected number of parts at the machining center?

Example 3 Solution. The finite buffer, Poisson arrivals, and exponential service times make this situation an M/M/1/6, where $k = 6$ is the size of the system (5 in buffer + 1 in service) and $\lambda = 1$ and $\mu = 1.2$ per minute. The desired performance measures are W_s and L_s.

```
┌─────────────────────────────────────────────────────────────────────┐
│ M/M/1/K: POISSON ARRIVALS TO A SPACE-LIMITED SINGLE EXPONENTIAL SERVER │
│                                                                         │
│   Input Parameters:                                                     │
│       Arrival rate (λ)                                    1.            │
│       Mean service time (1/μ)                        .833333           │
│       Maximum capacity of system (K > 1)                  6            │
│                                                                         │
│   Plot Parameters:                                                      │
│       Total time horizon for probability plotting        15.           │
│           [ALL PROBABILITIES ARE PLOTTED!]                              │
│                                                                         │
│   Results:                                                              │
│       Mean interarrival time (1/λ)                         1           │
│       Effective arrival rate (λ_eff)              0.92257607           │
│       Service rate (μ)                                   1.2           │
│       Traffic intensity (ρ)  [NEED NOT BE < 1]    0.83333333           │
│       Server utilization (ρ_eff)  [MUST BE < 100%]   76.88%            │
│       Fraction of time the server is idle (p0)    0.23118661           │
│       Probability that the system is full (pK)    0.07742393           │
│       Expected number turned away/unit time       0.07742393           │
│       Mean number of customers in the system (L)  2.29016258           │
│       Mean number of customers in the queue (Lq)  1.52134918           │
│       Mean waiting time (W)                       2.48235635           │
│       Mean waiting time in the queue (Wq)         1.64902302           │
└─────────────────────────────────────────────────────────────────────┘
```

■ **FIGURE 5-9**   Results for Example 3

Figure 5-9 presents the results using QTSPlus. Note that in this case, the effective arrival rate must be computed:

$$\lambda_e = \sum_{n=0}^{k-1} \lambda_n P_n = \sum_{n=0}^{k-1} \lambda P_n = \lambda \sum_{n=0}^{k-1} P_n = \lambda(1 - P_k)$$

Rearranging this formula yields $\lambda = \lambda_e + \lambda_{lost}$, where $\lambda_{lost} = \lambda P_k$ equals the mean number of customers turned away from the system because it is full. According to Figure 5-9, the average number of customers lost is about 0.077 per minute (or about 4.62 per hour).

Example 4. The university has a row of 10 parking meter–based spaces across from the engineering school. During peak hours, students arrive at the parking lot at a rate of 40 per hour according to a Poisson distribution, and the students use the parking space for approximately 60 minutes exponentially distributed. If all parking spaces are taken, it can be assumed that an arriving student does not wait (goes somewhere else to park). Suppose that the meters cost $w = \$0.03$ per minute, that is, \$2 per hour. How much income does the university potentially lose during peak hours on average because the parking spaces are full?

Solution Example 4. In this system, the parking spaces are the servers of the system. There are 10 parking spaces so that $c = 10$. In addition, there is no waiting for meters; thus, no queue forms. Therefore, the system size is also 10 ($k = 10$). Because of the Poisson arrival process and exponential service times, this can be considered an M/M/10/10 queuing system. In other words, the size of the system is the same as the number of servers, $c = k = 10$.

In the case of an M/M/c/c queuing system, P_c represents the probability that all servers in the system are busy. Thus, it also represents the probability that an arriving customer will be

```
M/G/c/c: PURE OVERFLOW MODEL
    Poisson input to multiple servers, with no queue (for any service distribution).

Input Parameters:
    Mean arrival rate (λ)                              40.
    Mean service time (1/μ)                             1.
    Number of available servers (c)                    10

Results:
    Mean interarrival time (1/λ)                     0.025
    Mean effective arrival rate (λeff)            9.692485
    Mean service rate (μ)                              1.0
    Individual server utilization (ρeff)            96.92%
    Fraction of time system is full (pc)          0.757688
    Rate of lost customers                       30.307515
    Expected system size (L)                      9.692485
```

■ **FIGURE 5-10**   Example 4 M/M/c/c results

turned away. The formula for the probability of a lost customer is called the Erlang loss formula:

$$P_c = \frac{\dfrac{r^c}{c!}}{\displaystyle\sum_{n=0}^{c}\dfrac{r^n}{n!}}$$

Customers arrive at the rate $\lambda = 20/hr$, whether the system is full or not. Thus, the expected number of lost customers per hour is λP_c. Since each customer brings $1/\mu$ service time charged at $w = \$2$ per hour, each arriving customer brings w/μ of income on average; however, not all arriving customers can park. Thus, the university loses $w \times \frac{\lambda}{\mu} \times P_c$ of income per hour. According to Figure 5-10, the university is losing about $\$2 \times 30.3 = \$60.6$ per hour because the metered lot is full during peak hours.

For this example, the service times are exponentially distributed. It turns out that for the case of $c = k$, the results for the M/M/c/c model are the same for the M/G/c/c model. In other words, the form of the distribution does not matter. The mean of the service distribution is critical in this analysis.

In the following example, a finite population of customers that can arrive, depart, and then return is considered. A classic and important example of this type of system is the machine interference or operator tending problem. A detailed discussion of the analysis and application of this model can be found in Stecke (1992). In this example, a set of machines are tended by 1 or more operators. The operators must tend to stoppages (breakdowns) of the machines. As the machines break down, waiting for the operator to complete service of other machines may occur. Thus, machine stoppages cause the machines to "interfere" with the productivity of the set of machines because of their dependence on a common resource, the operator. Let's consider an example of this system.

Example 5. Suppose a manufacturing system contains 5 machines, each subject to randomly occurring breakdowns. A machine runs for an amount of time that is an exponential random variable with a mean of 10 hours before breaking down. At present there are 2 operators to fix the broken machines. The amount of time that an operator takes to service the machines is exponential with a mean of 4 hours. An operator repairs only 1 machine at a time. If more machines are broken down than the current number of operators, the machines must wait for the next available operator for repair. They form a FIFO queue to wait for the next available operator. The number of operators required to tend to the machines in order to minimize down time in a cost-effective manner is desired. Assume that it costs the system $60 per hour for each machine that is broken down. Each operator is paid $15 per hour regardless of whether she or he is repairing a machine.

Solution Example 5. This situation can be modeled with a M/M/c/k/k queuing model, where c is the number of operators and k is the number of machines. Note that the size of the system is the same as the size of the calling population in this particular model. The arrival rate of an *individual machine* is $\lambda = 1/10$ per hour and the service rate of $\mu = 1/4$ per hour for each operator. Thus, the arrival and service rates are, respectively,

$$\lambda_n = \begin{cases} (k-n)\lambda & n = 0, 1, 2, \ldots k \\ 0 & n \geq k \end{cases}$$

and

$$\mu_n = \begin{cases} n\mu & n = 1, 2, \ldots c \\ c\mu & n \geq c \end{cases}$$

as illustrated in the state diagram of Figure 5-11 for the case of 2 operators and 5 machines. Thus, for this system, the arrival rate to the system decreases as more machines break down. When all the machines are broken down, the arrival rate to the system will be zero.

Note that the customer in this system is the machine. Even though the machines do not actually "move" to line up in a queue, they form a virtual queue in front of the operators to receive repair. In order to determine the most appropriate number of operators to tend the machines, a service criteria or a way to measure system cost is required. Since costs are given, let's formulate how much a given system configuration costs.

The easiest way to formulate a cost is to consider what a given system configuration costs on a per time basis, such as cost per hour. Clearly, the system costs $15 \times c$ ($/hour) to employ

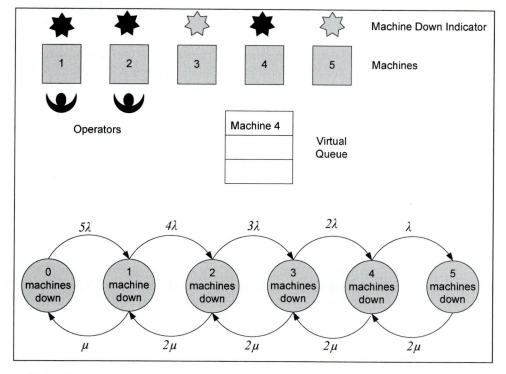

■ **FIGURE 5-11**   **Machine interference problem**

■ **TABLE 5-6**
Results for Example 5

Operator cost ($/hour)	15				
Machine cost ($/hour)	60				
Entity arrival rate	0.10	0.10	0.10	0.10	0.10
Service rate/channel	0.25	0.25	0.25	0.25	0.25
Number of servers	1	2	3	4	5
Maximum number in system	5	5	5	5	5
Number in population	5	5	5	5	5
Type	M/M/1/5/5	M/M/2/5/5	M/M/3/5/5	M/M/4/5/5	M/M/5/5/5
Mean number in system	2.674	1.661	1.457	1.430	1.429
Mean number in queue	1.744	0.325	0.040	0.002	0.000
Mean number in service	0.930	1.336	1.417	1.428	1.429
Operator utilization	0.930	0.668	0.472	0.357	0.286
Expected total cost	$175.460	$129.655	$132.419	$145.816	$160.714
Machine utilization	0.465	0.668	0.709	0.714	0.714

c operators. The problem also states that it costs $60/hour for each machine that is broken down. A machine is broken down if it is waiting for an operator or if it is being repaired by an operator. Therefore, a machine is broken down if it is in the queuing system. In terms of queuing performance measures, L_s machines can be expected to be broken down at any time in steady state. Thus, the expected steady-state cost of the broken-down machines is $60 \times L_s$ per hour. The total expected cost per hour, $E[TC]$, of operating a system configuration in steady state is thus

$$E[TC] = 60 \times L_s + 15 \times c$$

Consequently, the total expected cost, $E[TC]$, can be evaluated for various values of c and the system that has the lowest cost determined. Using the aforementioned spreadsheets from Jensen and Bard (2003), the necessary performance measures can be easily calculated. Table 5-6 shows results of the analysis.

Table 5-6 indicates that as the number of operators increase, the expected cost reaches its minimum at $c = 2$. As the number of operators is increased, the machine utilization increases but levels off. The machine utilization is simply the expected number of machines that are not broken down, divided by the number of machines:

$$\text{Machine Utilization} = \frac{k - L_s}{k} = 1 - \frac{L_s}{k}$$

5.1.3 Non-Markovian Queues and Approximations

Thus far, the queuing models that have been analyzed all assume a Poisson arrival process and exponential service times. For other arrival and service processes, only limited or approximate results are readily available. There are two cases worth mentioning here. The first case is the M/G/1 queue, and the second is an approximation for the GI/G/c queuing system. Recall that "G" means any general distribution. In other words, the results will hold regardless of the distribution. Also, "GI" refers to an arrival process in which the time between arrivals is an independent and identically distributed random variable with a distribution G.

For the M/G/1 model with a service distribution having a mean $E[ST] = 1/\mu$ and variance σ^2, the expected number in the system is

$$L_s = \frac{\lambda^2 \sigma^2 + r^2}{2(1 - r)} + r$$

Based on this formula for the expected number in the system, the other performance measures can be obtained via Little's formula. Note that only the mean and the variance of the service time distribution are necessary in this case.

For the case of the GI/G/c queue, a number of approximations have been influenced by an approximation for the GI/G/1 queue that first appeared in Kingman (1964). His single-server approximation is shown below:

$$W_q(GI/G/1) \approx \left(\frac{c_a^2 + c_s^2}{2}\right) W_q(M/M/1)$$

In this equation, $W_q(M/M/1)$ denotes the expected waiting time in the queue for the M/M/1 model, and c_a^2 and c_s^2 represent the squared coefficient of variation for the interarrival time and service time distributions, respectively. For a random variable, X, the squared coefficient of variation is given by $c_X^2 = Var[X]/(E[X])^2$. Whitt (1983) used a very similar approximation for the GI/G/c queue to compute the traffic congestion at each node in a queuing network for his work on the queueing network analyzer:

$$W_q(GI/G/c) \approx \left(\frac{c_a^2 + c_s^2}{2}\right) W_q(M/M/c)$$

A discussion of queuing approximations of this form as well as additional references can be found in Whitt (1993).

Thus, to approximate the performance of a GI/G/c queue, you need only the first two moments of the interarrival and service time distributions and a way to compute the waiting time in the queue for a M/M/c queuing system. These results are useful when trying to verify and validate a simulation model of a queuing system, especially in the case of a system that consists of more than one queuing system organized into a network. Before examining that more complicated case, some of the issues related to using Arena™ to simulate single-queue systems should be examined.

5.1.4 Simulating Single Queues in Arena™

This section presents a simulation model to analyze the M/M/c/k/k machine interference model. Naturally, the queuing formulas of the previous section could be used to calculate the performance of this system, but with only minor changes (see Exercise 5-14) the formulas cannot be applied. OptQuest will also be introduced in the Arena™ environment. OptQuest is an add-on to Arena™ that provides heuristic-based simulation optimization capabilities.

To develop this model, the standard modeling recipe can be used.

5.1.4.1 Example: Machine Interference Optimization Model. The system consists of machines and operators. The system must know the number of operators present, number of machines, how the machines break down (the running time to breakdown), repair time, costs associated with the broken-down machines, and the cost per hour for the operators. Thus, costs can be modeled as variables, running time as a distribution (expression), and repair time as a distribution (expression) in Arena™. As described in Example 5, the expected total cost per hour of operating the system must be estimated. This involves estimating the average number of machines that are broken down. In addition, to be consistent with Example 5, machine utilization needs to be estimated.

The potential entities and resources in this system can now be discussed. The operators clearly act as resources in the system. The most natural entity is the machine, since as in Example 5 the machines must wait in a virtual queue for operators to perform repairs. However, machine utilization is also required. Are the machines resources or entities? In this situation, it is useful to think a little harder about what is actually flowing in the system. When a machine stops running, a repair job is initiated for an operator. A repair job waits on a list for the next available operator. When the repair job is completed, the machine (associated with the repair

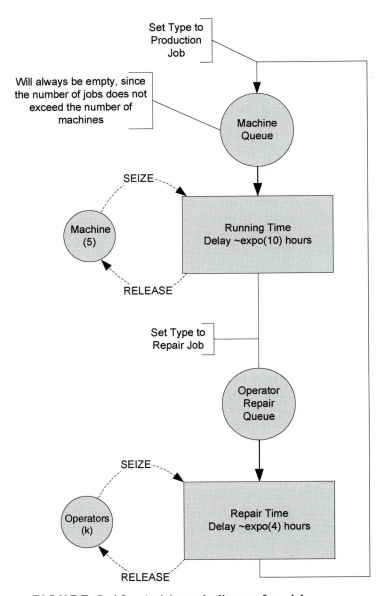

■ **FIGURE 5-12**   Activity cycle diagram for a job

job) is put back into service and essentially works on a production job. Thus, the modeling of the entity as a job might be useful from a conceptual standpoint. A production job requires a machine to run. A repair job requires the operator for repair. What is the maximum number of repair jobs possible in the system? There can never be more than k (#machines) broken down. Thus, there can never be more than k repair jobs. There can never be more than k (#machines) running. Thus, there can never be more than k production jobs. Because a machine is either broken down or running, it should be clear that at any time t,

Number of machines = Number of broken machines + Number of running machines

Number of machines = Number of repair jobs + Number of production jobs

If the jobs are modeled as entities, then the machines can be modeled as a resource (seized by production jobs). This will allow Arena™ to calculate the utilization of the machines automatically via its resource statistics.

Exhibit 5-1 Pseudo-Code for Machine Interference Model

```
 CREATE 5 jobs
A:   ASSIGN myType = 1        //1 for production
     vNumInProd = vNumInProd + 1
SEIZE 1 unit of machine
DELAY for running time = EXPO(10) hours
RELEASE 1 unit of machine
ASSIGN myType = 2         //2 for repair
     vNumInProd = vNumInProd - 1
     vNumInRepair = vNumInRepair + 1
SEIZE 1 unit of Operator
DELAY for repair time = EXPO(4) hours
RELEASE 1 unit of Operator
ASSIGN vNumInRepair = vNumInRepair - 1
GOTO A
```

Now, let's describe the process flow associated with the jobs. If a job can be thought of as simply changing type (between repair and production), then an activity cycle diagram can be developed for the life of a job. An activity cycle diagram is a special type of activity flow diagram in which the life of the entity "cycles." In Figure 5-12, this is illustrated by the loop. Note that in this case, there is no concept of "create" and "dispose." This conceptual model must be translated into something more suitable for Arena[TM].

The Arena-like pseudo-code is provided in Exhibit 5-1. In the pseudo-code, two variables have been used to keep track of the number of jobs in production and the number of jobs in repair. These concepts can now be translated into Arena[TM].

The building of the Arena[TM] model closely follows the pseudo-code. The completed model can be found in the file, *Ch5-MMckk-Example.doe*, that accompanies this chapter. You should first define the entities, variables, expressions, and resources in the model as illustrated in and Figures 5-13 and 5-14. Next, incorporate the model's flow modules as illustrated in Figure 5-15. The model closely follows the previously described activity cycle diagram and pseudo-code.

Figure 5-16 indicates that the CREATE module depends on the number of machines defined. There will be MR(Machines) jobs created at time 0.0 and then the CREATE module will stop its creation. *MR(resource name)* is a special-purpose function in Arena[TM] that returns the current number of units scheduled for a resource (Figure 5-17). MR jobs will be created because that is the most that can be running at any time. This logic assumes that the

Resource - Basic Process	Name	Type	Capacity
1	Machines	Fixed Capacity	5
2	Operators	Fixed Capacity	1

Entity - Basic Process	Entity Type	Initial Picture
1	ProdJob	Picture.Green Ball
2	RepairJob	Picture.Red Ball

■ **FIGURE 5-13** Entity and resource data modules

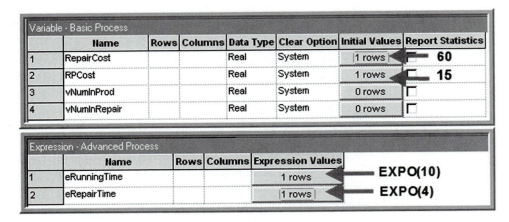

■ FIGURE 5-14   Variable and expression modules

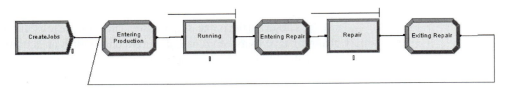

■ FIGURE 5-15   Machine interference flow chart modules

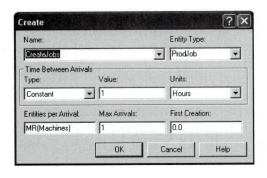

■ FIGURE 5-16   CREATE module for machine interference model

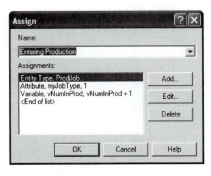

■ FIGURE 5-17   ASSIGN module for machine interference model

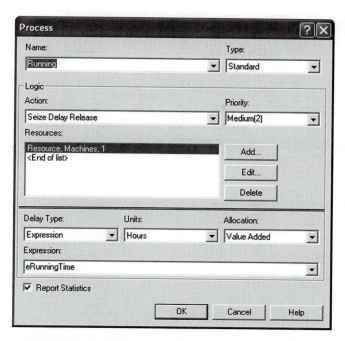

■ **FIGURE 5-18**   Machine-running-time PROCESS module

system starts with all the machines working on production jobs at the start of the simulation. Figure 5-17 indicates the ASSIGN module for assigning the type of job and for counting the current number of production jobs.

As the production job enters production on the machines, the number of jobs in production is incremented. A PROCESS module is then used to implement the SEIZE, DELAY, RELEASE logic for the running time of the machine as shown in Figure 5-18. After the machine completes the running time, the production job is changed to a repair job and the number of repair jobs in the system is incremented as shown in Figure 5-19. Finally, the repair job is processed and the number of repair jobs in the system is decremented as shown in Figures 5-20 and 5-21.

In order to collect the total cost on an hourly basis, an OUTPUT statistic should be defined as shown in Figure 5-22. A variable, *RPCost*, has been defined to represent the cost per hour for the operator (e.g., $15/hour). In addition, the repair cost is represented with a variable,

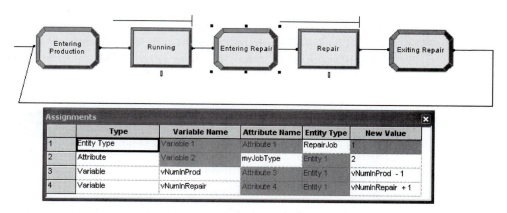

■ **FIGURE 5-19**   Entering repair ASSIGN module

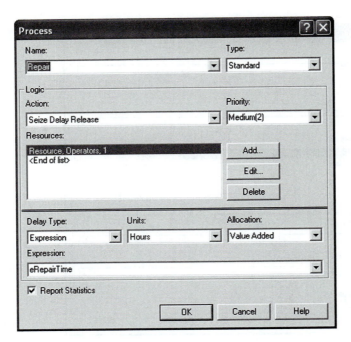

■ **FIGURE 5-20** Repair PROCESS module

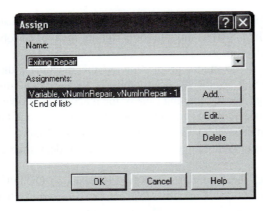

■ **FIGURE 5-21** ASSIGN for exiting repair

RepairCost, with an initial value of $60/broken machine. The statistic, *HourlyCostStat*, thus represents the following equation: $E[TC] = 15 \times c + 60 \times L_s$.

Figure 5-23 illustrates the results of the case of 1 operator and 5 machines. The statistics closely match results from the queuing analysis. The results are based on 21 replications of length 12,000 hours with a warm-up period of 2000 hours.

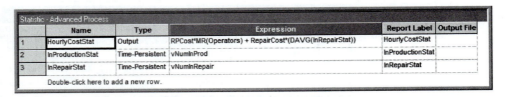

■ **FIGURE 5-22** Defining the total hourly cost

Time Persistent		
	Average	Half Width
InProductionStat	2.3179	0.04
InRepairStat	2.6821	0.04

Output		
	Average	Half Width
HourlyCostStat	175.93	2.23

Instantaneous Utilization		
	Average	Half Width
Machines	0.4636	0.01
Operators	0.9319	0.00

Number Busy		
	Average	Half Width
Machines	2.3179	0.04
Operators	0.9319	0.00

■ **FIGURE 5-23** Arena results for 1 operator and 5 machines

5.1.4.2 Using OptQuest[1] on the Machine Interference Model. As shown in Chapter 4, the expected total cost for this system can be easily evaluated by using Arena's Process Analyzer. The use of the Process Analyzer will be left as an exercise. Instead, the use of OptQuest for Arena™ will be illustrated. If you are following along with the model-building process, make sure that your model is open and go to Tools > OptQuest for Arena™. OptQuest will open and you should see a dialog asking whether you want to browse for an already formulated optimization run or start a new optimization run. Select the start new optimization run button. OptQuest will then read your model and start a new optimization model (Figure 5-24), which may

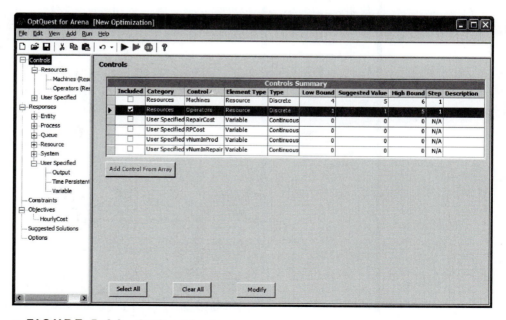

■ **FIGURE 5-24** OptQuest for Arena

---

[1]The use of OptQuest requires the professional license of Arena™.

take a few seconds, depending on the size of your model. OptQuest is similar to the Process Analyzer in some respects. It allows you to define controls and responses for your model. Then you can develop an optimization function that will be used to evaluate the system under the various controls. In addition, you can define a set of constraints that must not be violated by the final recommended solution to the optimization model. It is beyond the scope of this example to fully describe simulation optimization and all the intricacies of this technology. The interested reader should refer to April et al. (2001) and Glover et al. (1999) for more information on this technology. This example will simply illustrate the possibilities of using this technology. A tutorial on the use of OptQuest is available in the OptQuest help files.

Using the controls, double-click on the resource Operators and change the lower and upper bounds on the range for the control to 1 and 10, respectively. This defines the range of values that OptQuest will search to find a potential optimal solution. The controls act as the decision variables in the optimization model. Then, in the responses area, select the *Hourly-CostStat* as a response to include in the model. Finally, you should set up an objective function.

In OptQuest, you can build an objective function by using the controls and responses that were included in the optimization model. The objective function is quite simple here. The *HourlyCostStat* response should be selected as the objective function as shown in Figure 5-25. More general objective functions can be easily formed using the objective function editing capabilities of OptQuest. The objective function will be used by OptQuest to direct the search. OptQuest will use its intelligent search technology to run multiple replications at each

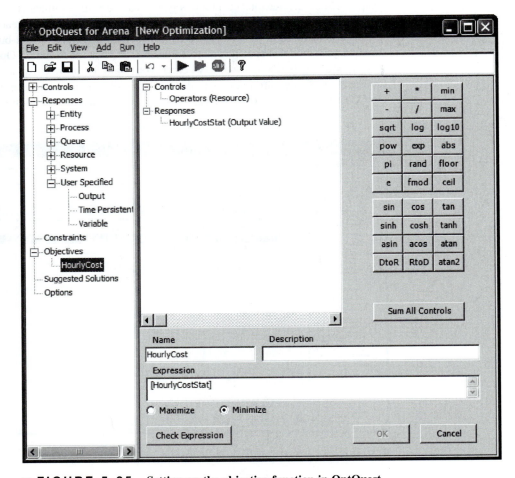

■ **FIGURE 5-25** Setting up the objective function in OptQuest

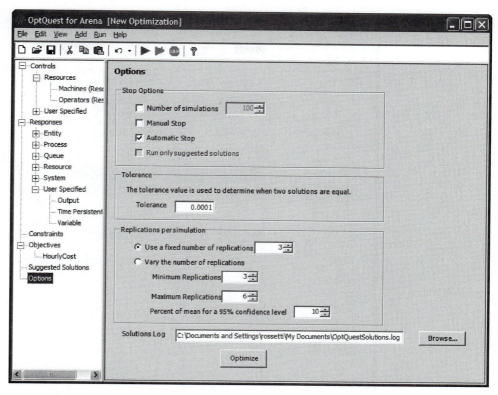

■ **FIGURE 5-26** OptQuest optimization options

combination of the decision variables. OptQuest will then evaluate the objective function and decide on new values for the decision variables in order to minimize (or maximize) its value (Figure 5-26). OptQuest uses a number of meta-heuristics (primarily based on Tabu Search; see Glover and Laguna 1997), to direct the search.

The OptQuest optimization options tab allows you to specify control parameters for how OptQuest controls the number of replications. The default settings for this simple search will be used. Press the triangle-like run button and execute the search. You should see a figure that tabulates the progress of the search at each simulation (Figure 5-27). In the figure, OptQuest has recommended a solution having 2 operators, which, based on the analytical results, is the optimal solution. Once you find a set of possibly optimal solutions, you can either use OptQuest's refine solution tab or you can use the Process Analyzer to statistically select the best system. This example is too limited to fully explore the capabilities of OptQuest. The use of OptQuest will be examined again when discussing a more complicated problem in Chapter 8.

The next example illustrates the modeling of a queuing situation for which an analytical solution is not readily available. This situation has multiple types of customers with different priorities, different service times, and in the case of one type of customer, the desire (or ability) to renege from the system.

5.1.4.3 Example: Walk-In Health Care Clinic.
After analyzing operations, the managers of a walk-in health-care clinic classified patients in three categories: high priority (urgent need of medical attention), medium priority (need standard medical attention), and low priority (non-urgent need of medical attention). On a typical day during the period of interest, there are about 15 arrivals per hour with (25% being high priority, 60% being medium priority, and the remaining being low priority). The managers are interested in understanding the

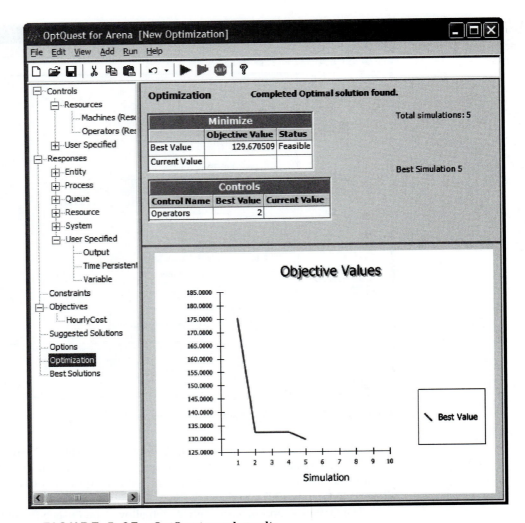

■ **FIGURE 5-27**   OptQuest search results

waiting time for patients at the clinic. Upon arrival at the clinic, the patients are triaged by a nurse into one of the three types of patients in only 2 to 3 minutes uniformly distributed. Then, the patients wait in the waiting room based on their priority. Patients with higher priority are placed at the front of the line. Patients with the same priority are ordered based on a first-come first-served basis. Service time distributions of the customers are given in Table 5-7.

The clinic has 4 doctors on staff to attend to the patients during the period of interest. Survey results indicated that if there are more than 10 people waiting for service, an arriving low-priority patient will exit before being triaged. Finally, non-urgent (low-priority) patients

■ **TABLE 5-7**
Clinic Service Time Distributions

Priority	Distribution (in minutes)
High	Lognormal(38, 8)
Medium	Triangular(16, 22, 28)
Low	Lognormal(12, 2)

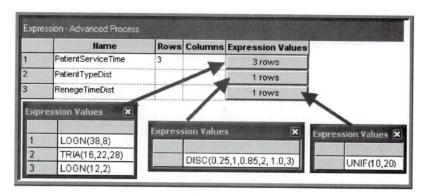

■ **FIGURE 5-28**   Expressions for walk-in clinic model

may depart if they have to wait longer than 15 ± 5 minutes after triage. That is, non-urgent patients may enter the clinic and begin waiting for a doctor, but if they have to wait more than 15 ± 5 minutes (uniformly distributed), they will decide to renege and leave the clinic without getting service. The clinic would like to estimate the following:

- Average system time of each type of patient
- Probability that low-priority patients balk
- Probability that low-priority patients renege
- Distribution of the number of customers waiting in the doctor queue

For this problem, the system is the walk-in clinic, which includes doctors and a triage nurse who serve three types of patients. The system must know how the patients arrive (time between arrival distribution), how to determine the type of the patient, the triage time, the service time by type of patient, and the amount of time that a low-priority patient is willing to wait. To determine the type of the patient, a discrete distribution (DISC) can be used. The service-time distributions depend on the patient type and can be easily held in an arrayed EXPRESSION as illustrated in Figure 5-28. In addition, the system must know the balk criteria for the low-priority patients. This information is modeled as variables as illustrated in Figure 5-29.

The entity for this model is the patient. Patients are created according to a particular type, enter the system, and then depart. Thus, a key attribute for patients should be their type. In addition, the required performance measures require that the arrival time of the patient be stored so that the total time in the system can later be calculated. There are two resources for this system: triage nurse and doctors, with 1 and 4 units, respectively. The process flow for this system is straightforward, except for the reneging of low-priority patients as illustrated in Exhibit 5-2.

Modeling the reneging of the patients in Arena™ requires additional effort. There are no "magic" dialogs or modules that directly handle reneging. The key to modeling the reneging described in this problem in Arena™ is to remember how Arena™ processes entities. In particular, as discussed in Chapter 2, no model logic is executed unless an entity is "moving"

	Name	Rows	Columns	Data Type	Clear Option	Initial Values	Report Statistics
1	vBalkCriteria			Real	System	1 rows	→ 10
2	vSearchNumb			Real	System	0 rows	☐

■ **FIGURE 5-29**   Walk-in clinic variable module

Exhibit 5-2 Pseudo-Code for Walk-In Clinic Model

```
CREATE patients
ASSIGN myType = PatientTypeDist
If myType == low priority
If NQ(Doctor's Q) >= balk criteria
      RECORD balk
      DISPOSE patient
      ELSE GOTO A
ELSE
A: SEIZE 1 unit of triage nurse
DELAY for UNIF(2,3) minutes
RELEASE 1 unit of triage nurse
Handle reneging patients
SEIZE 1 unit of Doctors
DELAY for service based on patient type
RELEASE 1 unit of Doctors
RECORD patient's system time
DISPOSE
```

through the model. In addition, you know that an entity in a queue is no longer moving. Thus, an entity in a queue cannot remove itself from the queue! While this presents a difficulty, it also indicates the solution to the problem.

If the entity that represents the low-priority patient cannot remove itself from the queue, then *some other* entity must do the removal. The only other entities in the system are other patients and it seems both unrealistic and illogical to have another patient remove the reneging patient. Thus, you can only conclude that another (logical) entity needs to be created to somehow implement the removal of the reneging patient. As you have learned, there are two basic ways to create entities: the CREATE module and the SEPARATE module. Since there is no clear pattern associated with when the patient will renege, this leaves the SEPARATE module as the prime candidate for implementing reneging. Recall that the SEPARATE module works by essentially "cloning" the entity that goes through it in order to create the specified number of clones. Now the answer should be clear. The low-priority patient should clone himself prior to entering the queue. The clone will be responsible for removing the cloned patient from the queue if the delay for service is too long.

Figure 5-30 illustrates the basic ideas for modeling reneging in the form of an activity diagram. Using a SEPARATE module, the patient entity can be duplicated prior to the entity entering the doctor's queue. Now the two flows can be conceptualized as acting in parallel. The (original) patient enters the doctor queue. If it has to wait, it stops being the active entity, stops moving, and gives control back the event scheduler. At this point, the duplicate (clone) entity begins to move. In the simulation, no actual simulation time has advanced. The clone then enters the delay module that represents the time until reneging. This schedules an event to represent this delay and the clone's progress stops. The event scheduler then allows other entities to proceed through their process flows. One of those other entities might be the original patient entity. If the original entity continues its movement and gets out of the queue before the clone finishes its delay, then it will automatically be removed from the queue by the doctor resource and be processed. If the clone's delay completes before the original exits the queue, then when the clone becomes the active entity, it will remove the original from the queue and proceed to being disposed. If the clone does not find the original in the queue, then the original must have proceeded and the clone can simply be disposed. One can think of this as the patient setting an alarm clock (the event scheduled by the duplicate) for when to renege. To implement

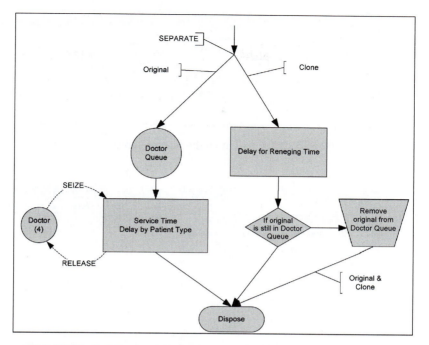

■ **FIGURE 5-30**   Activity diagram to model reneging

these ideas in Arena™, you will need to use the SEPARATE, SEARCH, and REMOVE modules. Now, let's take a look at the entire model in Arena™.

Figure 5-31 provides an overview of the walk-in clinic model that follows closely the previously discussed pseudo-code. The CREATE module has a time between arrival specified as Random(expo) with a mean of 6 minutes. The next ASSIGN module simply assigns the arrival time and the type of patient, and then changes the Entity Type and Entity Picture based on some sets that have been defined for each type of patient. This is shown in Figure 5-32.

The Triage and Doctor PROCESS modules are completed similarly to the implementation of many prior models. In order to implement the priority for the patients by type, the QUEUE module can be used. In this case, the *myType* attribute can be used with the lowest attribute value as shown in Figure 5-33. You should attempt to build this model or examine the supplied Arena™ file for details.

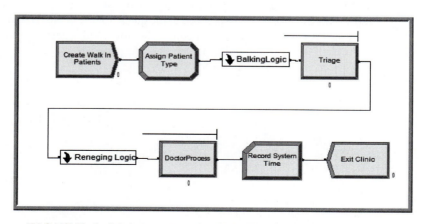

■ **FIGURE 5-31**   Overview of walk-in clinic model

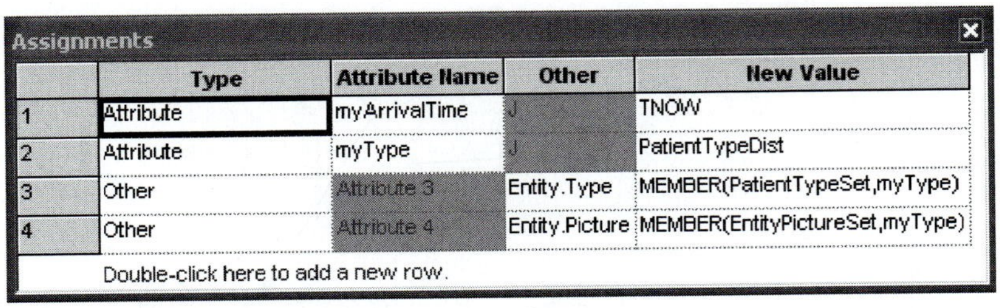

Assignments				✕
	Type	**Attribute Name**	**Other**	**New Value**
1	Attribute	myArrivalTime		TNOW
2	Attribute	myType		PatientTypeDist
3	Other	Attribute 3	Entity.Type	MEMBER(PatientTypeSet,myType)
4	Other	Attribute 4	Entity.Picture	MEMBER(EntityPictureSet,myType)
	Double-click here to add a new row.			

■ **FIGURE 5-32** Assigning the type of patient

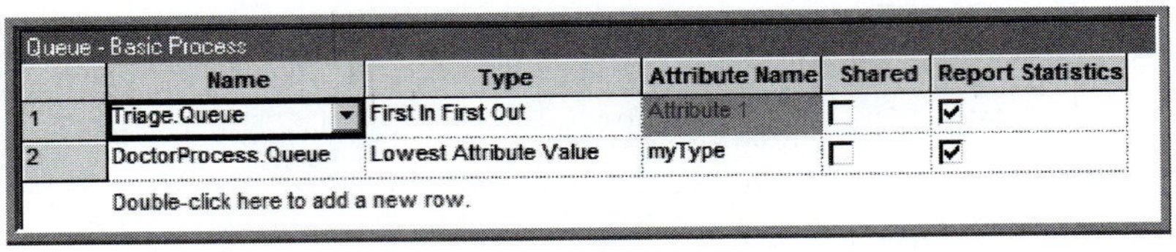

Queue - Basic Process					
	Name	**Type**	**Attribute Name**	**Shared**	**Report Statistics**
1	Triage.Queue	First In First Out	Attribute 1	☐	☑
2	DoctorProcess.Queue	Lowest Attribute Value	myType	☐	☑
	Double-click here to add a new row.				

■ **FIGURE 5-33** Using the QUEUE module to rank the queue

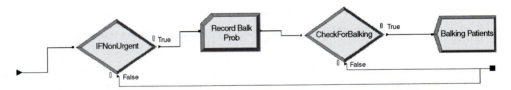

■ **FIGURE 5-34** Balking logic submodel

The balking and reneging logic have been placed inside two different submodels. Balking only occurs for nonurgent patients, so the type of patient is first checked. If the patient is of type non-urgent, whether a balk will occur can be recorded with the expression, NQ(DoctorProcess.Queue) $\geq$ vBalkCriteria. This expression evaluates to 1 for true or 0 for false. Thus, the probability of balking can be estimated. The patients who actually balk are then disposed.

The reneging logic is more complicated. Figure 5-35 presents an overview of the submodel logic for reneging. In the logic, the patient's type is checked to see whether the patient is of the non-urgent type. If so, the entity is sent to a SEPARATE module. The original entity simply exits the submodel (to proceed to the queue for the doctor resource). The duplicate then enters the DELAY module that schedules the time until reneging. After exiting the reneging

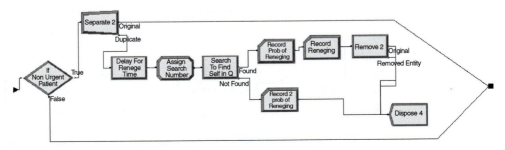

■ **FIGURE 5-35** Reneging Logic Submodel

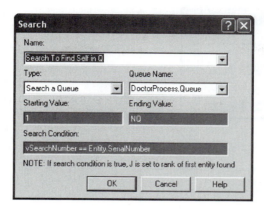

■ FIGURE 5-36   SEARCH module for reneging logic

delay, an ASSIGN module is used to set a variable (*vSearchNumber*) equal to *Entity.SerialNumber*. Recall from Chapter 2 that *Entity.SerialNumber* is a unique number given to each entity when created. When the original entity was duplicated by the SEPARATE module, the duplicate also has this same number assigned. Thus, you can use this number to have the clone search for the original entity in the queue. The SEARCH module allows the searching of a batch, queue, or expression for a particular result that may itself be the value from an expression.

As illustrated in Figure 5-36, the *DoctorProcess.Queue* is searched starting at the first entity in the queue (rank 1) to the last entity in the queue (the entity at rank NQ). The search proceeds to find the first entity where *vSearchNumber* $==$ *Entity.SerialNumber*. As the SEARCH module indicates, if the search condition is true, the global variable, J, is set to the rank of the first entity satisfying the condition. Thus, after the SEARCH module completes the variable, J, can be used to see whether an entity was found. An entity will have been found if $J > 0$ and not found if $J = 0$. The SEARCH module has two exit points: one for if the search found something, the other for the case of not finding something. If no entity was found, then the duplicate can simply be disposed because the original is no longer in the queue. If an entity was found, then the variable J can be used in the REMOVE module to remove the appropriate entity from the queue. The two RECORD modules on both paths after the SEARCH modules use the fact that the SEARCH module exit points indicate whether there was a renege. A one or a zero can be observed in the upper and lower path RECORD modules to collect the probability of reneging as illustrated in Figure 5-37.

If $J > 0$, then the entity at rank J can be removed from the *DoctorProcess.Queue* as illustrated in Figure 5-38. The rest of the model is relatively straightforward, and you are encouraged to explore the final dialog boxes.

Assuming that the clinic opens at 8 a.m. and closes at 6 p.m., the simulation was set up for 30 replications to yield the results shown in Figure 5-39. It appears that there is a relatively high chance (about 29%) that a non-urgent patient will renege. This may or may not be acceptable in light of the other performance measures for the system. The reader is encouraged to further

	Name	Type	Value	Recor	Counter Name	Tally Name
1	Record Reneging	Count	1	☐	Record Reneging	Record Reneging
2	Record Prob of Reneging	Expression	1	☐	Record Prob of Reneging	Record Prob of Reneging
3	Record2 Prob of Reneging	Expression	0	☐	Record2 Prob of Reneging	Record Prob of Reneging

Record - Basic Process

■ FIGURE 5-37   Recording whether reneging occurred

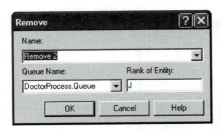

■ **FIGURE 5-38** REMOVE module for removing reneging entity

Tally		
Expression	Average	Half Width
Record Balk Prob	0.00222222	0.00
Record Prob of Reneging	0.2884	0.07

■ **FIGURE 5-39** Example output for walk-in clinic

explore this model in the exercises, including the implementation to collect statistics on the number of customers waiting in the doctor queue.

While some analytical work has been done for queuing systems involving balking and reneging, simulation allows for the modeling of more realistic types of queuing situations as well as even more complicated systems. The last two examples actually involved two single queue systems that are linked. The linking of a number of queuing systems is a very common situation, which is examined in the next section.

5.2   Networks of Queuing Stations

A network of queues can be thought of as a set of stations, where each station represents a service facility. In the simple case, each service facility might have a single queue feeding into a resource. In the most general case, customers may arrive at any station in the network from outside the system. Customers then proceed from station to station to receive service. After receiving all of their service requirements, the customers depart the system. In general, the customers do not take the same path through the network, and for a given customer type, the path may be deterministic or stochastic in nature. For example, consider the case of a job shop manufacturing setting. Each product that is manufactured may require a different set of manufacturing processes, which are supplied by particular stations (e.g., drilling, milling, etc.). As another example, consider a telecommunications network where packets are sent from some origin to some destination through a series of routers. In each of these cases, understanding how many resources to supply so that the customer can efficiently traverse the network at some minimum cost is important. As such, queuing networks are an important component of the efficient design of manufacturing, transportation, distribution, telecommunication, computer, and service systems. Figure 5-40 illustrates the concept of a network of queues for producing a vacuum cleaner.

The analytical treatment of the theory associated with networks of queues has been widely examined and remains an active area for theoretical research. Summarizing the enormous literature on queuing networks is beyond the scope of this text. For a starting point in the theory, see Gross and Harris (1998), Kelly (1979), Buzacott and Shanthikumar (1993), or Bolch et al. (2006).

A number of examples have already been examined that can be considered queuing networks (e.g., Chapter 2's Tie-Dye T-Shirts and Chapter 3's LOTR Makers, Inc.). The purpose of this section is to introduce some of the constructs in Arena[TM] that facilitate the simulation of

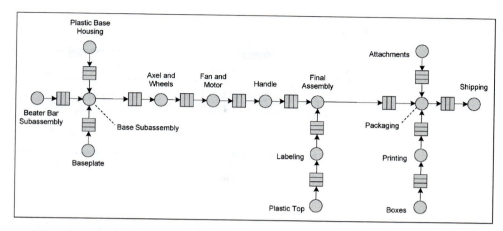

■ **FIGURE 5-40**   Network of queues

queuing networks. Since a queuing network involves the movement of entities between stations, the use of the STATION, ROUTE, and SEQUENCE modules from the Advanced Transfer template panel will be emphasized.

5.2.1   Example: Testing and Repair Shop

Consider a test and repair shop for computer parts (e.g., circuit boards, hard drives, etc.). The system consists of an initial diagnostic station through which all newly arriving parts must be processed. Currently, newly arriving parts arrive according to a Poisson arrival process with a mean rate of 3 per hour. The diagnostic station consists of 2 diagnostic machines that are fed the arriving parts from a single queue. Data indicate that the diagnostic time is quite variable and follows an exponential distribution with a mean of 30 minutes. Based on the results of the diagnostics, a testing plan is formulated for the parts. There are currently three testing stations {1, 2, 3} that consist of one machine each. The testing plan consists of an ordered sequence of testing stations that must be visited by the part before proceeding to a repair station. Because the diagnosis often involves similar problems, there are common sequences that occur for the parts. The company collected extensive data on the visit sequences for the parts and found that the following sequences constituted the vast majority of test plans for the parts.

For example, 25% of the newly arriving parts follow test plan 1, which consists of visiting test stations 2, 3, 2, and 1 before proceeding to the repair station. The testing of parts at each station takes time that may depend on the sequence that the part follows. That is, while parts that follow test plans 1 and 3 visit test station 1, data show that the time is not necessarily the same.

■ **TABLE 5-8**
Distribution of Test Plans

Test plan	% of parts	Sequence
1	25%	{2, 3, 2, 1}
2	12.5%	{3, 1}
3	37.5%	{1, 3, 1}
4	25%	{2, 3}

■ **TABLE 5-9**
Testing and Repair Time Distribution Parameters

Test plan	Testing time parameters	Repair time parameters
1	$\{(20, 4.1), (12, 4.2), (18, 4.3), (16, 4.0)\}$	$(30, 60, 80)$
2	$\{(12, 4), (15, 4)\}$	$(45, 55, 70)$
3	$\{(18, 4.2), (14, 4.4), (12, 4.3)\}$	$(30, 40, 60)$
4	$\{(24, 4), (30, 4)\}$	$(35, 65, 75)$

Data on the testing times indicate that the distribution is well modeled with a log-normal distribution with mean μ, and standard deviation σ in minutes. Table 5-9 presents the mean and standard deviation for each of the testing time distributions by each station in the test plan. For example, the first pair of parameters, (20, 4.1), for test plan 1 indicates that the testing time at test station 2 has a log-normal distribution with mean $\mu = 20$, and standard deviation $\sigma = 4.1$ minutes.

The repair station has 3 workers who attempt to complete the repairs based on the tests. The repair time also depends on the test plan that the part has been following. Data indicate that the repair time can be characterized by a triangular distribution with the minimum, mode, and maximum as indicated in Table 5-9. After the repairs, the parts leave the system. When the parts move between stations assume that there is always a worker available and that the transfer time takes between 2 to 4 minutes uniformly distributed. Figure 5-41 illustrates the arrangement of the stations and the flow of the parts following plan 2 in the test and repair shop.

The company is considering accepting a new contract that will increase the overall arrival rate of jobs to the system by 10%. They are interested in understanding where the potential bottlenecks are in the system and in developing alternatives to mitigate those bottlenecks so that they can still handle the contract. The new contract stipulates that 80% of the time the testing and repairs should be completed in 480 minutes. The company runs 2 shifts each day for each 5-day work week. Any jobs not completed at the end of the second shift are carried over to first shift of the next working day. Assume that the contract is going to last for 1 year (52 weeks). Build a simulation model that can assist the company in assessing the risks associated with the new contract.

Before implementing the model in Arena™, you should prepare by conceptualizing the process flow. Figure 5-42 illustrates the activity diagram for the test and repair system. Parts are created and flow first to the diagnostic station where they seize a diagnostic machine while the diagnostic activity occurs. Then, the test plan is assigned. The flow for the visitation of the parts

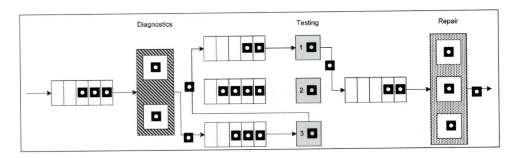

■ **FIGURE 5-41**   Test and repair shop

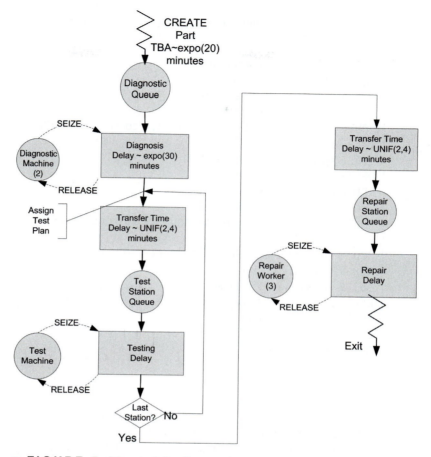

■ **FIGURE 5-42**   Activity diagram for test and repair system

to the test station is shown with a loop back to the transfer time between the stations. It should be clear that the activity diagram is representing any of the 3 test stations. After the final test station in the test plan has been visited, the part goes to the repair station, where 1 of 3 repair workers is seized for the repair activity. After the repair activity, the part leaves the system.

Exhibit 5-3 presents the pseudo-code for the test and repair model. This is a straightforward representation of the flow presented in the activity diagram. From the activity diagram and the pseudo-code, it should be clear that with previously discussed Arena™ modeling constructs you should be able to model this situation. In order to model the situation, you need some way to represent where the part is currently located in the system (e.g., the current station). Second, you need some way to indicate where the part should go to next. And finally, you need some way to model the transfer and the time of the transfer of the part between stations. This type of modeling is very common. Because of this, there are special modeling constructs in Arena™ that are specifically designed to handle situations like this.

5.2.2   STATION, ROUTE, and SEQUENCE Modules

To model the test and repair shop, three new modules and the concept of transferring entities must be discussed. Entities typically represent the things that are flowing through the system. In previously examined models, the *direct connect* method of transferring entities between

Exhibit 5-3 Pseudo-Code for Test and Repair System

```
CREATE part
SEIZE 1 diagnostic machine
DELAY for diagnostic time
RELEASE diagnostic machine
ASSIGN test plan sequence
ROUTE for transfer time by sequence to Test STATION
Test STATION
SEIZE appropriate test machine
DELAY for testing time
RELEASE test machine
If not at last station
ROUTE for transfer time by sequence to Test STATION
Else
ROUTE for transfer time by sequence to Repair STATION
Repair STATION
SEIZE repair worker from repair worker set
DELAY for repair time
RELEASE repair worker
Collect statistics
DISPOSE
```

modules in the model has been used. With the direct connect method two modules are directly connected via a connection line in the Arena™ model window and the entity "flow" along the connection line between the modules. The modeling constructs in this section allow entities to move between modules without a connection line.

Entity transfer refers to the various ways by which entities can move between modules. The STATION and ROUTE modules facilitate the transfer of entities between stations with a transfer delay.

Station: The STATION module represents a named location to which entities can be transferred. Initially, one can think of stations as the label part of the Go To – Label construct found standard programming languages; however, stations are more powerful than a simple label. Stations can be placed in sets and held in sequences. Stations are also necessary for mapping model logic transfers to a physical (spatial) representation of the world.

Route: The ROUTE module causes an entering entity to be transferred to a station with a possible time delay that accompanies the transfer. The entity leaves the route module and after the specified time delay reappears in the model at the specified station.

Entities have a number of special purpose attributes that are useful when modeling with the STATION and ROUTE modules. The attribute, *Entity.CurrentStation* is updated to the current station whenever the entity passes through the STATION module. This attribute is not user assignable, but can be used (read) in the model. It will return the station number that an entity is currently located at or 0 if the entity is not currently at a station. In addition, every entity has an *Entity.Station* attribute, which returns the entity's station or destination. The *Entity. Station* attribute is user assignable. It is (automatically) set to the intended destination station when an entity is transferred (e.g., via a ROUTE module). It will remain equal to the current station after the transfer or until either changed by the user or affected by another transfer-type module (e.g., ROUTE module). Thus, the modules attached to a STATION module are conceptually "at the station location."

In many modeling contexts, entities will follow a specific path through the system. In a manufacturing job shop, this is often called the process plan. In a bus system, this is called a route. In the test and repair system, this is referred to as the test plan. To model a specific path through the system, the SEQUENCE module can be used. A sequence consists of an ordered list of job steps. Each job step must indicate the STATION associated with the step, and may indicate a series of attribute assignments that must take place when the entity reaches the station associated with the job step. Each job step can have an optional name and can give the name of the next step in the sequence. Thus, a sequence is built by simply providing the list of stations that must be visited.

Each entity has a number of special purpose attributes that facilitate the use of sequences. The *Entity.Sequence* attribute holds the sequence that the entity is currently following or 0 if no sequence has been assigned. The ASSIGN module can be used to assign a specific sequence to an entity. In addition, the entity has the attribute, *Entity.JobStep*, which indicates the step in the sequence that the entity is currently on. *Entity.JobStep* is user assignable and is automatically incremented when a transfer type module (e.g., ROUTE) is used with the By Sequence option. Finally, the attribute (*Entity.PlannedStation*) is available and represents the number of the station associated with the next job step in the sequence. *Entity.PlannedStation* is not user assignable. It is automatically updated whenever *Entity.Sequence* or *Entity.JobStep* changes, or whenever the entity enters a station.

In the test and repair example, STATION modules will be associated with the diagnostic station, each of three test stations, and with the repair station. The work performed at each station will be modeled with a PROCESS module using the (SEIZE, DELAY, and RELEASE) option. The four different test plans will be modeled with four sequences. Finally, ROUTE modules, using the "By Sequence" transfer option will be used to transfer the entities between the stations with a UNIF(2, 4) transfer delay time distribution in minutes. Other than the new modules for entity transfer this model is similar to previous models. This model can be built by taking the following steps:

1. Define 5 different resources (*DiagnosticMachine, TestMachine1, TestMachine2, Test-Machine3, RepairWorkers*) with capacity (2, 1, 1, 1, 3), respectively.

2. Define 2 variables (*vContractLimit, vMTBA*) with initial values (480 and 20), respectively.

3. Define 3 expressions (*eDiagnosticTime, eTestPlanCDF*, and *eRepairTimeCDFs*). *eDiagnosticTime* should be expo(30), *eTestPlanCDF* should be DISC(0.25, 1, 0.375, 2, 0.75, 3, 1.0, 4), and *eRepairTimeCDFs* should be an arrayed expression with 4 rows. Each row should specify a triangular distribution according to the information given in Table 5-9 (e.g., TRIA(30, 60, 80)).

4. Use the SEQUENCE module on the Advanced Transfer panel to define four different sequences. This is illustrated in Figure 5-43. First, double-click on the SEQUENCE module to add a new sequence (row). The first sequence is called *TestPlan1Seq*. Then, add job steps to the sequence by clicking on the Steps button. Figure 5-43 shows the five job steps for test plan 1 (*TestStation2, TestStation3, TestStation2, TestStation1*, and *RepairStation*). Typing in a station name defines a station for use in the model window. Since every part must visit the repair station before exiting the system, *RepairStation* has been listed last in all the sequences. For each job step, define an assignment that will happen when the entity is transferred to the step. In the test and repair system, the testing time depends upon the job step. Thus, an attribute, *myTestingTime*, can be defined so that the value from the pertinent log-normally distributed test time distribution can be assigned. In the case illustrated, *myTestingTime* will be set equal to a random number

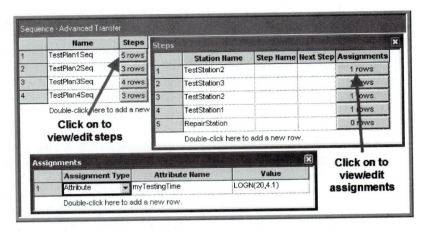

■ **FIGURE 5-43**   Defining sequences, job steps, and assignments

from a LOGN(20, 4.1) distribution as specified in Table 5-9. The *myTestingTime* attribute is used to specify the delay time for each of the PROCESS modules that represent the testing processes.

5. Finally, define a set to hold the sequences so that they can be randomly assigned to the parts after they visit the diagnostic machine. To define a Sequence set, you must use the Advanced Set module on the Advanced Process panel. Each of the sequences should be listed in the set as shown in Figure 5-44. The set type should be specified as *Other*. Unfortunately, the build expression option is not available here, and each sequence name must be carefully typed into the dialog box. The order of the sequences is important.

Now that all the data modules have been specified, you can easily build the model using the flow chart modules. Figure 5-45 presents an overview of the model. The pink-colored modules in the model window are the STATION and ROUTE modules. A CREATE module was used to generate the parts needing testing and repair with a mean time between arrivals of 20 minutes exponentially distributed. The entity then proceeds through an ASSIGN module where the attribute, *myArriveTime*, is set to TNOW. This will be used to record the job's system time. The next module is a STATION module that represents the diagnostic station (Figure 5-46).

After passing through the STATION module, the entity goes through the diagnostic process. Following the diagnostic process, the part is assigned a test plan. Figure 5-47 shows that first the test plan expression holding the DISC distribution across the test plans is used to assign a number {1, 2, 3, or 4} to the attribute *myTestPlan*. The attribute is then used to select the appropriate sequence from our previously defined set of test plan sequences. The sequence

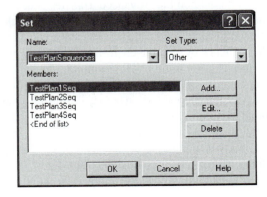

■ **FIGURE 5-44**   Advanced Set to hold sequences

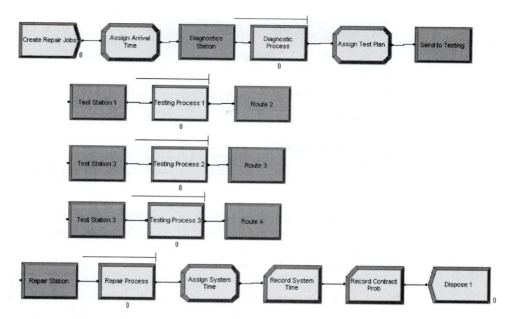

■ **FIGURE 5-45**   Overview of test and repair model

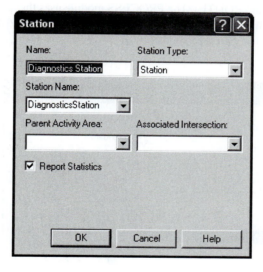

■ **FIGURE 5-46**   Diagnostics STATION module

returned from the set is assigned to the special-purpose attribute, *Entity.Sequence*, so that the entity can now follow this sequence.

The part then enters the ROUTE module for sending the parts to the testing stations. Figure 5-48 illustrates the ROUTE module. This module allows a time delay in the route time field and allows the user to select the method by which the entity will be transferred. Choosing

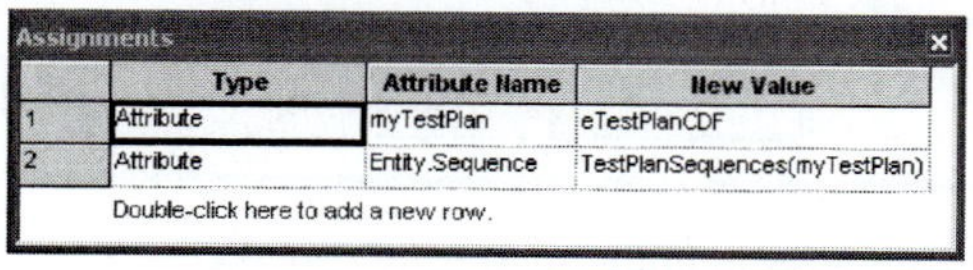

	Type	Attribute Name	New Value
1	Attribute	myTestPlan	eTestPlanCDF
2	Attribute	Entity.Sequence	TestPlanSequences(myTestPlan)
	Double-click here to add a new row.		

■ **FIGURE 5-47**   Test plan assignments

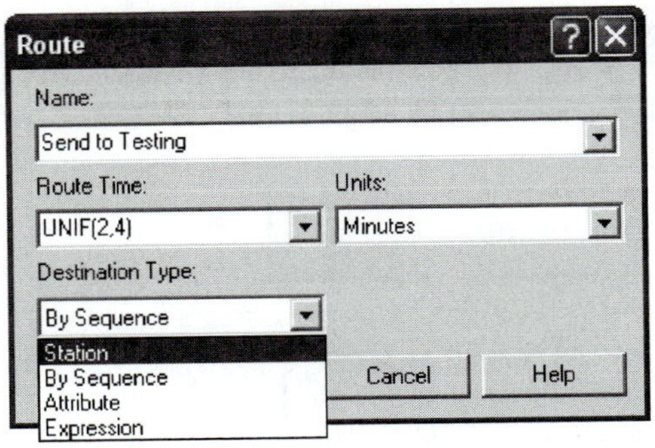

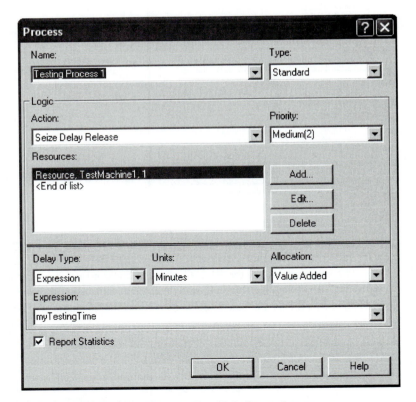

■ **FIGURE 5-48** **Route module**

the By Sequence option indicates that the entity should use its assigned sequence to determine the destination station for the route. The entity's sequence and its current job step are used. When the entity goes into the ROUTE module, *Entity.JobStep* is incremented to the next step. The entity's *Entity.Station* attribute is set equal to the station associated with the step and all attributes associated with the step are executed. The entity is then transferred (starts the delay associated with the transfer). After the entity completes the transfer and enters the destination station, the entity's *Entity.CurrentStation* attribute is updated.

In the example, the part is sent to the appropriate station on its sequence. Each of the stations used to represent the testing stations follow the same pattern (STATION, PROCESS (seize, delay, release), and ROUTE). The PROCESS module uses the attribute *myTestingTime* to determine the delay time for the testing process (Figure 5-49). This attribute was set when the entity's job step attributes were executed.

■ **FIGURE 5-49** **PROCESS module for testing process**

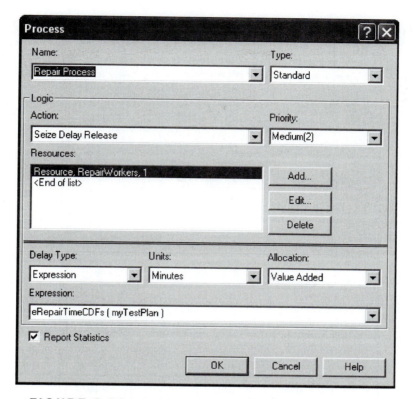

■ **FIGURE 5-50**   **Repair process module**

After proceeding through its testing plan, the part is finally routed to the *RepairStation*, since it was the station associated with the last job step. At the repair station, the part goes through its repair process by using the expression *eRepairTimeCDFs* and its attribute, *myTestPlan*, as shown in Figure 5-50.

The ASSIGN module after the repair process module simply computes the entity's total system time in the attribute, *mySysTime*, so that the following two RECORD modules can compute the appropriate statistics as indicated in Figure 5-51.

Now the model is ready to set up and run. According to the problem, the purpose of the simulation is to evaluate the risk associated with the contract. The life of the contract is specified as 1 year (52 weeks × 5 days/week × 2 shifts/day × 480 minutes/shift = 249,600 minutes). Since the problem states that any jobs not completed at the end of a shift are carried over to the next working day, it is as if there are 249,600 minutes or 4160 hours of continuous operation available for the contract. This is also a terminating simulation since performance during the life of the contract is the primary concern. The only other issue to address is how to initialize the system. The analysis of the two situations (current contract versus contract with 10% more jobs) can be handled via a *relative* comparison. To perform a relative comparison you

Record - Basic Process					
	Name	**Type**	**Value**	**Record into Set**	**Tally Name**
1	Record System Time	Expression	mySysTime	☐	SystemTimeStat
2	Record Contract Prob	Expression	mySysTime < vContractLimit	☐	ProbLTContractLimit

■ **FIGURE 5-51**   **Record modules**

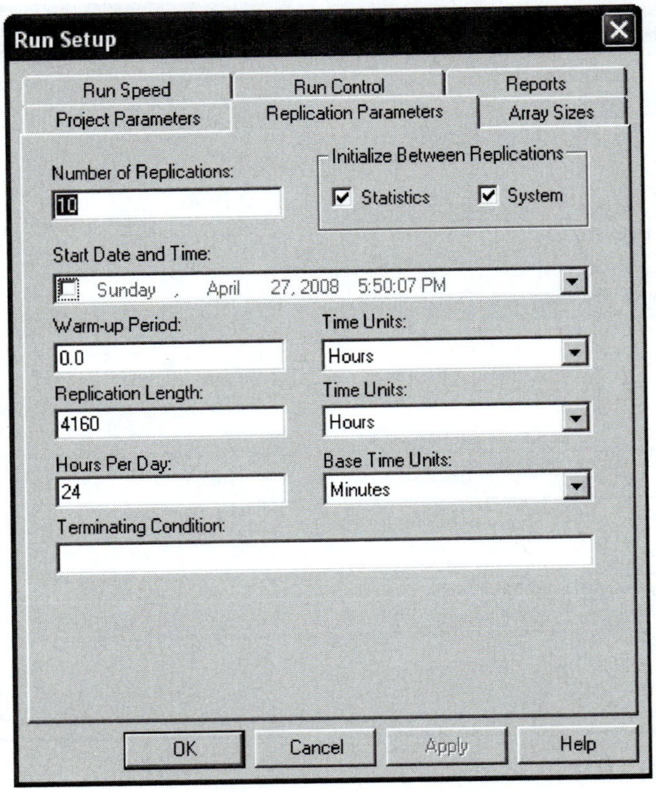

■ **FIGURE 5-52**   Test and repair shop run setup specification

need to ensure that both alternatives start under the same initial conditions. For simplicity, assume that the test and repair shop starts each contract alternative under empty and idle conditions. Let's assume that 10 replications of 4160 hours will be sufficient as illustrated in Figure 5-52.

As shown in Figure 5-53, for the current situation, the probability that a job completes its testing and repair in 480 minutes is about 82%. The addition of more jobs should increase the risk of not meeting the contract specification. You are asked to analyze the new contract's risks and make a recommendation to the company on how to proceed in the exercises.

In the test and repair example, the time that it took to transfer the parts between the stations was a simple stochastic delay (e.g., UNIF (2, 4) minutes). The STATION, ROUTE, and SEQUENCE modules make the modeling of entity movement between stations in this case very straightforward. In some systems, the modeling of the movement is much more important

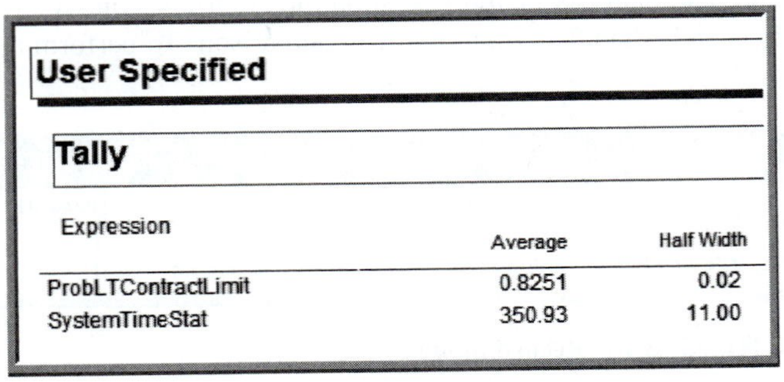

User Specified

Tally

Expression	Average	Half Width
ProbLTContractLimit	0.8251	0.02
SystemTimeStat	350.93	11.00

■ **FIGURE 5-53**   User-defined statistics for current contract because material handling devices (e.g., people to carry the parts, fork lifts, conveyors, etc.) may be required during the transfer. These will require an investigation of the modules in the Advanced Transfer Panel. This topic will be taken up in Chapter 6.

Another classic situation in which customers may wait to have their requests fulfilled will be examined in the remainder of this chapter: inventory systems.

5.3   Inventory Systems

In an inventory system, there are units of an item (e.g., computer printers, etc.) for which customers make demands. If the item is available (in stock), then the customer can receive the item and depart. If the item is not on hand when a demand from a customer arrives, then the customer may depart without the item (i.e., lost sales) or the customer may be placed on a list for when the item becomes available (i.e., backordered). In a sense, the item is like a resource that is consumed by the customer. Unlike the previous notions of a resource, inventory can be replenished. The proper control of the replenishment process is the key to providing adequate customer service. There are two basic questions that must be addressed when controlling the inventory replenishment process: (1) When to order? and (2) How much to order? If the system does not order enough or does not order at the right time, the system may not be able to fill customer demand in a timely manner. Figure 5-54 illustrates a simple inventory system.

There are a number of different ways to manage the replenishment process for an inventory item, known as *inventory control policies*. An inventory control policy must determine (at the very least) when to place a replenishment order and how much to order. This section examines the use of a reorder point (r), reorder quantity (Q) inventory policy. This is often denoted as an (r, Q) inventory policy. The modeling of a number of other inventory control policies will be explored as exercises. After developing a basic understanding of how to model an (r, Q) inventory system, the modeling can be expanded to study *supply chains*. A supply chain can be thought of as a network of locations that hold inventory in order to satisfy end customer demand.

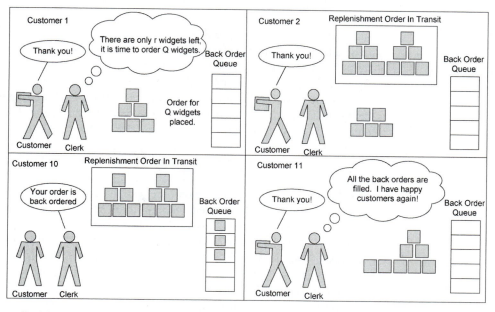

■ **FIGURE 5-54**   A simple (r, Q) inventory system where r = 6 and Q = 12

Just like the topic of queuing systems, the topic of inventory systems has been well studied. In the following, the analytical treatment of the (r, Q) inventory policy will be discussed. A full exposition of the topic of inventory systems is beyond the scope of this text, but the reader can consult a number of texts in the area, such as Hadley and Whitin (1963), Axsater (2006), Silver et al. (1998), or Zipkin (2000) for more details. For details on supply chain modeling, see Nahmias (2001), Askin and Goldberg (2002), Chopra and Meindl (2007), or Ballou (2004).

5.3.1   Modeling an (r, Q) Inventory Control Policy

This section develops a model of a continuous review (r, Q) inventory system with back-ordering. In a continuous review (r, Q) inventory control system, demand arrives according to some stochastic process. When a demand (customer order) occurs, the amount of the demand is determined, and then the system checks for the availability of stock. If the stock on hand is adequate for the order, the demand is filled and the quantity on hand is reduced. On the other hand, if the stock on hand is inadequate to fill the order, the entire order is backordered. The backorders are accumulated in a queue and they will be filled after the arrival of a replenishment order. Assume for simplicity that the backorders are filled on a first-come first-served basis. The inventory position (inventory on hand + on order − backorders) is checked each time after a regular customer demand and the occurrence of a backorder. If the inventory position reaches or falls under the reorder point, a replenishment order is placed. The replenishment order will require a possibly random amount of time to arrive and fill any backorders at the retailer and increase the on-hand inventory. The time from when a replenishment order is placed until the time that it arrives to fill backorders is often called the lead time for the item.

Three state variables are required to model this situation. Let $I(t)$, $IO(t)$, and $B(t)$ be the amount of inventory on hand, on order, and backordered, respectively, at time t. The net inventory, $IN(t) = I(t) - B(t)$, represents the amount (positive or negative) of inventory "available." Note that if $I(t) \geq 0$, then $B(t) = 0$, and that if $B(t) \geq 0$, then $I(t) = 0$. These variables compose the inventory position, $IP(t) = I(t) + IO(t) - B(t)$. The inventory position "remembers" the amount that has been previously ordered. Thus, when placing an order, the inventory position can be used to determine whether a replenishment order needs to be placed. In the continuous review (r, Q) policy, the inventory position must be checked against the reorder point as demands arrive to be filled. After filling (or backordering) a demand, either $I(t)$ or $B(t)$ will have changed (and thus $IP(t)$ will change). If $IP(t) \leq r$, then an order for the amount Q is placed.

Assume that each demand that arrives is for 1 unit to further simplify the discussion. The pseudo-code for this situation is given in Exhibit 5-4. Referring to the exhibit, the entity being created is a customer demand. The amount of the demand should be an attribute of the entity having value 1. If the demand can be satisfied ($I(t) >=$ amount demanded), the on-hand inventory is decremented and the inventory position updated. The inventory position is then checked against the reorder point. If the inventory position is less than or equal to the reorder point, then an order is placed. This is executed by sending the entity to the order-placing logic labeled B in the exhibit. Here, the on-order variable is incremented by the reorder quantity and the delay for the lead time started. Once the lead-time activity is completed, the on-order and on-hand variables are updated, and a signal is sent to the backorder queue. If there are demands waiting in the backorder queue, they are released to try to be filled again. If the demand cannot be satisfied, the demand is backordered by sending it to the backorder queue. Because this process may have changed the inventory position through $B(t)$, the inventory position is updated and checked against the reorder point.

The key performance measures for this type of system are the average amount of inventory on hand, the average amount of inventory backordered, the percentage of time that

Exhibit 5-4 Pseudo-Code for (r, Q) Inventory Model

```
Create the arrival of customer demand entity
Assign the amount demanded = 1
Label A: If I(t) >= amount demanded
            Decrement I(t) by the amount demanded
            Assign IP(t)=I(t)+IO(t)-B(t)
            If IP(t)<= r
Label B:          Increment IO(t) by Q
                  Delay the order entity for the lead time
                  Decrement IO(t) by Q
                  Increment I(t) by Q
                  Signal backorder queue to process backorders
        Else
                  No replenishment order, dispose the entity
        End If
    Else
        Increment B(t) by the amount demanded
        Assign IP(t)=I(t)+IO(t)-B(t)
        If IP(t)<= r
            Duplicate the entity
            Send original to Label B
            Send duplicate to backorder queue
        Else
            Send entity to backorder queue
    End If
When signal from the replenishment is received release all
entities from the queue
        Decrement B(t) by the amount demanded
        Send each entity to Label A
```

the system is out of stock, and the average number of orders made per unit of time. Let's discuss these performance measures before indicating how to collect them in a simulation. The average inventory on hand and the average amount of inventory backordered can be defined as follows:

$$\bar{I} = \frac{1}{T}\int_0^T I(t)dt \quad \bar{B} = \frac{1}{T}\int_0^T B(t)dt$$

As can be seen from the definitions, both $I(t)$ and $B(t)$ are time-based variables and their averages are time averages. Under certain conditions as T goes to infinity, these time averages will converge to the steady-state performance for the (r, Q) model. The percentage of time that the system is out of stock can be defined based on $I(t)$ as follows:

$$SO(t) = \begin{cases} 1 & I(t) = 0 \\ 0 & I(t) > 0 \end{cases} \quad \overline{SO} = \frac{1}{T}\int_0^T SO(t)dt$$

Thus, the variable $SO(t)$ indicates whether on hand equals zero at any time. A time average value for this variable can also be defined and interpreted as the percentage of time that the system is out of stock. One minus $\overline{SO}$ can be interpreted as the percentage of time that the system has stock on hand. Under certain conditions (but not always), this can also be interpreted as the fill rate of the system (i.e., the fraction of demands that can be filled immediately). The variables $\overline{SO}$ and $\bar{B}$ are measures of customer service. To understand the cost of operating the inventory policy, the average number of replenishment orders made per unit of time or the *average order frequency* needs to be measured. Let $N(t)$ be the number of replenishment orders placed in $(0, t]$, then the average order frequency over the period $(0, T]$ can be defined as

$$\overline{OF} = \frac{N(T)}{T}$$

Note that the average order frequency is a rate (units/time). A discussion of the technical issues related to these variables can be found in Zipkin (2000).

Before building the Arena^{TM} model for the (r, Q) inventory model, the analytical results for the case of Poisson demand and constant lead time will be summarized. Assuming that the customer demand process is Poisson with rate λ and that the lead time for the item is the constant L, Exhibit 5-5 summarizes the basic results for the performance measures.

The derivation of the results presented in Exhibit 5-5 can be found in Zipkin (2000). The following example should make this concrete and set up a problem for simulation.

Example. An inventory manager is interested in understanding the cost and service trade-offs related to the management of a particular item (e.g., computer printer). Suppose

Exhibit 5-5 Analytical Results for (r, Q) Inventory Model

---

r = reorder point
Q = reorder quantity
λ = mean customer demand rate in units/time
L = constant lead time (measured in time-units)
$g(x; t)$ = probability mass function for Poisson distribution with rate λ, representing the number of events that occur in an interval of length t

$$g(x; t) = \frac{(\lambda t)^x e^{-\lambda t}}{x!}$$

$G(x; t) = Pr\{X \leq x\} = \sum_{i=0}^{x} g(i; t) = $ Poisson cumulative distribution function

$G^0(x; t) = 1 - G(x; t) = $ Poisson complementary cumulative distribution function

$G^1(x; t) = -(x - \lambda t)G^0(x; t) + (\lambda t)g(x; t) = $ Poisson first-order loss function

$G^2(x; t) = (1/2)\{[(x - \lambda t)^2 + x]G^0(x; t) - (\lambda t)(x - \lambda t)g(x; t)\}$
$\qquad\qquad$ = Poisson second-order loss function

$$\overline{SO} = \frac{1}{Q}\left[G^1(r; L) - G^1(r + Q; L)\right]$$

$$\overline{B} = \frac{1}{Q}\left[G^2(r; L) - G^2(r + Q; L)\right]$$

$$\overline{I} = (1/2)(Q + 1) + r - \lambda L + \overline{B}$$

$$\overline{OF} = \frac{\lambda}{Q}$$

h = holding cost for the item in units ($/unit/time)
b = backordering cost for the item in units ($/unit/time)
k = order preparation cost in units ($/order)
TC = total cost of the policy per time, $TC = k\overline{OF} + h\overline{I} + b\overline{B}$, in units ($/time)

	A	B	C	D	E	F	G	H	I	J	K
1	Example										
2											
3	Continuous review (r, Q) Policy with backordering										
4	Poisson Demand										
5											
6	Demand Rate	3.6									
7	Lead Time	0.5									
8	Mean Lead Time Demand	1.8									
9											
10	Ordering Cost (k)	0.15									
11	Holding cost (h)	0.25									
12	BackOrder Cost (b)	1.75									
13											
14		r	Q	SO	I	B	OF	HC	BC	OC	TC
15		-1	2	0.917351	0.082649	1.382649	1.8	0.020662	2.419637	0.27	2.710299
16		0	2	0.685932	0.396717	0.696717	1.8	0.099179	1.219255	0.27	1.588435
17		1	2	0.403271	0.993446	0.293446	1.8	0.248362	0.513531	0.27	1.031893
18		2	3	0.138165	2.274149	0.074149	1.2	0.568537	0.129761	0.18	0.878298
19		3	2	0.072558	2.731845	0.031845	1.8	0.682961	0.055729	0.27	1.00869
20		4	2	0.023392	3.708453	0.008453	1.8	0.927113	0.014792	0.27	1.211906
21		5	2	0.006474	4.701979	0.001979	1.8	1.175495	0.003463	0.27	1.448958
22											

■ **FIGURE 5-55 Spreadsheet solution to (r, Q) inventory example**

customer demand occurs according to a Poisson process at a rate of $\lambda = 3.6$ per month and the lead time is 0.5 months. The manager has estimated that the holding cost for the item is approximately \$0.25 per unit per month. In addition, when a backorder occurs, the estimate cost will be \$1.75 per unit per month. Every time that an order is placed, it costs approximately \$0.15 to prepare and process the order.

Solution. The formulas of Exhibit 5-5 have been implemented in the *rQInventoryModel.xls* spreadsheet that accompanies this chapter. The spreadsheet is based on the use of VBA code that implements the formulas in the exhibit.

Figure 5-55 shows the performance of various (r, Q) inventory policies for different values of the reorder point and the reorder quantity based on the parameters of the example. The stock-out percentage, average inventory, and average number backordered are all computed. Note that the policy for r = 2 and Q = 3, appears to have the smallest total cost. In searching for the optimal values of r and Q, it is reasonable to start with the economic order quantity for Q, based on the assumption of deterministic demand:

$$EOQ = \sqrt{\frac{2k\lambda}{h}}$$

In addition, one can use a base-stock normal approximation for the initial value of r,

$$r = \lambda L + z\sqrt{\lambda L} \quad z = \Phi^{-1}\left(\frac{b}{b+h}\right)$$

where z is the standard normal value associated with the ratio $b/(b+h)$. The development of these approximations is discussed in Hopp and Spearman (2001). For the example problem, the approximations yield:

$$EOQ = \sqrt{\frac{2k\lambda}{h}} = \sqrt{\frac{2 \times 0.15 \times 3.6}{0.25}} = 2.08 \approx 3$$

$$z = \Phi^{-1}\left(\frac{b}{b+h}\right) = \Phi^{-1}\left(\frac{1.75}{1.75+0.25}\right) = \Phi^{-1}(0.88) = 1.15$$

$$r = \lambda L + z\sqrt{\lambda L} = 1.8 + 1.15 \times \sqrt{1.8} = 3.34 \cong 4$$

Variable - Basic Process						
	Name	**Rows**	**Columns**	**Data Type**	**Clear Option**	**Initial Values**	**Report Statistics**
1	vReorderPt ▼			Real	System	1 rows	☐
2	vReorderQty			Real	System	1 rows	☐
3	vOnHand			Real	System	0 rows	☑
4	vInvPos			Real	System	0 rows	☐
5	vOnOrder			Real	System	0 rows	☐
6	vBackOrdered			Real	System	0 rows	☑
7	vNumOrders			Real	System	0 rows	☐
8	vHoldingCost			Real	System	1 rows	☐
9	vBackOrderCost			Real	System	1 rows	☐
10	vOrderingCost			Real	System	1 rows	☐
	Double-click here to add a new row.						

■ **FIGURE 5-56**   **(r, Q) Inventory Arena model variables**

These are clearly approximate values and it would be important to experiment with values near these values as was done in the spreadsheet to better assess the situation. Now that the theoretical answers for this situation are known, let's build an Arena™ model to represent the problem.

The Arena™ model will follow closely the pseudo-code outlined in Exhibit 5-4. To develop this model, first define the variables to be used as per Figure 5-56. The reorder point and reorder quantity have been set to $(r = -1)$ and $(Q = 2)$, respectively. The costs have be set based on the example. Note that the report statistics checkboxes have been checked for the on hand inventory and the backordered inventory. This will cause time-based statistics to be collected on these variables.

Figure 5-57 illustrates the logic for handling incoming demands and the reprocessing of demands that were backordered. First, an entity is created to represent the demand. This occurs according to a time between arrivals with an exponential distribution with a mean of $(1.0/0.12)$ days $(3.6 \text{ units/month} * (1 \text{ month}/30 \text{ days}) = 0.12 \text{ units/day})$. In the following

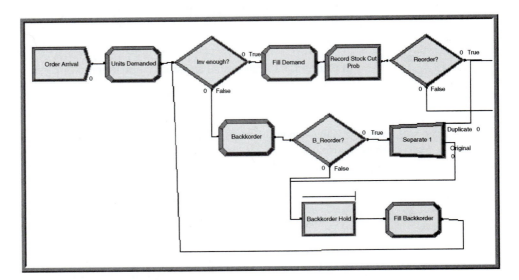

■ **FIGURE 5-57**   **Initial filling logic for (r, Q) inventory model**

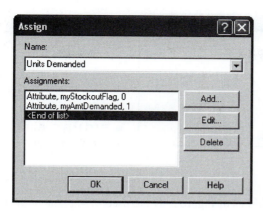

■ **FIGURE 5-58** Assigning the amount of demand

ASSIGN module (Figure 5-58), the amount demanded is set to 1 and a stock out indicator is set. This will be used to tally the probability that an incoming demand is not filled. In the DECIDE module, the amount on hand is checked to see if it can handle the amount demanded. If true, the demand is filled. If false, the demand is backordered.

The fill demand and backorder ASSIGN modules have the form shown in Figure 5-59. The amount on hand or the amount backordered is decreased or increased accordingly. In addition, the inventory position is updated. In the backordering ASSIGN module, the stock-out flag is set to 1 to indicate that this particular demand did not get an immediate fill.

When the demand is ultimately filled, it will pass through the RECORD module in Figure 5-57. Note that in the RECORD module (see Figure 5-60), the expression option is used to record on the attribute (*myStockoutFlag*), which has a value of 1 if the demand had been backordered upon arrival and the value 0 if it had not been backordered upon initial arrival. The value of this attribute is set initially to 0 in the ASSIGN module of Figure 5-58 and then updated to 1 if the demand is not filled as per Figure 5-59. This indicator variable will allow for the estimation of the probability that an arriving customer will be backordered. In inventory analysis parlance, this is called the probability of stock out. One minus the probability of stock out is called the fill rate for the inventory system (Figure 5-60). Under Poisson arrivals, it can be shown that the probability of an arriving customer being backordered will be the same as $\overline{SO}$, the percentage of time that the system is stocked out. This is due to a phenomenon called PASTA (Poisson arrivals see time averages) and is discussed in Zipkin (2000) as well as many texts on stochastic processes (e.g., Tijms 2003).

If the demand has been filled/backordered, the system then checks if the inventory position is less than or equal to the reorder point and if so a replenishment order must be placed.

Assignments   **Fill Demand ASSIGN Module**

	Type	Variable Name	New Value
1	Variable	vOnHand	vOnHand - myAmtDemanded
2	Variable	vInvPos	vOnHand + vOnOrder - vBackOrdered
	Double-click here to add a new row.		

Assignments   **Backorder ASSIGN Module**

	Type	Variable Name	Attribute Name	New Value
1	Attribute	Variable 1	myStockoutFlag	1
2	Variable	vBackOrdered	Attribute 2	vBackOrdered + myAmtDemanded
3	Variable	vInvPos	Attribute 3	vOnHand + vOnOrder - vBackOrdered
	Double-click here to add a new row.			

■ **FIGURE 5-59** ASSIGN modules for filling and backordering demand

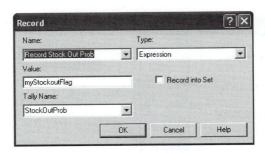

■ **FIGURE 5-60** **Recording demand initial-fill condition**

The two DECIDE modules of Figure 5-57 are used to implement the reorder point–checking logic. A demand that has been backordered must wait in the back order queue. Thus, if a replenishment order is required in the backordering case, the demand must be cloned, with the duplicate acting as the replenishment order and the original flowing into the backorder queue. The backorder queue is represented by a HOLD module (Figure 5-61). The demands that enter the HOLD module will wait until they get a signal (value = 1) before they are released. After being released, the demands will attempt to be filled again. The release will be triggered after a replenishment order arrives.

The logic representing the replenishment-ordering process is given in Figure 5-62. First, the amount of the order is determined, and then the order entity experiences the delay representing the lead time. After the lead time, the order has essentially arrived to replenish the inventory. The corresponding ordering and replenishment ASSIGN modules and the subsequent SIGNAL modules are given in Figure 5-63 and Figure 5-64, respectively.

The basic model structure is now completed; however, there are a few issues related to the collection of the performance measures that must be discussed. The average order frequency must be collected. Recall that $\overline{OF} = N(T)/T$. Thus, the number of orders placed in the time interval of interest must be observed. This can be done by creating a logic entity that will observe the number of orders that have been placed since the start of the observation period. In Figure 5-63, the variable called *vNumOrders* is incremented each time an order is placed. This is, in essence, $N(T)$. Figure 5-65 shows the order-frequency collection logic. First, an entity is created every T time units. In this case, the interval is monthly (every 30 days).

Then the RECORD module observes the value of *vNumOrders* (Figure 5-66). The following ASSIGN module sets *vNumOrders* equal to zero. Thus, *vNumOrders* represents the number of orders accumulated during the 30-day period.

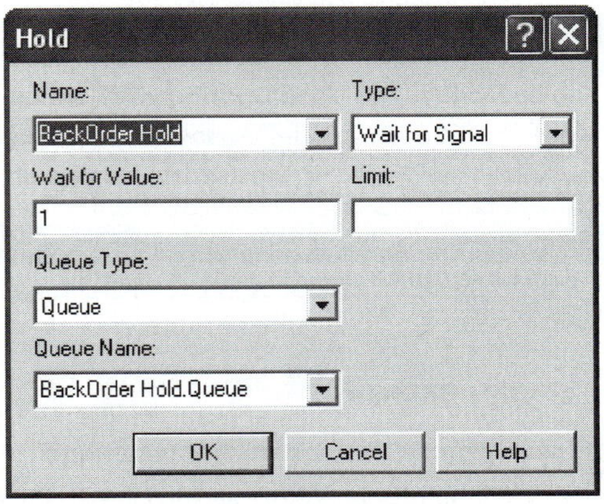

■ **FIGURE 5-61** **Backorder queue as a HOLD module**

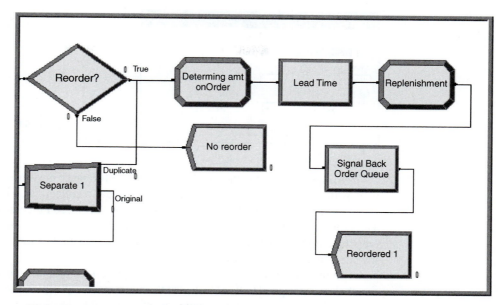

■ **FIGURE 5-62**   Replenishment logic

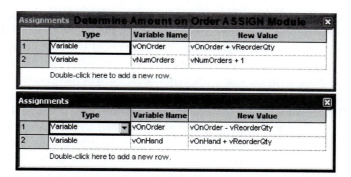

■ **FIGURE 5-63**   Ordering and replenishment ASSIGN modules

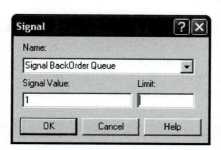

■ **FIGURE 5-64**   Signaling the backorder queue

■ **FIGURE 5-65**   Order frequency collection logic

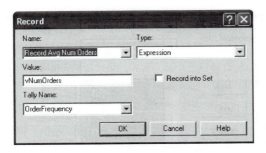

■ **FIGURE 5-66** Recording the number of orders placed

To close out the statistical collection of the performance measures, the collection of $\overline{SO}$ as well as the cost of the policy needs to be discussed. To collect $\overline{SO}$, a time-persistent statistic needs to be defined using the Statistic module of the Advanced Process panel as in Figure 5-67. The expression (vOnHand == 0) is Boolean, which evaluates to 1.0 if true and 0.0 if false. By creating a time-persistent statistic on this expression, the proportion of time that the on-hand inventory is equal to 0.0 can be tabulated.

To record the costs associated with the policy, three output statistics are needed. Recall that these are designed to be observed at the end of each replication. For the inventory holding cost, the time average of the on-hand inventory variable, DAVG(*vOnHand* value), should be multiplied by the holding cost per unit per time. The backordering cost is done in a similar manner. The ordering cost is computed by taking the cost per order times the average order frequency. Recall that this was captured via a RECORD module every month. The average from the RECORD module is available through the TAVG() function. Finally, the total cost is tabulated as the sum of the backordering cost, the holding cost, and the ordering cost. In Figure 5-67, this was accomplished by using the OVALUE() function for OUTPUT statistics. This function returns the last observed value of the OUTPUT statistic. In this case, it will return the value observed at the end of the replication. Each of these statistics will be shown on the Arena™ reports as user-defined statistics.

The case of (r = −1, Q = 2) with the cost parameters indicated in Figure 5-55 was run for 30 replications with a warm-up period of 3600 days and a run length of 39,600 days. As seen in Figure 5-68, these results match very closely the performance calculated in Figure 5-55. Note that the stock-out probability is essentially the same as the percentage of time that the system was out of stock. This is an indication that the PASTA property of Poison arrivals is working according to theory.

This example should give you a solid basis for developing more sophisticated inventory models. While analytical results were available for the example, small changes in the assumptions necessitate the need for simulation. For example, what if the lead times are stochastic or the demand is not in units of 1. In the latter case, the filling of the backorder queue should be done in different ways. For example, suppose that customers wanting 5 and 3 items, respectively, were waiting in the queue. Now suppose that a replenishment of 4 items comes in. Do you give 4 units of the item to the customer wanting 5 units (partial fill), or do you choose the

	Name	Type	Expression	Report Label	Output File
1	TimeOutOfStock	Time-Persistent	(vOnHand == 0)	TimeOutOfStock	
2	HoldingCost	Output	vHoldingCost * DAVG(vOnHand Value)	HoldingCost	
3	BackorderCost	Output	vBackOrderCost * DAVG(vBackOrdered Value)	BackorderCost	
4	OrderingCost	Output	vOrderingCost * TAVG(OrderFrequency)	OrderingCost	
5	TotalCost	Output	OVALUE(BackorderCost) + OVALUE(HoldingCost) + OVALUE(OrderingCost)	TotalCost	
	Double-click here to add a new row.				

■ **FIGURE 5-67** Statistic module for (r, Q) inventory model

Tally

Expression	Average	Half Width
OrderFrequency	1.8022	0.01
StockOutProb	0.9173	0.00

Time Persistent

Time Persistent	Average	Half Width
TimeOutOfStock	0.9177	0.00

Variable	Average	Half Width
vBackOrdered	1.3849	0.01
vOnHand	0.08226392	0.00

Output

Output	Average	Half Width
BackorderCost	2.4236	0.02
HoldingCost	0.02056598	0.00
OrderingCost	0.2703	0.00
TotalCost	2.7145	0.02

■ **FIGURE 5-68**   Results for the (r, Q) inventory model

customer wanting 3 units and fill their entire backorder and give only 1 unit to the customer wanting 5 units? The more realism needed, the lower the likelihood that analytical models can be applied, and the more simulation becomes useful.

In the next section, the previous inventory system is expanded to handle a supply chain. Because of the complexity, only an overview of the modeling issues will be described. In addition, an Arena™ model capable of modeling a system with a supplier, a warehouse, and multiple retailers having only one product will be presented.

5.3.2   Modeling a Multiechelon Inventory System

This section describes how to model a simple multiechelon inventory system. Figure 5-69 illustrates the basic system that will be modeled. The system consists of an external supplier, a warehouse, and a set of retailers. Assume that there is only 1 item type stocked at each location in

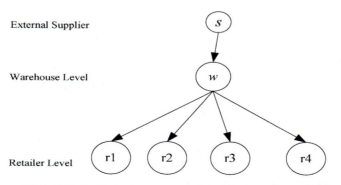

■ **FIGURE 5-69**   A simple multiechelon inventory system

the system. The warehouse supplies each of the four retailers when they make a replenishment request. In this model, each of the four retailers will be identical. They have the same control policy settings for their items and they experience the same customer demand. The arrangement of suppliers and customers in this manner results in a tree structure as illustrated in the figure.

In the model, a reorder point reorder quantity (r, Q) inventory policy is utilized at each location. If the location does not have sufficient stock to satisfy demand, then the order gets backordered. The retail level experiences customer demand according to a Poisson process. The warehouse experiences replenishment requests for the order quantities, Q, that are associated with the retailers. The time between the placement of a replenishment order by a retailer and the arrival of the replenishment from the warehouse is called the lead time. The lead time can consist of the waiting time to fill the order if backordered plus a transport time to move the order from the warehouse to the retailer. This example assumes that the transport time is deterministic. The warehouse, in turn, will order its replenishment from the external supplier. The external supplier can satisfy any order placed on it, with the order being satisfied after a corresponding delay for the lead time. Conceptually, the external supplier's lead time is the production and transport time for the order.

In order to develop an Arena™ model for this situation, the previous example can be expanded. Note that each location follows the (r, Q) inventory policy. Thus, the basic logic can be repeated for each retailer and for the warehouse. Then, the retailers must send their orders to the warehouse and the warehouse must send its filled orders to the appropriate retailer. Rather than cutting and pasting the basic logic 5 times (once for the warehouse and four times for the retailers), the model can be made more generic by using arrays. This will enable the expansion of the model to more retailers if required in the future.

Figure 5-70 shows the logic for the retailer level in the multiechelon inventory system. As you can see, this logic is essentially the same as that used in the previous example. Because of this similarity, all details of the model are not presented. A summary of the changes will be emphasized.

In Figure 5-71, the variables are defined for the model. The key difference is the representation of the on-hand, on order, backordered, and inventory position for each retailer using Arena™ arrays.

Variables are listed in the following.

vNumR: This variable indicates the total number of retailers. It is used in the Order Arrival of the CREATE Module to determine the aggregate arrival rate for all retailers (i.e., EXPO (v*MTBA*/v*NumR*)).

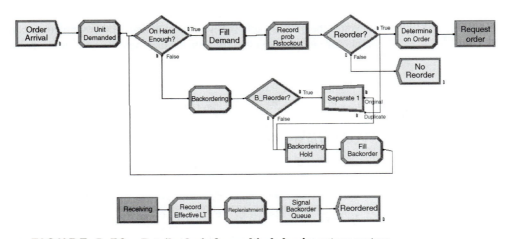

■ **FIGURE 5-70**   Retailer logic for multiechelon inventory system

Variable - Basic Process						
Name	**Rows**	**Columns**	**Data Type**	**Clear Option**	**Initial Values**	**Report Statistics**
1 vNumR			Real	System	1 rows	☐
2 vMTBA			Real	System	1 rows	☐
3 vRetailerLT			Real	System	1 rows	☐
4 vWhsLT			Real	System	1 rows	☐
5 vQw			Real	System	1 rows	☐
6 vRw			Real	System	1 rows	☐
7 vQr			Real	System	1 rows	☐
8 vRr			Real	System	1 rows	☐
9 vOnHand	4		Real	System	1 rows	☐
10 vOnOrder	4		Real	System	0 rows	☐
11 vBackOrdered	4		Real	System	0 rows	☐
12 vInvPos	4		Real	System	0 rows	☐
13 vSignalValue	4		Real	System	4 rows	☐
14 vWhsOnHand			Real	System	1 rows	☑
15 vWhsOnOrder			Real	System	0 rows	☐
16 vWhsBackOrdered			Real	System	0 rows	☑
17 vWhsInvPos			Real	System	0 rows	☐

Double-click here to add a new row.

■ **FIGURE 5-71** **Variable module**

vMTBA: This variable is the mean time between the arrivals of customer demand. If the demand rate is λ per period, the mean interarrival time is 1/λ.

vRetailerLT: This variable represents the transport lead time for the retailer from the warehouse. In the model, it is a constant set at 1 day.

vWhsLT: This variable represents the replenishment lead time for the warehouse from the supplier. It is assumed to be constant with the value of 1 day.

vQw: This variable is the replenishment reorder quantity for the warehouse.

vRw: This variable is the reorder point at the warehouse.

vQr: This variable is the replenishment reorder quantity for the retailers.

vRr: This variable is the reorder point of the retailers.

vOnHand(vNumR), vOnOrder(vNumR), vBackOrdered(vNumR), vInvPos(vNumR): These variable arrays represent the amount on hand, on order, backordered, and the inventory position of each retailer.

vSignalValue(vNumR): This variable array represents the signal value used at each retailer to release the orders that are held in the backorder queue.

vWhsOnHand, vWhsOnOrder, vWhsBackOrdered, vWhsInvPos: These variables represent the amount of inventory on hand, on order, and backordered, and the inventory position at the warehouse.

If each retailer experiences Poisson demand at rate λ_i, then the overall demand to the system is $\lambda = \sum_{i=1}^{4} \lambda_i$. This overall demand rate can be probabilistically split so that each retailer experiences a Poisson process with rate λ_i. See Tijms (2003) for more on the merging and splitting properties of the Poisson process. Let $p_i = \lambda_i/\lambda$ be the probability that an arriving demand should be sent to the ith retailer. Thus, an arriving demand can be created, and then the retailer can be determined according to $p_i = \lambda_i/\lambda$. Figure 5-72 illustrates the ASSIGN module after creating the arriving demand.

Assignments

	Type	Attribute Name	
1	Attribute	myRetailerNum	ANINT(UNIF(0,1)*vNumR+0.5)
2	Attribute	myAmtDemanded	1
3	Attribute	myRetailerSOFlag	0
	Double-click here to add a new row.		

■ **FIGURE 5-72** **Assigning the retailer number**

Since each retailer is identical in the example, the first line randomly generates a discrete uniform number over the range 1 to 4. After the retailer number has been assigned, this attribute will be used as the index into the arrays representing the retailer's on hand, on-order, amount backordered, and inventory position. Figure 5-73 shows an example of how the arrays are used in the ASSIGN module for filling demand. This makes it possible to have any number of retail locations, by simply increasing the size of the arrays and appropriately assigning to the attribute *myRetailerNum*.

The rest of the logic for modeling the retail level is the same as described for the previous example, except for the two pink ROUTE and STATION modules. When a retailer decides to place a replenishment order, the entity representing the order is sent via the ROUTE module to a station module that denotes the warehouse logic as per Figure 5-74.

When the warehouse fills a demand from a retailer, it is sent to the STATION module labeled Receiving. The pseudo-code associated with the Receiving station is as follows:

```
Receiving replenishment Order from warehouse
Record the effective lead time
Replenishment assignment
Decrement the vOnOrder(myRetailerNum) by vQr
Increment the inventory vOnhand (myRetailerNum) by vQr
Send signal, vSignalValue(myRetailerNum), to release backlog hold queue
Dispose entity
```

Assignments

	Type	Other	New Value
1	Other	vOnHand(myRetailerNum)	vOnHand(myRetailerNum) - myAmtDemanded
2	Other	vInvPos(myRetailerNum)	vOnHand(myRetailerNum) + vOnOrder(myRetailerNum) - vBackOrdered(myRetailerNum)
	Double-click here to add a new row.		

■ **FIGURE 5-73** **Assignments for filling demand**

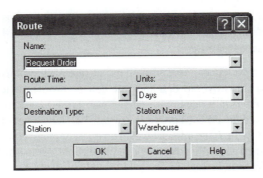

■ **FIGURE 5-74** **Sending the order to the warehouse**

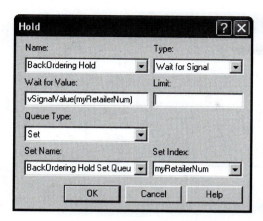

■ **FIGURE 5-75** HOLD module for queuing backorders at the retailers

The signal sent is based on an array indexed by the retailer number. Arena's HOLD module has the ability to hold entities until they receive a specific signal and then to release the entities associated with that signal number. Figure 5-75 shows the corresponding HOLD module associated with the signal. Note that the Set option has been used for the type of queuing. This allows a separate backorder queue to be defined for each retailer. The set contains four queues, one for each retailer, and it is indexed by the attribute *myRetailerNum*. Thus, each backordered demand waits in a queue specific to the retailer that is currently processing the demand.

The logic associated with the warehouse is essentially the same as previously described. The key difference is that the demand sent to the warehouse will require *vQr* units of inventory from the warehouse (no longer a single unit). Assume that backordered demands are satisfied on a first in first out basis. Once a demand has been filled, it must be sent to the proper retailer. This is accomplished by the ROUTE module denoted as Ship Order in Figure 5-76. The ROUTE module is shown in Figure 5-77. The entity is sent directly to the station with the name *Receiving* using a routing time of 1 day (*vRetailerLT*).

Collecting statistics on this model is a bit more tedious. Figure 5-78 illustrates the primary problem. Because the state variables, such as *on hand*, are represented as arrays, time-persistent statistics on each element of the array cannot be directly defined in the VARIABLE module. Thus, in order to collect statistics on the average inventory and average amount backordered, a specific time-persistent statistic must be defined as shown in Figure 5-78. Other statistics need to be handled in a similar fashion. These extensions are left for you to pursue as exercises.

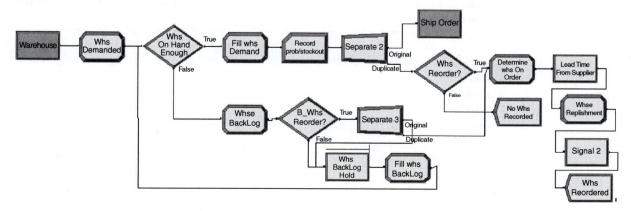

■ **FIGURE 5-76** Warehouse demand processing logic

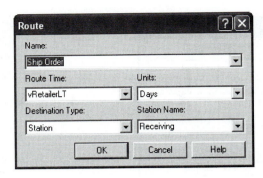

■ **FIGURE 5-77** **Shipping the order to the retailer**

	Name	Type	Expression	Report Label	Output File
1	onHandR1	Time-Persistent	vOnHand(1)	onHandR1	
2	onHandR2	Time-Persistent	vOnHand(2)	onHandR2	
3	backOrderR1	Time-Persistent	vBackOrdered(1)	backOrderR1	
4	backOrderR2	Time-Persistent	vBackOrdered(2)	backOrderR2	
5	backOrderR3	Time-Persistent	vBackOrdered(3)	backOrderR3	
6	backOrderR4	Time-Persistent	vBackOrdered(4)	backOrderR4	
7	onHandR3	Time-Persistent	vOnHand(3)	onHandR3	
8	onHandR4	Time-Persistent	vOnHand(4)	onHandR4	

Double-click here to add a new row.

■ **FIGURE 5-78** **Statistic module for multiechelon inventory model**

This completes the discussion of the multiechelon inventory system. To run the model, you must set the reorder point and reorder quantity for the warehouse and retailers. As an example, the model was run with the settings shown in Table 5-10.

Figure 5-79 shows the results for running the multiechelon inventory system. The average inventory on hand and on backorder for each of the retailers is shown. As an extension, one might want to collect statistics at the aggregate level (e.g., average amount of inventory in the system or at the retailer level).

Although this is one of the more complicated models, it should be evident that extending the model to include such elements as shipping via trucks or multiple items at

■ **TABLE 5-10**
Parameters for Running Multiechelon Inventory System

Parameter	Variable name	Value
Number of retailers	vNumR	4
Mean time between demand arrivals	vMTBA	10 days
Retailer lead time	vRetailerLT	1 days
Warehouse lead time	vWhsLT	1 days
Warehouse reorder quantity	vQw	4
Warehouse reorder point	vRw	−1
Retailer reorder quantity	vQr	1
Retailer reorder point	vRr	0
Number of replications		10
Warm-up period		360 days
Replication length		3600 days

Tally		
Expression	Average	Half Width
Record probRstockout	0.1223	0.01
Record probWstockout	0.3488	0.01
Interval	Average	Half Width
Record EffectiveLT	1.3006	0.00

Time Persistent

Time Persistent	Average	Half Width
backOrderR1	0.00893928	0.00
backOrderR2	0.00875987	0.00
backOrderR3	0.00974924	0.00
backOrderR4	0.00872368	0.00
onHandR1	0.8766	0.01
onHandR2	0.8782	0.00
onHandR3	0.8743	0.01
onHandR4	0.8800	0.00
Variable	Average	Half Width
vWhsBackOrdered	0.1219	0.01
vWhsOnHand	1.2164	0.02

■ **FIGURE 5-79** **Results for multiechelon inventory system**

each location can be accomplished using Arena^{TM}. Consider exploring some of these extensions in the exercises.

5.4   Summary

This chapter provided an introduction to two very important application areas for simulation: queuing and inventory models. For queuing models, you learned how to analyze situations involving a single station using analytical formulas. This analysis involves the following:

- Identifying the arrival and service processes
- Identifying the size of the arriving population and the size of the system
- Specifying the appropriate queuing model and its input parameters
- Identifying the desired performance measures
- Computing the required performance measures

Two key concepts were also covered: Little's formula and steady-state analysis. Little's formula represents a conservation principle that relates the average waiting time in the system to the average number in the system. The steady-state performance of the queuing system is readily available based on the steady-state probability distribution for the number of customers in the system. This distribution can be thought of as the probability of having a given number of customers in the system at some arbitrary point in the infinite future. Under certain conditions, analytical formulas are available; however, when they are not available simulation can be used. Even for simple single-station models, small changes to assumptions or the structure of the system necessitate the use of simulation. For example, if the priority of the customer depends on the amount of resource available, then simulation is needed.

Queuing systems often involve a network of stations. While analytical models exist for certain classes of system structure, simulation quickly becomes the most viable method for analysis. Arena™ has important constructs that facilitate the modeling of queuing networks:

> *STATION:* Allows the marking in the model for a location to which entities can be directed for processing
>
> *SEQUENCE:* Allows for prespecified routes of stations to be defined and attributes to be assigned when entities are transferred to the stations
>
> *ROUTE:* Facilitates the movement between stations with a time delay

In addition to queuing models, you learned about inventory models and the basic operation of the (r, Q) inventory policy. Again, the application of analytical models quickly becomes limited. When this occurs, simulation provides a robust and useful tool for analyzing systems involving these elements. In particular, a supply chain can be conceptualized as a network of inventory locations. Again, the modeling of these kinds of systems can be implemented using Arena™ and the STATION and ROUTE modules. The analysis of these types of systems has only been introductory. You are encouraged to seek out a number of the excellent texts mentioned in the references to learn more about these topics.

The modeling journey is not yet complete. Through five chapters, you should have developed a good working knowledge on how to use Arena™ to model a number of interesting situations. Chapter 6 takes this modeling even further by studying constructs in Arena™ that facilitate the modeling of systems involving material movement. These constructs include transporters, conveyors, and automated guided vehicles.

Exercises

For problems 5-1 to 5-12, specify the appropriate queuing models needed to solve the problem using Kendall's notation. Specify the parameters of the model, such as λ_e, μ, c, N, and so on. Specify how and what you would compute to solve the problem. Be as specific as possible by specifying the equation needed. Then, compute the desired quantities. You might also try to use Arena™ to solve the problems via simulation.

5.1 Consider a single-pump gas station where the arrival process is Poisson with a mean time between arrivals of 10 minutes. The service time is exponentially distributed with a mean of 6 minutes.

(a) What is the probability that you have to wait for service?

(b) What is the mean number of customers at the station?

(c) What is the expected time waiting in the line to get a pump?

5.2 Suppose an operator has been assigned the responsibility of maintaining 3 machines. For each machine the probability distribution of the running time before a breakdown is exponentially distributed with a mean of 9 hours. The repair time also has an exponential distribution with a mean of 2 hours.

(a) What is the probability that the operator is idle?

(b) What is the expected number of machines that are running?

(c) What is the expected number of machines that are not running?

5.3 SuperFastCopy wants to install self-service copiers, but cannot decide whether to put in one or two machines. Managers indicate that arrivals are Poisson with a rate of 30 per hour, and the time spent copying is exponentially distributed with a mean of 1.75 minutes. Because the shop is small, they want the probability of 5 or more customers in the shop to be small, say less than 7%. Make a recommendation based on queuing theory to SuperFastCopy.

5.4 Two machines are being considered for a certain processing job. The first machine has an exponentially distributed processing time with a mean of 10 minutes. For the second machine the vendor has indicated that the mean processing time is 10 minutes but with a standard deviation of 6 minutes. Assume that jobs arrive according to a Poisson process with an arrival rate of 4.5 jobs per hour. Using queuing theory, which machine is better in terms of the average waiting time of the jobs?

5.5 Customers arrive at a one-window drive-in bank according to a Poisson distribution with a mean of 10 per hour. The service time for each customer is exponentially distributed with a mean of 5 minutes. There are 3 spaces in front of the window including that for the car being served. Other arriving cars can wait outside these 3 spaces.

(a) What is the probability that an arriving customer can enter one of the 3 spaces in front of the window?

(b) What is the probability that an arriving customer will have to wait outside the 3 spaces?

(c) How long is an arriving customer expected to wait before starting service?

(d) How many spaces should be provided in front of the window so that an arriving customer can wait in front of the window at least 20% of the time? In other words, the probability of at least one open space must be greater than 20%.

5.6 Joe Rose is a student at Big State U. He does odd jobs to supplement his income. Job requests come every 5 days on the average, but the time between requests is exponentially distributed. The time for completing a job is also exponentially distributed with a mean of 4 days.

(a) What would you compute to find the chance that Joe will not have any jobs to work on?

(b) What would you compute to find the average value of the waiting jobs if Joe gets about $25 per job?

5.7 Sly's convenience store operates a two-pump gas station. The lane leading to the pumps can hold at most 5 cars, including those being serviced. Arriving cars go elsewhere if the lane is full. The distribution of the arriving cars is Poisson with a mean of 20 per hour. The time to fill up and pay for the purchase is exponentially distributed with a mean of 6 minutes.

(a) Specify using queuing notation, exactly what you would compute to find the percentage of drivers who will seek business elsewhere.

(b) Specify using queuing notation, exactly what you would compute to find the utilization of the pumps.

5.8 An airline ticket office has two ticket agents answering incoming phone calls for flight reservations. In addition, two callers can be put on hold until one of the agents is available to take the call. If all four phone lines (both agent lines and the hold lines) are busy, a potential customer gets a busy signal, and it is assumed that the call goes to another ticket office and that the business is lost. The calls and attempted calls occur randomly (i.e., according to Poisson process) at a mean rate of 15 per hour. The length of a telephone conversation has an exponential distribution with a mean of 4 minutes.

(a) Specify using queuing notation, exactly what you would compute to find the probability of losing a potential customer?

(b) What would you compute to find the probability that an arriving phone call will not start service immediately but will be able to wait on a hold line?

5.9 SuperFastCopy has three identical copying machines. When a machine is being used, the time until it breaks down has an exponential distribution with a mean of 2 weeks. A repair person is kept on call to repair the machines. The repair time for a machine has an exponential distribution with a mean of 0.5 week. The downtime cost for each copying machine is $100 per week.

(a) Let the state of the system be the number of machines not working. Construct a state diagram for this queuing system.

(b) Write an expression using queuing performance measures to compute the expected downtime cost per week.

5.10 NWH Cardiac Care Unit (CCU) has 5 beds, which are virtually always occupied by patients who have just undergone major heart surgery. Two registered nurses (RNs) are on duty in the CCU in each of the three 8-hour shifts. About every 2 hours following an exponential distribution, one of the patients requires a nurse's attention. The RN will then spend an average of 30 minutes (exponentially distributed) assisting the patient and updating medical records regarding the problem and care provided. Because immediate service is critical to the 5 patients, two important questions follow:

(a) What would you compute to find the average number of patients being attended by the nurses?

(b) What would you compute to fine the average time that a patient spends waiting for one of the nurses to arrive?

5.11 Rick managers a small barber shop at Big State U, and employs one barber. Rick is also a barber and he works only when he has more than one customer in the shop. Customers arrive randomly at a rate of 3 per hour. Rick takes 15 minutes on the average for a haircut, but his employee takes 10 minutes. Assume that there are only 2 chairs available with no waiting room in the shop.

(a) Let the state of the system be the number of customers in the shop. Construct a state diagram for this queuing system.

(b) What is the probability that a customer is turned away?

(c) What is the probability that the barber shop is idle?

(d) What is the steady-state mean number of customers in the shop?

5.12 Consider the M/G/1 queue with the following variation. The server works continuously as long as there is at least one customer in the system. The customers are processed FIFO. When the server finishes serving a customer and finds the system empty, the server goes away for a length of time called a "vacation." At the end of the vacation, the server returns and begins to serve the customers, if any, who have arrived during the vacation. If the server finds no customers waiting at the end of a vacation, it immediately takes another vacation, and continues in this manner until it finds at least one waiting customer upon return from a vacation. Assume that the time between customer arrivals is exponentially distributed with mean of 3 minutes. The service distribution for each customer is a gamma distribution with a mean of 4.5 seconds and a variance of 3.375. The length of a vacation is a random variable uniformly distributed between 8 and 12 minutes. Estimate the average number of customers waiting in the queue upon return from the vacation. Run the model for 1 replication of 20000 minutes with a warm up of 5000 minutes. In addition, develop an empirical distribution for the number of customers waiting upon the return of the server from vacation. In other words, estimate the probability that $j = 0$, 1, 2, and so on, where j is the number of waiting customers in the queue upon the return of the server from a vacation. Hint: Use the FREQUENCIES option of the STATISTIC module, see Chapter 7. This queue has many applications, for example, consider how a bus stop operates.

5.13 Reconsider Problem 5-8. Suppose the airline ticket office has expanded the number of lines to 8 lines, still with two agents. In addition, they have instituted an automated caller identification system that automatically places first class business (FCB) passengers at the head of the queue, waiting for an agent. Of the original 15 calls per hour, they estimate that roughly one-third of these come from FCB customers. They have also noticed that FCB customers take approximately 3 minutes on average for the phone call, still exponentially distributed. Regular customers still take on average 4 minutes, exponentially distributed. Simulate this system with and without the new prioritization scheme, and compare the average waiting time for the two types of customers.

5.14 Consider a system having 6 machines tended by 2 operators. In this system, there are two types of stoppages. Type 1 stoppage occurs after a fixed constant amount of machine running time, $1/\lambda_1 = 30$ minutes, and has a constant value of $1/\mu_1 = 10$ minutes as the service time for the stoppage. Type 2 stoppages occur after random intervals of time, are negatively exponentially distributed, with a mean of $1/\lambda_2 = 10$ minutes. Service times for type 2 stoppages are negative exponentially distributed with a mean of 4 minutes. Both of the operators have the same skills and can handle either type of stoppage. The machines wait for service from the operators in a first come first served queue with no priority given to either type of stoppage. Assume that the chance of a type 1 stoppage occurring next is 30%; otherwise, the next stoppage will be of type 2. Simulate this system for 21 replications of 12000 minutes with a 2000 minute warm up period to estimate the average number of waiting machines by type of stoppage, the average utilization of the operator, the average utilization of the machines, and the average waiting time of the machines by type of stoppage.

5.15 Consider the walk-in health care clinic example earlier in this chapter. Clinic managers are interested in improving the handling of low-priority patients. To that end, they are considering hiring a special registered nurse practitioner who is licensed to provide minor medical care under the supervision of doctors. In changing the system, the low-priority patients will be handled by the nurse practitioner after triage. The balking and reneging will still occur as before, except that the patients now wait in a dedicated area for the nurse practitioner and do not wait in the main line for the doctor. Management predicts that 80% of the low-priority patients can be handled solely by the nurse practitioner; however, after seeing the nurse practitioner 20% of the patients must also see the a doctor. The service time at the nurse practitioner is the same as it would have been if the low-priority patient had visited the doctor. Those patients who must see the doctor are placed at the head of the line for the doctor queue. When these patients visit the doctor, their service time is log-normally distributed with a mean of 6 minutes and a standard deviation of 1 minute. Compare the operation of the proposed system to the current operation of the clinic based on 30 replications. Do you think that the dedicated nurse practitioner is a good idea?

5.16 Suppose a service facility consists of two stations in series (tandem), each with its own FIFO queue. Each station consists of a queue and a single server. A customer completing service at station 1 proceeds to station 2, while a customer completing service at station 2 leaves the facility. Assume that the interarrival times of customers to station 1 are IID exponential random variables with a mean of 1 minute. Service times of customers at station 1 are IID exponential random variables with a mean of 0.7 minute, and at station 2 are IID exponential random variables with mean 0.9 minute. Develop an Arena^{TM} model for this system using the STATION and ROUTE modules.

(a) Run the simulation for 20 replications of 200000 minutes with a 20,000 minute warm up and estimate for each station the expected average delay in queue for the customer, the expected time-average number of customers in queue, and the expected utilization. In addition, estimate the average number of customers in the system and the average time spent in the system.

(b) Use the results of queuing theory to verify and validate your results for part (a).

(c) Suppose now there is a travel time from the exit of station 1 to the arrival at station 2. Assume that this travel time is distributed uniformly between 0 and 2 minutes. Modify your simulation and rerun it under the same conditions as in part (a).

5.17 Reconsider Problem 5-16. Suppose that there is limited space at the second station. In particular, there is room for 1 customer to be in service at the second station and room for only 1 customer to wait at the second station. A customer completing service at the first station will not leave the service area at the first station unless there is a space available for it to wait at the second station. In other words, the customer will not release the server at the first station unless a move to the second station is possible. Reanalyze this situation assuming an arrival rate of 1 customer every 2 minutes, with and without the transfer delay.

(a) Use a resource to model the waiting space at the second station. Ensure that a customer leaving the first station does not release its resource until it is able to seize the space at the second station.

(b) Use a HOLD module with the wait-and-signal option to model this situation. Compare your results to those obtained in part (a).

(c) Use a HOLD module with the scan for condition option to model this situation. Compare your results to parts (a) and (b).

(d) Discuss the effectiveness of each of the methods. Discuss the efficiency of each the methods.

5.18 Consider the simple three-workstation flow line in Figure 5-80.[2] Parts entering the system are placed at a staging area for transfer to the first workstation. The staging area can be thought as the place where the parts enter the system prior to going to the first workstation. No processing takes place at the staging area, other than preparation to be directed to the appropriate stations. After the parts have completed processing at the first workstation, they are transferred to a paint station attended by a second worker, and then to a packaging station where they are packed by a third worker, and then to a second staging area where they exit the system.

The time between the parts' arrivals at the system is exponentially distributed with a mean of 28 minutes (stream 1). The processing time at the first workstation is uniformly distributed between 21 and 25 minutes (stream 2). The paint time is log-normally distributed with a mean of 22 minutes and a standard deviation of 4 (stream 3). The packing time follows a triangular distribution with a minimum of 20, mode of 22, and a maximum of 26 (stream 4). The transfers are unconstrained, in that they do not require a vehicle or resource, but all transfer times are exponential with a mean of 2 or 3 minutes (stream 5). Transfer times from the staging to the workstation and from pack to exit are 3 minutes. Transfer times from the workstation to paint and from paint to pack are 2 minutes. The performance measures of interest are the utilization and work in progress at the workstation, paint, and packaging operations.

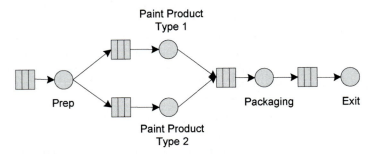

■ **FIGURE 5-80**   **Paint shop flow line**

---

[2]This problem is based on an example in Pegden et al. (1995), p. 209 and continuing on p. 217.

Suppose that statistics on the part flow time, that is, the total time a part spends in the system, need to be collected. However, before simulating this process, it is discovered that a new part needs to be considered. This part will be painted a different color. Because the new part replaces a portion of the sales of the first part, the arrival process remains the same, but 30% of the arriving parts are randomly designated as the new type of part (stream 10). The remaining parts (70% of the total) are produced in the same manner as described above. However, the new part requires the addition of a different station with a painting time that is log-normally distributed with a mean of 49 minutes and a standard deviation of 7 minutes (stream 6). Assume that an additional worker is available at the new station. The existing station paints only the old type of part and the new station paints only the new parts. After the painting operation, the new part is transferred to the existing packing station and requires a packing time that follows a triangular distribution with a minimum value of 21, a mode of 23, and a maximum of 26 (stream 7). Develop models for the original system and for the system involving the new part type. Run the models for 20 replications of 600,000 minutes with a 50,000 minute warm up period. If you were to select a resource to add capacity, which would it be?

5.19 Consider the testing and repair shop. Suppose instead of increasing the overall arrival rate of jobs to the system, the new contract will introduce a new type of component into the system that will require a new test plan sequence. Tables 5-11 and 5-12 provide specifics associated with the new testing plan.

Managers wish to improve their understanding of potential system bottlenecks and to develop alternatives to mitigate those bottlenecks so that they can handle the contract. The contract stipulates that 80% of the time the testing and repairs should be completed in 480 minutes. The company runs 2 shifts each day for each 5-day work week. Any jobs not completed at the end of the second shift are carried over to first shift of the next working day. Assume that the contract is going to last for 1 year (52 weeks). Build a simulation model that can assist the company in assessing the risks associated with the new contract.

5.20 Parts arrive at a 4-workstation system according to an exponential interarrival distribution with a mean of 10 minutes. The workstation A has 2 machines. Three workstations (B, C, D) each have a single machine. There are 3 part types, each with an equal probability of arriving.

■ **TABLE 5-11**
Distribution of Test Plans

Test Plan	% of Parts	Sequence
1	20%	{2, 3, 2, 1}
2	12.5%	{3, 1}
3	37.5%	{1, 3, 1}
4	20%	{2, 3}
5	10%	{2, 1, 3}

■ **TABLE 5-12**
Testing and Repair Time Distribution Parameters

Test Plan	Testing Time Parameters	Repair Time Parameters
1	{(20, 4.1), (12, 4.2), (18, 4.3), (16, 4.0)}	(30, 60, 80)
2	{(12, 4), (15, 4)}	(45, 55, 70)
3	{(18, 4.2), (14, 4.4), (12, 4.3)}	(30, 40, 60)
4	{(24, 4), (30, 4)}	(35, 65, 75)
5	{(20, 4.1), (15, 4), (12, 4.2)}	(20, 30, 40)

The process plans for the part types are given below. The entries are for exponential distributions with the mean processing time parameter given.

Part Type	Workstation, Processing Time	Workstation, Processing Time	Workstation, Processing Time
Part 1	A, 9.5	C, 14.1	B, 15
Part 2	A, 13.5	B, 15	C, 8.5
Part 3	A, 12.6	B, 11.4	D, 9.0

Assume that the transfer time between arrival and the first station, between all stations, and between the last station and the system exit is 3 minutes. Using the ROUTE, SEQUENCE, and STATION modules, simulate the system for 30,000 minutes and discuss potential bottlenecks in the system.

5.21 A particular stock-keeping unit (SKU) has demand that averages 14 units per year and is Poisson distributed. That is, the time between demands is exponentially distributed with a mean a 1/14 years. Assume that 1 year = 360 days. The inventory is managed according to a (r, Q) inventory control policy with r = 3 and Q = 4. The SKU costs $150. An inventory carrying charge of 0.20 is used and the annual holding cost for each unit has been set at 0.2 ∗ $150 = $30 per unit per year. The SKU is purchased from an outside supplier and it is estimated that the cost of time and materials required to place a purchase order is about $15. It takes 45 days to receive a replenishment order. The cost of backordering is very difficult to estimate, but a guess has been made that the annualized cost of a backorder is about $25 per unit per year.

(a) Using the analytical results for the (r, Q) inventory model compute the total cost of the current policy.

(b) Using Arena^TM simulate the performance of this system using Q = 4 and r = 3. Report the average inventory on hand, the cost for operating the policy, the average number of backorders, and the probability of a stock out for your model.

(c) Now suppose that the lead time is stochastic and governed by a log-normal distribution with a mean of 45 days and a standard deviation of 7 days. What assumptions do you have to make to simulate this situation? Simulate this situation and compare/contrast the results with parts (a) and (b).

5.22 Simulate problem 5-21 part (b) using Arena^TM and OptQuest to recommend optimal values for the (r, Q). Use a range of 1 to 6 for the reorder point and a range of 1 to 12 for the reorder quantity.

5.23 Suppose that the amount demanded for each customer is governed by a geometric random variable with parameter p = 0.9. What assumptions do you have to make to simulate this situation? How will you handle the possibility of a customer order taking the inventory position significantly below the reorder point? What if ordering Q does not get the inventory position above the reorder point? How will you process a demand for 2 items when you only have 1 item on the shelf?

5.24 Reconsider Problem 5-21. Suppose backordering is not allowed. That is, customers who arrive when there is not enough to fill their demand are lost. Suppose the amount demanded for each customer is governed by a geometric random variable with parameter p = 0.9. Develop an Arena^TM model for this situation. Estimate the expected number of lost sales per year for this situation.

5.25 Simulate Problem 5-24 using Arena^TM and OptQuest to recommend optimal values for the (r, Q) when the value of a lost sale is set at $40.

5.26 In a continuous review (s, S) inventory control system, the system operates essentially the same as an (r, Q) control policy except that when the reorder point, s, is reached, the system instead orders enough inventory to reach a maximum inventory level of S. Thus, the amount of the order will vary with each order. Repeat the analysis of Problem 5-21 (b) assuming that the SKU is controlled with a (s, S) policy. Assume that s = 4 and S = 7. Develop and simulate this system and compare the cost results to the results of Problem 5-21 (b).

5.27 Suppose in problem 5-26 that the amount demanded for each customer is governed by a geometric random variable with parameter p = 0.9. In addition, assume that the backorder queue is filled on a first come first served basis. Develop a model and simulate this system.

5.28 In a periodic review (R, S) inventory control system, the time between reviews of the inventory position is given by the parameter R. For example, if R = 1 week, at the end of each week, the inventory position is reviewed. If the inventory position is less than S, the amount necessary to bring the inventory position back up to S is ordered. Repeat the analysis of Problem 5-21 (b) assuming the SKU is controlled with a (R, S) policy. Use R = 0.5 year and S = 7. Develop and simulate this system and compare the cost results to the results of Problem 5-21 (b).

5.29 This problem considers the periodic review (R, s, S) inventory control system with back ordering. This control policy is a combination of the (s, S) and (R, S) policies. See Problems 5-26 and 5-28. The time between reviews of the inventory position is given by the parameter R. When the inventory position is reviewed and the inventory position is less than s, then the amount necessary to bring the inventory position back up to S is ordered. Repeat the analysis of problem 5-21 (b) assuming the SKU is controlled with a (R, s, S) policy. Use R = 0.5 year, s = 3, and S = 7. Develop and simulate this system and compare the cost results to the results of Problem 5-21 (b).

5.30 Consider problem 5-21. Suppose that the replenishment orders are served by a production facility that acts like a single server queuing system. The production time is log-normally distributed with a mean of 30 days and a standard deviation of 7 days. The transport delay is a constant of 1 day. Simulate this situation and compare/contrast the results with part (b) of Problem 5-21. What happens if the demand for the item doubles?

5.31 Consider a shop that produces and stocks items. The items are produced in lots. The demand process for each item is Poisson with the rate given below. Backordering is permitted. The pertinent data for each of the items is given in the following table. The carrying charge on each item is 0.20. The lead time is considered constant with the value provided in the table. Assume that there are 360 days in each year.

Item	1	2	3	4	5	6
Demand rate (units per year)	1000	500	2000	100	50	250
Unit cost (dollars per unit)	20	100	50	500	1000	200
Lead-time mean (days)	2	5	5	6	14	8

Suppose that management desires to set the optimal (r, Q) policy variables for each of the items so as to minimize the total holding cost and ordering costs. Assume that it costs approximately $5 to prepare and process each order. In addition, they wish to have the system fill rate to be no lower than 80% and no individual item's fill rate to be below 60%. They also wish to have the average customer wait time (for any part) to be less than or equal to 3 days. Simulate this system and use OptQuest to find an optimal policy recommendation for the items. You might start by analyzing the parts individually to have initial starting values for the optimization search.

5.32 Consider the model for the multiechelon inventory system presented in the chapter. Suppose that the retail locations do not allow backordering. That is, customers who arrive when there is not enough to fill their demand are lost. In addition, suppose that the amount demanded for each customer is governed by a geometric random variable with parameter $p = 0.9$. The warehouse and the retailers control the SKU with (s, S) policies as shown in the following table.

	Policy (s, S)	Demand Poisson (λ)
Retailer 1	(1, 3)	(0.2)
Retailer 2	(1, 5)	(0.1)
Retailer 3	(2, 6)	(0.3)
Retailer 4	(1, 4)	(0.4)
Warehouse	(4, 16)	

The overall rate to the retailers is 1 demand per day, but individually they experience the rates as shown in the table. In other words, the retailers are no longer identical in terms of the policy or the demand that they experience. The lead times remain as 1 day for the warehouse and for each retailer. The warehouse allows backordering. Develop an Arena™ model for this situation. Estimate the expected number of lost sales per year for each retailer, the average amount of inventory on hand at the retailers and at the warehouse, the fill rate for each retailer and for the warehouse, and the average number of items backordered at the warehouse. Run the model for 10 replications of 3960 days with a warm-up of 360 days.

5.33 Consider the multiechelon inventory system presented in this chapter. Suppose multiple types of items can be located at each retailer and at the warehouse. Describe the basic model changes that would need to be made to address this situation.

5.34 Consider the multiechelon inventory system presented in the chapter. Suppose that the system is being expanded so that there are multiple item types (as in problem 5-31) and there are now 2 warehouse locations. Each warehouse supplies 2 of the retailers. In addition, each warehouse is supplied by a national distribution center. Describe the basic model changes that would need to be made to address this situation.

5.35 This problem examines a single-item, multistage, serial production system using a kanban control system. The word *kanban* is Japanese for card. A kanban is simply a card that authorizes production. Kanban-based control systems are designed to limit the amount of inventory in production. While there are many variations of kanban control, this problem considers a simplified system with the following characteristics:

- A series of work centers.
- Each work center consists of one or more identical machines with an input queue feeding the machines that is essentially infinite.
- Each work center also has an output queue where completed parts are held to await being sent to the next work center.
- Each work center has a schedule board where requests for new parts are posted.
- Customer orders arrive randomly at the last work center to request parts. If there are no parts available, then the customers will wait in a queue until a finished part is placed in the output buffer of the last work center.

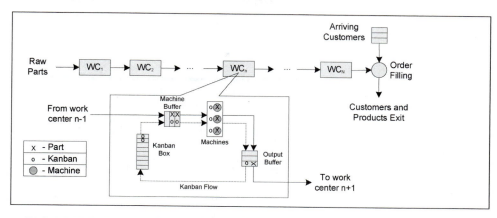

■ **FIGURE 5-81** Kanban production flow line

Figure 5-81 illustrates a production flow line consisting of kanban-based work centers. Raw parts enter at the first work center and proceed, as needed, through each work center until eventually being completed as finished parts at the last work center. Finished parts are used to satisfy end-customer demand.

A detailed view of an individual kanban-based work center is also shown in the figure. Here parts from the previous work center are matched with an available kanban to proceed through the machining center. Parts completed by the machining center are placed in the output buffer (along with their kanban) to await demand from the next work center.

An individual work center functions as follows:

- Each work center has a finite number of kanbans, $k_n \geq 1$.

- Every part must acquire one of these kanbans in order to enter a given work center and begin processing at the machining center. The part holds the kanban throughout its stay in the work center (including the time it is in the output buffer).

- Unattached kanbans are stored on the schedule board and are considered as requests for additional parts from the previous work center.

- When a part completes processing at the machines, the system checks to see if there is a free kanban at the next work center. If there is a free kanban at the next work center, the part releases its current kanban and seizes the available kanban from the next work center. The part then proceeds to the next work center to begin processing at the machines. If a kanban is not available at the next work center, the part is placed in the output buffer of the work center to await a signal that there is a free kanban available at the next station.

- Whenever a kanban is released at the work center it signals any waiting parts in the output buffer of its feeder work center. If there is a waiting part in the output buffer when the signal occurs, the part leaves the output buffer, releasing its current kanban, and seizing the kanban of the next station. It then proceeds to the next work center to be processed by the machines.

For the first work center in the line, whenever it needs a part (i.e., when one of its kanbans is released), a new part is created and immediately enters the work center. For the last work center, a kanban in its output buffer is not released until a customer arrives and takes away the finished part to which the kanban was attached. Consider a three work center system. Let k_n be the number of kanbans and let c_n be the number of machines at work center n, $n = 1, 2, 3$. Let's assume that each machine's service time distribution is an exponential distribution with service rate μ_n. Demands from customers arrive at the system according to a Poisson process with rate λ. As an added twist, assume that the queue of arriving customers can be limited. In other words, the customer queue has a finite waiting capacity, say C_q. If a customer arrives when there are C_q customers already in the customer queue they do not enter the system. Build a simulation model to simulate the following case:

c_1	c_2	c_3
1	1	1
k_1	k_2	k_3
2	2	2
μ_1	μ_2	μ_3
3	3	3

where $\lambda = 2$ and $C_q = \infty$, with all rates per hour. Estimate the average hourly throughput for the production line, that is, the number of items produced per hour. Run the simulation for 10 replications of length 15,000 hours with a warm-up period of 5000 hours. Suppose $C_q = 5$. What is the effect on the system?

CHAPTER 6

Entity Movement and Material Handling Constructs

LEARNING OBJECTIVES

After completing this chapter, you should be able to:

- Model constrained entity transfer with resources
- Model constrained entity transfer with transporters
- Model systems involving conveyors
- Model systems involving automatic guided vehicles
- Perform basic animation for entity transfer situations

Chapter 5 introduced the concept of unconnected entity movement through the use of the ROUTE and STATION modules. The ROUTE module models entity movement between stations as a simple time delay. The entities can be imagined as "having little feet" that allow them to move from one station to another. That is, the entity is able to move itself. If the entity is a person (e.g., a patient in a clinic), this representation makes sense; however, if the entity is a part in a manufacturing system, this representation begins to break down. For example, in the test-and-repair situation of Chapter 5, the parts moved between stations with a time delay. But how did they physically move? One way to think about this is that workers were always available to move the parts between the stations. If there is *always* a worker available, then it is as if there is an infinite supply of workers. Thus, whenever a part must be moved from one station to another, the part uses one of the workers to make the movement. Since there is an infinite supply of workers, this is the same as the part moving itself (i.e., having little feet), and only the time delay for moving between the stations is relevant.

In many situations, modeling transfers with a delay is perfectly reasonable, especially if you are not interested in how the entities moved (only that they moved). However, in many situations, the movement of entities can become constrained by a transfer mechanism. For example, the movement of parts may require that the parts be placed on a pallet and that a fork lift be used to move the parts. At any point in time, the quantity of available fork lifts may be insufficient, and thus the parts may have to wait for a fork lift to become available. When the waiting-for-transport potential is significant, you might want to model with more detail the "how" behind the entity transfer. In addition, since movement can be a significant part of an operation, the design of the material movement system may be the main focus of the modeling effort.

This chapter explores the various constructs available in Arena^{TM} to facilitate the modeling of the physical movement of entities between stations. The chapter begins by describing how to model transfers using resources. In this case, the transfer delay is accompanied by the use of a resource. Then, Section 6.2 presents how Arena^{TM} facilitates resource-constrained movement using the TRANSPORTER module and its accompanying constructs. A transporter is a special kind of resource in Arena^{TM} that can move. Since not all movement is as freely moving through space as people walking or fork lifts moving, Arena^{TM} provides constructs for modeling entity movement when the space between the locations becomes an important aspect of the modeling. Section 6.3 indicates how conveyors can be represented in Arena^{TM} and how they represent the space between stations. Then, in Section 6.4, the modeling of transporters will be revisited to understand how to model the situation where the transporters may compete for space while moving. This will involve the modeling of the space between stations as a fixed path (i.e., like a road network with intersections, etc.). As usual, these concepts will be illustrated with example models using Arena^{TM}.

6.1   Constrained Transfer with Resources

When an entity requires the use of something to complete the movement between stations, Arena's RESOURCE module can be used to model the situation. In this situation, the physical (e.g., distance) aspects of the movement are not of interest, only the time that may be constrained by the availability of the transport mechanism. To illustrate this modeling, let's revisit the test-and-repair shop from Chapter 5.

Recall that in the test-and-repair shop, parts follow one of four test plans through the shop. Each part first goes to the diagnostic station where it determines the sequence of stations that it will visit via an assignment of a test plan. After being diagnosed, it then proceeds to the first station in its test plan. In the example of Chapter 5, it was assumed that a worker (from somewhere) was always available to move the entity to the next station and that the transfer time took between 2 to 4 minutes uniformly distributed. The diagnostic station had two diagnostic machines and each test station had one testing machine. Finally, the repair station had three workers who performed the necessary repairs on the parts after testing. An *implicit* assumption in the model was that a worker staffed each of the two diagnostic machines (one worker for each machine) and that a worker was assigned to each test station. The modeling of the workers was not a key component of the modeling, so such an implicit assumption was fine.

In this section, the use of workers to move the entities and staff the stations will be explicitly modeled. Assume that there are two workers at the diagnostic station, one worker per testing station, and three workers at the repair station. Thus, there are a total of eight workers in the system. For simplicity's sake, assume that any of these eight workers are capable of moving

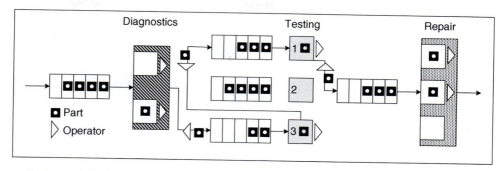

■ **FIGURE 6-1**   **Test-and-repair shop with workers providing the movement**

parts between the stations. For example, when a part completes its operation at the diagnostic station, any worker in the system can carry the part to the next station. In reality, it may be useful to assign certain workers to certain transfers (e.g., diagnostic workers move parts to the part's first station); however, for the sake of simplicity these issues will be ignored and any worker will be allowed to do any transport in this example. This also requires that any worker is capable of noticing that a part needs movement. For example, perhaps the part is put in a basket and a light goes on indicating that the part needs movement. When a part requires movement, it will wait for the next available idle worker to complete the movement. In this situation, a worker may be busy tending to a part in process at a station or the worker may be busy moving a part between stations. Figure 6-1 illustrates the new situation for the test-and-repair shop involving the use of workers.

Since workers are required for the processing of parts at the stations and they might have to perform the movement of parts between stations, the workers must be shared by two activities. Thus, when a worker completes one activity, a mechanism is needed to indicate which activity should proceed next. A simple mechanism is to assign a priority to one of the tasks. Thus, it seems reasonable to assume that parts waiting for processing at a station are given priority over parts that require movement between stations.

Figure 6-2 illustrates an activity diagram for the situation in which workers are called for transport. In the figure, each worker is individually labeled. For example, DW1 refers to worker 1 at the diagnostic station. In the figure, the visitation of each part at a test station is illustrated with a loop back to the transfer time after the testing delay. Thus, the figure represents all three test stations with the test station queue, testing delay, and test machine combination. Unfortunately, this does not explicitly indicate that a test worker is assigned to each test station individually. In particular, the other resources marked TW2 and TW3 should technically have seize and release arrows associated with them for the testing activity at a particular station.

It should be clear from the figure that three sets of resources will be required in this model. A resource set should be defined for the diagnostic workers with two members. A resource set should be defined for the three repair workers. Finally, a resource set should be defined to hold each of the workers {DW1, DW2, TW1, TW2, TW3, RW1, RW2, RW3} available for transporting parts. When a part requires diagnostics, it will seize one of the workers in the diagnostic workers set. When a part requires testing, it will seize the appropriate test worker for its current station. Finally, at the repair station, the part will seize one of the repair workers. In order to receive transport, the part will seize from the entire worker set.

6.1.1   Implementing Resource-Constrained Transfer

Based on Figure 6-2, the basic pseudo-code for resource-constrained transfer should be something like that shown in Exhibit 6-1. This logic assumes that each worker has been placed in the appropriate sets. At the diagnostic station, both the machine and a diagnostic worker are

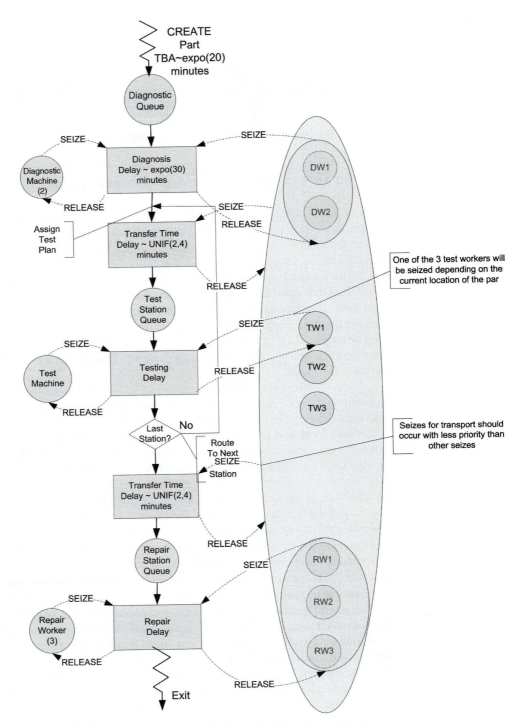

■ **FIGURE 6-2** **Activity diagram for revised test-and-repair situation**

required. When the part completes the processing at the diagnostic station, the part seizes a worker from the overall set of workers and routes with a transfer delay to the appropriate testing station.

After arriving at the test station, the part releases the worker who performed the transport and proceeds with the seizing of the test machine and worker. Again, prior to routing to the next

Exhibit 6-1 Pseudo-Code for Resource-Constrained Transfer

```
CREATE part
SEIZE 1 diagnostic machine
SEIZE 1 diagnostic worker from diagnostic worker set
DELAY for diagnostic time
RELEASE diagnostic machine
RELEASE diagnostic worker
ASSIGN test plan sequence
SEIZE 1 worker from worker set
ROUTE for transfer time by sequence to Test STATION
Test STATION
RELEASE worker
SEIZE appropriate test machine
SEIZE appropriate test worker
DELAY for testing time
RELEASE test machine
RELEASE test worker
SEIZE 1 worker from worker set
If not at last station
    ROUTE for transfer time by sequence to Test STATION
Else
    ROUTE for transfer time by sequence to Repair STATION
Repair STATION
RELEASE worker
SEIZE repair worker from repair worker set
DELAY for repair time
RELEASE repair worker
Collect statistics
DISPOSE
```

station, a worker from the overall set of workers is seized. Note also that the worker is released after the part is transferred to the repair station.

The combination of logic (SEIZE, ROUTE, STATION, RELEASE) is very common in entity-transfer modeling. While this logic can be directly implemented with the corresponding Arena™ modules, Arena™ has provided two additional modules that enable a wide range of transfer modeling options through dialog box entries. These modules are the ENTER and LEAVE modules of the Advanced Transfer panel. The ENTER module represents a number of different modules based on how its dialog box entries are completed. Essentially, an ENTER module represents the concepts of a STATION, a delay for unloading, and the releasing of the transfer method. In the LEAVE module, the user can specify a delay for loading, how the entity will be transferred out (e.g., resource, conveyor, transporter), and routing the entity. Now let's take a look at how to modify the test-and-repair model to use these new constructs.

Starting with a copy of the *RepairShop.doe* file from Chapter 5, the first step is to define 8 individual resources for the diagnostic workers (2), the test workers (1 for each station), and the repair workers (3). This is illustrated in Figure 6-3, where the workers have been added to the RESOURCE module. Note that the repair workers have been changed from having a single resource with capacity 3 to three resources each with capacity 1. This will enable all 8 workers represented as resources to be placed in an overall worker (resource) set.

Resource - Basic Process

	Name	Type	Capacity
1	DiagnosticMachine	Fixed Capacity	2
2	TestMachine1	Fixed Capacity	1
3	TestMachine2	Fixed Capacity	1
4	TestMachine3	Fixed Capacity	1
5	RepairWorker1	Fixed Capacity	1
6	RepairWorker2	Fixed Capacity	1
7	RepairWorker3	Fixed Capacity	1
8	DiagnosticWorker1	Fixed Capacity	1
9	DiagnosticWorker2	Fixed Capacity	1
10	TestWorker1	Fixed Capacity	1
11	TestWorker2	Fixed Capacity	1
12	TestWorker3	Fixed Capacity	1

Double-click here to add a new row.

■ **FIGURE 6-3** **Resources for test-and-repair shop**

Use the SET module on the Basic Process panel to define each of the three sets required for this problem. Figure 6-4 illustrates the *Workers* set in the SET module. The *DiagnosticWorkers* and *RepairWorkers* sets are defined in a similar manner using the appropriate resources. Since there was no preference given in the use of the resources in the sets just list them as shown. Recall that for the preferred resource selection rule, the order of the resources matters. The cyclical rule is used in this model.

Set - Basic Process

	Name	Type	Members
1	DiagnosticWorkers	Resource	2 rows
2	RepairWorkers	Resource	3 rows
3	Workers	Resource	8 rows

Double-c

Members

	Resource Name
1	TestWorker1
2	TestWorker2
3	TestWorker3
4	DiagnosticWorker1
5	DiagnosticWorker2
6	RepairWorker1
7	RepairWorker2
8	RepairWorker3

Double-click here to add a new row

■ **FIGURE 6-4** **Resource sets for test-and-repair shop**

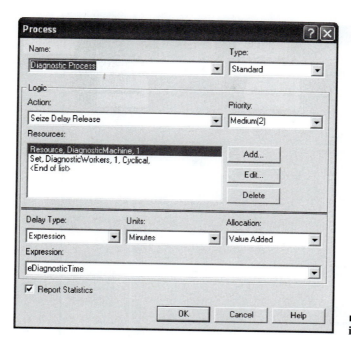

■ **FIGURE 6-5** Seizing in the diagnostic process

Now that the new resources and their sets have been defined, you need to update each of the PROCESS modules in order to ensure that the appropriate resources are seized and released. This is illustrated in Figure 6-5. In the Diagnostic Process dialog box, the SEIZE-DELAY-RELEASE option has been used with the list of resources required. In this case, the part needs both one unit of the diagnostic machine and one unit from the *DiagnosticWorkers* set. Each of the testing stations is done in a similar manner by first seizing the test machine and then seizing the test worker. The repair station is done in a similar fashion, except there is only the seizing of one unit from the *RepairWorkers* set. In all cases, the cyclical rule is used with the level of medium for the seize priority. If you are following along with the model building process, update each of the PROCESS modules as described. The completed model is found in the file *RepairShopResourceConstrained.doe* that accompanies this chapter.

Now the ENTER and LEAVE modules can be specified. The STATION modules associated with the three testing stations and the repair station will be replaced with appropriately configured ENTER modules. In addition, the ROUTE modules will be replaced with LEAVE modules for routing to the testing stations, between the testing stations, and to the repair station. Let's start with specifying how to leave the diagnostic station. Delete the ROUTE module associated with the diagnostic station and drag a LEAVE module into its place. Then fill out the LEAVE module as shown in Figure 6-6. When you first open the LEAVE module, it will not look like the figure. As seen in the figure, the module is divided into three components: Delay, Logic (Transfer Out), and Connect Type. The Delay text field allows a time delay to be specified after getting the transfer-out option. This can be used to represent a loading time. In this case, the loading time will be ignored (the time used by the worker to pick up the part is negligible). The transfer-out logic allows the specification of the use of a transporter, a conveyor, a resource, or no constrained transfer (none). In this case, the seize resource option is required. After selecting one of the resource-constrained transfer options, additional options will become available. Because the transfer is constrained, the entity must have a place to wait for the transfer if the resource is not immediately available. Thus, there is the option to specify a queue for the entity. You can think of this queue as the output bin for the station. Just like in the

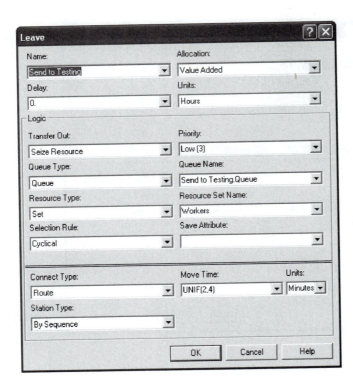

■ **FIGURE 6-6** **LEAVE module for test-and-repair shop example**

PROCESS module, you can directly seize a resource or seize a resource from a resource set. In this case, one worker will be seized from the *Workers* set. Note that a low seize priority has been specified. Since this priority is lower than that used by the parts when seizing with the PROCESS modules, the PROCESS modules will have higher priority when parts are waiting for processing over those waiting for transfer.

Finally, the Connect Type option in the LEAVE module is just like the corresponding options in the ROUTE module. In this case, the entity can be routed by sequence with a routing delay of UNIF(2, 4) minutes. Each of the other ROUTE modules in the previous model should be replaced with LEAVE modules like that shown in Figure 6-6. To do this replacement quicker, you can simply delete the ROUTE modules and copy/paste LEAVE modules as per Figure 6-6. You should make sure to change the name of the modules after the cut/paste operation, since no two Arena™ modules can have the same name.

Now the ENTER module can be specified. The ENTER module for test station 1 is given in Figure 6-7. Again, the ENTER module allows the usage of a number of other Arena™ constructs via the Station Name and Logic options. In the Logic options, you can specify a delay and the transfer-in options. The Delay option represents a delay that occurs before the transfer-in option. This is commonly used to represent a delay for unloading the transferring device. In this model, the time needed by the worker to unload (put down the part) is assumed to be negligible.

The transfer-in options indicate how the entity should react with respect to the transfer mechanism: Free Transporter, Exit Conveyor, Release Resource, and None. A simple unconstrained transfer uses the None option. In this model, the Release Resource option should be used. Because the worker seized for the transfer was part of a set, you can select the Set option for the Resource Type. The Release Rule specifies how the resource will be released. In this case, the Last-Member-Seized option should be used. This means that the most recent member of the *Workers* set that the part seized will be released.

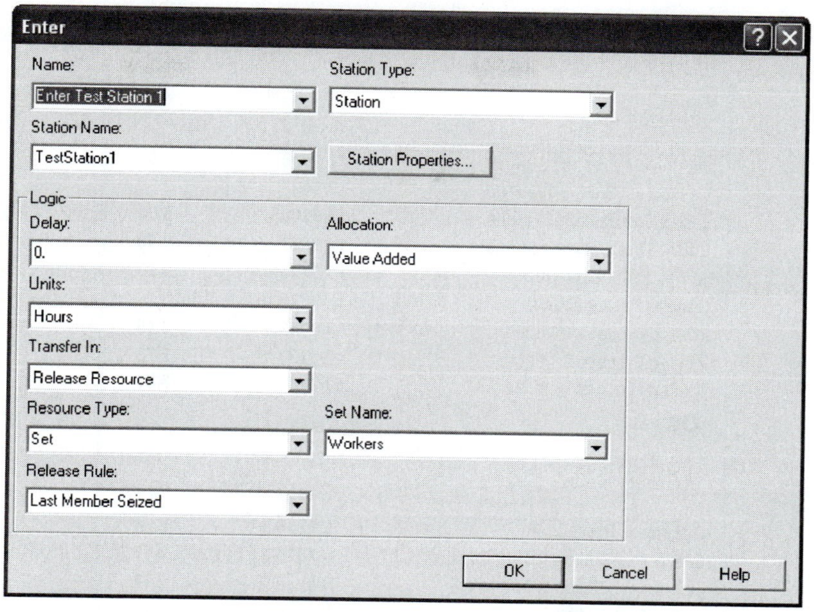

■ FIGURE 6-7 ENTER module for test-and-repair shop example

Similar to what was done in Chapter 5, the model was run for 10 replications of 4160 hours to examine the effect on the performance measures. As seen in Figure 6-8, the probability of meeting the contract specification has been reduced by about 10% (from 82.51% in the Chapter 5 example to 72.04% in this example). This is due to the increase in average system time. As seen in Figure 6-9, the time spent waiting in the station's output buffer for pick-up ranges from 13 to 15 minutes. Since there are four queues, this time has added significantly to the system time of the parts. In Figure 6-10, the utilization of the resources is relatively high (near 90% in most cases).

Because of the increased risk of not meeting the contract specifications for the system modeled with the more realistic use of workers to transfer parts, it might be useful to consider alternative ways to transfer parts between stations. The next section examines the use of dedicated workers modeled as transporters to perform this work. Before proceeding with that modeling, the animation needs to be enhanced.

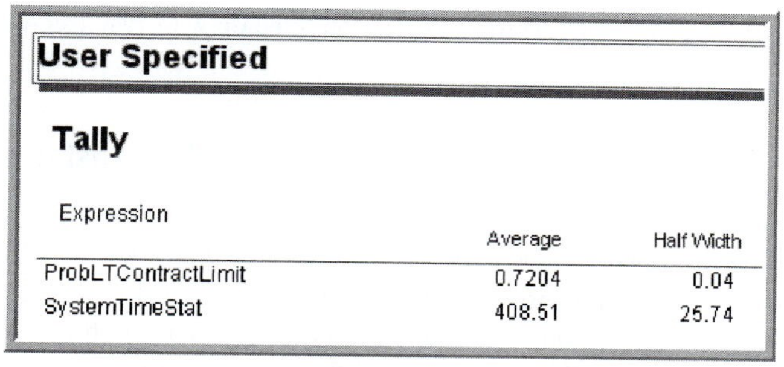

■ FIGURE 6-8 Probability of meeting contract requirements

Time

Waiting Time	Average	Half Width
Diagnostic Process.Queue	41.9538	2.54
Leave Test Station 1.Queue	13.0437	2.88
Leave Test Station 2.Queue	14.0260	3.04
Leave Test Station 3.Queue	13.3562	2.98
Repair Process.Queue	25.3725	3.23
Send to Testing.Queue	14.9525	3.16
Testing Process 1.Queue	53.0058	3.49
Testing Process 2.Queue	40.5793	2.73
Testing Process 3.Queue	53.7482	6.07

Other

Number Waiting	Average	Half Width
Diagnostic Process.Queue	2.0944	0.14
Leave Test Station 1.Queue	0.7325	0.17
Leave Test Station 2.Queue	0.5285	0.12
Leave Test Station 3.Queue	0.6676	0.15
Repair Process.Queue	1.2663	0.17
Send to Testing.Queue	0.7478	0.16
Testing Process 1.Queue	2.9724	0.21
Testing Process 2.Queue	1.5285	0.11
Testing Process 3.Queue	2.6842	0.32

■ **FIGURE 6-9** Queue statistics for test-and-repair with resource-constrained transfer

Instantaneous Utilization	Average	Half Width
DiagnosticMachine	0.7484	0.01
DiagnosticWorker1	0.8399	0.01
DiagnosticWorker2	0.8399	0.01
RepairWorker1	0.9223	0.01
RepairWorker2	0.9228	0.01
RepairWorker3	0.9224	0.01
TestMachine1	0.8545	0.01
TestMachine2	0.7773	0.01
TestMachine3	0.8607	0.01
TestWorker1	0.9225	0.01
TestWorker2	0.8735	0.01
TestWorker3	0.9229	0.01

■ **FIGURE 6-10**   Utilization for test-and-repair with resource-constrained transfer

6.1.2   Animating Resource-Constrained Transfer

If you ran the model described in the previous section with the animation turned on you would note the queues for the PROCESS modules and for the ROUTE modules. Unfortunately, you cannot see the entities during the transfer between stations. In this section, you will augment the basic flow chart animation with route animation, so that the flow of the entities between the stations will be visible. The animation of the stations will also be updated. Unfortunately, to effectively represent the use of the resources in the animation, more effort would need to be

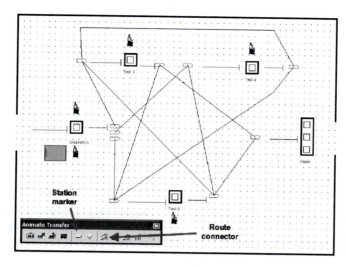

■ **FIGURE 6-11** Final animation for resource-constrained transfer

expended on the animation than is useful at this time. For example, workers can be busy tending a process or walking. To properly represent this, the definition of additional resource states and their animation would need to be discussed. As discussed in the next section, the use of transporters will greatly assist in that regard. Thus, for sake of simplicity, the movement of the entities between stations will be a rather crude representation of resource usage.

Figure 6-11 shows the final animation for the model. The completed model is available in the file, *RepairShopResourceConstrainedTransferWithAnimation.doe*. Note that in the figure, the Animate Transfer toolbar has been detached. The two key buttons to be used are the Station Marker button and the Route Connector button.

An overview of how to create this animation, beginning from the file *RepairShopResourceConstrainedTransfer.doe*, follows:

1. Use Arena's drawing toolbar to place the dark lines for the outline of the repair shop and for the dark rectangles to represent the stations. Use the Text Label button to label each of the station areas.

2. Use the Resource button on the Basic Animate toolbar to place each of the resources (diagnostic machines, test machines, repair workers, diagnostic workers, and test workers). In the figure, the standard animation for resources was used to represent the diagnostic machines, test machines, and repair workers. The picture library (as described in Chapter 1) was used to assign a seated worker to the idle picture and a worker who appears to be walking to the busy state picture icon of the diagnostic and test workers.

3. Delete the queues from the flow chart area and re-add them using the Queue button of the Basic Animate toolbar in the locations indicated in the figure. This process is similar to the one described in Chapter 1.

4. Use the Variable button on the Basic Animate toolbar to place a variable animation showing the current number of busy diagnostic machines.

Now the animation of the routes taken by the parts as they move from station to station can be described. Recall that the test plan sequences are as shown in Table 6.1.

From the test plans, a from/to indicator table can be developed, which shows whether a part will go from a given station to another station. In Table 6.2, "Yes" means that the station sends a part from the origin station to the destination station. This table indicates whether a route needs to be placed between the indicated stations.

■ **TABLE 6-1**
Distribution of Test Plans

Test plan	% of parts	Sequence
1	25%	{2, 3, 2, 1}
2	12.5%	{3, 1}
3	37.5%	{1, 3, 1}
4	25%	{2, 3}

■ **TABLE 6-2**
From/To Indicator Table

		Destination			
	Station	Test 1	Test 2	Test 3	Repair
	Diagnostics	Yes	Yes	Yes	No
	Test 1	No	No	Yes	Yes
Origin	Test 2	Yes	No	Yes	No
	Test 3	Yes	Yes	No	Yes

To animate the routes between these stations, you need to place station markers and connect the appropriate stations using the route connectors. To place a station marker, click on the station button of the Animate Transfer toolbar. Your cursor will turn into a crosshair, and you can place the marker in the model window where appropriate. As indicated in Figure 6-12, you can place a station marker for entering and exiting the station, which helps in visualizing the flow through the station. Station markers and connections will have to be placed for all the connections indicated in Table 6.2. If you miss a connection, you will not see the part moving between the associated stations. To connect existing station markers, click on the Route button. The cursor will turn into a crosshair and you can select the appropriate station markers. Note that the station marker will become highlighted as you connect. If station markers do not yet exist, after clicking the Route button your first click will lay down a station marker and then you

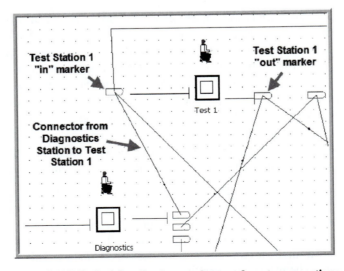

■ **FIGURE 6-12** Station markers and route connections

can move your cursor to the location for the second station marker. This process takes a little patience and practice. If you are having trouble selecting stations consider using the View toolbar to zoom in to help with the process. In addition, if you have to repeat the same step, you can right-click on the mouse after a step (e.g., placing a station marker) and choose the Repeat Last Action item from the pop-up menu. Thus, you do not have to repeatedly press the toolbar button.

If you run the model with the animation on, you will see the parts moving along the route connectors between the stations, the resource states changing, and the queues increasing and decreasing. If you watch the animation carefully, you should note that the repair workers become busy when either working on a part or transporting the part. As previously mentioned, it would be nice to separate these states in the animation. This can be accomplished by using the STATESET module in conjunction with the resources. Consult the *Arena*^{TM} *Users Guide* or help system for more information on the STATESET module.

Having the workers tend to the processing on the machines and move the parts between the stations has caused system time to increase. The next section examines the use of dedicated workers to transport the parts between stations using the TRANSPORTER module.

6.2   Constrained Transfer with Transporters

A *transporter* refers to one or more identical transfer devices that can be allocated to an entity for the purpose of transferring entities between stations. Two types of transporters are available in Arena^{TM}:

> *Free path transporter:* The travel time between stations depends on distance and speed.
>
> *Guided-path transporter:* The travel time between stations depends on distance and speed, and the movement of the transporter is restricted to a predefined network of intersections and links.

The standard delay option of the ROUTE or LEAVE modules assumes no resource-constrained transfer. As indicated in the last section, the situation of a general resource being used for the transfer can be easily modeled. In this situation, the time of the transfer was specified. In the case of transporters, the resource-constrained transfer is still being modeled, but in addition, the physical distance of the transfer must be modeled. In essence, rather than specifying a time to transfer, you must specify velocity and distance for the transfer. From the velocity and distance associated with the transfer, the time of the transfer can be computed. In these situations, the physical device and its movement (with or without the entity) through the system are of key interest in the modeling.

To model with transporters, Arena^{TM} provides the following modules on the Advanced Transfer Panel:

> *Allocate:* Attempts to allocate a unit of the transporter to the entity. This is like a seize module when using resources. If the transporter is not immediately available for the request, the entity must wait in a queue. ALLOCATE only gives a unit of the transporter to the entity. It does not move the transporter to the entity's location, nor does it change the status from idle to busy. This is useful when you want to perform other activities before or during the time that the empty transporter is moved to the entity's location.
>
> *Move:* Causes an allocated transporter to move to a particular location. The controlling entity is not moved during this process.

Request: Attempts to allocate a transporter and then move the transporter to the location of the requesting entity. REQUEST has the net effect of having an ALLOCATE followed by a MOVE module.

Transport: Causes the entity to be transported to its destination. The entity must have possession of the transporter unit. To contrast TRANSPORT and MOVE, MOVE permits unloaded movement and TRANSPORT is moving loaded with the entity. Before the transporter can be reallocated to another requesting entity the transporter must be de-allocated using the FREE module.

Free: Causes the entity to de-allocate the transporter. This is similar to releasing a resource.

Halt: Causes the transporter to stop moving and become inactive. If the transporter is currently busy at the time when an entity enters the Halt module, the status of the transporter is considered busy and inactive until the entity that controls the transporter frees the unit. If the transporter is idle at the time when an entity halts the transporter, it is set to inactive immediately. This is useful in modeling breakdowns during movement. Once halted, the transporter cannot be allocated until it has been activated.

Activate: This module causes a halted transporter to become active so that it can then continue. It increases the capacity of a previously halted transporter or a transporter that was initially inactive (as defined in the TRANSPORTER module). The transporter unit that is activated will reside at the station location at which it was halted until it is moved or requested by an entity. If an allocation request is pending for the transporter at the time the unit is activated, the requesting entity will immediately gain control of the transporter.

Transporter: This data module allows the user to define the characteristics of the transport device, such as number of transporter units, default velocity, and initial location. In addition, the DISTANCE module associated with the transporter must be specified.

Distance: This data module allows the user to define the distances (in a consistent unit of measure, such as meters) for all the possible moves that the transporter may make. The values for the distances must be non-negative integers. Thus, if the distance is given in decimals (e.g., 20.2 meters), the distance should be rounded (e.g., 20 meters) or scaled to integers (202 decameters). In any case, the units of measure for the distances must be consistent. The DISTANCE module essentially creates a from/to matrix for the distances between stations. These are the traveling distances for the *transporters*, not the entities, and should include both loaded and empty moves. For n station locations, you will have to specify at most $n(n-1)$ possible distances. The distances in the distance matrix do not have to by symmetrical. In addition, you do not need to specify a distance if it never occurs (either when the transporter is loaded or empty). If a distance is not given, it is assumed to be zero. If the distance is symmetric, (the same in both directions) only one pair needs to be specified.

Figure 6-13 illustrates the general case of using a transporter and how it relates to the Arena™ modules. Since transporters are used to move entities between two stations, two stations have been indicated in the figure. At the first station, the entity performs the standard SEIZE, DELAY, and RELEASE logic. The entity then attempts to allocate a unit of the transporter via the REQUEST module. Note the queue for awaiting transporter arrival. The REQUEST also causes the transporter to move to the entity's location, as indicated with the travel time to origin activity. A loading activity can then occur. After the loading activity has been completed, the entity experiences the delay for the transport to the desired location. As seen in the figure, the LEAVE module facilitates this modeling. Once the entity arrives at its

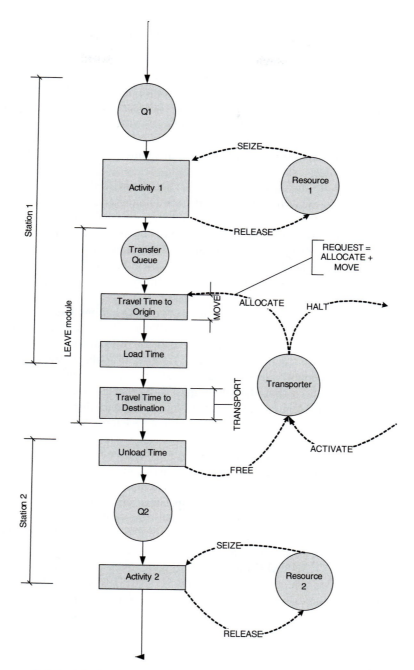

■ **FIGURE 6-13**   Activity diagram for general transporter case

destination station, there may be a delay for unloading the transfer device and then the transporter is freed. As shown in the figure, conceptually the HALT and ACTIVATE modules affect whether the transporter is available to be requested/allocated.

The activity diagram in Figure 6-13 represents how the test-and-repair system will need to operate with transporters. The next section discusses how to use transporters to model the dedicated transport workers in the test-and-repair shop.

6.2.1   Test-and-Repair Shop with Workers as Transporters

The results of the constrained-resource analysis indicated that constrained transfer for the parts in the test-and-repair system can have a significant effect on the system's ability to meet the contract requirements. This may be due to having the workers share the roles of tending machines and transporting parts. The following example investigates whether a set of dedicated workers would make sense for this system. In particular, the number of workers to dedicate to the transport task must be determined. Since there is a lot of walking involved, the model needs to be more precise in the physical modeling of the situation. Various layout configurations can also be simulated.

To model this situation using Arena's transporters, the distance between the stations and a way to model the velocity of transport are required. Since the workers have a natural variability in walking speed, a model for human walking speed is needed. Based on time-study data, the velocity of a worker walking in the facility has been found to be distributed according to a triangular distribution with a minimum of 22.86, a mode of 45.72, and a maximum of 52.5, all in meters per minute. Since this distribution will be used in many locations in the model, an EXPRESSION should be defined, called *eWalkingTimeCDF*, which is equal to TRIA(22.86, 45.72, 52.5).

Based on measuring the distance between the stations, the approximate distance between the stations has been determined as given in Table 6-3. Recall that both the loaded and unloaded distances for the transporter must be specified. For example, even though no parts are routed from repair to diagnostics, the distance from repair to diagnostics must be given because the worker (transporter) may be at the repair station when something needs to be moved from the diagnostic station. Thus, the worker must walk from the repair station to the diagnostic station (unloaded) in order to pick up the part. Note also that the distances do not have to be symmetric (i.e., the distance from test 1 to test 2 does not have to be the same as the distance from test 2 to test 1).

Starting with the finished model for the resource-constrained example (without the animation), you can make small changes in order to use transporters. The first step is to define the distances. Figure 6-14 shows the DISTANCE module from the Advanced Transfer panel. The spreadsheet view allows easier specification of the distances as shown in the figure. Multiple distance sets can be defined and used in the same model. The distance set must be given a name so that the name can be referenced by the TRANSPORTER module. The transporters for the model can now be defined using the TRANSPORTER module.

The TRANSPORTER module (Figure 6-15) allows the transporter to be given a name and various other attributes. The TRANSPORTER module defines a fleet of identical mobile resources (e.g., vehicles). You can specify the number of units of the transporter. In this example, there will be three workers (all identical) who can transport the parts.

■ **TABLE 6-3**
Transporter Distances between Stations

	Station	Destination				
		Diagnostics	**Test 1**	**Test 2**	**Test 3**	**Repair**
	Diagnostics	–	40	70	90	100
	Test 1	43	–	10	60	80
origin	Test 2	70	15	–	65	20
	Test 3	90	80	60	–	25
	Repair	110	85	25	30	–

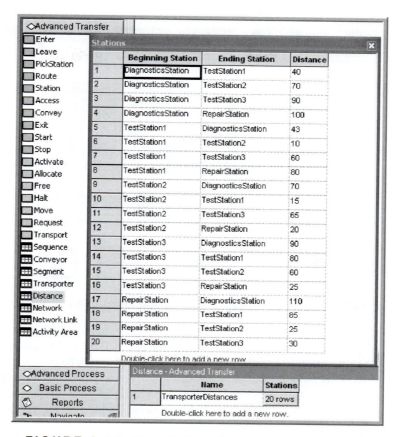

■ FIGURE 6-14   DISTANCE module

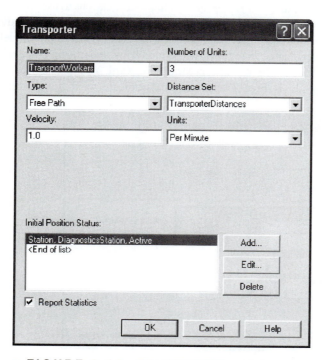

■ FIGURE 6-15   TRANSPORTER module

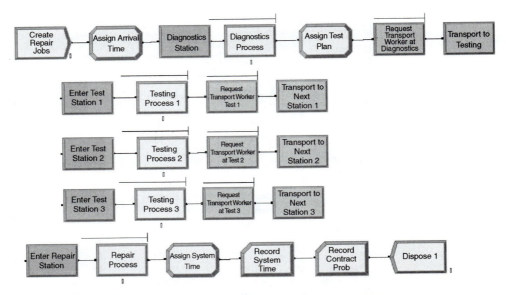

■ **FIGURE 6-16**   **Overall test-and-repair model with transporters**

The type of transporter in this case is Free Path, and the Distance Set has been specified with the name used in the appropriate distance set in the DISTANCE module. The velocity in the TRANSPORTER module must be a real number (it cannot be an expression). The default value of the velocity will be used in the dialog box because the velocity will be specified for each movement of the transporter in the model. The initial position specifies where the transporter will be located at the beginning of the replication. In this case, the workers start active at the diagnostics station.

Now that the data modules are defined, let's take a look at the overall model. Figure 6-16 shows the overall test-and-repair model after modifications to use transporters. If you have the model open, note the new blue-colored modules. These are from the Advanced Transfer panel and have replaced the LEAVE modules of the previous example model.

The other change from the previous model involves the updating of the ENTER modules. The ENTER module change is very straightforward. You just need to specify that the type of transfer-in option is Free Transporter and indicate what transporter to free. Figure 6-17

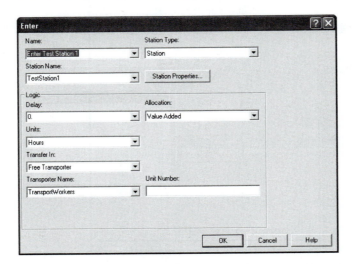

■ **FIGURE 6-17**   **ENTER module with Free Transporter option**

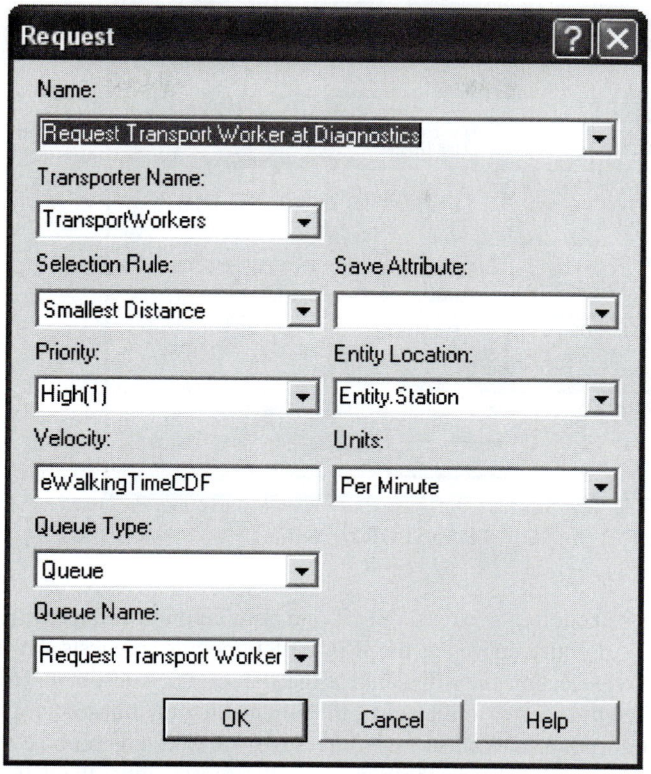

■ **FIGURE 6-18** **REQUEST module**

illustrates the changes to the ENTER module for entering test station 1. By default, the transporter unit that the entity currently has is freed.

The changes to the LEAVE module are not as simple. Because a random velocity value is needed whenever the worker begins moving, the LEAVE module cannot be used. The LEAVE module relies on the transporter default velocity as defined in the TRANS-PORTER module. Since the default velocity in the TRANSPORTER module must be a real number (not an expression), the REQUEST and TRANSPORT modules are used in this example.

Figure 6-18 shows the REQUEST module for the example. In the REQUEST module, you must specify which transporter to request as well as the selection rule. The selection rules are similar in concept to those for resources. The transporter selection rules follow:

Cyclical: Cyclic priority selects the first available transporter beginning with the successor of the last transporter allocated (cycles through the transporters)

Largest Distance: Select the available transporter farthest from the requesting station

Preferred Order: Select the available transporter unit that has the lowest unit number

Random: Select randomly from the available transporter units

Smallest Distance: Select the nearest available transporter

Specific Member: Specific unit of the transporter

ER(Index): Select based on rule number Index defined in experiment frame

UR(Index): Select transporter index UR, where UR is computed in a user-coded function UR(LENT, NUR)

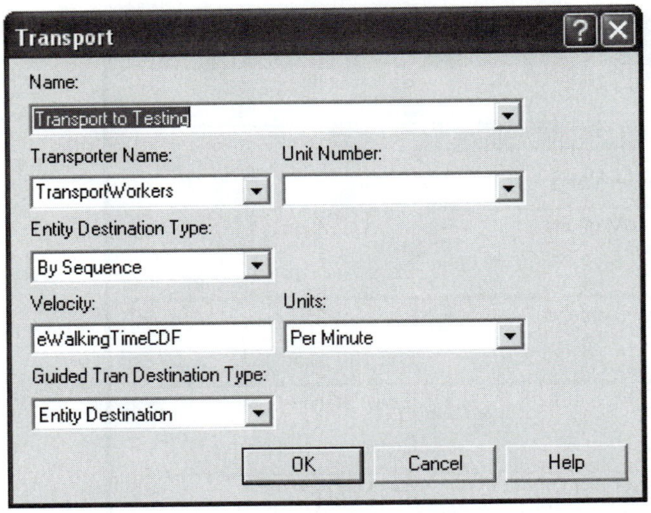

■ **FIGURE 6-19**   **TRANSPORT module**

The selection rule is used only if one or more units of the transporter are available to be allocated. An entity arrives at the REQUEST module, requests the transporter according to the specified selection rule. If a unit of the transporter is available (active and idle), ARENA allocates the transporter to the entity, thereby making the transporter busy (and allocated). The selected transporter can be saved in the Save Attribute combo box of Figure 6-18. Arena™ calculates the travel time from the transporter's present station to the station of the requesting entity, and imposes a time delay representing the travel time using the velocity argument. If a transporter is not available, the entity waits in queue. The REQUEST module gets the transporter to the entity. At this point, the transporter is used to transport the entity via the TRANSPORT module.

Figure 6-19 shows the TRANSPORT module for the test-and-repair example. The TRANSPORT module is very similar to the ROUTE module, except that the user specifies the transporter to use by name (and/or unit number). In most cases, this will be the same transporter as allocated in the REQUEST module. The user can then specify the type of destination (by sequence, attribute, station, or expression). In this case, the By Sequence option should be used. In addition, the velocity of the transport can be specified. The Guided Tran Destination Type field allows the user to send the transporter to a location different than the one specified by the entity's destination. This is useful in guided-path transporters to send the transporter through specific intersections on its way to the final destination. This option will not be used in this text.

After making the previous changes, you can run the model with the default animation and find out how the system operates with the transporters. Recall that in Figure 6-15, there is a little check-box to indicate whether statistics on the transporter will be collected. If this is checked, and the Transporters check-box is checked on Arena's Project Parameters tab on the Run Setup dialog box, Arena™ will automatically report statistics on the number that are busy and on transporter utilization.

As seen in Figure 6-20, utilization of the three transporters is very low. Fewer than three workers are likely needed for the transport task. The reader is asked to explore this issue as an exercise.

6.2.2   Animating Transporters

A significant advantage of using transporters in this situation rather than resource-constrained transfer is the animation available with transporters. This section converts

Transporter

Usage

Number Busy	Average	Half Width
TransportWorkers	0.4660	0.00

Number Scheduled	Average	Half Width
TransportWorkers	3.0000	0.00

Utilization	Average	Half Width
TransportWorkers	0.1553	0.00

■ **FIGURE 6-20**   Transporter statistics

the previous animation so that it uses transporters. The basic steps in converting the model follow:

1. Cut and paste the animation from the file *RepairShopResourceConstrainedTransferWith-Animation.doe* to the file *RepairShopWithTransportersNoAnimation.doe*, and rename it.
2. Select and delete each route connector and station marker from the previous animation.
3. Delete each animation queue from the flow chart modules. Redefine each animation queue with the appropriate name in the copied animation.
4. Use the Add Distance button on the Animate Transfer toolbar to place station markers and distance connectors for every from/to combination in the distance set as shown in Figure 6-21. For every station marker, make sure that the "Parking" check-box is checked. This indicates that the transporter will be shown at that station when idle (parked).

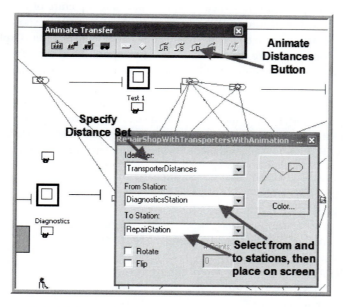

■ **FIGURE 6-21**   Placing distance animation elements

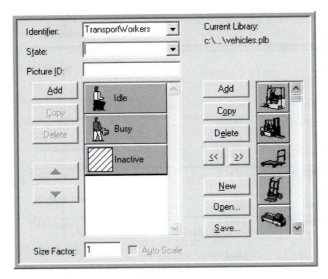

■ **FIGURE 6-22** Defining the transporter picture

Add a variable animation to show the current number of busy transporters. This can be easily done using the expression builder to find the transporter variables (NT(*Transport-Workers*)). To learn more about the variables associated with transporters, look up "transporter variables" in Arena's help system.

Use the transporter button in the Animate Transfer toolbar to define a transporter picture. The process is essentially the same as that for defining a resource animation picture, as shown in Figure 6-22.

Now the animation involving the transporters is complete. If you run the model with animation turned on, you will see that there is only about one unit of the transporter busy at any time. In addition, you will see that the idle transporters become parked after their transport when no other transport requests are pending. They remain at the station where they last dropped off a part until a new transport request comes in. In the animation, a picture depicting a person sitting was used to indicate the idle state and a picture of a person walking/carrying something was used to depict the busy state. Another way to show that the transporter is carrying something is to enable the "ride point" for the transporter. To place a ride point in the busy transporter picture, select the Ride Point option from the Object menu after double-clicking on the busy picture in the Transporter Animation dialog box. If no ride point is defined, the entity picture will not appear during transport.

In this model, the worker stays at the location where they drop off the part if no part requires transport. Instead, the worker could go to a common staging area (e.g., in the middle of the shop) to wait for the next transport request. The "trick" to accomplishing this is to not use the ENTER module. The basic idea would be implemented as follows:

```
STATION
If requests are waiting
   FREE the transporter
   Continue with normal station processing
Else
   SEPARATE 1 duplicate into the station
   MOVE to staging area
   FREE transporter
```

Note that when there are no waiting requests, the original entity is used to continue control of the transporter to move it to the staging area. The reader is asked to implement this idea as an extension to this problem in the exercises. The notion of having an entity control or "drive" the transporter around is very useful in modeling systems that have complicated transporter allocation rules or systems that naturally have a driver, such as bus systems.

A distance module allows the physical transport aspects of systems to be modeled. An important aspect of this type of modeling is the movement of entities through space. The conveyor constructs in Arena^{TM} allow for a finer representation of the usage of space between stations. That topic is taken up in the next section.

6.3   Modeling Systems with Conveyors

A conveyor is a track, belt, or some other device that provides movement over a fixed path. Typically, conveyors are used to transport high-volume items over short- to medium-range distances. Conveyance speeds typically range from 20 to 80 feet per minute to as fast as 500 feet per minute. Conveyors can be gravity based or powered. Conveyor modeling can be roughly classified as follows:

Accumulating: Items on the conveyor continue to move forward when there is a blockage on the conveyor.

Nonaccumulating: Items on the conveyor stop when the conveyor stops.

Fixed spacing: Items on the conveyor have a fixed space between them or the items ride in a bucket or bin.

Random spacing: Items are placed on the conveyor in no particular position and take up space on the conveyor.

You have likely experienced conveyors in everyday life. For example, the belt conveyor at a grocery store resembles an accumulating conveyor. An escalator is a fixed-spaced nonaccumulating conveyor (the steps are like a "bucket" for the person to ride in). People movers in airports resemble nonaccumulating, random-spacing conveyors. When designing systems with conveyors, a number of performance measures should be considered:

Throughput capacity: Number of loads processed per time

Delivery time: Time taken to move the item from origin to destination

Queue lengths: Queues to get on the conveyor and for blockages when the conveyor accumulates

Number of carriers used or space used: Number of spaces on the conveyor used

Modeling conveyors does not necessarily require specialized simulation constructs. For example, gravity-based conveyors can be modeled as a PROCESS or a DELAY with a deterministic delay. The delay is set to model the time that it takes the entity to "fall" or slide from one location to another. One simple way to model multiple items moving on a conveyor is to use a resource to model the "front" of the conveyor with a small delay to load the entity on the conveyor. This allows for spacing between the items on the conveyor. Once on the conveyor, the entity releases the front of the conveyor and delays for its move time to the end of the conveyor. At the end of the conveyor, there could be another resource and delay to unload the conveyor. As long as the delays are deterministic, the entities will not "pass" each other on the conveyor. If you are not interested in modeling the space taken by the conveyor, then this type of modeling is very reasonable. Henriksen and Schriber (1986) discuss a number of ways to approach the simulation modeling of conveyors. When space becomes an important element of the modeling, then the simulation language constructs for conveyors become useful.

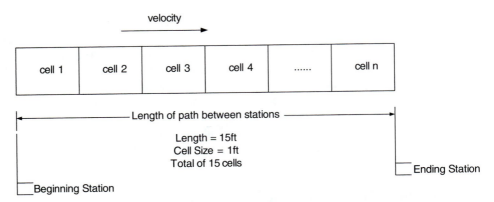

■ FIGURE 6-23   **A conveyor conceptualized as a set of contiguous cells**

For example, if there actually is a conveyor between two stations and the space allocated for the length of the conveyor is important to system operation, you might want to use Arena's conveyor constructs.

In Arena™, a conveyor is a material-handling device for transferring or moving entities along a predetermined path having fixed predefined loading and discharge points. Each entity to be conveyed must wait for sufficient space on the conveyor before it can gain entry and begin its transfer. In essence, simulation modeling constructs for conveyors model the travel path via a mapping of space or distance to resources. Space along the path is divided into units of resources called cells. The conveyor then is essentially a set of moving cells of equal length. Whenever an entity reaches a conveyor entry point, it must wait for a predefined amount of unoccupied and available consecutive cells in order to get on the conveyor.

Figure 6-23 illustrates the idea of modeling a conveyor as a set of contiguous cells representing the space on the conveyor. One way to think of this is an escalator with each cell being a step.

In Figure 6-23, if each cell represents 1 foot and the total length of the conveyor is 15 feet, then there will be 15 cells. The cell size of the conveyor is the smallest portion of a conveyor that an entity can occupy. In modeling with conveyors, the size of the entity also matters. For example, think of people riding an escalator. Some people have a suitcase and fill two steps and others do not have a suitcase and require only one step. An entity must acquire enough contiguous cells to hold its physical size in order to be conveyed.

The key modules in Arena™ for modeling the use of conveyors follow:

Access: When an entity enters an ACCESS module, it will wait until the appropriate amount of contiguous cells are available at the access point. Once the entity has control of the cells, it can then be conveyed from its current station to a station associated with the conveyor. This is similar in concept to the ALLOCATE module for transporters.

Convey: The CONVEY module causes the entity to move from its origin station to the next station. The time to move is based on the velocity of the conveyor and the distance between the stations. This is similar in concept to the MOVE module for transporters.

Exit: The EXIT module causes the entity to release the cell(s) that it holds on the conveyor. If another entity is waiting in queue for the conveyor at the same station where the cells are released, the waiting entity will then access the conveyor. This is like releasing a resource or freeing a transporter.

Start: The START module changes the status of the conveyor to active. Entities may reside on the conveyor while it is stopped. These entities will maintain their respective positions on the conveyor once it is started. The entity using the START module does not have to be on the conveyor.

Stop: The STOP module changes the status of the conveyor to inactive. The conveyor will stop immediately, regardless of the number of entities on the conveyors. This is useful for modeling conveyor failures or blockages on the conveyor. The entity using the STOP module does not have to be on the conveyor.

Conveyor: This data module defines whether the conveyor is an accumulating or nonaccumulating conveyor, cell size (in physical units e.g., feet), and velocity. A conveyor consists of a sequence of segments as defined by the SEGMENT module. The sum of the distances among the stations on the conveyor (specified by the SEGMENT module) must be divisible by the cell size. If not, an error will occur when the model is checked.

Segment: This data module defines the segment network that makes up a conveyor. Each segment is a directed link between two stations. The conveyor path is defined by a beginning station and a set of next station–distance pairs. The beginning station of the segment is the beginning station number or name, and the next station defines the next station name that is a length of distance units from the previously defined station. The length of the segment must be specified in integer distance units (feet, meters, etc.). No station should be repeated in the segment network unless the conveyor is a loop conveyor. In that case, the last ending station should be the same as the beginning station for the segment network.

As was the case for transporters, the ENTER and LEAVE modules of the advanced transfer template will also provide some conveyor functionality. LEAVE provides for ACCESS and CONVEY. ENTER provides for EXIT. The modules work slightly differently for accumulating and nonaccumulating conveyors.

Nonaccumulating conveyors travel in a single direction, and the spacing remains the same between the entities. When an entity is placed on the conveyor, the entire conveyor is actually disengaged or stopped until instructions are given to transfer the entity to its destination. When the entity reaches its destination, the entire conveyor is again disengaged until instructions are given to remove the entity from the conveyor, at which time it is engaged or started. As previously mentioned, the conveyor is divided into a number of cells. To get on the conveyor and begin moving, the entity must have its required number of contiguous cells. For example, in Figure 6-24, the circle needed one cell to get on the conveyor. Suppose the hexagon was trying to get on the conveyor. As cell 2 became available, the hexagon would seize it. Then it would wait for the next cell (cell 1) to become available and seize it. After having both required cells, it is "on" the conveyor. But moving on a conveyor is a bit more continuous than

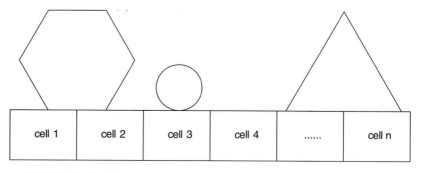

■ **FIGURE 6-24**   **Various entity sizes on a conveyor**

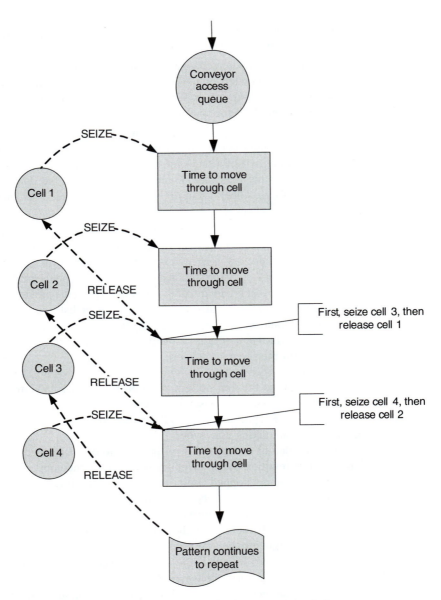

■ **FIGURE 6-25**   Activity diagram for nonaccumulating conveyor

this. Conceptually, you can think of the hexagon starting to move on the conveyor when it gets cell 2. When one cell moves, all cells must move in lock step. Since entities are mapped on to the cells by their size, when a cell moves the entity moves. Reversing this analogy becomes useful. Suppose that the cells are fixed, but the entities can move. In order for an entity to "take a step" it needs the next cell. As it crosses over to the next cell, it releases the previous cell that it occupied. It is as if the entity's "step size" is one cell at a time.

Figure 6-25 illustrates an *approximate* activity flow diagram for an entity using a nonaccumulating conveyor. Suppose that the entity size is 2 feet, and each cell is 1 foot on the conveyor. There is a queue for the entities waiting to access the conveyor. Cells of the conveyor act as resources modeling the space on the conveyor. Movements of the entity through the space of each cell are activities. For the 2-foot entity, it first seizes the first cell and then moves through the distance of the cell. After the movement, it seizes the next cell and moves through its length. Because it is 2 feet (cells) long, it seizes the third cell and then releases the first cell before delaying for the move time for the third cell. As cells become released, other entities can seize them and continue their movement. The repeated pattern of overlapping SEIZE and RELEASE modules that underlies conveyor modeling clearly depends on the number of cells and entity size. Thus, the activity diagram would actually vary by entity size (number of cells required).

The larger the number of cells to model a segment of a conveyor, the more slowly the model will execute; however, a larger number of cells allows for a more "continuous" representation of how entities actually exit and leave the conveyor. Larger cell sizes force entities to delay longer to move; thus, waiting entities must delay longer for available space. A smaller mapping of cells to distance allows then entity to "creep" onto the conveyor in a more continuous fashion. For example, suppose that the conveyor was 6 feet long and space was modeled in inches. That is, the length of the conveyor is 72 inches. Next, suppose that the entity required 1 foot of space while on the conveyor. If the conveyor is modeled with 72 cells, the entity starts to get on the conveyor after only 1 inch (cell) and requires 12 cells (inches) when riding on the conveyor. If the mapping of distance was 1 cell equals 1 foot, the entity would have to wait until it could access an entire 1-foot cell before it moved onto the conveyor.

You are now ready to put these concepts into action.

6.3.1   Test-and-Repair Shop with Conveyors

The test-and-repair shop will be used for illustrating the use of conveyors. Figure 6-26 shows the test-and-repair shop with the conveyors and the distance between stations. The distances from each station in clockwise order follow:

- Diagnostic station to test station 1, 20 feet
- Test station 1 to test station 2, 20 feet
- Test station 2 to repair, 15 feet
- Repair to test station 3, 45 feet
- Test station 3 to diagnostics, 30 feet

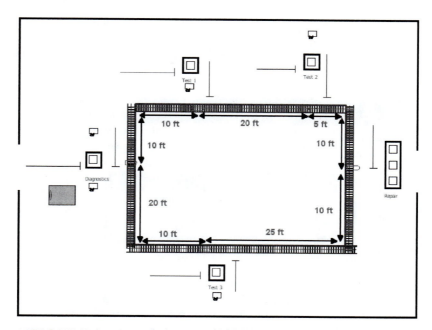

■ **FIGURE 6-26**   **Test-and-repair shop with conveyors**

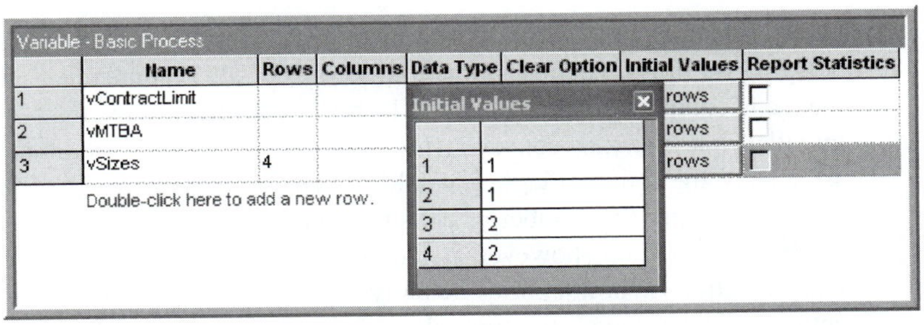

Variable - Basic Process							
	Name	**Rows**	**Columns**	**Data Type**	**Clear Option**	**Initial Values**	**Report Statistics**
1	vContractLimit					rows	☐
2	vMTBA					rows	☐
3	vSizes	4				rows	☐

Double-click here to add a new row.

Initial Values		✖
1	1	
2	1	
3	2	
4	2	

■ **FIGURE 6-27**   **Array for part sizes**

Assume that the conveyor's velocity is 10 feet per minute. These figures have been provided in this example; however, in modeling a real system, you will have to tabulate this information. To illustrate the use of entity sizes, assume the following concerning the parts following the four test plans:

- Test plan 1 and 2 parts require 1 foot of space while riding on the conveyor
- Test plan 3 and 4 parts require 2 feet of space while riding on the conveyor.

The conveyor logic will first be implemented without animation, and then the animation for the conveyors will be added. To convert the previous model so that it uses conveyors, change the model *RepairShopWithTransportersNoAnimation.doe* as follows:

1. Delete the TRANSPORTER and DISTANCE modules defined for using the transporter.
2. Delete the REQUEST and TRANSPORT modules for leaving the stations and replace them with LEAVE modules. The LEAVE modules will be filled out in a moment.
3. Add a variable array called *vSizes*() of dimension 4 to hold the sizes of the parts following each of the test plans as shown in Figure 6-27.
4. Save your file. The completed model is named, *RepairShopWithConveyorsWithNoAnimation.doe*, which can be found on the CD that accompanies the text.

You are now ready to define the conveyor and its segments. First, you will define segments for the conveyor and then the conveyor itself. Figure 6-28 shows the segments to be used on the conveyor. First, give the segment a name and then specify the station that represents

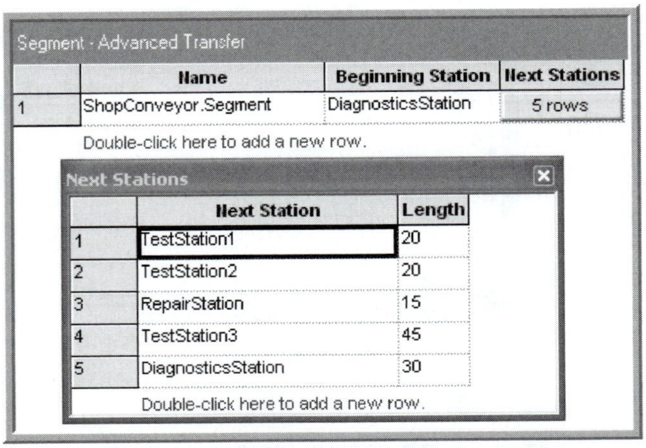

Segment - Advanced Transfer			
	Name	**Beginning Station**	**Next Stations**
1	ShopConveyor.Segment	DiagnosticsStation	5 rows

Double-click here to add a new row.

Next Stations			
		Next Station	**Length**
1		TestStation1	20
2		TestStation2	20
3		RepairStation	15
4		TestStation3	45
5		DiagnosticsStation	30

Double-click here to add a new row.

■ **FIGURE 6-28**   **SEGMENT module for test-and-repair example**

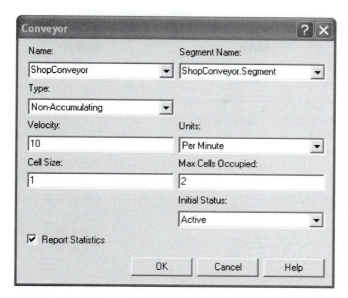

■ **FIGURE 6-29**   Conveyor module

the "beginning" station. Since this is a loop conveyor, the beginning station is arbitrary. The diagnostic station is used here since this is where the parts enter the system.

The next station and the distance to the next station are given for each part of the conveyor. Note that the distances have been added as shown in Figure 6-26. The lengths specified in Figure 6-28 are the actual physical distances (e.g., feet). The mapping to cells occurs in the CONVEYOR module.

Figure 6-29 shows the CONVEYOR module. To associate the conveyor with a segment, use the Segment Name drop-down box. The next decision is the type of conveyor. In this case, a nonaccumulating conveyor is used.

Conveyor velocity is 10 feet per minute. The cell size and maximum cells occupied text boxes must now be completed. The cell size is the most important decision. Thinking back to Figure 6-25, it should be clear that space must be represented by discrete cells of a specified integral length. Thus, the total length of the conveyor must be integer valued and the cell size must result in an integer number of cells *for each* segment. The lengths of each distance specified in the SEGMENT module for the example are all divisible by 5. Thus, you could choose a cell size of 5. This would mean that each cell would be 5 feet in length. Since entities access the conveyor by grabbing cells, this hardly seems appropriate if the parts are 1 and 2 feet in length, respectively. In the case of a 5-foot cell size, there would be a lot of space between the parts as they ride on the conveyor. Instead, the cell size will be specified as 1 foot. This will always work, since every number is divisible by 1. It also works well in this case, since the smallest part is 1 foot, and the distances in the SEGMENT module are based on foot units. The maximum number of cells text field is easy. This field represents the size of the biggest entity in terms of cells, which is 2 in this case because parts that follow test plans 3 and 4 are 2 feet long.

With the CONVEYOR and SEGMENT modules defined, changing the ENTER and LEAVE modules to use the conveyors is easy. The first item to address is to make sure that each created part knows its size so that the size can be used in the LEAVE module. You can do this with the *Assign Test Plan* ASSIGN module as shown in Figure 6-30.

The LEAVE modules must be updated to use the conveyor to transfer the entity and to indicate the number of cells required by the entity. This is done in Figure 6-31 using the entity's *mySize* attribute. Each of the other LEAVE modules for the other stations should be updated in a

	Type	Attribute Name	New Value
1	Attribute	myTestPlan	eTestPlanCDF
2	Attribute	Entity.Sequence	TestPlanSequences(myTestPlan)
3	Attribute	Entity.Picture	MEMBER(PartPictures,myTestPlan)
4	Attribute	mySize	vSizes(myTestPlan)

Double-click here to add a new row

■ **FIGURE 6-30** **Assigning part sizes**

similar fashion. You must now update the ENTER module so that the entity can exit the conveyor after arriving at the desired station.

In Figure 6-32, the Exit Conveyor option is used when transferring into test station 1. In both of the ENTER and LEAVE modules, the option of delaying the entity is available. This can be used to represent a loading or an unloading time for the conveyor.

With the conveyor statistics check-box clicked, the conveyor statistics will appear on the summary reports. Utilization is calculated for the conveyor, regardless of conveyor type. This utilization represents the length of entities on the conveyor in comparison to the length of the entire conveyor over the simulation run. Thus, conveyor utilization is articulated in terms of space used. In addition, there is very little queuing for getting on the conveyor. The statistics for the conveyor in this problem (Figure 6-33) indicate that the conveyor is not highly utilized; however, the chance of meeting the contract limit is up to 80%. The production rates in this example may make it difficult to justify the use of conveyors for the test-and-repair shop. Ultimately, the decision would come down to a cost/benefit analysis.

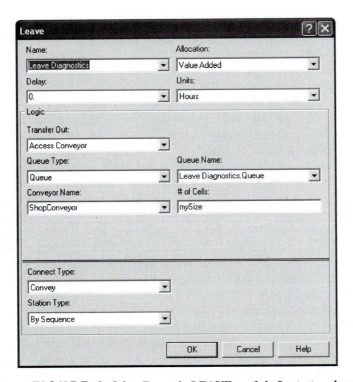

■ **FIGURE 6-31** **Example LEAVE module for test-and-repair shop**

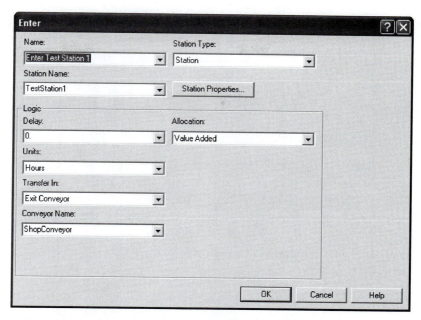

■ **FIGURE 6-32**   **Enter module for test station 1**

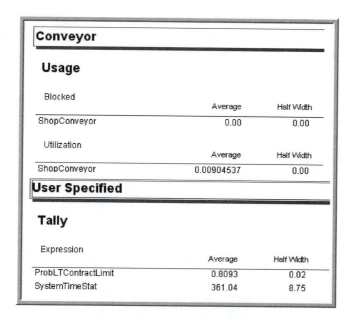

■ **FIGURE 6-33**   **Statistics for test-and-repair with conveyors**

6.3.2   Animating Conveyors

To augment the model with animation, the Animate Transfer toolbar can be used. To visibly discern the plan that an entity is following in the animation, you can define a picture set, with appropriately drawn pictures for each type of plan. This is illustrated in Figures 6-34 and 6-35. Assignment to the *Entity.Picture* attribute is shown in Figure 6-30.

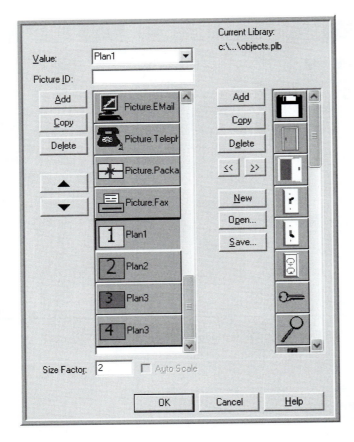

■ **FIGURE 6-34** Entity pictures for test plans

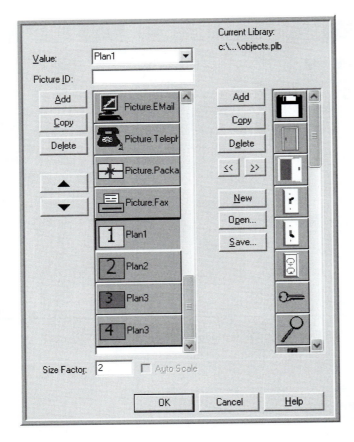

■ **FIGURE 6-35** Picture set for test plans

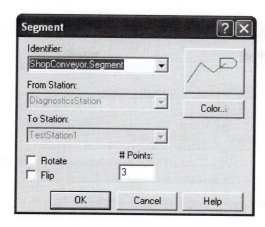

■ **FIGURE 6-36**   **Segment Animation dialog**

To update the previous transporter animation, take the following steps:

1. Copy and paste the animation elements from the file, *RepairShopWithTransportersWithAnimation.doe* to your own model file. The completed Arena™ model is called *RepairShopWithConveyorsWithAnimation.doe*.

2. Delete the transporter picture, station markers, and distance connectors.

3. Using the Animate Transfer toolbar's Segment button, make station markers and segment connectors to represent the conveyors as shown in Figure 6-26. The Segment Animation dialog box is shown in Figure 6-36. It works very much like the Route and Distance Animation dialog.

After the segments have been laid down, you might want to place some lines to represent the conveyors. This was done in Figure 6-36 where the roller pattern from the line patterns button on the drawing toolbar was also used. That's it! Animating conveyors is very easy. When you run the model with the animation on, you will see that the entities move along the conveyor segments, getting on and off at the appropriate stations. You will also clearly see that most of the time the conveyor does not have very many parts. This section concentrated on nonaccumulating conveyors. The next section describes how accumulating conveyors operate and discusses a number of other conveyor modeling issues.

6.3.3   Miscellaneous Issues in Conveyor Modeling

This section discusses accumulating conveyors, how to model merging conveyors and diverging conveyors, how to allow processing to occur on the entity while it is still on the conveyor, and how to model recirculation conveyors.

6.3.3.1 Accumulating Conveyors. An accumulating conveyor can be thought of as always running. When an entity stops moving on the conveyor (e.g., to unload), other entities are still allowed on the conveyor since the conveyor continues moving. When entities on the conveyor meet up with the stopped entity, they stop, and a queue accumulates behind the stopped entity until the original stopped entity is removed or transferred to its destination. As an analogy, imagine wearing roller blades and being on a people mover in an airport. (Don't ask how you got your roller blades through security!) While on the people mover, you aren't skating (you are standing still, resting, but still moving with the people mover). Ahead of you, a seriously deranged simulation book author places a bar across the people mover. When you reach the bar, you grab on. Since you are on roller blades, you remain stationary at the bar, while your roller blades are turning like mad. Now imagine all the people on the mover having roller blades. The people following you will continue approaching the bar until they bump into you.

Everyone will pile up behind the bar until the bar is removed. You are on an accumulating conveyor! Note that while you were on the conveyor and there was no blockage, the spacing between the people remained the same, but that when the blockage occurred, the spacing between the entities decreased until they bump up against each other. To summarize:

- Accumulating conveyors are always moving.
- If an entity stops to exit or receives processing on the conveyor, other entities behind it are blocked and begin to queue up.
- Entities in front of the blockage continue moving.
- When a blockage ends, blocked entities may continue, but must first wait for the entities immediately in front of them to move forward.

In modeling accumulating conveyors in Arena™, the main differences occur in the actions of the ACCESS and CONVEY modules. Instead of disengaging the conveyor as with nonaccumulating conveyors, the conveyor continues to run. ACCESS allocates the required number of cells to any waiting entities as space becomes available. Any entities that are being conveyed continue until they reach a blockage. If the blocking entity is removed or conveyed, the accumulated entities only start to move when the required number of cells becomes available.

In the ACCESS module, the number of cells still refers to the number of cells required by the entity while moving on the conveyor; however, you must indicate what the space requirements will be when the entities experience a blockage. In the CONVEYOR module, if the accumulating conveyor option is selected, the user must decide how to fill in the Accumulation Length text field. The accumulation length specifies the amount of space required by the entity when accumulating. It does not need to be the same as the amount of space implied by the number of cells required when the entity is being conveyed. Also, it does not have to be divisible into an even number of cells. In addition, as seen in Figure 6-37, the accumulation length can also be an entity (user-defined) attribute. Thus, the accumulation size can vary by entity. Typically, the space implied by the number of cells required while moving will be larger than that required when accumulating. This will allow spacing between the entities when moving and allow them to get closer to each other when accumulating.

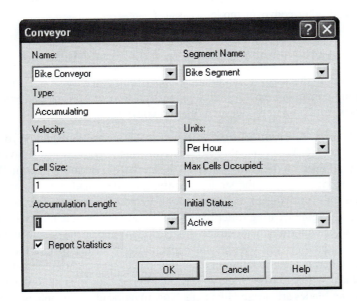

■ **FIGURE 6-37**   **Accumulating conveyor dialog**

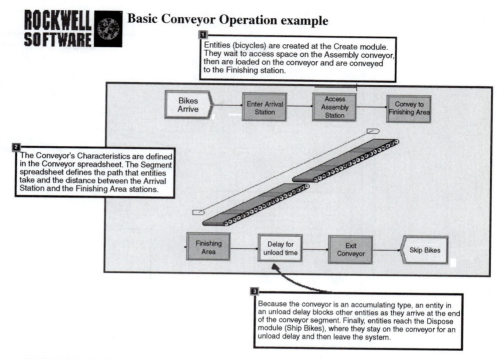

■ FIGURE 6-38   Arena's *Smarts101.doe* model for accumulating conveyors

The test-and-repair example barely required nonaccumulating conveyors. Thus, to illustrate accumulating conveyors, one of Arena's SMART files (*Smarts101.doe*) (Figure 6-38) will be used. The SMART files can be found in the Arena™ folder in your installation of Arena™ (typically *C:\Program Files\Rockwell Software\Arena™\Smarts*).

In this example, there is one segment for the conveyor of length 10, which (while not specified in the model) can be assumed to be in meters. Velocity is only 1 meter per hour, the conveyor cell size is 1, and the maximum size of an entity is 1. The entities, which are bicycles, enter at the arrival station and access the conveyor, where they are conveyed to the finishing area. You can think of the bicycles as being assembled as they move along the conveyor; in fact, this type of system often occurs in assembly lines. Once the bicycles reach the finishing area, there is an unload delay *prior* to exiting the conveyor. Because the unload delay occurs prior to exiting the conveyor, the entity's movement is blocked and an accumulation queue will occur behind the entity on the conveyor. Run the model to convince yourself that the bicycles are accumulating.

When modeling with accumulating conveyors, additional statistics are collected on the average number of accumulating entities if the conveyors check-box is checked on the Project Parameters tab of the Run Setup dialog box. You can check out the help system to see all the statistics.

6.3.3.2 Merging and Diverging Conveyors. In conveyor modeling a common situation involves one or more conveyors feeding another conveyor. For example, in a distribution center, there may be conveyors associated with each unloading dock, which are then attached to a main conveyor that takes the products to a storage area. To model this situation in Arena™, the feeding conveyors and the main conveyor can be modeled with different CONVEYOR modules and separate SEGMENT definitions. In this distribution center example, an item accesses the unloading dock's conveyor, rides the length of the

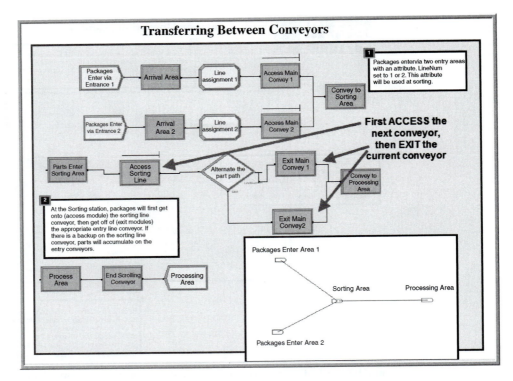

■ **FIGURE 6-39** **Transferring between conveyors**

conveyor, and then attempts to get on the main conveyor. After it has accessed the main conveyor, the entity exits its unloading dock conveyor.

Arena™ has two SMART files that illustrate these concepts. Let's first take a look at *Smarts107.doe*. Figure 6-39 illustrates the overall model. There are two conveyor lines for areas 1 and 2, respectively. When the entities arrive at an area, an attribute called *LineNum* is assigned to indicate where it came from. The entities access the appropriate conveyor and then are conveyed to the sorting area. The sorting area in this example is the station where the main conveyor begins. Once the parts arrive at the sorting station, they first access the conveyor to the processing area. The entity then exits its current conveyor. The DECIDE module is used to select the correct conveyor to exit based on the *LineNum* attribute. By first accessing the next conveyor, the entity may have to wait on its current conveyor. Because of this, it is very natural to model this situation with accumulating conveyors. The feeder conveyors will back up as entities try to get on the main conveyor.

A similar example involves the merging of one conveyor with another, as in the SMART file *Smarts110.doe* (Figure 6-40). In this example, there are two conveyors. The main conveyor is associated with two segment lengths. For example, the main conveyor goes from Entry A Station to the Midpoint Station to the EndPoint station, with 5-feet lengths, respectively. The second conveyor is a single conveyor from the Entry B station to the MidPoint Station with a length of 10 feet. Note that the MidPoint station is associated with segments related to the two conveyors. In other words, they share a common station. The entities arrive at both the Entry A and Entry B stations. Entities that arrive at the Entry A station first convey to the MidPoint station and then immediately convey to the EndPoint station. They do not exit the conveyor at the MidPoint station. Entities from the Entry B station are conveyed to the MidPoint station. Once at the MidPoint station they first ACCESS the main conveyor and then EXIT their conveyor, causing entities arriving from Entry B to potentially wait for space on the main

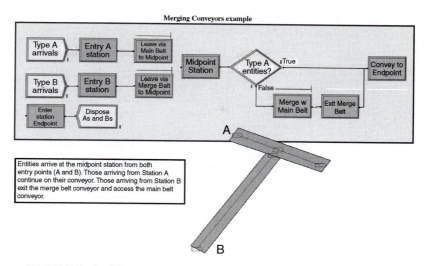

■ **FIGURE 6-40**   **One conveyor merging into another**

conveyor. This basic idea can be expanded to any number of feeder conveyors along a lengthier main conveyor.

Diverging conveyors are often used in systems that sort items. The items arrive on a main conveyor and are transferred to any number of other conveyors based on their attributes (e.g., destination). Arena's *Smarts108.doe* file illustrates the basic concepts behind diverging conveyors. In this example, as shown in Figure 6-41, the parts arrive at the arrival area and access a conveyor to the sorting area. Once an entity reaches the sorting station, the entity first exits the incoming conveyor and then there is a small delay for the sorting. The module labeled

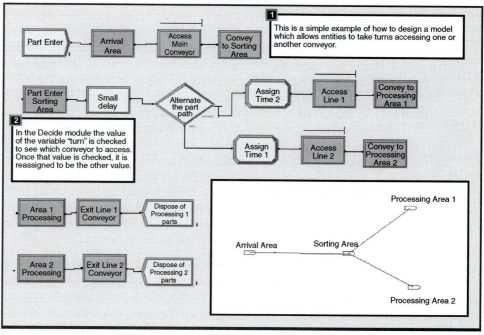

■ **FIGURE 6-41**   **Simple alternating conveyor**

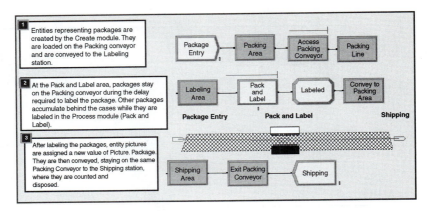

■ **FIGURE 6-42**   Processing while on a conveyor

Parts Enter Sorting Area is an ENTER module with the transfer-in option used to exit the conveyor. The entity is then shunted down one of the lines based on a DECIDE module. In this model, a variable is used to keep track of the last line that an entity was sent to so that the other line will be used in the DECIDE for the next entity. This implements a simple alternating logic. After being shunted down one of the conveyors, the entity uses the standard ACCESS, CONVEY, STATION, EXIT logic before being disposed.

With a little imagination, a more sophisticated sorting station can be implemented. For example, a number of conveyors that handle the parts by destination might be used in the system. When the parts are created, they are given an attribute indicating their destination station. When they get to the sorting station, a DECIDE module can be used to pick the correct conveyor handling their destination.

6.3.3.3 Processing while on the conveyor. In the bicycle example from *Smarts101.doe*, workers can be conceptualized as moving along with the conveyor as they assemble the bicycles. In many systems, it is common for a machine to be positioned to use the surface of the conveyor as its work area. For example, in circuit board manufacturing, the circuit boards ride on a conveyor through chip placement machines. While in the machine, the conveyor stops to do the insertion. These types of situations can be easily modeled in Arena^{TM} by not exiting the conveyor while undergoing a PROCESS module.

Arena's SMART file, *Smarts103.doe*, contains an example of this type of modeling. In this example, as shown in Figure 6-42, packages are created for the packing area, where they access and convey on a conveyor to the pack and label station. The conveyor consists of two segments attached at the pack and label station. Once at the Labeling Area station, the entity enters a PROCESS module, which requires a delay and the labeling machine. Note that the entity did not exit the conveyor. The *Labeling Area* module is a simple STATION module. After completing the processing, the entity proceeds to be conveyed to the shipping area, where it exits the conveyor. In this example, the conveyor is a nonaccumulating conveyor. If you run the model, you will see that the whole conveyor stops when the entity is being processed. If an accumulating conveyor is used, the packages would queue up behind the labeling machine.

6.3.3.4 Recirculation Conveyors. A recirculation conveyor is a loop conveyor in which the entities may ride on the conveyor until there is adequate space at their desired location. In essence, the conveyor is used as space to hold the entities that cannot get off due to inadequate space at a given station. For example, suppose that in the test-and-repair shop, there was only space for one waiting part at test station 2. Thus, the size of queue for test station 2 can be at most 1.

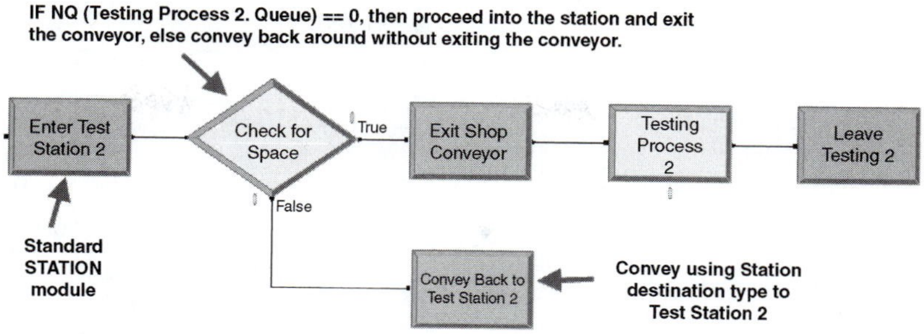

IF NQ (Testing Process 2. Queue) == 0, then proceed into the station and exit the conveyor, else convey back around without exiting the conveyor.

■ **FIGURE 6-43**   **Test-and-repair system with recirculation conveyor**

The previous model can be easily modified to handle this situation by not using an ENTER module at test station 2. Figure 6-43 shows the changes to the logic for test station 2 to handle this situation. As seen in the figure, the ENTER module has been replaced with a combination of STATION, DECIDE, EXIT, and CONVEY modules. The DECIDE module checks the queue at the Testing Process 2 module. If a part is not in the queue, then the arriving part can exit the station and try to seize the testing resources. If there is a part waiting, then the arriving part is sent to the CONVEY module to be conveyed back around to test station 2. As long as there is a path back to the station, this is perfectly fine (as in this case). The modified model is available with the files for this chapter in the file called *RepairShopWithRecirculatingConveyorsWithAnimation.doe*. If you run the model and watch the animation, you will see many more parts on the conveyor because of the recirculation.

A number of issues related to conveyor modeling have not been discussed, including the use of specialized variables defined for conveyors. Refer to Arena's Variables Guide or the help system under "conveyor variables" for more information on this topic. Arena™ also has a number of other SMART files that illustrate the use of conveyors. For example, you might want to explore SMART file *Smarts105.doe* to better understand entity size and how it relates to cells on the conveyor. Conveyors allow the modeling of space through the use of cells. The next section examines how Arena™ models space when it is relevant for transporter modeling.

6.4   Modeling Guided-Path Transporters

This section presents Arena's modeling constructs for automated guided vehicle (AGV) systems. An AGV is an autonomous powered vehicle that can be programmed to move between locations along paths. A complete discussion of AGV systems is beyond the scope of this text. See standard texts on material handling or manufacturing systems design for a more complete introduction to the technology. For example, Askin and Standridge (1993) and Singh (1996) have complete chapters dedicated to AGV system modeling. For purposes of this section, the discussion will be limited to AGV systems that follow a prescribed path (e.g., tape, embedded wires, etc.) to make things simpler. There are newer AGVs that are not limited to following a path, but rather are capable of navigating via sensors in buildings (see, for example, Rossetti and Seldanari, 2001).

When modeling with guided transporters, the most important issue to understand is that the transporters can now compete with each other for the space along their paths. The space along the path must be explicitly modeled (like it was for conveyors) and the size of the transporter matters. Conceptually, a guided transporter is a mobile resource that moves along a fixed path and competes with other mobile resources for space along the path. When using a guided transporter, the travel path is divided into a network of links and intersections. This is

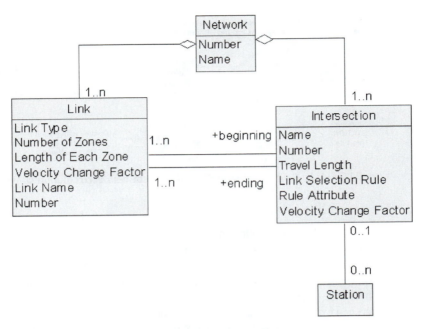

called the vehicle's path network. Figure 6-44 illustrates the major concepts in guided vehicle modeling: networks, links, intersections, and stations.

A link is a connection between two points in a network. Every link has a beginning intersection and an ending intersection. The link represents the space between two intersections. The space on the link is divided into an integer number of zones. Each zone has a given length as specified in a consistent unit of measure. All zones have the same length. The link can also be of a certain type (bidirectional, spur, unidirectional). A bidirectional link implies that the transporter is allowed to traverse in both directions along the link. A spur is used to model "dead ends," and unidirectional links restrict traffic to one direction. A velocity change factor can also be associated with a link to indicate a change in the basic velocity of the vehicle while traversing the link. For example, for a link going through an area with many people present, you may want the velocity of the transporter to automatically be reduced.

Figure 6-45 illustrates a simple three-link network. In the figure, Link 1 is bidirectional with a beginning intersection label I1 and an ending intersection labeled I2. Link 1 consists of four zones. There can also be a direction of travel specified. The beginning and ending directions of the link can be used to define the direction of the link (in degrees) as it leaves the beginning intersection and as it enters the ending intersection. The direction entering the ending intersection defaults to the direction leaving the link's beginning intersection. Thus, the beginning direction for Link 1 in the figure is 60 degrees. Zero and 360 both represent the east or right. The beginning direction for Link 2 is 300 degrees. Think of it this way: The vehicle has to turn to go onto Link 2. It would have to turn 60 degrees to get back to zero and then another 60 to head down Link 2. Since the degrees are specified from the right, this means that Link 2 is at 300 degrees relative to an axis going horizontally through intersection I2. The direction of travel is only relevant if the acceleration or deceleration of the vehicles as they make turns is important to the modeling.

In general, an intersection is simply a point in the network; however, Arena's representation of intersections can be more detailed. An intersection also models space: the space between two or more links. Thus, as indicated in Figure 6-44, an intersection may have a length, a velocity change factor, and link selection rule. For more information on the modeling of

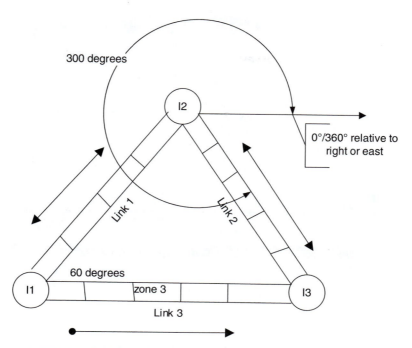

■ **FIGURE 6-45**   Example three link network

intersections, see the INTERSECTIONS element in Arena's Elements template. In addition, Banks et al. (14) and Pegden et al. (15) cover the SIMAN blocks and elements that underlie guided-path transporter modeling in Arena™.

A fixed path network consists of a set of links and associated intersections. Figure 6-44 also shows that a station may be associated with an intersection in the network and an intersection may be associated with many stations. Transporters in a guided-path network can be sent to specific stations, intersections, or zones. Only the use of stations will be illustrated here.

The new Arena™ constructs to be discussed include the NETWORK and NETWORK LINK modules on the Advanced Transfer template. The NETWORK module allows the user to specify a set of links for a path network. Figure 6-46 shows the NETWORK module. To add

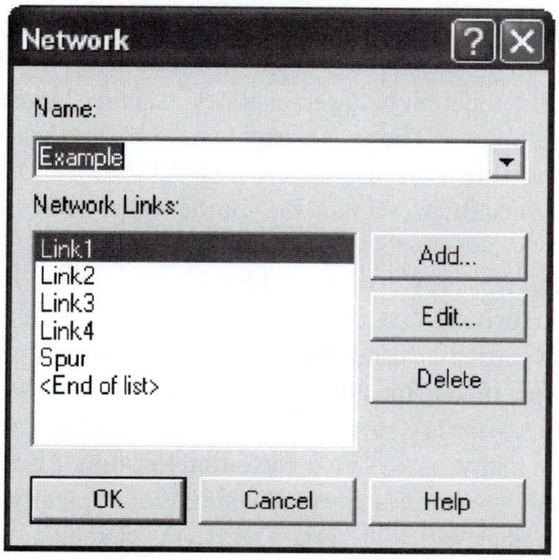

■ **FIGURE 6-46**   **NETWORK module for guided-path modeling**

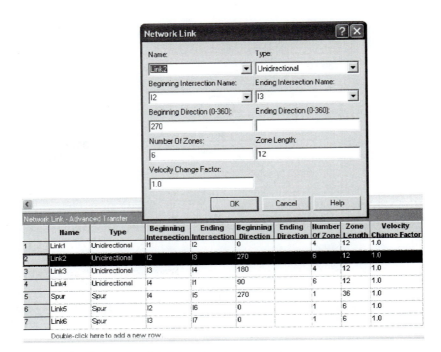

■ **FIGURE 6-47** NETWORK LINK module

links, the user presses the Add button or uses the Spreadsheet dialog box entry. The links added to the NETWORK module must have already been defined via the NETWORK LINK module. The NETWORK LINK module permits the definition of the intersections on the links, the directional characteristics of the links, number of zones, zone length, and velocity change factor. Figure 6-47 illustrates the NETWORK LINK module. Note that the spreadsheet view of the module makes editing multiple links relatively easy.

Now, let's take a look at a simple example. In this example, parts arrive at an entry station every 25 minutes, where they wait for one of two available AGVs. Once loaded onto the AGV (taking 20 minutes), they are transported to the exit station. At the exit station, the item is unloaded, and again, this takes 20 minutes. The basic layout of the system is shown in Figure 6-48. There are seven intersections and seven links in this example. The distances between the intersections are shown in the figure.

In modeling this situation, you need to make a number of decisions regarding transporter characteristics, the division of space along the links, directions of the links, and zone control of the links. Let's assume that the item being transported is rather large and heavy. That's why it takes so long to load and unload. The transporter is a 6-foot-long cart that travels 10 feet per minute. Two carts are available for transport. Figure 6-49 shows the TRANSPORTER module for using guided-path transporters. When the Guided type is selected, additional options become available.

Because transporters occupy space on the path, the choice of initial position can be important. A transporter can be placed at a station that has been associated with an intersection, at an intersection, or on a particular zone on a link. The default option will place the transporter at the first network link specified in the NETWORK module associated with the transporter. As long as you place the transporters so that deadlock (to be discussed shortly) is prevented, everything will be fine. In many real systems, the initial location of the vehicles will be obvious. For example, since AGVs are often battery powered, they typically have a charging station. Because of this, most systems will be designed with a set of spurs to act as a "home base" for

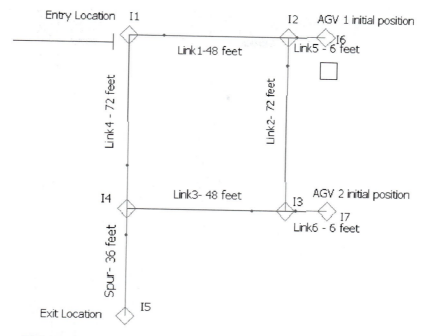

■ **FIGURE 6-48** **Layout for simple AGV example**

the vehicles. For this simple system, intersections I6 and I7 will be used as the initial position of the transporters. Remaining elements to specify include indicating that the transporter is active and specifying the length for the transporter of 6 (feet). To be consistent, velocity is specified as 10 feet per minute. Before discussing the zone control rule, (governs how the transporter moves between zones) how to specify the zones for the links needs to be discussed.

Guided-path transporters move from zone to zone on the links. The size of the zone, the zone control, and the size of the transporter govern how close the transporters can get to each other while moving. Zones are similar to the cell concept for conveyors. The selection of the size of the zones is governed by the physical characteristics of the actual system. In this example, assume that the carts should not get too close to each other during movement (perhaps because a large part overhangs the cart). In this example, let's choose a zone size of 12 feet. Thus, while the cart is in a zone, it essentially takes up half of the zone. If you think of the cart as sitting at the midpoint of the zone, then the closest the carts can be is 6 feet (3 feet to the front and 3 feet to the rear).

In Figure 6-47, Link 1 consists of 4 zones, each 12 feet long. Remember that the length is whatever unit of measure you want, as long as you are consistent in your specification. With these values, the maximum number of vehicles that could be on Link 1 is four. The rest of the zone specifications are also given in Figure 6-47. Now, you need to decide on the direction of travel permitted on the links. In this system, let's assume that the AGVs will travel clockwise around the figure. Figure 6-47 indicates that the direction of travel from I1 along Link1 is zero degrees (to the east). Then, Link 2 has a direction of 270 degrees, Link 3 has a direction of 180 degrees, and Link 4 has a direction of 90 degrees. The spur link to the exit station has a direction of 270 degrees (to the south). Thus, a clockwise turning of the vehicle around the network is achieved. In general, unless you specify the acceleration/deceleration and turning factor for the vehicle, the specification of these directions is not necessary. It has been illustrated here to indicate what the modeling entails. When working with a real AGV system, these factors can be discerned from the specification of the vehicle and the physical requirements of the system.

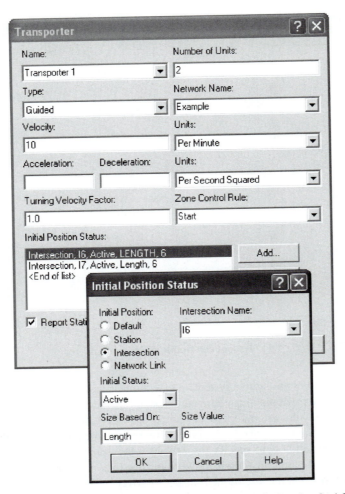

■ **FIGURE 6-49**   **TRANSPORTER module for simple AGV example**

To understand why the direction of travel and spurs are important, you need to understand how the transporters move from zone to zone and the concept of deadlock. Zone control governs how the transporter moves from zone to zone. There are three types of control provided by Arena[TM]: Start (of next zone), End (of next zone), or distance units k (into next zone). The START rule specifies that the transporter will release its backward-most zone when it starts into its next zone. This allows following vehicles to immediately begin to move into the released zone. Figure 6-50 illustrates this concept. In this case, the vehicle is contained in a single zone. In general, a vehicle could cover many zones. The END rule specifies that the transporter will release its backward most zone after it has moved into the next zone (or reached the end of the next zone). This prevents a following vehicle from moving into the transporter's current zone until it is fully out of the zone. The specification of the zone control rule is not critical to most system operations, especially if the vehicle size is specified in terms of zones. The release at end form of control allows for more separation between the vehicles when they are moving. In the distance units k (into next zone) rule, the transporter releases its backward-most zone after traveling k distance units through the next zone. In general, intersections can be modeled with a traversal distance. These rules also apply to how the vehicles move through intersections and onto links. For more details about the operation of these rules, see Pegden et al. (1995) for a discussion of the underlying SIMAN constructs.

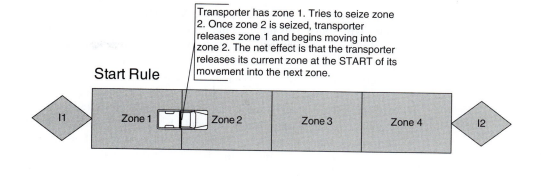

Transporter has zone 1. Tries to seize zone 2. Once zone 2 is seized, transporter releases zone 1 and begins moving into zone 2. The net effect is that the transporter releases its current zone at the START of its movement into the next zone.

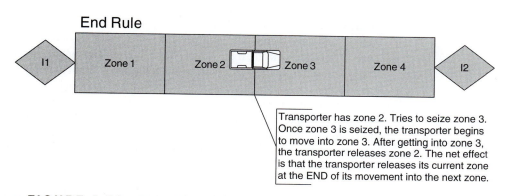

Transporter has zone 2. Tries to seize zone 3. Once zone 3 is seized, the transporter begins to move into zone 3. After getting into zone 3, the transporter releases zone 2. The net effect is that the transporter releases its current zone at the END of its movement into the next zone.

■ **FIGURE 6-50**   **Illustrating zone control**

As indicated in the zone control discussion, guided-path transporters move by seizing and releasing zones. Now consider the possibility of a bidirectional link with two transporters moving in opposite directions. What will happen when the transporters meet? They will crash! No, not really, but they will cause the system to get in a state of deadlock. Deadlock occurs when one vehicle requires a zone currently under the control of a second vehicle, and the second vehicle requires the zone held by the first vehicle. In some cases, Arena™ can detect when this occurs during the simulation run. If it is detected, then Arena™ will stop with a run-time error. In some cases, Arena™ cannot detect deadlock. For example in Figure 6-48, suppose a station was attached to I4 and a transporter (T1) was sent directly to that station for processing. Also, suppose that the part stays on transporter (T1) during the processing so that transporter (T1) is not released at the station. Next, suppose that transporter (T2) has just dropped off something at the exit point, I5, and that a new part has just come in at I1. Transporter (T2) will want to go through I4 on its way to I1 to pick up the new part. Transporter (T2) will wait until transporter (T1) finishes its processing at I4, and if transporter (T1) *eventually* moves there will be no problem. However, suppose transporter (T1) is released at I4 and no other parts arrive for processing. Transporter (T1) will become idle at the station attached to I4 and will continue to occupy the space associated with I4. Since, no arriving parts will ever cause transporter (T1) to move, transporter (T2) will be stuck. There will be the one part waiting for transporter (T2), but transporter (T2) will never get there to pick it up.

Obviously, this situation has been concocted to illustrate a possible undetected deadlock situation. The point is that care must be taken when designing guided-path networks so that deadlock is prevented by design or that special control logic is added to detect and react to various deadlock situations. For novice modelers, deadlock can seem troublesome, especially if you don't think about where to place the vehicles and if you have bidirectional links. In most real systems, you should not have to worry about deadlock because you should design the system to prevent it in the first place! In the simple example, all unidirectional links were used along with spurs. This prevents many possible deadlock situations. To illustrate what is meant by design, consider the concocted undetectable deadlock situation. A simple solution to prevent it would be to provide a spur for transporter (T1) to receive the processing that was occurring at the station associated with I4, and associate the processing station with the intersection at the end of the new spur. Thus, during the processing the vehicle will be out of the way. If you were designing a real system, wouldn't you adopt this common-sense practice anyway? Another simple solution would be to always send idle vehicles to a staging area or home base that gets them out of the way of the main path network.

To conclude this discussion, let's briefly discuss the operation of spurs and bidirectional links. This discussion is based in part on Pegden et al. (1995: 394). Note that the link from I4 to I5 is a spur in the previous example. A spur is a special type of link designed to help prevent some deadlock situations. A spur models a "dead end" and is bidirectional. If a vehicle is sent to the intersection at the end of the spur (or to a station attached to the end intersection), then the vehicle will need to maintain control of the intersection associated with the entrance to the spur so that it can get back out. If the vehicle is too long to fit on the spur, then the vehicle will naturally control the entering intersection. It will block vehicles from moving through the intersection. If the spur link is long enough to accommodate the size of the vehicle, the entering intersection will be released when the vehicle moves down the spur; however, any other vehicles that are sent to the end of the spur will not be permitted to gain access to the entering intersection until all zones in the spur link are available. This logic only applies to vehicles that are sent to the end of the spur. Any vehicles sent to or through the entering intersection of the spur are still permitted to gain access to the spur's entering intersection. This allows traffic to continue on the main path while vehicles are down the spur. Unfortunately, if a vehicle stops at the entering intersection and then never moves, the situation described in the concocted example occurs.

Bidirectional links work similarly to unidirectional links, but a vehicle moving on a bidirectional link must have control of both the entering zone on the link and the direction of travel. If another vehicle tries to move on the bidirectional link it must be moving in the same direction as any vehicles currently on the link; otherwise, it will wait. Unfortunately, if a vehicle (wanting to go down a bidirectional link) gains control of the intersection that a vehicle traveling on the bidirectional link needs to exit the link, deadlock may occur. Because it is so easy to get in deadlock situations with bidirectional links, you should proceed carefully when using them in your models.

The animation of guided-path transporters is very similar to regular transporters. On the Animation Transfer toolbar, the Network Link button is used to lay down intersections and the links between the intersections as shown in Figure 6-51. Arena™ will automatically recognize and fill the pull-down text boxes with the intersections and links that have been defined in the NETWORK LINK module. Although the diagram is not to scale, the rule/scale button is very handy when laying out the network links to try to get a good mapping to the physical distances in the problem. Once the network links have been placed, the animation works just like you are used to for transporters. The file *SimpleAGVExample.doe* is available for this problem. In the model, the parts are created and sent to the entry station, where they request a transporter, delay for the loading, and then transport to the exit station. At the exit station, there is a delay for the unloading before the transporter is freed. The REQUEST, TRANSPORT, and FREE modules all work in the same way as for free path transporters.

Try running the model. You will see that the two transporters start at the specified intersections, and you will clearly see the operation of the spur to the exit point, as it prevents deadlock. The second transporter will wait for the transporter already on the spur to exit the spur before heading down to the exit point. A handy technique is to use the View > Layers menu to view the transfer layers during the animation. This will allow the intersections and links to remain visible during the model run.

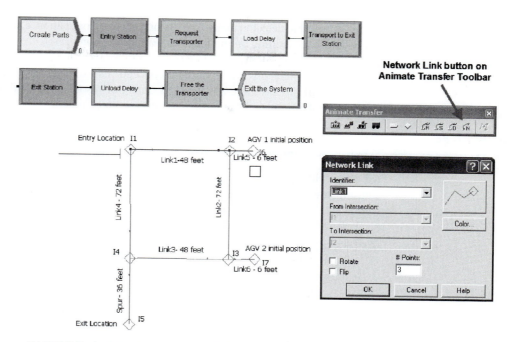

■ **FIGURE 6-51** **Animating guided-path transporters**

This section has provided a basic introduction to using guided-path transporters in Arena™ models. Many issues related to the use of these constructs that have not been covered for example:

- Construction and use of the shortest path matrix for the guided-path network. Since there may be multiple paths through the network, Arena™ uses standard algorithms to find the shortest path between all intersections in the network. The shortest path matrix is used when dispatching the vehicles.

- Specialized variables used in guided-path networks. These variables assist with indicating where the vehicle is in the network, accessing information about the network in the simulation (e.g., distances, beginning intersection for a link, etc.), status of links, intersections, and so on. Arena's help system under "transporter variables" discusses these variables, and a discussion of their use can also be found in Pegden et al. (1995).

- Changing vehicle speed, handling failures, changing vehicle size, redirecting vehicles through portions of the network, and link selection rules.

Details on the these topics can be found in Banks et al. (1995) and Pegden et al. (1995). Advanced examples are also given in both of those texts, including the possible use of guided-path constructs for modeling automated storage and retrieval systems (AS/RS) and overhead cranes in Pegden et al. (1995).

6.5   Summary

This chapter provided an introduction to modeling entity transfer. In particular, the use of resource-constrained transfer, transporters, and conveyors from an Arena™ modeling perspective were all discussed. Many of these concepts are especially useful in modeling manufacturing and distribution systems. The animation of entity transfer was also discussed through a number of embellishments to the test-and-repair model. While still somewhat crude, the animation concepts presented in this chapter are very effective for validating and selling an

Arena™ model to decision makers. With more time, effort, patience, and artistic skill, you can easily create compelling animations for your simulations.

A large portion of the modeling concepts in Arena™ have now been covered. There are a number of miscellaneous (yet important) concepts that still need to be discussed. In particular, Chapter 7 will discuss more advanced modeling of resources (including staffing and break-downs) and the basics of Arena's cost-modeling constructs.

Exercises

6.1 Reconsider problem 5.16 of Chapter 5. Analyze this situation assuming an arrival rate of 1 customer every 2 minutes. Assume now that the travel time between the two systems is more significant. The travel time is distributed according to a triangular distribution with parameters (1, 2, 4) minutes.

(a) Model the system assuming that the worker from the first station moves the parts to the second station. The movement of the part should be given priority if there is another part waiting to be processed at the first station.

(b) Model the system with a new worker (resource) to move the parts between the stations.

(c) From your models, estimate the total system time for the parts, the utilization of the workers, and the average number of parts waiting for the workers. Run the simulation for 20 replications of 200000 minutes with a 20,000 minute warm up period.

6.2 Reconsider part (b) of problem 6.1. Instead of immediately moving the part, the transport worker waits until a batch of 5 parts has been produced at the first station. When this occurs, the worker moves the batch of parts to the second station. The parts are still processed individually at the second station. From your model, estimate the total system time for the parts, utilization of the workers, and average number of parts waiting for the workers. Run the simulation for 20 replications of 200000 minutes with a 20,000 minute warm up period.

6.3 Redo problem 5.19 of Chapter 5 using resource-constrained transfer. Assume that there are 2 workers at the diagnostic station, 1 worker per testing station, and 3 workers at the repair station. Thus, there are a total of 8 workers in the system. Furthermore, assume that any of these 8 workers are capable of moving parts between the stations. For example, when a part completes its operation at the diagnostic station, any worker in the system can carry the part to the next station. When a part requires movement, it will wait for the next available idle worker to complete the movement. Assume that parts waiting for processing at a station will be given priority over parts that require movement between stations. Build a simulation model that can assist the company in assessing the risks associated with the new contract under this resource-constrained situation.

6.4 Redo problem 5.20, assuming that a pool of 3 workers performs the transport between the stations. Assume that the transport time is triangularly distributed with parameters (2, 4, 6) all in minutes. Make an assessment for the company for the appropriate number of workers to have in the transport worker pool.

6.5 In Section 6.2.1, the test-and-repair system was analyzed with three transporters. Re-analyze this situation and recommend an appropriate number of transport workers.

6.6 Reconsider part (b) of problem 6.1. Instead of modeling the problem with a resource, the problem should be modeled with a transporter. Assume that there are 3 transporters (forklifts) available to move the parts between the two stations. The distance between the two stations is 100 meters and the fork lift's velocity between the stations is triangularly distributed with parameters (25, 50, 60) in meters per minute.

(a) Model the system and estimate the total system time for the parts, utilization of the workers, and average number of parts waiting for the fork lift. Run the simulation for 20 replications of 200000 minutes with a 20,000 minute warm up period.

■ **TABLE 6-4**

Distance Table for Problem 6.7

	Station	Destination				
		A	**B**	**C**	**D**	**Exit**
	A	—	50	60	80	90
	B	55	—	25	55	75
origin	C	60	20	—	70	30
	D	80	60	65	—	35
	Exit	100	80	35	40	—

(b) Instead of immediately moving the part, a fork lift is not called until a batch of 5 parts has been produced at the first station. When this occurs a fork lift is called to move the batch of parts to the second station. The parts are still processed individually at the second station. Redo the analysis of part (a) for this situation.

6.7 Reconsider problem 6.4 using transporters. The distances between the four stations (in feet) are given in Table 6.4. After the parts finish processing at the last station of their sequence, they are transported to an exit station, where the shipping process begins.

There are 2 fork lifts in the system. The velocity of the fork lifts varies in the facility due to various random congestion effects. Assume that the fork lift's velocity between the stations is triangularly distributed with parameters (3, 4, 5) miles per hour. Simulate the system for 30,000 minutes and discuss the potential bottlenecks in the system.

6.8 Reconsider the test-and-repair example in Section 6.2.1. In this problem, the workers have a home base that they return to whenever there are no more waiting requests rather than idling at their last drop-off point. The home base is located at the center of the shop. The distances from each station to the home base are given in Table 6.5.

Update the animation so that the total number of waiting requests for movement is animated by making it visible at the home base of the workers. Rerun the model using the same run parameters as the chapter example and report the average number of requests waiting for transport. Provide an assessment of the risk associated with meeting the contract specifications based on this new design.

6.9 Reconsider problem 5.35 with transporters to move the parts between the stations. The stations are arranged sequentially in a flow line with 25 meters between each station; distances are shown in Table 6.6.

■ **TABLE 6-5**

Transporter Distances between Stations in Problem 6.8

	Station	Destination					
		Diagnostics	**Test 1**	**Test 2**	**Test 3**	**Repair**	**Home base**
	Diagnostics	—	40	70	90	100	20
	Test 1	43	—	10	60	80	20
origin	Test 2	70	15	—	65	20	15
	Test 3	90	80	60	—	25	15
	Repair	110	85	25	30	—	25
	Home base	20	20	15	15	25	—

■ **TABLE 6-6**
Distance Table for Problem 6.9

		Destination			
	Station	**1**	**2**	**3**	**Exit**
origin	1	—	25	50	75
	2	25	—	25	50
	3	50	25	—	25
	Exit	75	50	25	—

Assume that there are two transporters available to move the parts between the stations. Whenever a part is required at a down stream station and a part is available for transport the transporter is requested. If no part is available, no request is made. If a part becomes available, and a part is need, then the transporter is requested. The velocity of the transporter is considered to be TRIA(22.86, 45.72, 52.5) in meters per minute. By using the run parameters of problem 5.35, estimate the effect of the transporters on the throughput of the system.

6.10 Three independent conveyors deliver 1-foot parts to a warehouse. Once inside the warehouse, the conveyors merge onto one main conveyor to take the parts to shipping. Parts arriving on conveyor 1 follow a Poisson process with a rate of 6 parts per minute. Parts arriving on conveyor 2 follow a Poisson process with a rate of 10 parts per minute. Finally, parts arriving on conveyor 3 have an interarrival time uniformly distributed between 0.1 and 0.2 minutes. Conveyors 1 and 2 are accumulating conveyors and are 15 and 20 feet long respectively. They both run at a velocity of 12 feet per minute. Conveyor 3 is an accumulating conveyor and is 10 feet long. It runs at a velocity of 20 feet per minute. The main conveyor is 25 feet long and is an accumulating conveyor operating at a speed of 25 feet per minute. Consider the parts as being disposed after they reach the end of the main conveyor. Simulate this system for 480 minutes and estimate the mean time to access and the mean number of waiting parts for the conveyors.

6.11 A 140-foot, nonaccumulating loop conveyor is used to feed jobs to 7 workers that are evenly spaced around a loop. Jobs are loaded onto the conveyor 5 feet prior to worker 1. An incomplete job will continue to move around the conveyor until an idle worker is encountered. Once encountered, the job is unloaded from the conveyor, processed, and the completed job is reloaded on the conveyor. Completed jobs are unloaded from the conveyor 5 feet after worker 7. Jobs arrive exponentially with a mean time between arrivals of 2.2 minutes. Job processing is log-normally distributed with a mean of 14 minutes and a variance of 3.6 minutes squared for all workers. The loop conveyor has a velocity of 15 feet per minute. Arriving jobs wait in a storage buffer until space on the loop conveyor becomes available.

(a) Simulate for 1 day of continuous operation and determine the utilization of the conveyor, the average number of jobs on the conveyor, the average time spent in the system, and the utilization of each worker.

(b) Suppose that the conveyor is subject to breakdowns. The time between breakdowns is exponentially distributed with a mean time of 9 hours. The time to repair is also exponentially distributed with a mean of 7 minutes. Simulate for 1 day of continuous operation and determine the utilization of the conveyor, average number of jobs on the conveyor, average time spent in the system, and utilization of each worker. Hint: Consider the use of the STOP and START modules.[1]

---

[1]This problem is based on Banks et al. (1995), problem E10.6. Reprinted with permission of John Wiley & Sons, Inc.

6.12 Reconsider problem 5.35 with conveyors to move the parts between the stations. Suppose that a single nonaccumulating conveyor of length 75 meters, with 25-meter segments between each of the stations is used in the system. When a part is needed by a downstream station, it is loaded onto the conveyor if available. The load time takes between 15 and 45 seconds, uniformly distributed. The speed of the conveyor is 5 meters per minute. If a part is produced and the downstream station requires the part, it is loaded onto the conveyor. By using the run parameters of problem 5.35, estimate the effect of conveyors on system throughput.

6.13 This problem considers the use of AGVs for the test-and-repair system in the current chapter. The layout of the proposed system is given in Figure 6-52 with all measurements in meters. The only AGV in this system is 1 meter in length and moves at a velocity of 30 meters per minute. Its home base is at the dead end of the 9-meter spur. Since there will be only one AGV, all links are bidirectional. The design associates the stations with the nearest intersection on the path network.

(a) Simulate the system for 10 replications of 4160 hours. Estimate the chance that the contract specifications are met and the average system time of the jobs. In addition, assess the queuing for and utilization of the AGV.

(b) Consider the possibility of having a two-AGV system. What changes to your design do you recommend? Be specific enough that you could simulate your design.

6.14 Reconsider problem 5-18 of Chapter 5 with the use of transporters. Assume that all parts are transferred by using two fork lifts that travel at an average speed of 150 feet per minute. The distances (in feet) between the stations are provided in Table 6.7. The distances are symmetric. Both the drop-off and pickup points at a station are at the same physical location. Once the truck reaches the pickup/drop-off station, it requires a load/unload time of two minutes.

Analyze this system to determine any potential bottleneck operations. Report on the average flow times of the parts as a whole and individually. In addition, obtain statistics representing the average number of parts waiting for allocation of a transporter and the number of busy

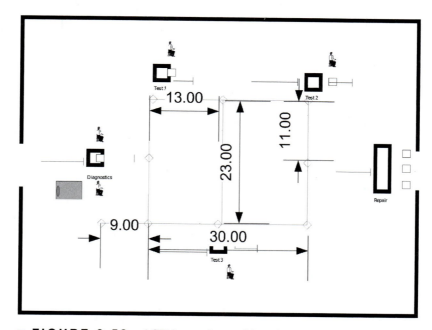

■ **FIGURE 6-52** **AGV layout for problem 6.13**

■ **TABLE 6-7**
Distance Table for Problem 6.14

	Enter	Workstation	Paint	New Paint	Pack	Exit
Enter	—	325	445	455	565	815
Workstation		—	120	130	240	490
Paint			—	250	120	370
New paint				—	130	380
Pack					—	250
Exit						—

transporters. Run the model for 600,000 minutes with a 50,000-minute warm-up period. Develop an animation for this system that illustrates the transport and queuing in the system.[2]

6.15 Reconsider problem 6.14 with the use of AGVs. There are now three AGVs that travel at a speed of 100 feet per minute to transport the parts in the system. The guided-path network for the AGVs is given in Figure 6-53.

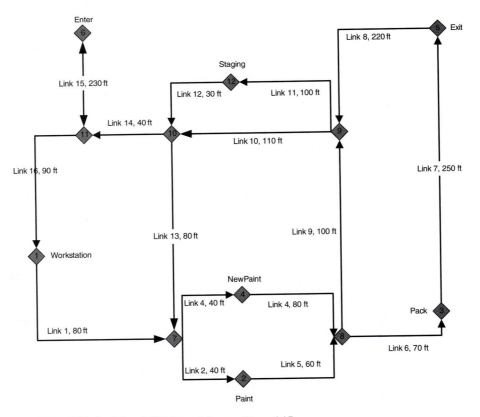

■ **FIGURE 6-53** AGV layout for problem 6.15

[2]This problem is based on an example in Pegden et al. (1995: 223).

To avoid deadlock situations the guided-path network has been designed for one way travel. In addition, to prevent an idle vehicle from blocking vehicles performing transports, a staging area (Intersection 12) has been placed on the guided-path network. Whenever a vehicle completes a transport, a check should be made of the number of requests pending for the transporters. If there are no requests pending, the AGV should be sent to the staging area. If there are multiple vehicles idle, they should wait on Link 11. Link 15 is a spur from intersection 6 to intersection 11. The stations for the operations on the parts should be assigned to the intersections as shown in the figure. The links should all be divided into zones of 10 feet each. The initial position of the vehicles should be along Link 11. Each vehicle is 10 feet in length or 1 zone. The release at start form of zone control should be used.

Analyze this system to determine any potential bottleneck operations. Report on the average flow times of the parts as a whole and by part type. Also obtain statistics representing the average number of parts waiting for allocation of a transporter and the number of busy transporters. Run the model for 600,000 minutes with a 50,000-minute warm-up period. Develop an animation for this system that illustrates the transport and queuing in the system.[3]

---

[3]This problem is based on an example in Pegden et al. (1995: 381).

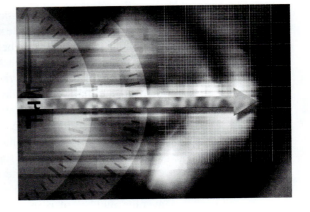

C H A P T E R   7

Miscellaneous Topics in Arena^{TM} Modeling

LEARNING OBJECTIVES

After completing this chapter, you should be able to:

- Model the scheduling and failure/repair of resources
- Use Arena's entity and resource costing constructs
- Model with advanced concepts involving picking amongst stations, picking up entities into groups, dropping off entities from groups, and generic station modeling
- Understand and use some of Arena's advanced programming options

This chapter tackles a number of miscellaneous topics that can enhance your modeling capabilities with Arena^{TM}. The chapter first examines in more detail how to utilize resources in your models by incorporating resource schedules. This will enable you to better model and tabulate the effects of realistic resource changes. For example, you might want to vary the resource capacity via a resource schedule in order to better match staffing levels to customer arrivals. In addition, with enhanced resource modeling, you will want to better model the costs associated with entities using the resources. Thus, a discussion of how Arena^{TM} tabulates entity and resource costs is also presented.

As entities move through the model, they often are formed into groups. Arena's constructs for grouping and ungrouping entities will also be presented. Finally, an introduction to the programming capabilities (e.g., Visual Basic for Applications, etc.) that Arena^{TM} provides will be discussed. Let's get started with enhanced resource modeling.

7.1 Advanced Resource Modeling

A resource can be something that an entity uses as it flows through the system. If the required units of the resource are available for the entity, then the units are allocated to the entity. If the required units are unavailable for allocation, the entity's progress through the model stops until the required units become available. So far, the only way in which the units could become unavailable was for the units to be seized by an entity. The seizing of at least one unit of a resource causes the resource to be considered busy. A resource is considered to be in the idle state when all units are idle (no units are seized). Thus, in previous modeling a resource could be in one of two states—busy or idle. This section discusses the other two default states that Arena^{TM} permits for resources—failed and inactive.

When specifying a resource, its capacity must be given. Capacity represents the maximum number of units of the resource that can be seized at any time. In previous modeling, the capacity of the resource was *fixed*. That is, the capacity did not vary as a function of time; however, in many situations, the capacity of a resource can vary with time. For example, in the pharmacy model, you might want to have two or more pharmacists staffing the pharmacy during peak business hours. This type of capacity change is predictable and can be managed via a resource schedule. Alternatively, the changes in capacity of a resource might be unpredictable or random. For example, a machine being used to produce products on a manufacturing line may experience random failures that require repair. During the repair time, the machine is not available for production. The situations of scheduled capacity change and random capacity change define the inactive and failed states in Arena^{TM} modeling.

There are two key resource variables for understanding resources and their states.

MR(Resource ID): Resource capacity. MR returns the number of capacity units currently defined for the specified resource. This value can be changed by the user via the resource schedule or through the use of the ALTER block.

NR(Resource ID): Number of busy resource units. Every time an entity seizes or preempts capacity units of a resource, the NR variable changes accordingly. NR is an integer value, and is not user assignable.

The index Resource ID should be the unique name of the resource or its construct number. These variables can change over time for a resource. Changing of these variables along with the possibility of a resource failing defines the possible states of a resource. The states of a resource are:

Idle: A resource is in the idle state when all units are idle and the resource is not failed or inactive. This state is represented by the Arena^{TM} constant, IDLE_RES, which evaluates to the number (-1).

Busy: A resource is in the busy state when it has one or more busy (seized) units. This state is represented by the Arena^{TM} constant, BUSY_RES (-2).

Inactive: A resource is in the inactive state when it has zero capacity and is not failed. This state is represented by the Arena^{TM} constant, INACTIVE_RES (-3).

Failed: A resource is in the failed state when a failure is currently acting on the resource. This state is represented by the Arena^{TM} constant, FAILED_RES (-4).

The special function STATE(Resource ID) indicates the current state of the resource. For example, STATE(Pharmacist) = = BUSY_RES will return true if the resource named *Pharmacist* is currently busy. Thus, the STATE() function can be used in conditional statements and in the Arena^{TM} environment (e.g., variable animation) to check the state of a resource.

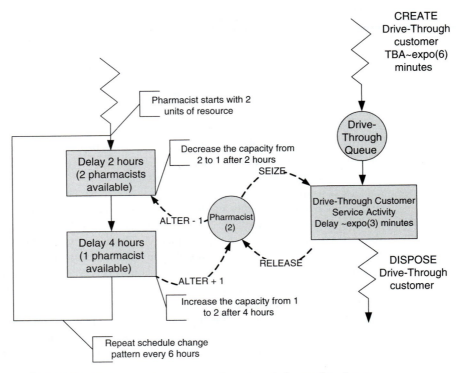

■ **FIGURE 7-1** **Activity diagram for scheduled capacity changes**

To vary the capacity of a resource, you can use the SCHEDULE module on the Basic Process panel, the Calendar Schedules builder, or the ALTER block on the Blocks panel. Only resource schedules will be presented in this text. The Calendar Schedule builder (Edit > Calendar Schedule) allows the user to define resource capacities in relation to the Gregorian calendar. For more information on this, refer to Arena's help system. The ALTER block can be used in the model window to invoke capacity changes based on specialized control logic. The ALTER block allows the capacity of the resource to be increased or decreased by a specified amount.

In the pharmacy model, suppose that two pharmacists are available at the beginning of the day to serve the drive-through window. During the first 2 hours of the day, there are two pharmacists, and then one pharmacist during the next 4 hours. In addition, suppose that this on-and-off pattern of availability repeated itself every 6 hours. This situation can be modeled with an activity diagram by indicating that the resource capacity is decreased or increased at appropriate times. Figure 7-1 illustrates this concept. The diagram has been augmented with the ALTER keyword to indicate the capacity change. Conceptually, a "schedule entity" is created in order to cycle through the pharmacist's schedule. At the end of the first 2-hour period, capacity is reduced to one pharmacist. Then, after the 4-hour delay, the capacity is increased to two pharmacists. This is essentially how you use the ALTER block in a model. Refer to Pegden et al. (1995) and the Arena™ help system for further information on the ALTER block. As seen in Figure 7-1, this pattern can be easily extended for multiple capacity changes and durations. This is what the SCHEDULE module in Arena™ does in an easier-to-use form.

As an example, consider the following simple illustration of the SCHEDULE module based on a modified version of an Arena™ SMART file, *Smarts139 ResourceScheduleEx-ample.doe*. In this system, customers arrive according to a Poisson process at a rate of one customer per minute. The service times of the customers are distributed according to a triangular distribution with parameters (3, 4, 5) in minutes. Each arriving customer requires one

■ **TABLE 7-1**
Example Staffing Schedule

Shift	Start time	Stop time	# Available	Duration
1	0	60	2	60
1	60	180	3	120
1	180	360	9	180
1	360	480	6	120
2	480	540	2	60
2	540	660	3	120
2	660	840	9	180
2	840	960	6	120

unit of a single resource when receiving service; however, the capacity of the resource varies with time. Let's suppose that two 480-minute operation shifts will be simulated. Assuming that the first shift starts at time zero, each shift is staffed with the same pattern that varies according to the values shown in Table 7-1. With this information, you can easily set up the resource in Arena[TM] to follow the staffing plan for the two shifts. This is done by first defining the schedule using the SCHEDULE module and then indicating that the resource should follow the schedule in the appropriate RESOURCE module.

To set up the schedule in Arena[TM], a number of schedule editing formats are available in Arena[TM]. Figure 7-2 shows the schedule spreadsheet editor for this example. In the spreadsheet, you can enter (value, duration) pairs that represent the capacity and time duration that the capacity will be available. The schedule must be given a name, format type (Duration), type (Capacity for resources), and time units (hours in this case). The scale factor field does not apply to capacity-type schedules.

Alternatively, you can use the graphical schedule editor to define the schedule as shown in Figure 7-3. The graphical schedule editor allows you to use the mouse to draw the capacity changes. The help button describes its use. Note that capacity goes up and down in the figure according to the staffing plan. The Schedule dialog box is also useful in entering a schedule. The method of schedule editing is entirely your preference.

In the SCHEDULE module, only the schedule for the first 480-minute shift was entered. What happens when the durations specified by the schedule are completed? The default action is to repeat the schedule indefinitely until the simulation run length is completed. If you want a specific capacity to remain for the entire run, you can enter a blank duration, which defaults the duration to infinite. The schedule will not repeat, so the capacity will remain at the specified value when the duration is invoked.

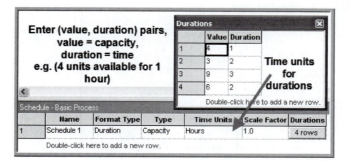

■ **FIGURE 7-2** Schedule value, duration spreadsheet

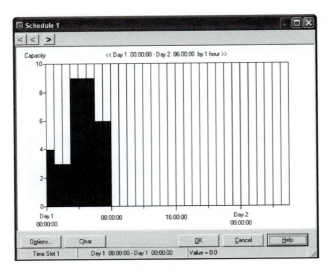

■ **FIGURE 7-3** **Graphical schedule editor**

Once the schedule has been defined, you need to specify that the resource will follow the schedule by changing its Type to Based on Schedule as shown in Figure 7-4. In addition, you must specify the rule that the resource will use if there is a schedule change and it is currently seized by entities. This is called the Schedule Rule. The following three rules are available in Arena™.

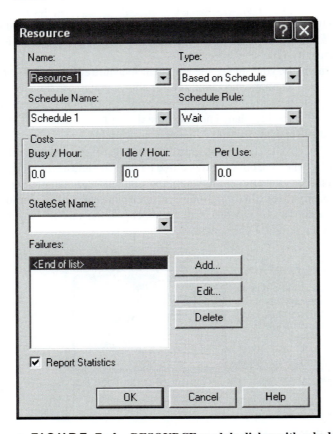

■ **FIGURE 7-4** **RESOURCE module dialog with scheduled capacity**

Ignore: Starts the time duration of the schedule change or failure immediately, but allows the busy resource to finish processing the current entity before affecting the capacity change.

Wait: Waits until the busy resource has finished processing the current entity before changing the resource capacity or starting the failure and starting the time duration of the schedule change/failure.

Preempt: Interrupts the currently processing entity, changes the resource capacity, and starts the time duration of the schedule change or failure immediately. The resource will resume processing the preempted entity as soon as the resource becomes available (after schedule change/failure).

The wait rule has been used in Figure 7-4. Let's discuss a bit more about how these rules work with an example.

Let's suppose that your simulation professor has office hours throughout the day, except for lunch time at 12 to 12:30 p.m. In addition, since your professor teaches simulation using Arena™, he or she follows Arena's rules for capacity changes. What happens under the conditions of each rule if you arrive at 11:55 a.m. with a question?

7.1.1 Ignore

Case 1.

- You arrive at 11:55 a.m. with a 10-minute question and begin service.
- At noon, the professor hangs a "lunch in progress" sign on his door.
- Your professor continues to answer your question for the remaining service time of 5 minutes.
- You leave at 12:05 p.m.
- Regardless of whether there are students waiting in the hallway, the professor starts lunch.
- At 12:30 p.m., the professor finishes lunch and removes the sign. If there are students waiting in the hallway, they can begin service at this time.

The net effect of case 1 is that the professor lost 5 minutes of lunch time. During the 30-minute scheduled break, the professor was busy for 5 minutes and inactive for 25 minutes.

Case 2.

- You arrive at 11:55 a.m. with a 45-minute question and begin service.
- At noon, the professor hangs a "lunch in progress" sign.
- Your professor continues to answer your question for the remaining service time of 40 minutes.
- At 12:30 p.m., the professor removes the sign and continues to answer your question.
- You leave at 12:40 p.m.

The net effect of case 2 is that the professor did not get to eat lunch that day. During the 30-minute scheduled break, the professor was busy for 30 minutes.

This rule is called Ignore since the scheduled break may be *ignored* by the resource if the resource is busy when the break occurs. Technically, the scheduled break actually occurs and the time that was scheduled is considered as unscheduled (inactive) time. Let's assume that the professor is at work for 8 hours every day (including the lunch break). But because of the scheduled lunch break, the total time available for useful work is 450 minutes. In case 1, the professor worked for 5 minutes more than scheduled. In case 2, the professor worked for

30 minutes more than scheduled. As indicated shortly, this extra work time must be factored into how the utilization of the resource (professor) is computed. Now, let's examine what happens if the rule is Wait.

7.1.2 Wait

Case 1.

- You arrive at 11:55 a.m. with a 10-minute question and begin service.
- At noon, the professor's lunch reminder rings on the computer. The professor recognizes the reminder but doesn't act on it, yet.
- Your professor continues to answer your question for the remaining service time of 5 minutes.
- You leave at 12:05 p.m. The professor recalls the lunch reminder and hangs a "lunch in progress" sign. Regardless of whether there are students waiting in the hallway, the professor hangs the sign and starts a 30-minute lunch.
- At 12:35 p.m., the professor finishes lunch and removes the sign. If there are students waiting in the hallway, they can begin service.

Case 2.

- You arrive at 11:55 a.m. with a 45-minute question and begin service.
- At noon, the professor's lunch reminder rings. The professor recognizes the reminder but doesn't act on it, yet.
- Your professor continues to answer your question for the remaining service time of 40 minutes.
- You leave at 12:40 p.m. The professor recalls the lunch reminder and hangs a "lunch in progress" sign. Regardless of whether there are students waiting in the hallway, the professor hangs the sign and starts a 30-minute lunch.
- At 1:10 p.m., the professor finishes lunch and removes the sign. If there are students waiting in the hallway, they can begin service.

The net effect of both these cases is that the professor does not miss lunch (unless the student's question takes the rest of the afternoon!). Thus, in this case the resource will experience the scheduled break after *waiting* to complete the entity in progress. Again, in this case, the tabulation of the amount of busy time may be affected by when the rule is invoked. Now, let's consider the situation of the Preempt rule.

7.1.3 Preempt

Case 1.

- You arrive at 11:55 a.m. with a 10-minute question and begin service.
- At noon, the professor's lunch reminder rings. The professor pushes you into the corner of the office, hangs the "lunch in progress" sign on the door, and begins a 30-minute lunch. If students are waiting in the hallway, they continue to wait.
- You wait patiently in the corner of the office until 12:30 p.m.
- At 12:30 p.m. the professor finishes lunch, removes the sign, and tells you that you can get out of the corner. You continue your question for the remaining 5 minutes.
- At 12:35 p.m., you finally finish your 10-minute question and depart the system, wondering what the professor had to eat that was so important.

As you can see from the handling of Case 1, the professor always gets the lunch break. The customer's service is preempted and resumed after the scheduled break. The result for Case 2 is essentially the same, with the student finally completing service at 12:55 p.m. While this rule may seem a bit rude in a service situation, it is quite reasonable for many situations where the service can be restarted (e.g., parts on a machine).

The example involving the professor involved a resource with one unit of capacity. But what happens if the resource has a capacity of more than one, and what happens if the capacity change is more than one? The rules work essentially the same. If the scheduled change (decrease) is less than or equal to the current number of idle units, then the rules are not invoked. If the scheduled change will require busy units, then any idle units are first taken away and then the rules are invoked. In the case of the ignore rule, the units continue serving, the inactive sign goes up, and whichever unit is released first becomes inactive first.

7.1.4   Running the Scheduled Capacity Change Model

Now that you understand the consequences of the rules, let's run the example model. The final model including the animation is shown in Figure 7-5. When you run the model, the variable animation boxes will display the current time (in minutes), the current number of scheduled resource units (MR), the current number of busy resource units (NR), and the state of the resource (as per the resource state constants). The resource statistics results after running the model for 960 minutes appear in Figure 7-6. You should note something new. In all previous work, the instantaneous utilization was the same as the scheduled utilization. In this case, these quantities are not the same. Tabulation of the amount of busy time as per the scheduled rules accounts for the difference. Let's take a more detailed look at how these quantities are computed.

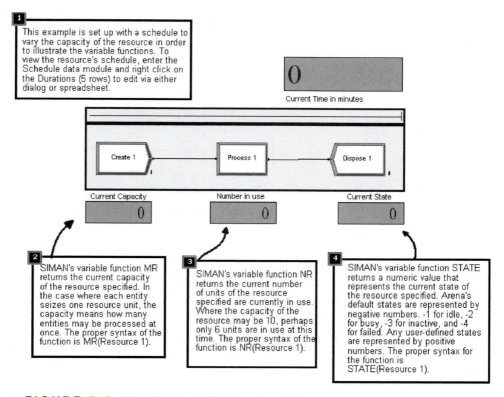

■ **FIGURE 7-5** **Scheduled capacity change model**

Resource

Usage

Instantaneous Utilization	Average	Half Width
Resource 1	0.7586	(Correlated)

Number Busy	Average	Half Width
Resource 1	4.1850	0.462899358

Number Scheduled	Average	Half Width
Resource 1	6.1154	(Insufficient)

Scheduled Utilization	Value
Resource 1	0.6843

Total Number Seized	Value
Resource 1	1007.00

■ **FIGURE 7-6 Resource statistical results for scheduled capacity change example**

The calculation of instantaneous and scheduled utilization depends on the two variables NR and MR for a resource. Instantaneous utilization is defined as the time weighted average of the ratio of these variables. The formulas for instantaneous and scheduled utilization are as follows:

$$NR(t) = \text{number of busy resource units at time } t$$

$$\overline{NR} = \frac{1}{T}\int_{o}^{T} NR(t)dt = \text{time average number of busy resources}$$

$$MR(t) = \text{number of scheduled resource units at time } t$$

$$\overline{MR} = \frac{1}{T}\int_{o}^{T} MR(t)dt = \text{time average number of scheduled resource units}$$

$$IU(t) = \begin{cases} 0.0 & NR(t) = 0 \\ 1.0 & NR(t) \geq MR(t) \\ NR(t)/MR(t) & \text{otherwise} \end{cases}$$

$$\overline{IU} = \frac{1}{T}\int_{o}^{T} IU(t)dt = \text{(time average) instantaneous utilization}$$

The scheduled utilization is the time average number of busy resources divided by the time average number of scheduled resources. As seen in the formula for scheduled utilization, this is the same as the total time spent busy divided by the total time available for all resource units.

$$\overline{SU} = \frac{\overline{NR}}{\overline{MR}} = \frac{\frac{1}{T}\int_{0}^{T} NR(t)dt}{\frac{1}{T}\int_{0}^{T} MR(t)dt} = \frac{\int_{0}^{T} NR(t)dt}{\int_{0}^{T} MR(t)dt} = \text{scheduled utilization}$$

■ **TABLE 7-2**

Example Calculation for Professor Utilization

Time interval	Busy time	Scheduled time	NR(t)	MR(t)	IU(t)
8 a.m.–noon	240	240	1.0	1.0	1.0
12–12:30 p.m.	30	0	1.0	0	1.0
12:30–4 p.m.	210	210	1.0	1.0	1.0
	480	450	$\overline{NR} = 1.0$	$\overline{MR} = 450/480 = 0.9375$	$\overline{IU} = 1.0$

If $MR(t)$ is constant, then $\overline{IU} = \overline{SU}$. The Arena™ function, $RESUTIL(Resource\ ID)$, returns $IU(t)$. Caution should be used in interpreting $\overline{IU}$ when $MR(t)$ varies with time, because instantaneous utilization does not adequately reflect the time working when not scheduled.

Now let's return to the example of the professor holding office hours. Let's suppose that the Ignore option is used and consider Case 2 and the 45-minute question. Let's also assume for simplicity's sake that the professor had assigned the class a large homework assignment due the next day and was therefore busy answering student questions all day long. What would be the average instantaneous utilization and the scheduled utilization of the professor? Table 7-2 illustrates the calculations.

Table 7-2 indicates that $\overline{IU} = 1.0$ (or 100%) and $\overline{SU} = 1.06$ (or 106%). Who says that professors don't give their all! Thus, with scheduled utilization, a schedule can result in the resource having its utilization higher than 100%. There is nothing wrong with this result, and if the resource is busy for more time than it is scheduled, you would definitely want to know.

Arena™ help files contain an excellent discussion of how it calculates instantaneous and scheduled utilization (as well as what they mean) (see Figure 7-7). The interested reader should search for, "resource statistics: instantaneous utilization vs. scheduled utilization."

Consider the following examples, all with an eight-hour period split into two four-hour shifts.

Case A: We have 150 workers available during an 8-hour period, and 50 of these are busy the entire 8 hours. Note that since the number scheduled is constant, both utilization statistics are identical.

Case B: We have 200 workers available for 4 hours, then 100 workers are available for the next 4 hours. All 100 of the workers are idle for the first 4 hours, then busy for the last 4 hours.

Case C: We have 200 workers available for 4 hours, then 100 workers are available for the next 4 hours. 50 of the workers are busy for the entire 8 hours.

Case D: We have no workers available for the first 4 hours, then 100 workers busy of the 300 available for the last 4 hours. In this last case, you could certainly advance the argument that period 1 should be excluded, and that the utilization is undefined during a period when no resources are scheduled. But Arena does not exclude selected periods (or states) from time persistent statistics (see Frequencies Element). So Arena does include this period and defines it as a utilization of 0%.

Scenario		Case A		Case B		Case C		Case D	
	4 Hour Shift	1	2	1	2	1	2	1	2
	# Busy # Scheduled Utilization	50 150 33.33%	50 150 33.33%	0 200 0%	100 100 100%	50 200 25%	50 100 50%	0 0 0%	100 300 33.33%
Total Hours Busy		(50*8) = 400		((0*4)+(100*4)) = 400		(50*8) = 400		(100*4) = 400	
Total Hours Scheduled		(150*8) = 1200		((200*4)+ (100*4)) = 1200		((200*4)+ (100*4)) = 1200		(300*4) = 1200	
Average Number Busy		(50*8)/8 = 50		((0*4)+ (100*4))/8 = 50		(50*8)/8 = 50		((0*4)+(100*4))/8 = 50	
Average Number Scheduled		(150*8)/8 = 150		((200*4)+ (100*4))/8 = 150		((200*4)+ (100*4))/8 = 150		((0*4)+(300*4))/8 = 150	
Scheduled Utilization		50/150 = 33.33%		50/150 = 33.33%		50/150 = 33.33%		50/150 = 33.33%	
Instantaneous Utilization		(33.33%*8)/8 = 33.33%		((0%*4)+ (100*4))/8 = 50%		((25%*4)+(50% *4))/8 = 37.5%		((0%*4)+(33.33% *4))/8 = 16.67%	

■ **FIGURE 7-7**   Utilization example from help files

7.1.5 Modeling Resource Failure

As mentioned, a resource has four states: idle, busy, inactive, and failed. The SCHEDULE module is used to model planned capacity changes. The FAILURES module is used to model unplanned or random capacity changes that place the resource in the failed state. The failed state represents the situation of a breakdown for the *entire* resource. When a failure occurs for a resource, all units of the resource become unavailable and the resource is placed in the failed state. For example, imagine standing in line at a bank with three tellers. Assume that the computer system goes down and all tellers are thus unable to work. This situation is modeled with the FAILURES module of the Advanced Process panel. A more common application of failure modeling occurs in manufacturing settings to model the failure and repair of production equipment.

There are two ways in which a failure can be initiated for a resource: time based and usage (count) based. The concept of time-based failures is illustrated in Figure 7-8. Note the similarity to scheduled capacity changes. You can think of this situation as a failure "clock" ticking. The uptime delay is governed by the time to failure distribution and the downtime delay is governed by the time to repair distribution. When the time of failure occurs, the resource is placed in the failed state. If the resource has busy units, then the rules for capacity change (ignore, wait, and preempt) are invoked.

Usage-based failures occur *after* the resource has been released. Each time the resource is released, a count is incremented; when it reaches the specified number or failure value, the resource fails. Again, if the resource has busy units, the capacity change rules are invoked. A resource can have multiple failures (both time and count based) defined to govern its behavior. In the case of multiple failures that occur before the repair, the failures "queue up" and cause the repair times to be acted on consecutively.

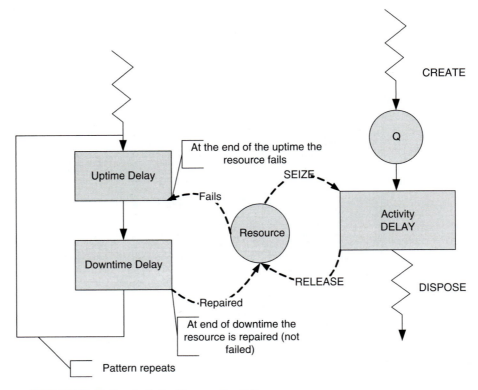

■ **FIGURE 7-8** **Activity diagram for failures**

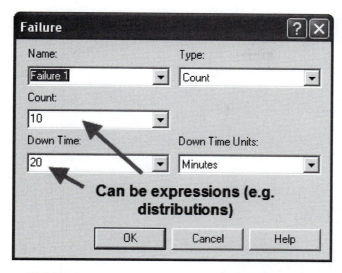

■ **FIGURE 7-9**   Count-based failure dialog

Figures 7-9 and 7-10 illustrate how to define count-based and time-based failures in the FAILURES module. In both cases, the uptime, downtime, or count fields can all be expressions. In the case of the time-based failure (Figure 7-10), the field Uptime in this State only requires special mention. A time-based failure defines a failure "clock" that ticks until the failure occurs. After the failure is repaired, the clock is reset and the ticking to failure starts again. The uptime in this state field forms the basis for the clock. If nothing is specified, then simulation clock time is used as the basis.

However, you can specify a state (from a STATESET or auto state [Idle, Busy, etc.]) to use as the basis for accumulating the time until failure. The most common situation is to choose the Busy state. In this fashion, the resource only accumulates time until failure during its busy time. Figures 7-11 and 7-12 illustrate how to attach the failures to resources.

Note that in the case of Figure 7-12, you can clearly see that a resource can have multiple failures. In addition, the same failure can be attached to different resources. These figures are all

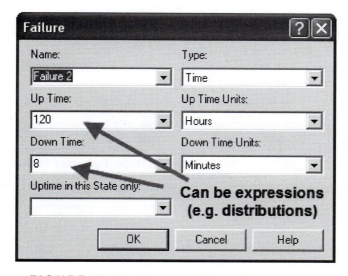

■ **FIGURE 7-10**   Time-based failure dialog

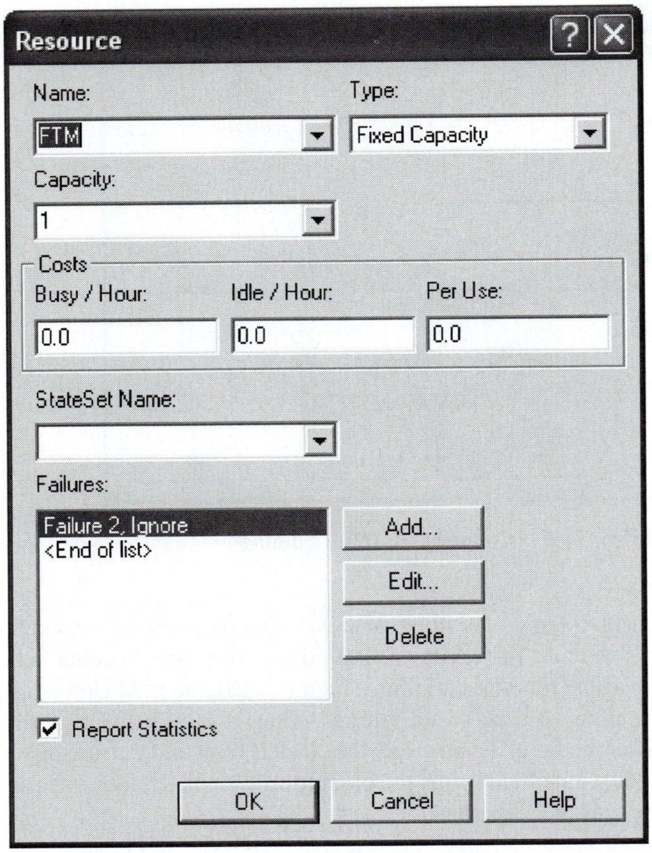

■ **FIGURE 7-11** Resource with failure attached

based on the file, *Smarts112-TimeAndCountFailures.doe*, which accompanies this chapter. In this file, resource animation is used to indicate that the resource is in the failed state. You are encouraged to run the model. You might also look at the *Smarts123.doe* file for other ideas about displaying failures.

When resources can be in a variety of states (idle, busy, inactive, and failed), it is often useful to be able to tabulate the time (and percentage of time) spent in the states. The next section presents how Arena™ facilitates the tabulation of these quantities using the frequency option of the STATISTIC module.

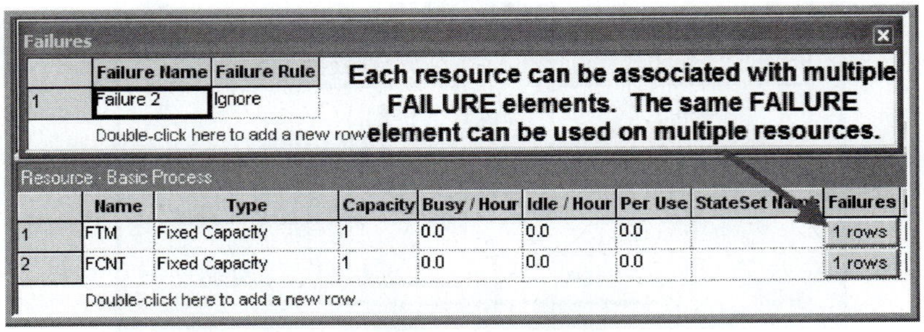

■ **FIGURE 7-12** Resources with failures spreadsheet view

7.2   Tabulating Frequencies Using the STATISTIC Module

Time-based variables in a simulation take on a particular value for a duration of time. For example, consider a variable like NQ(Queue Name) which represents the number of entities currently waiting in the named queue. If the variable NQ equals zero, the queue is considered empty. You might also want to tabulate the time (and percentage of time) that the queue was empty. This requires that you record when the queue becomes empty and when it is not empty. From this, you can summarize the time spent in the empty state. The ratio of the time spent in the state to the total time yields the percentage of time spent in the state and under certain conditions, this percentage can be considered the probability that the queue is empty.

To facilitate statistical collection of these quantities, use the frequencies option of the STATISTIC module in Arena™. With the frequency option, you specify either a value or a range of values over which you want frequencies tabulated. For example, suppose that you want to know the percentage of time that the queue is empty, that it has 1 to 5 customers, and 6 to 10 customers. In the first case (empty), you need to tabulate the total time for which NQ equals 0. In the second case, you need to tabulate the total time for which there were 1, 2, 3, 4, or 5 customers in the queue. This second case can be specified by a range. In Arena's frequency statistics, the range is specified such that it does not include the lower limit. For example, if LL represents the lower limit of the range and UL represents the upper limit of the range, the time tabulation occurs for the values in (LL, UL]. Thus, in order to collect the time that the queue has 1 to 5 customers, you would be specify a range (0, 5]. Frequencies can also be tabulated over the states of a resource, essentially using the STATE(Resource Name) variable.

The file, *Smarts112-TimeAndCountFailuresWithFrequencies.doe* that accompanies this chapter illustrates the use of the frequencies option. Figure 7-13 shows how to collect frequencies on a queue and on a resource. In the figure, the categories for the frequency tabulation on the queue have been specified. There have been three categories defined (empty, 1 to 5, and 6 to 10). Since the empty category is defined solely on a single value, it is considered as a constant. The other two categories have been defined as ranges. The user needs to provide the value in the case of the constant or a range of values, as in the case of a range. Note how the range for 1 to 5 has a low value specified as 0 and a high value specified as 5. This is because the lower value of the range is not included in the range. The specification of categories is similar to specifying the bins in a histogram.

Arena™ will tabulate the following quantities for a frequency (as per Arena's help files).

FAVG (Frequency ID, Category): Average time in category. FAVG is the average time that the frequency expression has had a value in the specified category range. FAVG equals FRQTIM divided by FCOUNT.

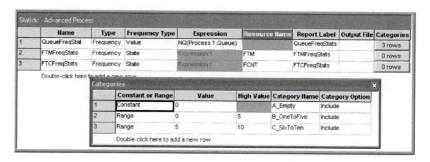

■ FIGURE 7-13   Frequency option in STATISTIC module

FCATS (Frequency ID): Number of categories. FCATS returns the number of categories of a frequency, including the out-of-range category. FCATS is an integer value.

FCOUNT (Frequency ID, Category): Frequency category count. FCOUNT is the number of occurrences of observations for the Frequency Number of values in the Category number; it is an integer value. Only occurrences of time > 0 are counted.

FHILIM (Frequency ID, Category): Frequency category high limit. FHILIM is the upper limit of a category range or simply the value if no range is defined for the particular Category number of Frequency Number. FHILIM is user-assignable.

FLOLIM (Frequency ID, Category): Frequency category low limit. FLOLIM defines the lower limit of a frequency category range. Values equal to FLOLIM are not included in the Category; all values larger than FLOLIM and less than or equal to FHILIM for the category are recorded. FLOLIM is user-assignable.

FSTAND (Frequency ID, Category): Standard category percent. FSTAND calculates the percent of time in the specified category compared to the time in all categories.

FRQTIM (Frequency ID, Category): Time in category. FRQTIM stores the total time of the frequency expression value in the defined range of the Category number.

FRESTR (Frequency ID, Category): Restricted category percent. FRESTR calculates the percent of time in the specified category compared to the time in all restricted categories.

FTOT (Frequency ID): Total frequency time. FTOT records the total amount of time that frequency statistics have been collected for the specified Frequency Number.

FTOTR (Frequency ID): Restricted frequency time. FTOTR records the amount of time that the specified Frequency Number has contained values in nonexcluded categories (i.e., categories that have a value in the restricted percent column).

FVALUE (Frequency ID): Last recorded value. FVALUE returns the last recorded value for the specified frequency. When animating a frequency histogram, it is FVALUE, not the FAVG, which is typically displayed.

Note that the number of occurrences or number of times that the category was observed is tallied as well as the time spent in the category. The user has the opportunity to include or exclude a particular category from the total time. This causes two types of percentages to be reported—standard and restricted. The restricted percentage is computed based on the total time that has removed any excluded categories.

Figure 7-14 illustrates the results from running the example model for 480 hours. Recall that in this model the FTC resource subject to random failures and that the FTM resource is

Replication 1	Start Time:	0.00	Stop Time:	480.00	Time Units: Hours
FTCFreqStats	Number Obs	Average Time	Standard Percent	Restricted Percent	
BUSY	456	0.08859380	8.42	8.42	
FAILED	48	0.3333	3.33	3.33	
IDLE	435	0.9738	88.25	88.25	
FTMFreqStats	Number Obs	Average Time	Standard Percent	Restricted Percent	
BUSY	343	1.1675	83.43	93.74	
FAILED	73	0.1077	1.64	1.84	
IDLE	46	0.4100	3.93	4.42	
INACTIVE	358	0.1476	11.01	--	
QueueFreqStats	Number Obs	Average Time	Standard Percent	Restricted Percent	
A_Empty	46	1.3034	12.49	12.88	
B_OneToFive	104	2.5785	55.87	57.60	
C_SixToTen	68	2.0211	28.63	29.52	
OUT OF RANGE	9	1.6058	3.01	--	

■ **FIGURE 7-14**   **Results when using Frequency option**

subject to both a schedule and random failures. From the figure, you can see that the FTC resource failed 48 times, with an average time spent failed of 0.3333 hours per failure. This resulted in about 3.33% of the time being failed or $480 \times 0.03333 = 15.9984$ hours. In the case of the FTM resource, the resource spent about 11% (or about 52.848 hours) of the total time inactive. When tabulating frequencies for a resource, the restricted percentage column automatically restricts the inactive state from its calculations. Thus, the value 93.74 for the restricted percent busy represents 93.74% of $(480 - 52.848 = 427.151$ hours) or about 400 hours. This is the same as 83.43% of 480. For the frequency tabulation on the number in queue, there are four not three categories listed. The last category, OUT OF RANGE, records frequencies when the value of the variable is not in one of the previously defined categories. Note that the categories do not have to define contiguous intervals. As indicated in the figure, the OUT OF RANGE category is automatically excluded from the restricted percentage column.

When using frequencies, Arena^{TM} will tabulate the frequencies for each replication. It does not automatically tabulate frequency statistics across replications. To tabulate across replications, you can define an OUTPUT statistic and use one of the previously discussed frequency functions (e.g., FRQTIM, FAVG, etc.). The frequency element allows finer detail on the time spent in user-defined categories. This can be very useful, especially when analyzing the effectiveness of resource staffing schedules.

Besides the tabulation of time, Arena^{TM} also facilitates the tabulation of cost. Recall from Figure 7-11 that there are three fields related to tabulating the cost of a resource. The next section discusses the use of these fields as well as how costs can be associated with entities.

7.3   Entity and Resource Costing

There are two types of cost-related information that can be used in Arena^{TM}: resource cost and entity cost. Both cost models support the tabulation of costs that can support an activity-based costing (ABC) analysis in an Arena^{TM} model. Activity-based costing is a cost management system that attempts to better allocate costs to products based on the activities they undergo within the firm rather than a simple direct and indirect cost allocation. The details of activity-based costing will not be discussed in this text; however, how Arena^{TM} tabulates costs will be examined so that you can understand how to utilize the cost estimates available from an Arena^{TM} model.

Arena's resource costing allows costs to be tabulated based on the time a resource is busy or idle. In addition, a cost can be assigned to each time the resource is used. Let's take a look at a simple example to illustrate these concepts. In the Arena^{TM} SMARTS file, *Smarts019.doe*, the processing of bill payments is modeled. The bills arrive according to a Poisson process with a mean of one bill every minute. The bills are first sent to a worker who handles the bill processing, which requires processing time that is distributed according to a triangular distribution with parameters (0.5, 1.0, 1.5) minutes. Then the bills are sent to a worker who handles bill mailings. The processing time for the mailing also occurs according to a triangular distribution with parameters (0.5, 1.0, 1.5) minutes. Both workers are paid an hourly wage regardless of whether they are busy. The bill processing worker is paid $7.75/hour and the mailing worker is paid $5.15/hour. In addition, the workers are paid an additional $0.02 for each bill that they process. Management would like to tabulate the cost of this processing over an 8-hour day.

It should be clear that you can get the wage cost by simply multiplying the hourly wage by 8 hours because the workers are paid regardless of whether they are busy or idle during the period. The number of bills processed will be random and thus the total cost must be simulated. Arena^{TM} can also facilitate calculation of workers getting paid differently if they are busy or

	Name	Type	Capacity	Busy / Hour	Idle / Hour	Per Use	StateSet Name	Failures	Report Statistics
1	Biller	Fixed Capacity	1	7.75	7.75	.02		0 rows	✓
2	Mailer	Fixed Capacity	1	5.15	5.15	.02		0 rows	✓
	Double-click here to add a new row.								

■ **FIGURE 7-15** Specifying resource costs

idle. In this example, Arena^{TM} can do all the calculations if the costs in the RESOURCE module are specified. The model is very simple (CREATE, PROCESS, PROCESS, DISPOSE). See the *Smarts019.doe* file for details of the various dialog boxes. To implement the resource costing, you must specify the RESOURCE costs as per Figure 7-15.

In the Run Setup dialog, make sure that the Costing check-box is enabled on the Project Parameters tab. This will ensure that the statistical reports include the cost reports. Arena^{TM} will tabulate the costs and present a summary across all resources and allow you to drill down to get specific cost information for each resource for each category (busy, idle, and usage) as shown in Figure 7-16.

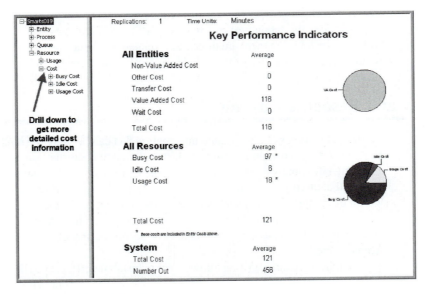

■ **FIGURE 7-16** Resource cost summary

■ **TABLE 7-3**

Tabulation of Resource Costs

Resource cost	$/Hour	Hours	Cost	Totals
Biller	7.75	8	62	
Mailer	5.15	8	41.2	
			103.2	
Usage Cost	$/Usage	# Uses		103.2
Biller	0.02	456	9.12	
Mailer	0.02	456	9.12	
				18.24
			Total cost	121.44

■ TABLE 7-4
Tabulation of Busy and Idle Costs

	Util	Time busy	$/Hour	Cost	Total
Biller	0.9374	7.50	7.75	58.12	
Mailer	0.9561	7.65	5.15	39.39	
				97.51	
	Idle	Time idle			97.51
Biller	0.0626	0.50	7.75	3.88	
Mailer	0.0439	0.35	5.15	1.81	
					5.69
				Total Cost	103.20

Tables 7-3 and 7-4 indicate how the costs are tabulated from Arena's statistical reports for the resources. In Table 7-3, the number of uses is known because there were 435 entities (bills) processed by the system. In Table 7-4, the utilization from the resource statistics was used to estimate the percentage of time busy and idle. From this, the amount of time can be calculated and from that the cost. Arena^{TM} tabulates the actual amount of time in these categories. The value-added cost calculation will be discussed shortly.

To contrast this consider *Smarts049.doe* in which the billing worker and mailing workers follow a schedule as shown in Figures 7-17 and 7-18. Because the resources have less time available to be busy or idle the costs are less for this model as shown in. In this case, the workers are not paid when they are not scheduled.

	Name	Type	Schedule Name	Schedule Rule	Busy / Hour	Idle / Hour	Per Use
1	Biller	Based on Schedule	Schedule 1	Wait	7.75	7.75	.02
2	Mailer	Based on Schedule	Schedule 1	Wait	5.15	5.15	.02

Double-click here to add a new row.

■ FIGURE 7-17 *Smarts049* with resources based on schedule

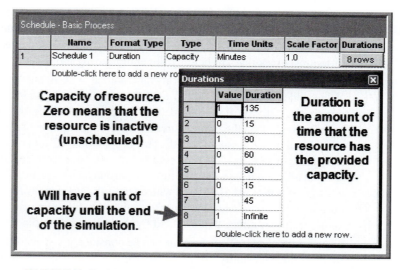

■ FIGURE 7-18 Schedule for *Smarts049*

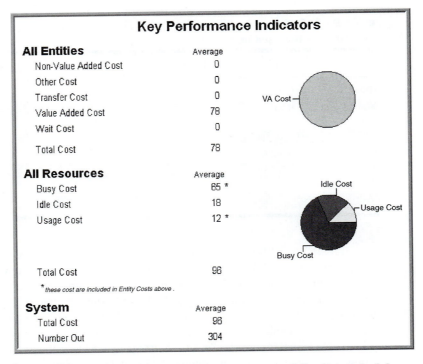

■ FIGURE 7-19 Summary costs for resources following a schedule

As seen in Figure 7-19, Arena™ also tabulates costs for the entities. Let's take a look at another example to examine how entity costs are tabulated. Entity costing is slightly more complex than resource costing. Arena™ assigns entity costs into five different activity categories.

- Waiting time/cost: Wait time is any time designated as Wait. Waiting time in queues is by default allocated as waiting time. Waiting cost is the cost associated with an entity's designated wait time. This cost includes the value of the time the entity spends in the waiting state and the value of the resources that are held by the entity during the time.

- Value added time/cost: Value-added time is any time designated as Value added. Value-added time directly contributes value to the entity during the activity. For example, the insertion of components on a printed circuit board adds value to the final product. Value-added cost is the cost associated with an entity's designated value-added time. This cost includes the value of the time the entity spends in the value-added activity and the value of the resources that are held by the entity during the time.

- Non–value-added time/cost: Non–value-added time is any time designated as Non–Value added. Non–value-added time does not directly contribute value to the entity during the activity. For example, the preparation of the materials needed to insert the components on the printed circuit board (e.g., organizing them for insertion) can be considered as non–value added. The designation of non–value-added time can be subjective. As a rule, if the time could be reduced or eliminated and efficiency increased without additional cost, then the time can be considered as non–value added. Non–value-added cost is the cost associated with an entity's designated non–value-added time. This cost includes the value of the time the entity spends in non–value-added activity and the value of the resources that are held by the entity during the time.

- Transfer time/cost: Transfer time can be considered a special case of non–value-added time in which the entity experiences a transfer. By default, all time spent using material-handling devices from the Advanced Transfer panel (such as a conveyor or transporter) is

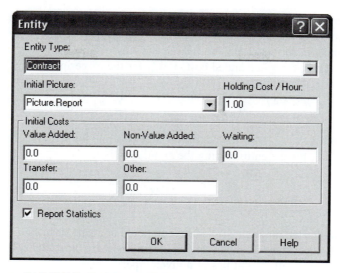

■ **FIGURE 7-20** **ENTITY module costing terms**

specified as Transfer time. Transfer cost is the cost associated with an entity's designated transfer time. This cost includes the value of the time the entity spends in the designated transfer activity and the value of the resources that are held by the entity during the time.

■ Other time/cost: This category is a catch-all for any activities that do not naturally fit in the other four categories.

As mentioned, Arena™ designates wait and transfer time automatically based on the constructs being used. In the case of the transfer time category, you can also designate specific delays as transfer time, even if a material handling device is not used. In the modeling, it is incumbent on the modeler to properly designate the various activities in the model; otherwise, the resulting cost tabulations will be meaningless. Thus, Arena™ does not do the cost modeling for you; however, it tabulates the costs automatically for you. Don't forget the golden rule of computing: garbage in = garbage out. When you want to use Arena™ for cost modeling, you need to *carefully* specify the cost elements for the *entire* model. Let's take a closer look at how Arena™ tabulates the costs.

Entity cost modeling begins by specifying the cost rates for the types of entities in the ENTITY module. In addition, to get a meaningful allocation of the resource cost for the entity, you need to specify the costs of using the resource as per the RESOURCE module. As seen in Figure 7-20, the ENTITY module allows the user to specify the holding cost of the entity as well as initial costs for each of the five categories. Let's ignore the initial costs for a moment and examine the meaning of the holding cost/hour field. The holding cost per hour represents the cost of processing the entity *anywhere* in the system. This is the dollar value of having the entity in the system expressed as a time rate. For example, in an inventory model, you might consider this the cost of holding one unit of the item in the system for 1 hour. This cost can be difficult to estimate. It is typically based on a cost of capital argument; see Silver et al. (1998) for more on estimating this value. Provided the holding cost per hour for the entity is available and the resource costs are specified for the resources used by the entity, Arena™ can tabulate the costs.

Let's consider tabulating the value-added cost for an entity. As an entity moves through the model, it will experience activities designated as value added. Let n be the number of value-added activity periods experienced by the entity while in the system. Let VAT_i be the value-added time for period i and h be the holding cost rate for the entity's type. While the entity is experiencing the activity, it may be using various resources. Let R_i be the set of resources used by the entity in activity period i. Note that the cardinality of R_i must be less than or equal to the total number of resources in the model. Let b_j be the busy cost per hour for the jth resource held by the entity in period i. Let u_j be the usage cost associated with the jth resource used during the activity. Thus, the value-added cost, VAC_i, for the ith period is

$$VAC_i = h \times VAT_i + \left(\sum_{j \in R_i} b_j \right) \times VAT_i + \sum_{j \in R_i} u_j = \left(h + \sum_{j \in R_i} b_j \right) \times VAT_i + \sum_{j \in R_i} u_j$$

The quantity, $\sum_{j \in R_i} b_j$, is called the resource cost rate for period i. Thus, the total value-added cost, $TVAC$, for the entity during the simulation is

$$TVAC = \sum_{i=1}^{n} VAC_i$$

The costs for the other categories are computed in a similar fashion by noting the number of periods for the category and the associated time spent in the category by period. These costs are then totaled for all the entities of a given type.

The initial costs as specified in the ENTITY module are treated in a special manner by Arena's cost reports. Arena's help system has this to say about the initial costs:

The initial VA cost, NVA cost, waiting cost, transfer cost and other cost values specified in this module are automatically assigned to the entity's cost attributes when the entity is created. These initial costs are not included in the system summary statistics displayed in the Category Overview and Category by Replication reports. These costs are considered to have been incurred outside the system and are not included in any of the All Entities Cost or the Total System Cost. These initial costs are, however, included in the cost statistics by entity type.

To illustrate entity cost modeling, consider SMARTS file *Smarts047.doe*. In this model contracts arrive according to a Poisson process with a mean rate of one contract per hour. There are two value-added processes on the contract: an addendum is added to the contract and a notary signs the contract. These processes each take TRIA(0.5, 1, 1.5) hours to complete. The contract processor has an $8/hour busy/idle cost. The notary is paid on a per usage basis of $10 per use. The holding cost is $1 per hour for the contracts. These values are shown in Figures 7-20 and 7-21. Figure 7-22 illustrates how to allocate the value-added time in the contract addendum process. By running this model with one entity, you can more easily see how the value-added costs are tabulated. If you run the model, the value-added cost for one entity is about $19.

Figure 7-23 indicates that the value-added processing time for the contract at the addendum process was 55.4317 minutes. This can be converted to hours as shown in Table 7-5. Then the cost per hour for the entity in the system (holding cost plus resource cost) is tabulated (e.g., $1 + $8 = $9). This is multiplied by the value-added time in hours to get the cost of the

	Name	Type	Capacity	Busy / Hour	Idle / Hour	Per Use
1	Processor ▼	Fixed Capacity	1	8.00	8.00	0.0
2	Notary	Fixed Capacity	1	0.0	0.0	10.00

Double-click here to add a new row.

■ **FIGURE 7-21**   Resource costs for *Smarts047.doe*

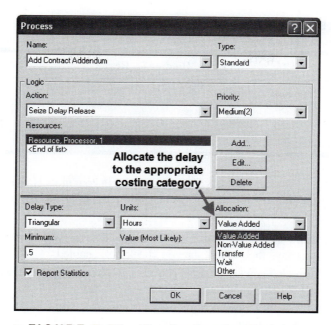

■ **FIGURE 7-22**   Allocating the value-added time

process. If the resource for the process has a resource cost then it must be included as shown in Table 7-5 and as explained in the equation for VAC_i.

To obtain the cost for the entire simulation, this calculation would be done for each contract entity for every value-added activity experienced by the entity. In the case of the single entity, the total cost of $19.16 for the addendum and notary activities is shown in Table 7-5.

Process

Time per Entity

VA Time Per Entity

	Average
Add Contract Addendum	55.4317
Sign and Notarize Contract	50.6456

Wait Time Per Entity

	Average
Add Contract Addendum	0.00
Sign and Notarize Contract	0.00

Total Time Per Entity

	Average
Add Contract Addendum	55.4317
Sign and Notarize Contract	50.6456

■ **FIGURE 7-23**   Processing time

■ **TABLE 7-5**
Tabulation of Entity Cost

Process	VAT (minutes)	VAT (hours)	Holding Cost ($/hr)	Resource Cost ($/hr)	Cost ($/hr)	Usage Cost	Total Cost	Cost
Addendum	55.4317	0.9238617	$1.00	$8.00	$9.00	$8.31	$0.00	$8.31
Notary	50.6456	0.8440933	$1.00	$0.00	$1.00	$0.84	$10.00	$10.84
								$19.16

Again, when using Arena's costing construct in your models you should be extra diligent in entering the information and in understanding the effect of specifying entity types. The following issues should be carefully considered:

- Make sure that only those DISPOSE modules for which you want entity statistics tabulated have their entity statistic check-box enabled.
- Make sure that you use the ENTITY data module and specify the type of entity to create in your CREATE modules.
- Carefully specify the allocation for your activities. For example, if you want an entity's value-added cost, then you must allocate to value added for every pertinent value-added activity that the entity may experience in the model.
- Be sure to read carefully how the BATCH and SEPARATE module's handle the assignment of entity attributes so that the proper attribute values are carried by the entities. Specify the attribute assignment criteria that best represents your situation.
- Read carefully how Arena™ handles PROCESS submodels and regular submodels in terms of cost tabulation if you use submodels. See "submodel processing" in Arena's help system.
- Remember also that you don't have to use Arena's costing models. You can specify your cost tabulations directly in the model and tabulate only the things that you need. In other words, you can do it yourself.

The next section outlines a number of other constructs in Arena™ that can enhance your models.

7.4   Miscellaneous Modeling Concepts

This section discusses a number of additional constructs available in Arena™ that facilitate common modeling situations that have not been previously described. The first topic allows for finer control of the routing of entities to stations for processing. The second situation involves improving the scalability of your models by making stations generic. Lastly, the use of the active entity to pick up and drop off other entities will be presented. This allows for another way to form groups of entities that travel together that is different than the functionality available in the BATCH module.

7.4.1   Picking Stations

Imagine that you have just finished your grocery shopping and that you are heading towards the check-out area. You look ahead and scan the checkout stations in order to pick what you think will be the line that will get your check-out process completed the fastest. This is a very common situation in many modeling instances: the entity is faced with selecting from a number

ALLEN-BRADLEY · **ROCKWELL SOFTWARE** · DODGE · RELIANCE ELECTRIC   **Rockwell Automation**

The PickStation Module example

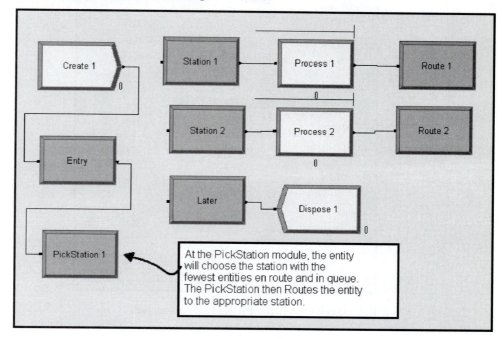

■ **FIGURE 7-24   PICKSTATION module example**

of available stations based on the state of the system. While this situation can be handled with the use of DECIDE modules, the implementation of many criteria to check can be tedious (and error prone). In addition, there are often a set of common criteria to check (e.g., shortest line, least busy resource, etc.). Because of this, the PICKSTATION module is available on the Advanced Transfer panel. The PICKSTATION module combines the ability to check conditions prior to transferring the entity out of a station.

Arena's SMARTS file, *Smarts113.doe*, is a good example of this situation. In the model, entities are created according to a Poisson process with a rate of one every minute. The arriving entities then must choose between two stations for service. In this case, their choice should be based on the number of entities en route to the stations and the current number of entities waiting at the stations. After receiving processing at the stations, the entities then depart the system.

As seen in Figure 7-24, the entity is created and then immediately enters the Entry station. From the Entry station, the entity will be routed via the PICKSTATION module to one of the two processing stations. The processing is modeled with the classic SEIZE, DELAY, RELEASE logic. After the processing, the entity is routed to the Later (exit) station where it is disposed. The only new modeling construct required for this situation is the PICKSTATION module.

The PICKSTATION module is very similar to a ROUTE module, as shown in Figure 7-25. The top portion of the module defines the criteria by which the stations will be selected. The bottom portion of the module defines the transfer out mechanism. In the figure, the module has been set to select based on the minimum of the number in queue and the number en route to the station. The Test Condition field allows the user to pick according to the minimum or the maximum. The Selection Based On section specifies the constructs that will be included in the selection criterion. The user must provide the list of

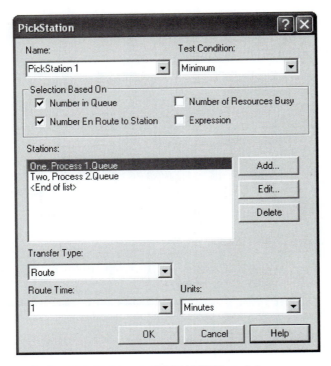

■ **FIGURE 7-25** **PICKSTATION module**

stations and the relevant construct to test at each station (e.g., the process queue). The selection of the construct is dependent on criteria that were selected in the Selection Based On area. Figure 7-26 illustrates the dialog box for adding a station to the list. Note that the specification is based on the station. In the figure, the user can select the station, a queue at

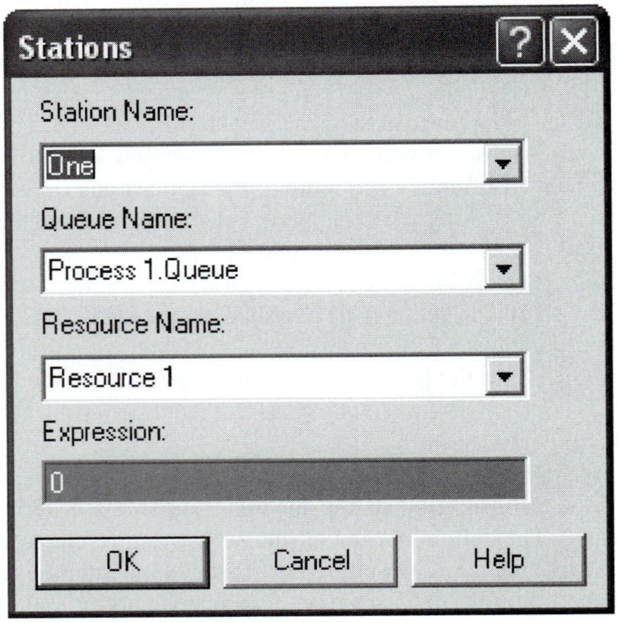

■ **FIGURE 7-26** **Station criteria specification**

the station, a resource at the station or a general expression to be evaluated based on the station. Arena^{TM} scans the list to find the station that has the minimum (or maximum) according to the criteria selected. In the case of ties, Arena^{TM} will pick the station that is listed first according the Stations list. After the station has been selected, the entity will be routed according to the standard route options (transport, route, convey, and connect). In the case of the connect option, the user is given the option to save the picked station in an attribute, which can then be used to specify the station in a transfer module (e.g., CONVEY). In addition, since the choice of transport or convey requires that the entity has control over a transporter or has accessed the conveyor, the entity must first have used the proper modules to gain control of the material-handling construct (e.g., REQUEST) prior to entering the PICKSTATION.

As you can see, the specification and operation of the PICKSTATION module is relatively straightforward. The only difficulty lies in properly using the Expression option.

Let's consider modifying the example to illustrate the use of an expression. Assume that the two resources in the example follow a schedule such that every hour the resources take a 10-minute break, with the first resource taking the break at the beginning of the hour and the second resource taking a break at the middle of the hour. This ensures that both resources are not on break at the same time. In this situation, arriving customers would not want to choose the station with the resource on break; however, if the station is picked based solely on the number in queue and the number en route, the entities may be sent to the station on break. This is because the number in queue and the number en route will be small (if not zero) when the resource is on break. Therefore, some way is needed to ensure that the station on break is not chosen during the break time. This can be done by testing to see if the resource is inactive. If it is inactive, the criteria can be multiplied by a large number so that it will not be chosen.

To implement this idea, two schedules were created, one for each resource to follow. Figure 7-27 shows the two schedules. In the schedule for resource 1, the capacity is 0 for the first 10 minutes and is then 1 for the last 50 minutes of the hour. The schedule then repeats. For the second resource's schedule, the resource is available for 30 minutes and then the break starts and lasts for 10 minutes. The resource is then available for the rest of the hour. This allows the schedule to repeat hourly.

Now the PICKSTATION module needs to be adjusted so that the entity does not pick the station with the resource that is on break. This implementation can be accomplished with the expression: *(STATE(resource name) = INACTIVE_RES)*vBigNumber*. In this case, if the named resource's state is inactive the Boolean expression yields a 1 for true. Then the 1 is multiplied by a big number (e.g., 10,000), which would yield the value 10,000 if the state is inactive. Since the station with minimum criteria is to be selected, this will penalize any station that is currently inactive. The implementation of this in the PICKSTATION module is

Durations			
		Value	Duration
1		0	10
2		1	50
Double-click here to			

Durations			
		Value	Duration
1		1	30
2		0	10
3		1	20

Schedule for Resource 1   **Schedule for Resource 2**

■ **FIGURE 7-27** Schedules for PICKSTATION extension

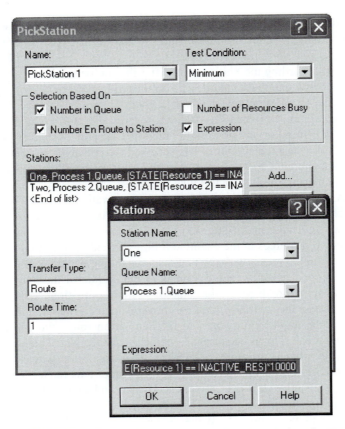

■ **FIGURE 7-28**   **Preventing a station from being chosen while inactive**

shown in Figure 7-28. If you run the model, *ch7Smarts113-PickStation-ResourceStaffing.doe*, that accompanies this chapter, you can see that the entity does not pick the station that is on break.

In many models (including this one), the station logic is essentially the same, SEIZE, DELAY, RELEASE. The next section considers the possibility of replacing many stations with a generic station that can handle any entity intended for a number of stations.

7.4.2   Generic Station Modeling

A generic station is a station (and its accompanying modules) that can handle the processing for any number of designated stations. To accomplish this, sets must be used to their fullest. Generic stations allow a series of commonly repeating modules to be replaced by a set of modules, sort of like a subroutine. However, the analogy to subroutines is not very good because of the way Arena™ handles variables and attributes. The trick to converting a series of modules to generic stations is to properly index into sets. An example is probably the best way to demonstrate this idea.

Let's reconsider the test-and-repair shop from Chapter 5 and shown in Figure 7-29. Note that the three testing stations in the middle of the model all have the same structure. In fact, the only difference between them is that the processing occurs at different stations. In addition, the parts are routed via sequences and have their processing times assigned in the SEQUENCE module. These two facts will make it very easy to replace those nine modules with four modules that can handle any part intended for any of the test stations.

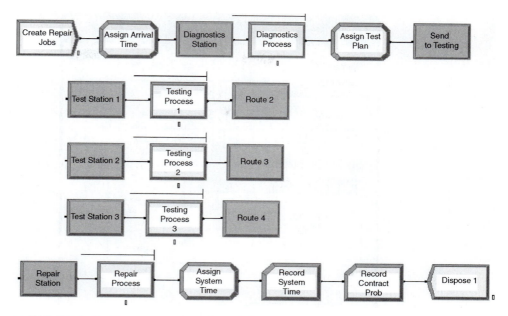

■ **FIGURE 7-29** **Test-and-repair model**

Figure 7-30 shows the generic version of the test-and-repair model. As shown in the figure, the testing stations have been replaced by the station name *Generic Station*. Note that the generic modeling comprises STATION, SEIZE, DELAY, RELEASE, ROUTE. The real trick of this model begins with set modeling.

Three sets must be defined for this model: a set to hold the test stations, a set to hold the test machine resources, and a set to hold the processing queues for the test machines. There are two ways to define a set of stations: using the Advanced Set module or using the set option in the STATION module. Since a generic station is being built, it is a bit more convenient to use the set option in the STATION module. Figure 7-31 illustrates the generic station module. Note how the station type field has been specified as Set. This allows a set to be defined to hold the stations to be represented by this station set. The station set members for the test stations have all been added. Arena™ will use this one module to handle any entity that is transferred to any of the listed stations. A very important issue in this modeling

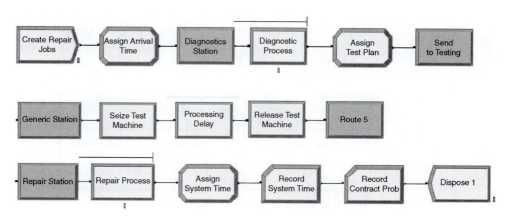

■ **FIGURE 7-30** **Generic test-and-repair model**

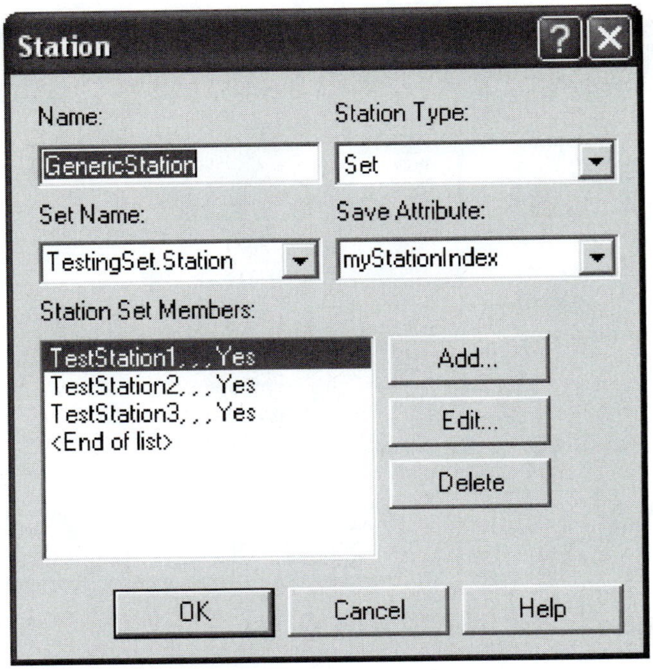

FIGURE 7-31 Generic station module with Set option

is the order of the stations listed. In the other sets that will be used, you need to make sure that the resources and queues associated with the stations are listed in the same order.

When an entity is sent to one of the stations listed, the station that the generic station currently represents must be remembered. This is accomplished with the Save Attribute. The saved attribute will record the *index* of the station in its set. That's important enough to repeat. It does not hold the station selected. It holds the index for the station in the station set. This index can be used to reference into a queue set and a resource set to make the processing generic. Figure 7-32 illustrates the resource and queue sets needed for the model. The resource set is defined using the SET module on the Basic Process panel, and the queue set is defined using the Advanced SET module on the Advanced Process panel. With these sets defined, they can be used in the SEIZE and RELEASE modules.

In the original model, a PROCESS module was used to represent the testing processes. In this generic model, the PROCESS module will be replaced with the SEIZE, DELAY, and RELEASE modules from the Advanced Process panel. The primary reason for doing this is so that the queue set can be accessed when seizing the resource.

Figure 7-33 illustrates the SEIZE module for the generic test-and-repair model. When seizing the resource, the Set option has been used by selecting a specific member from the

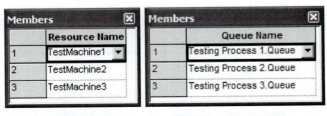

TestMachineSet   **TestMachineQSet**

FIGURE 7-32 Resource and queue sets for generic model

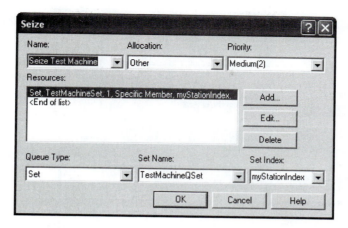

■ **FIGURE 7-33** SEIZE module for generic model

TestMachineSet as specified by the *myStationIndex*. Thus, if the entity had been sent to the test station 2, the *myStationIndex* would have the value 2. This value is then used as the index into the *TestMachineSet*. Now you should see why the order of the sets matters. In addition, if the entity must wait for the resource it needs to know which queue to wait in. The SEIZE module has an option to allow the queue to be selected from a set. In this case, the *TestMachineQSet* can be accessed by using the *myStationIndex*. Since there are multiple queues being handled by this one SEIZE module Arena™ takes away the default queue animation. If the animation of these queues is important to your modeling, then you can use the Animate toolbar to place animation queues in the model window.

The DELAY module for the generic station is shown in Figure 7-34. The DELAY module is simple to make generic because the SEQUENCE module is being used to assign to the *myTestingTime* attribute when the entity is routed to the station.

The RELEASE module for the generic station is shown in Figure 7-35. In this case, the set option has been used to specify which resource to release. The *myStationIndex* is used to index into the *TestMachineSet* to release a specific member. The ROUTE module is just like the ROUTE module used in the original test-and-repair model. It routes the part using the sequence option.

If you run this model, found in the file *GenericRepairShop.doe*, the results will be duplicates of those seen in Chapter 5. Now, you might be asking yourself: "This generic stuff seems like a lot of work. Is it worth it?" Well, in this case, four flow chart modules were saved (five in the generic versus nine in the original model). Plus, three additional sets were added in the generic model. This doesn't seem like much of a savings. Now, imagine a realistically sized model with many more test stations, such as 10 or more. In the original model, to add more

■ **FIGURE 7-34** DELAY module for generic model

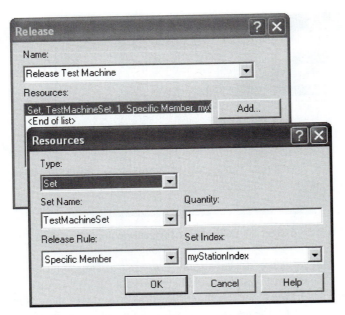

■ **FIGURE 7-35** **RELEASE module for generic model**

testing stations, you could "cut and paste" the STATION, PROCESS, ROUTE combination as many times as needed, updating the modules accordingly. In the generic model, all you need to do is add more members to the sets. You do not need to change the flow chart modules. All the changes required to run a bigger model would be made in the data modules (e.g., sets, sequences, etc.). This can be a big advantage when experimenting and running different model configurations. The generic model will be much more scalable.

Until you feel comfortable with generic modeling, I recommend the following. First, you should develop the model without the generic features. Once the model has been tested and is working as you intended, then look for opportunities to make the model generic. Consider making it generic if you think you can take advantage of the generic components during experimentation, and if you think reuse of the model is an important aspect of the modeling effort. If the model is only going to be used once to answer a specific question, then you probably don't need to consider generic modeling. Even in the case of a one-time model, if the model is very large, replacing parts of it with generic components can be a real benefit when working with the model. In this simple example, the generic modeling was easy because of the model's structure. This will not always be the case. Thus, generic modeling will often require more thinking and adjustments than illustrated in this example; however, I have found it very useful.

7.4.3 Picking Up and Dropping Off Entities

In many situations, entities must be synchronized so that they flow together in the system. Chapter 2 presented how the BATCH, SEPARATE, and MATCH modules can be used to handle these types of situations. Recall that the BATCH module allows entities to wait in queue until the number of desired entities enter the BATCH module. Once the batch is formed, a single representative entity leaves the BATCH module. The batch representative entity can be either permanent or temporary. In the case of a temporary entity, the SEPARATE module is used to split the representative entity into its constituent parts.

What exactly is a "representative entity"? The representative entity is an entity created to hold the entities that form the batch in an *entity group*. An entity group is simply a list of entities that have been attached to a representative entity. The BATCH module is not the only way to

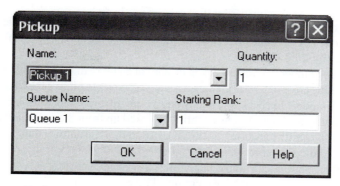

■ **FIGURE 7-36**   **PICKUP module**

form entity groups. In the case of a BATCH module, a new entity is created to hold the group. The newly created entity serves as the representative entity. In contrast, the PICKUP module causes the active entity (the one passing through the PICKUP module) to add entities from a queue to its entity group, thereby becoming a representative entity. To get entities out of the entity group, the DROPOFF module can be used. When the active entity passes through the DROPOFF module, the user can specify which entities in the group are to be removed (dropped off) from the group. Since the entities are being removed from the group, the user must also specify where to send the entities once they have been removed. Since the entities that are dropped off were represented in the model by the active entity holding their group, the user can also specify how to handle the updating/specification of the attributes of the dropped-off entities.

The PICKUP and DROPOFF modules are found on the Advanced Process Panel. Figures 7-36 and 7-37 illustrate the PICKUP and DROPOFF modules. In the PICKUP module, the

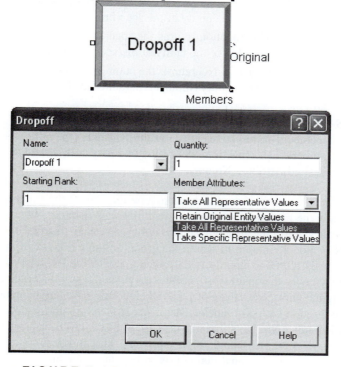

■ **FIGURE 7-37**   **DROPOFF module**

quantity field specifies how many entities to remove from the specified queue starting at the given rank. For example, to pick up all the entities in the queue, you can use NQ(queue name) in the quantity field.

The number of entities picked up cannot be more than the number of entities in the queue or a run time error will result. Thus, it is very common to place a DECIDE module in front of the PICKUP module to check how many entities are in the queue. This is very useful when picking up one entity at a time. When an entity is picked up from the specified queue, it is added to the group of the entity executing the PICKUP module. The entities that are picked up are added to the end of the list representing the group.

Since entities can hold other entities in a group, you may need to get access to the entities in the group during model execution. There are a number of special-purpose functions/ variables that facilitate working with entity groups. Information on the following functions is available in the Arena™ help system.

AG(Rank, Attribute Number): AG returns the value of the general-purpose attribute Attribute Number of the entity at the specified Rank in the active entity's group. The function NSYM may be used to translate an attribute name into the desired Attribute Number.

ENTINGROUP(Rank [, Entity Number]): ENTINGROUP returns the entity number (i.e., IDENT value) of the entity at the specified Rank in the group of entity representative Entity Number.

GRPTYPE[(Entity Number)]: When referencing the representative of a group formed at a BATCH module, GrpType returns 1 if it is a temporary group and 2 if it is a permanent group. If there is no group, then a 0 will be returned.

ISG(Rank): This function returns the special-purpose jobstep (Entity.Jobstep, or IS) attribute value of the entity at the specified Rank of the active entity's group.

MG(Rank): This function returns the special-purpose station (Entity.Station, or M) attribute value of the entity at the specified Rank of the active entity's group.

NSG(Rank): This function returns the special-purpose sequence (Entity.Sequence, or NS) attribute value of the entity at the specified Rank of the active entity's group.

NG[(Entity Number)]: NG returns the number of entities in the group of representative Entity Number. If Entity Number is defaulted, NG returns the size of the active entity's group.

SAG(Attribute Number): SAG adds the values of the specified Attribute Number of all members of the active entity's group. The data type of the specified attribute must be numeric. The function NSYM may be used to translate an attribute name into the desired Attribute Number.

Note that many of the functions require the rank of the entity in the group. That is, the location in the list holding the group of entities. For example, AG(1, NSYM(myArrivalTime)) returns the value of the attribute named *myArrivalTime* for the first entity in the group. Often a SEARCH module is used to search the group to find the index of the required entity.

The DROPOFF module is the counterpart to the PICKUP module. The DROPOFF module removes the quantity of entities starting at the provided rank from the entity's group. The original entity exits the exit point labeled Original. Any members dropped off exit the module through the point labeled Members. If there are no members in the group, then the original simply exits. Note that like the case of SEPARATE, you can specify how the entities being dropped off will handle their attributes. The option to Take Specific Representative Values causes an additional dialog box to become available that allows the user to specify which attributes the entities will get from the representative entity. The functions discussed earlier can be used on the members of the group to find the rank of specific members to be dropped off. Thus, entities can be dropped off singly or in the quantity specified. The representative entity will continue to hold the other members in the group. The special variable NG is very useful in determining how many entities are in the group. Again, the SEARCH module can be useful to search the group to find the rank of the entities that need to be dropped off.

   To illustrate the PICKUP and DROPOFF modules, the modeling of a simple shuttle bus system will be examined. In this system, there are two bus stops and one bus. Passengers arrive to bus stop 1 according to a Poisson arrival process with a mean rate of 6 per hour. Passengers arrive to bus stop 2 according to a Poisson arrival process with a mean rate of 10 per hour. Passengers arriving at bus stop 1 desire to go to bus stop 2, and passengers arriving to bus stop 2 desire to go to bus stop 1. The shuttle bus has 10 seats. When the bus arrives to a stop, it first drops off all passengers for the stop. The passenger disembarking time is distributed according to a lognormal distribution with a mean of 8 seconds and a standard deviation of 2 seconds per passenger. After all the passengers have disembarked, the passengers waiting at the stop begin the boarding process. The per passenger boarding time (getting on the bus and sitting down) is distributed according to a lognormal distribution with a mean of 12 seconds and a standard deviation of 3 seconds. Any passengers that cannot get on the bus remain at the stop until the next bus arrival. After the bus has completed the unloading and loading of the passengers, it travels to the next bus stop. The travel time between bus stops is distributed according to a triangular distribution with parameters (16, 20, 24) in minutes. This system should be simulated for 8 hours of operation to estimate the average waiting time of the passengers, the average system time of the passengers, and the average number of passengers left at the stop because the bus was full.

   On first analysis of this situation, it might appear that the way to model this situation is with a transporter to model the bus; however, Arena's mechanism for dispatching transporters does not facilitate the modeling of a bus route. In standard transporter modeling, the transporter is requested by an entity. As you know, most bus systems do not have the passengers request a bus pick up. Instead, the bus driver drives the bus along its route stopping at the bus stops to pick up and drop off passengers. The "trick" to modeling this system is to model the bus as an entity. The bus entity will execute the PICKUP and DROPOFF modules in Arena™. The passengers will also be modeled as entities. This will allow the tabulation of the system time for the passengers. In addition, since the passengers are entities, a HOLD module with an "infinite hold" option can be used to model the bus stops. That is really all there is to it! Actually, since the logic that occurs at each bus stop is the same, the bus stops can also be modeled using generic stations. Let's start with the modeling of the passengers.

   Figure 7-38 presents an overview of the bus system model. The entire model can be found in the file *ch7BusSystem.doe* associated with this chapter. In the figure, the top six modules represent the arrival of the passengers. The pseudo-code is straightforward:

CREATE passenger with time between arrivals exponential with mean 10 minutes
ASSIGN arrival time
HOLD until removed

The next four modules in Figure 7-38 represent the creation of the bus. The bus entity is created at time zero and assigned the sequence that represents the bus route. A sequence module was used called *BusRoute*. It has two job steps representing the bus stops, *BusStation1* and *BusStation2*. Then the bus is routed using a ROUTE module according to the By Sequence option. Since these are all modules that have been previously covered, all the details will not be presented; however, the completed dialogs can be found in the accompanying model file. Now let's discuss the details of the bus station modeling.

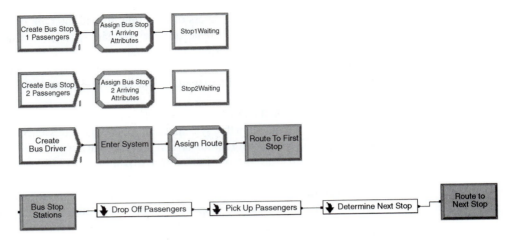

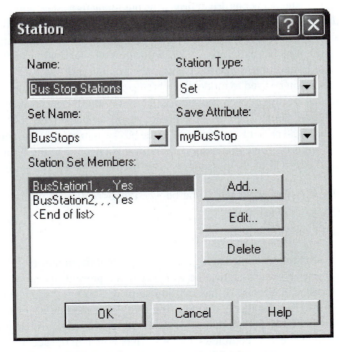

■ **FIGURE 7-38** Bus system overview

As previously mentioned, a generic station can be used to represent each bus stop. Figure 7-39 shows the generic station for the bus stops. The station type has been defined as "Set" and the stations representing the bus stops, *BusStation1* and *BusStation2* have been added to the station set. The key to the generic modeling will be the attribute, *myBusStop*, which will hold the index of the current station. In Figure 7-38, three submodels have been used to represent the drop off logic, the pick up logic, and the logic to decide on the next bus stop to visit. The logic for dropping off passengers is shown in Figure 7-40. This is a perfect place to utilize the WHILE-ENDWHILE blocks. Since the variable NG represents the number of entities in the active entity's group, it can be used in the logic test to decide whether to continue to drop off entities. This works especially well in this case since all the entities in the group will be dropped off. If the active entity (bus) has picked up any entities (passengers), then NG will be

■ **FIGURE 7-39** Generic station for bus stops

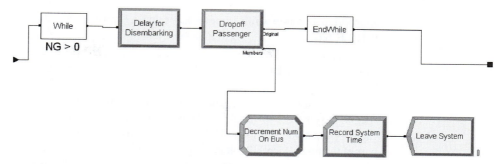

■ **FIGURE 7-40**   Drop off passengers submodel

greater than zero and the statements in the while construct will be executed. The bus experiences a DELAY to represent the disembarking time of the passenger and then the DROPOFF module (Figure 7-41) is executed. Note that in this case, since the system time of the passengers must be captured, the Retain Original Entity Values option can be used. The details of the rest of the modules can be found in the accompanying model file. In the case of a bus system with more stops, logic would be needed to search the group for passengers that want to get off at the current stop. The reader will be asked to explore this idea in the exercises.

The logic to pick up passengers is very similar to the logic for dropping off the passengers. In this case, a WHILE-ENDWHILE construct will again be used, as shown in Figure 7-42; however, there are a couple of issues that need to be handled carefully in this submodel.

In the drop off passengers submodel, the bus stop queues did not need to be accessed; however, when picking up the passengers, the passengers must be removed from the

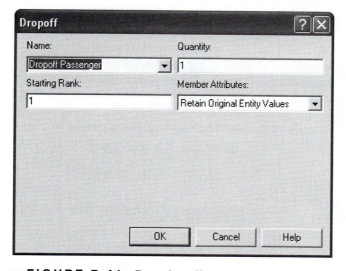

■ **FIGURE 7-41**   Dropping off passengers

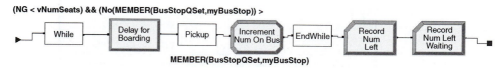

■ **FIGURE 7-42**   Pick up passengers submodel

appropriate bus stop waiting queue. Let's take a careful look at the logical test condition for the WHILE block:

> (NG < vNumSeats) && (NQ(MEMBER(BusStopQSet, myBusStop)) > 0)

A variable, *vNumSeats*, has been defined which represents the capacity of the bus. The special purpose group variable, NG, is compared to the capacity, as long as NG is less than the capacity the next passenger can be picked up from the bus stop queue. What about the second part of the test? Recall that it is a logical error to try to pick up an entity from an empty queue. Thus, the statement, NQ(queue name) > 0, tests to see if the queue holds any entities. Since generic modeling is being used here, the implementation must determine which queue to check based on the bus stop that the bus is currently serving. A Queue Set called *BusStopQSet* has been defined to hold the bus stop waiting queues. The queues have been added to this set in the same order as the bus stop stations were added to the bus stop station set. This ensures that the index into the set will return the queue for the appropriate station.

The function MEMBER(set name, index) returns the member of the set for the supplied index. Thus, in the case of *MEMBER(BusStopQSet,myBusStop))*, the queue for the current bus stop will be returned. In this case, both NG and NQ need to be checked so that passengers are picked up as long as there are waiting passengers and the bus has not reached its capacity. If a passenger can be picked up, the bus delays for the boarding time of the passenger and then the PICKUP block is used to pick up the passenger from the appropriate queue. The PICKUP block, shown in Figure 7-43, is essentially the same as the PICKUP module from the Advanced Process panel; however, with one important exception. The PICKUP block allows an expression to be supplied in the Queue ID dialog field that evaluates to a queue. The PICKUP module from the Advance Process panel does not allow this functionality, which is critical for the generic modeling.

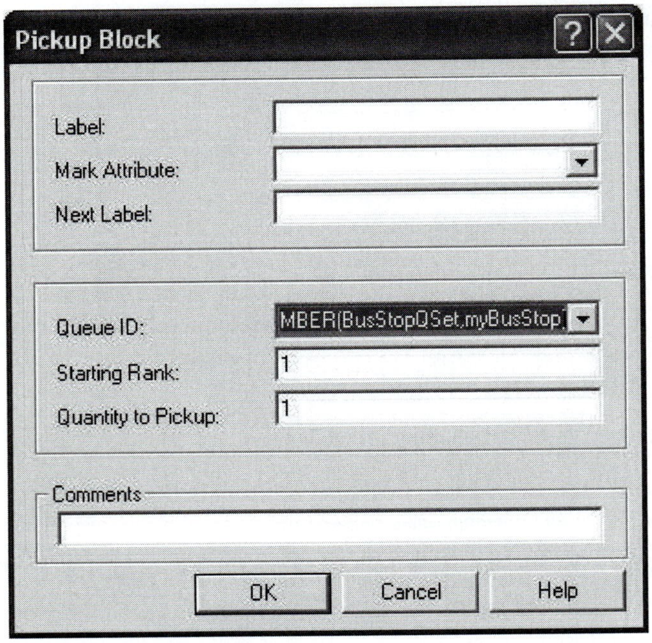

■ **FIGURE 7-43** PICKUP block

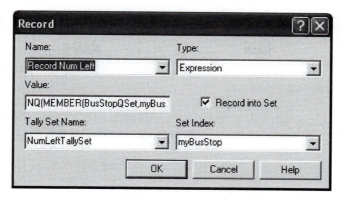

■ **FIGURE 7-44**   **Recording the number of passengers not picked up**

In this situation, the *MEMBER(BusStopQSet,myBusStop)* is used to return the proper queue in order to pick up the passengers. Each time the PICKUP block is executed, it removes the first passenger in the queue.

After the passengers are picked up at the stop, the logic to record the number left waiting is executed. In this case, a snap shot view of the queue at this particular moment in time is required. This is an observation-based statistic on the number in queue. Figure 7-44 shows the appropriate RECORD module. In this case, a tally set can be used so that the statistics can be collected by bus stop. The expression to be observed is NQ*(MEMBER(BusStopQSet, myBusStop))*.

Running the model for 10 replications of 8 hours yields the results shown in Figure 7-45. Note that the waiting time for the passengers is a little over 20 minutes for the two queues. In

Waiting Time	Average	Half Width
Stop1Waiting.Queue	20.7487	1.24
Stop2Waiting.Queue	23.2777	2.90

Other

Number Waiting	Average	Half Width
Stop1Waiting.Queue	2.0739	0.21
Stop2Waiting.Queue	3.9714	0.80

User Specified

Tally

Expression	Average	Half Width
Record Num Left Waiting	0.2567	0.29

Interval	Average	Half Width
Record System Time	42.9741	1.65

Usage

None	Average	Half Width
NumLeftQ1	0.00	0.00
NumLeftQ2	0.5182	0.58

■ **FIGURE 7-45**   **Bus system results**

addition, the results indicate that on average, there is less than one passenger left waiting because the bus is full. The reader will be asked to explore this model further in the exercises.

Based on this example, modeling with the PICKUP and DROPOFF modules is not too difficult. Additional functions (e.g., AG(), etc.) that have been not illustrated are especially useful when implementing more complicated decision logic that can select specific entities to be picked up or dropped off. In addition, with the generic representation for the bus system, it should be clear that adding additional bus stops can be readily accomplished.

The coverage of the major modeling constructs in Arena™ has been completed. The next section closes out this chapter with some additional insights into how to use some of the programming aspects of Arena™ in your modeling projects.

7.5   Programming Concepts in Arena™

In Arena™, programming support comes in two forms—laying down flow chart modules and computer language integration (e.g., VBA, C, etc.). This section presents some common programming issues that are helpful to understand when trying to get the most out of your models. The discussion involving input and output started in Chapter 2 will be continued. Then, the use of Visual Basic for Applications (VBA) in the Arena™ environment will be introduced.

7.5.1   Working with Files, Excel, and Access

As discussed in Chapter 2, Arena™ allows the user to directly read from or write to files in a model during a simulation run by using the READWRITE Module located on the Advanced Process panel. Using this module, the user can read from the keyboard, read from a file, write to the screen, or write to a file. When reading from or writing to, the user must define an Arena™ File Name and an optionally a file format. The Arena™ File Name is the internal name (within the model) for the external file defined by the operating system. The internal file name is defined with the FILE data module. Using the FILE data module, the user can specify the name and path to the external file. The file format can be specified either in the FILE data module or in the READWRITE module to override the specification given in the FILE data module. The format can be free format, a valid C or FORTRAN format, WKS for Lotus spreadsheets, Microsoft Excel, Microsoft Access, and ActiveX Data Objects Access types. In order for the READWRITE module to operate, an entity must pass through the module. Thus, as demonstrated in Chapter 2, it is typical to create a logic entity at the appropriate time of the simulation to read from or write to a file. This section presents examples of the use of the READWRITE module. Then, the pharmacy model is extended to read parameters from a database and write statistical output to a database.

7.5.2   Reading from a Text File

In this section, Arena's SMART file, *Smarts162.doe*, is used to show how to read from a text file (Figure 7-46). Open the file named *Smarts162Revised.doe*.

In this model, entities arrive according to a Poisson process, the type of entity and thus the resulting path through the processing is determined via the values in the *simdat.txt* file, shown in Figure 7-47. The first number in the file is the type (1 or 2). Then, the following two numbers are the station 1 and station 2 processing times, respectively. In the READWRITE module reads directly from the SIMDAT file using a free format. Each time a read occurs, the attributes (*myType*, *myStation1PT*, *myStation2PT*) are read in. The *myType* attribute is tested in the

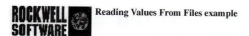

Reading Values From Files example

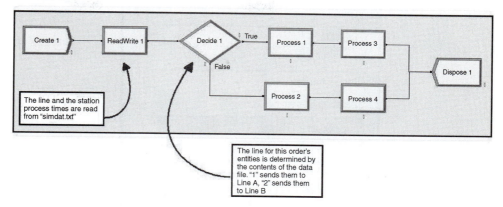

■ FIGURE 7-46   Arena Smarts 162 model

■ FIGURE 7-47   Sample processing times in *simdat.txt* file

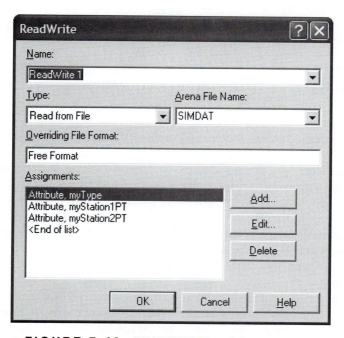

■ FIGURE 7-48   READWRITE module

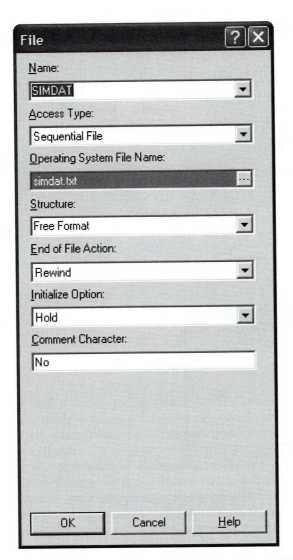

■ **FIGURE 7-49** FILE module for *Smarts162Revised.doe*

DECIDE module and the attributes *myStation1PT* and *myStation2PT* are used in the PROCESS modules.

In Figure 7-49, the end-of-file action specifies what to do with the entity when the end of the file is reached. The Error option can be used if an unexpected EOF condition occurs. The Dispose option may be used to dispose of the entity, close the file or ADO record set and stop reading from the file. The Rewind option may be used so that every time you reach an EOF, you start reading the file or record set again from Record 1. Finally, the Ignore option can be used if you expect an EOF and want to determine your own action (such as reading another file). With the Ignore option when the EOF is reached, all variables read in the READWRITE module will be set to 0.

The READWRITE module can be followed with a DECIDE module to ensure that if all values are 0, the desired action is taken. The "Comment Character" is useful to embed lines in the file that are skipped. These lines can add columns headers, comments, etc. to the file for easier readability of the file. Suppose the comment character was a semicolon (;):

```
; This would be a comment followed by a header comment
; Num Servers      Arrival Rate
5        10
```

The Initialize option indicates what Arena™ should do at the beginning of a replication. The file can stay at the current position (Hold), be rewound to the beginning of the file (Rewind), or the file can be closed (Close).

The rest of the model is straightforward. In this case, the values from the file are read into the attributes of an entity. The values of an array can also be read in using this technique.

7.5.3   Reading a Two-Dimensional Array

The Smarts file, *Smarts164.doe*, shows how to read into a two-dimensional array. The CREATE module creates a single entity and the ASSIGN module initializes the array index. Then, iterative looping, see Figure 7-50, is performed using a DECIDE module as previously discussed in Chapter 2.

In this particular model, the entity delays for 10 minutes before looping to read in the next values for the array. The assignments for the READWRITE module are shown in Figure 7-51. You can easily see how the use of two WHILE-ENDWHILE loops could allow for reading in the size of the array. You would first read in the number of rows and columns and then loop through each row/column combination.

7.5.4   Reading from an Excel Named Range

So far the examples have focused on text files and for simple models these will often suffice. For more user friendly input of the data to files, you can read from Excel files. The Smarts file *Smarts185.doe* (Figure 7-52) shows how to read in values from an Excel named range.

In order to read or write to an Excel named range, the named range must already be defined in the spreadsheet. Open the Excel file *Smarts185.xls*. Select cells C5:E6 and look in the

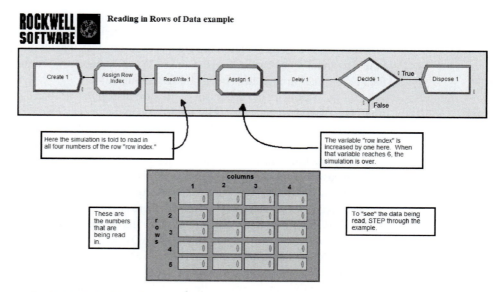

■ **FIGURE 7-50**   *Smarts164.doe* **reading in a 2-D array**

	Type	Other
Assignments		
1	Other ▼	DupNum(Row Index,1)
2	Other	DupNum(Row Index, 2)
3	Other	DupNum(Row Index, 3)
4	Other	DupNum(Row Index, 4)
	Double-click here to add a new row.	

■ **FIGURE 7-51**   **READWRITE assignments module using arrays**

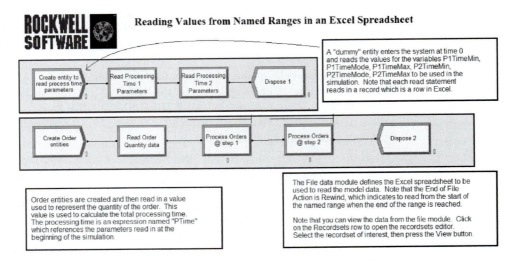

■ **FIGURE 7-52**   *Smarts185.doe* **reading from an Excel named range**

upper left corner of the Excel input bar. You will see the named range. By selecting any cells in Excel you can name them by typing the name into the named range box as indicated in Figure 7-53. You can organize your spreadsheet in any way that facilitates the understanding of your data input requirements and then name the cells required for input. The named ranges are then accessible through the Arena™ FILE module.

In this example, the processing times are distributed according to a triangular distribution. The named spreadsheet cells hold the parameters of the triangular distribution (min, mode, max) for each of the two PROCESS modules. An EXPRESSION module was used to define expressions with the variables to be read in indicating the parameter values. The order quantities are how much the customer has ordered. The lower CREATE module creates 50 arriving customers, where the order quantity is read and then used in the processing times. In the use of named ranges, essentially the execution of each READWRITE module causes a new row to be read. As indicated in Figure 7-52, there are two back-to-back READWRITE modules attached to the upper CREATE module that implements logic to read in each of the rows associated with the processing times. The READWRITE module attached to the "Create Order entities" CREATE module represents logic to allow each new entity to read in a new row from the named range.

After setting up the spreadsheet and defining the named ranges in Excel, you must then indicate to Arena™ how to associate a file to the named range. This is accomplished through

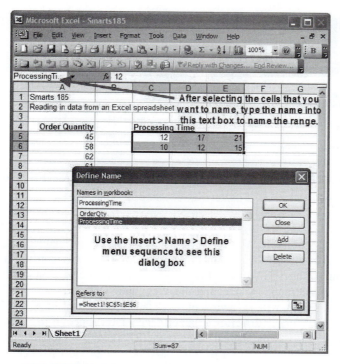

■ **FIGURE 7-53**   Named range in Excel

the use of the FILE module. Select the FILE module in *Smarts185.doe*. The FILE module is already defined for you (shown in Figure 7-54), but the steps will be indicated here. In the spreadsheet data view, double-click to add a new row and fill in AccessType, Operating System File Name, End of File Action, and Initialize Option the same as in the previous row. Then, click on the define Recordsets row button.

   You will see a dialog box similar to Figure 7-55. Arena™ is smart enough to connect to Excel and to populate the Named Range text box with the named ranges that you defined for

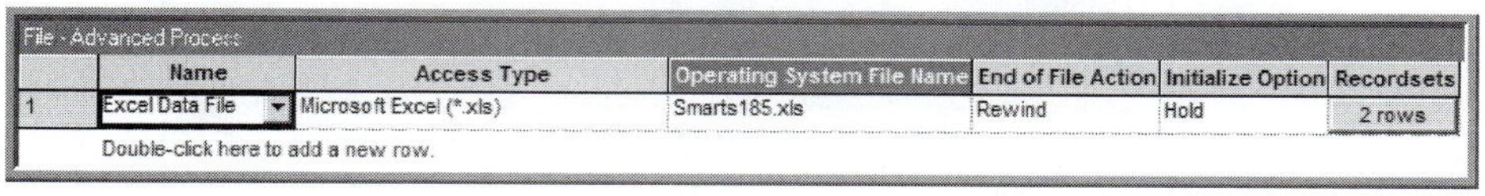

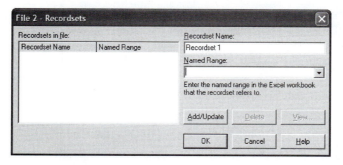

■ **FIGURE 7-54**   Datasheet view of FILE module

■ **FIGURE 7-55**   Record set defining for FILE module

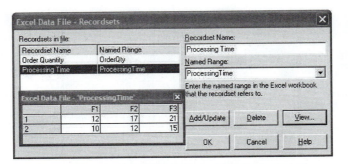

■ **FIGURE 7-56** Viewing the *ProcessingTime* named range in Arena

your Excel workbook. You need only select your desired named ranges and click on Add/ Update. This will add the named range as a record set in the Recordsets file area. You should add both named ranges as shown in Figure 7-56. Once you select a named range you can also view the data in the range by selecting the View button. If you try this with the *ProcessingTime* named range, you will see a view of the data similar to that shown in Figure 7-56. As you can see, it is very simple to define a named range and connect it to Arena™. Now, you have to indicate how to use the named range in the READWRITE module. Open the READWRITE module (see Figure 7-57) for reading the processing time 1 parameters. After selecting read from file and the associated Arena™ file name, Arena™ will automatically recognize the file as a named range-based Excel file. You can then choose the record set that you want to read from and then the module works essentially as it did in our text file examples. You can also use these procedures to write to Excel named ranges and to Excel using active data objects (ADOs). See for example SMARTS files 189 and 190.

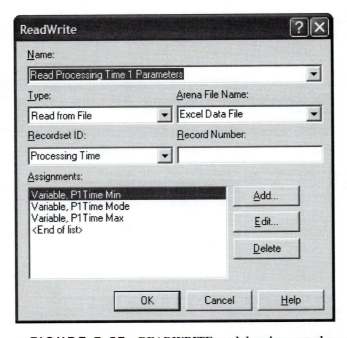

■ **FIGURE 7-57** READWRITE module using named range

Making the simulation "file driven" requires special planning focused on how to organize the model to take advantage of the data input. Using sets and arrays appropriately is often necessary when making a model "file driven." Some ideas for using files follow.

- Reading in the parameters of distributions
- Reading in model parameters (e.g., reorder points, order quantities, etc.).
- Reading in specified sequences for entities to follow. Define a set of sequences and read in the index into the set for the entity to follow. You can then define different sequences based on a file.
- Reading in different expressions to be used. Define a set of expressions and read in the index into the set to select the appropriate expression.
- Reading in a number of entities to create, and then creating them using a SEPARATE module.
- Creating entities and assigning their attributes based on a file.
- Reading in each entity from a file to create a trace-driven simulation.

7.5.5   Reading Model Variables from Microsoft Access

This section uses the pharmacy model and augments it to allow the parameters of the simulation to be read in from a database. In addition, the simulation will write out statistical data at the end of each run. This example involves the use of Microsoft Access, if you do not have Access then just follow along with the text. In addition, the creation of the database shown in Figure 7-58 will not be discussed. The database has two tables—one to hold the inputs and one to hold the output across simulation replications. Each row in the table *InputTable* has three fields. The first field indicates the current replication, the second field indicates the mean time between arrivals for the experiment and the last field indicates the mean service time for the experiment. This example will only have a total of three "replications"; however, each replication is *not* going to be the same. At the beginning of each replication, the parameters for the replication will be read in and then the replication will be executed. At the end of the replication, some of the statistics

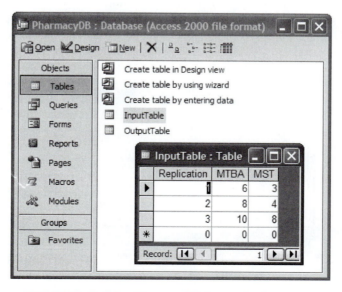

■ **FIGURE 7-58** *PharmacyDB* **Access database**

■ **FIGURE 7-59**   **Read/write logic for example**

related to the simulation will be written out to the table called *OutputTable*. For simplicity, this table also only has three columns. The first column is the replication number, the second column will hold the average waiting time in the queue and the third column will hold the half-width reported from Arena™.

Open the *PharmacyModelRWv1.doe* file and examine the VARIABLE module. Note that variables have been defined to hold the mean time between arrivals and the mean service time. Also, you should look at the CREATE and PROCESS modules to see how the variables are being used. The logic to read in the parameters at the beginning of each replication and to write the statistics at the end of the replication must now be implemented. The logic will be quite simple as indicated in Figure 7-59. An entity should be created at time zero, read the parameters from the database, delay for the length of the simulation, and then write out the data.

You should lay down the modules indicated in Figure 7-59. Now, the Arena™ FILE module must be defined. The process for using Access as a data source is very similar to the way Excel named ranges operate. Figure 7-60 shows the basic setup in the spreadsheet view. Opening up the record set rows allows you to define the tables as record sets. You should define the record sets as indicated in the figure.

The two READWRITE modules are quite simple. In the first READWRITE module, the variables RepNum, MTBA, and MST are assigned values from the file (Figure 7-61). This will occur at time zero and thereby set the parameters for the run. The second READWRITE module writes out the value of the RepNum and uses the Arena™ functions TAVG() and THALF() to get the values of the statistics associated with the waiting time in queue (Figure 7-62).

Now, the setup of the replications must be considered. In this model, at the beginning of each replication the parameters will be read in. Since there are three rows in the database input table, the number of replications will be set to 3 so that all rows are read in. In addition, the run length is set to 1000 hours and the base time unit to minutes, as shown in Figure 7-63.

Only one final module is left to edit. Open the DELAY module. The entity entering this module should delay for the length of the simulation. Arena™ has a special purpose variable called TFIN which holds the length of the simulation *in base time units*. Thus, the entity should

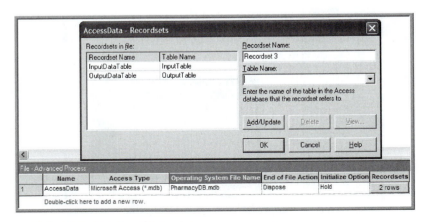

■ **FIGURE 7-60**   **FILE module for Access database**

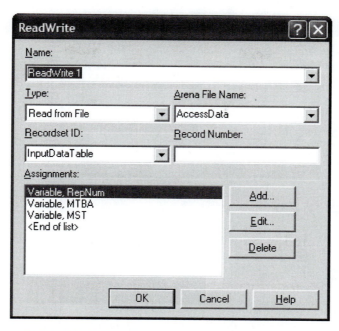

■ **FIGURE 7-61**   READWRITE assignments in first READWRITE module

delay for TFIN units as shown in Figure 7-64. Make sure that the units match the base time units specified in the Run Setup dialog.

After the entity delays for TFIN time units, it enters the READWRITE module where it writes out the values of the statistics at that time. After running the model, you can open the Microsoft Access database and view the *OutputTable* table. You should see the results for each of the three replications as shown in Figure 7-65.

Suppose that now you want to replicate each run involving the parameter settings three times. All you would need to do would be to set up your Access input table as shown in Figure 7-66 and change the number of replications to 9 in the Run Setup dialog.

This same approach can be easily repeated for larger models. This allows you to specify a set of experiments, say according to an experimental design plan, execute the experiments and easily capture the output responses to a database. You can then use any of your favorite statistical analysis tools (e.g., MINITAB, SAS, etc.) to analyze your experiments. I have found that for very large experiments where I want fine-grained control over the inputs and outputs, the approach outlined here works quite nicely.

Assignments

	Type	Variable Name	Other
1	Variable	RepNum	
2	Other	Variable 2	TAVG(Get Medicine.Queue.WaitingTime)
3	Other	Variable 3	THALF(Get Medicine.Queue.WaitingTime)

ReadWrite - Advanced Pro

	Name	Type	Arena File Name	Recordset ID	Record Number	Assignments
1	ReadWrite 1	Read from File	AccessData	InputDataTable		3 rows
2	ReadWrite 2	Write to File	AccessData	OutputDataTable		3 rows

■ **FIGURE 7-62**   READWRITE assignments in second READWRITE module

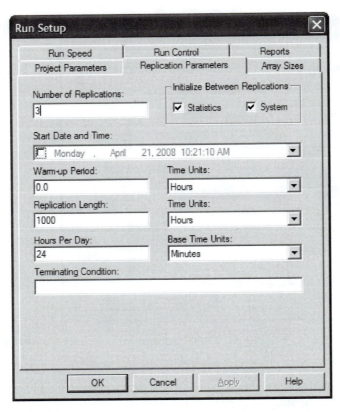

■ **FIGURE 7-63**   **Run setup parameters for example**

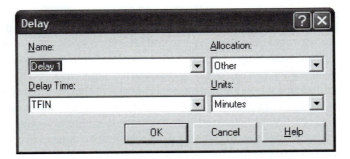

■ **FIGURE 7-64**   **Delaying for the length of the simulation**

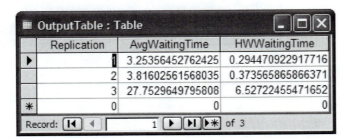

■ **FIGURE 7-65**   **Output in Access database**

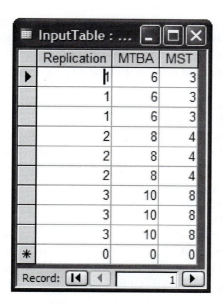

■ **FIGURE 7-66**   **Making repeated replications**

7.5.6   Using Visual Basic for Applications

This section discusses the relationship between Arena™ and Visual Basic for Applications (VBA). VBA is Microsoft's "macro" language for applications built on top of the Windows operating system. As its name implies, VBA is based on the visual basic (VB) language. VBA is a full featured language that has all the aspects of modern computer languages, including the ability to create objects with properties and methods. A full discussion of VBA is beyond the scope of this text. For an introduction to VBA, see Albright (2001).

The section assumes that the reader has some familiarity with VBA or is, at the very least, relatively computer-language literate. This topic is relatively advanced and typical usage of Arena™ often does not require the user to delve into VBA. The interaction between Arena™ and VBA will be illustrated through a discussion of the VBA block and the use of Arena's user defined function. Through VBA, Arena™ also has the ability to create models (e.g., lay down modules, fill dialogs, etc.) via Arena's VBA automation model. This advanced topic is not discussed in this text, but extensive help is available on the topic via Arena's on-line help system.

To understand the connection between Arena™ and VBA, you must understand the user event oriented nature of VBA. Visual basic was developed as an augmentation of the BASIC computer language to facilitate the development of visual user interfaces. User interface interaction is inherently event driven. That is, the user causes various events to occur (e.g., move the mouse, clicks a button, etc.) that the user interface must handle. To build a program that interacts significantly with the user involves writing code to react to specific events. This paradigm is quite useful, and as VB became more accepted, the event-driven model of computing was expanded beyond that of directly interacting with the user. Additional non-user interaction "events" can be defined and can be called at specific places in the execution of the program. The programmer can then place code to be called at those specific times in order to affect the actions of the program. Arena™ is integrated with VBA through a VBA event model.

There are a number of VBA events that are predefined in Arena's VBA interaction model. The following key VBA events will be called automatically if an event routine is defined for these events in VBA in Arena™.

DocumentOpen, DocumentSave: These events are called when the model is opened or saved. The SIMAN object is not available, but the Arena™ object can be used. The SIMAN object will be discussed in some examples.

RunBegin: This event is called prior to the model being checked. When the model is checked, the underlying SIMAN code is compiled and translated to executable machine code. You can place code in the RunBegin event to add or change modules with VBA automation so that the changes get complied into the model. The SIMAN object is not available, but the Arena™ object can be used. This event occurs automatically when the user invokes a simulation run. The Arena™ object will not be discussed in this text, but plenty of help is available in the help system.

RunBeginSimulation: This event is called prior to starting the simulation (i.e., before the first replication). This is a perfect location to set data that will be used by the model during all replications. The SIMAN object is active when this event is fired and remains active until after RunEndSimulation fires. Because the simulation is executing, changes using the Arena™ object via automation are not permissible.

RunBeginReplication: This event is called prior to the start of each replication.

OnClearStatistics: This event is called if the clear statistics option has been selected in run setup. This event occurs prior to each replication and at the designated warm up time if a warm up period has been supplied.

RunEndReplication: This event is called at the end of each replication. It represents a perfect location for capturing replication statistical data. It is only called if the replication completes naturally with no interruption from errors or the user.

RunEndSimulation: This event is called after all replications are completed (i.e., after the last replication). It represents a perfect place to close files and get across replication statistical information.

RunPause, RunRestart, RunResume, RunStep, RunFastForward, RunBreak: These events occur when the user interacts with Arena's run control (e.g., pauses the simulation via the pause button on the VCR control). These events are useful for prompting the user for input.

RunEnd: This event is called after RunEndSimulation and after the run has been ended. The SIMAN object is no longer active in this event.

OnKeystroke, OnFileRead, OnFileWrite, OnFileClose: These events are fired by the SIMAN runtime engine if the named event occurs.

UserFunction, UserRule: The UserFunction event is fired when the UF function is used in the model. The UserRule event is fired when the UR function is called in the model.

SimanError: This event is called if SIMAN encounters an error or warning. The modeler can use this event to trap SIMAN related errors.

A VBA event is defined when a subroutine is written that corresponds to the VBA event naming convention in the ThisDocument module accessed through the VBA editor. In VBA, a file that holds code is called a *module*. This should not be confused with an Arena™ module.

7.5.7   Using VBA

Let us take a look at how to write a VBA event for responding to the RunBegin event. Open the Arena™ file, *VBAEvents.doe*, available with this chapter. The Arena™ model is a simple CREATE, PROCESS, DISPOSE combination to create entities and has a model to illustrate VBA. The specifics of the model are not critical to the discussion here. In order to write a

■ **FIGURE 7-67**   **Showing the Visual Basic Editor**

subroutine to handle the RunBegin event, you must use the VBA Editor. In the Arena™ environment use the Tools > Macro > Show Visual Basic Editor menu option (as shown in Figure 7-67).

This will open the VBA Editor as shown in Figure 7-68. If you double-click on the ThisDocument item in the VBA projects tree as illustrated in Figure 7-68, you will open a VBA module that is specifically associated with the current Arena™ model.

A number of VBA events have already been defined for the model. Let's insert an event routine to handle the *RunBegin* event. This is illustrated in Figure 7-69. Place your cursor on a line in the *This Document* module and go to the event drop down box called (General). In this drop-down box, select Model Logic. Now go to the adjacent drop down box that lists the

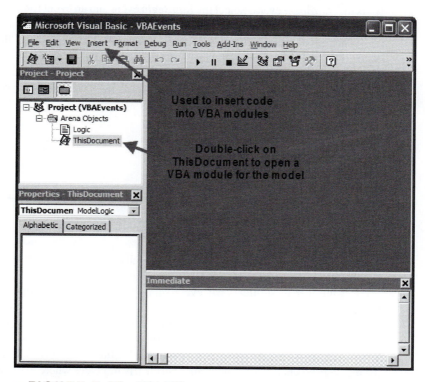

■ **FIGURE 7-68**   **VBA Editor**

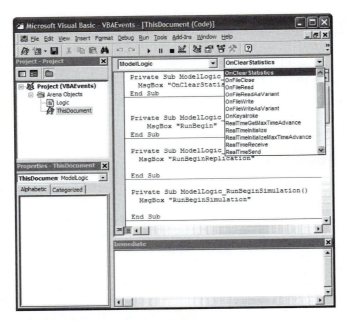

■ **FIGURE 7-69**   **Arena's VBA events**

available VBA events. Clicking on this drop down list will reveal all the possible Arena™ VBA events that are available. The event routines that have already been written are in bold. As seen in Figure 7-39, the *OnClearStatistics* event is indicating that it has been written and this can be confirmed by looking at the code associated with the *ThisDocument* module. At the end of the list is the *RunBegin* event. Select the *RunBegin* event and an event routine will automatically be placed in the *ThisDocument* module at your current cursor location.

The event procedure has a very special naming convention so that it can be properly called when the VBA integration mechanism needs to call it during the execution of the model. The code will be very simple. It will just open a message box that indicates that the *RunBegin* VBA event has occurred. The code to open a message box when the event procedure "fires" follows.

```
Private Sub ModelLogic_RunBegin()
MsgBox "RunBegin"
End Sub
```

The use of the *MsgBox* function in this example is just to illustrate that the subroutine is called at the proper time. You will quite naturally want to put more useful code in your VBA event subroutines.

A number of similar VBA event subroutines have been defined with similar message boxes. Go back to the Arena™ Environment and press the run button on the VCR run control toolbar. As the model executes a series of message boxes will open (Figure 7-70). After each message box appears, press okay and continue through the run. As you can see, the code in the VBA event routines is executed when Arena™ fires the corresponding event. The use of the message box here is a bit annoying, but clearly indicates where in the sequence of the run the event occurs.

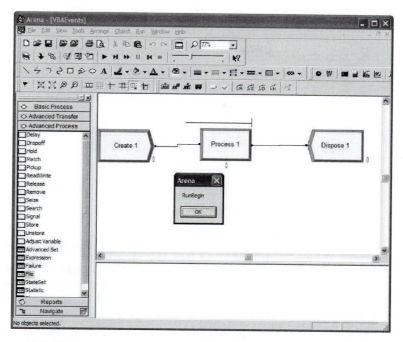

■ **FIGURE 7-70**   **Message box for *RunBegin* example**

The Arena™ Smart files are a good source of examples related to VBA. The following models are located in the Smarts folder in your main Arena™ folder (e.g., *Program Files\\Rockwell Software\\Arena™\\Smarts*).

Smarts 001: VBA—VariableArray Value

Smarts 016: Displaying a Userform

Smarts 017: Interacting with Variables

Smarts 024: Placing Modules and Filling in Data

Smarts 028: Userform Interaction

Smarts 081: Using a Shape Object

Smarts 083: Ending a Model Run

Smarts 086: Creating and Editing Resource Pictures

Smarts 090: Manipulating a Module's Repeat Groups

Smarts 091: Creating Nested Submodels Via VBA

Smarts 098: Manipulating Named Views

Smarts 099: Populating a Module's Repeat Group

Smarts 100: Reading in Data from Excel

Smarts 109: Accessing Information

Smarts 121: Deleting a Module

Smarts 132: Executing Module Data Transfer

Smarts 142: VBA Submodels

Smarts 143: VBA—Animation Status Variables

Smarts 155: Changing and Editing Global Pictures

Smarts 156: Grouping Objects

Smarts 159: Changing an Entity Attribute

Smarts 161: User Function

Smarts 166: Inserting Entities into a Queue

Smarts 167: Changing an Entity Picture

Smarts 174: Reading/Writing Excel Using VBA Module

Smarts 175: VBA Builds and Runs a Model

Smarts 176: Manipulating Arrays

Smarts 179: Playing Multimedia Files within a Model

Smarts 182: Changing Model Data Interactively

The Smarts files 001, 024, 090, 099, and 109 are especially illuminating. In the next section, the *UserFunction* event and the VBA block are illustrated.

7.5.8 The VBA Module and User-Defined Functions

This example discusses how to use a user form to set the value of a variable, call a user-defined function that uses the value of an attribute to compute a quantity, and use a VBA block to display information at a particular point in the model. Since the purpose of this example is to illustrate VBA, the model is a simple single-server queuing system as illustrated in Figure 7-71. The model consists of CREATE, ASSIGN, PROCESS, VBA, and DISPOSE modules.

The VBA block is found on the Blocks panel. To attach the Blocks panel, you can right-click in the Basic Process Panel area and select Attach, and then find the *Blocks.tpo* file. Now you are ready to lay down the modules as shown in Figure 7-71. The information for each module is given below:

CREATE: Choose Random(expo) with mean time between arrivals of 1 hour and set the maximum number of entities to 5.

ASSIGN: Use an attribute called *myPT*, and assign a U(10,20) random number to it via the UNIF(10,20) function.

PROCESS: Use the SEIZE, DELAY, RELEASE option. Define a resource and seize 1 unit of the resource. In the Delay Type area, choose Expression and type in UF(1) for the expression.

Using the VARIABLE data module to define a variable called *vPTFactor* and initialize it with the value 1.0.

No editing is necessary for the VBA block and the DISPOSE module. If you have any difficulty completing these steps you can look at the module dialogues in the Arena™ file called *Ch7VBAExample.doe*.

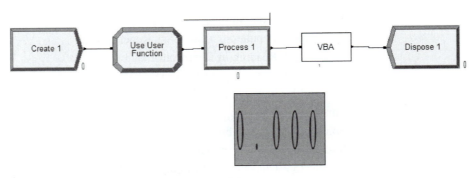

■ **FIGURE 7-71** **VBA example model**

Exhibit 7-1   Defining Variables in VBA

```
Option Explicit

'' Declare global object variables to refer
'' to the Model and SIMAN objects
'' Public allows them to be accessed anywhere in this module
'' or any other vba module
Public gModelObj As Model
Public gSIMANObj As SIMAN

'' Variables can be accessed via their uniquely assigned
'' symbol number. An integer variable is needed to hold the index
'' It is declared public here so it can be used throughout this module
'' and other vba modules
Public vPTFactorIndex As Integer

'' Index for the myPT attribute
'' It is declared private here so it can be used throughout this module
Private myPTIndex As Integer
```

Now you are ready to enter the world of VBA. Use Alt-F11 to open the VBA Editor. Double-click on the ThisDocument object and enter the code as shown in Exhibit 7-1. If you don't want to retype this code, then you can access it in the file *Ch7VBAExample.doe*. Let's walk through what this code means. The two public variables *gModelObj* and *gSIMANObj* are global object reference variables of type Model and SIMAN, respectively. These variables allow access to the properties and methods of these two objects once the object references have been set. The Model object is part of Arena's VBA Object model and allows access to the features of the Model as a whole. The details of Arena's Object model can be found in the Arena™ help system by searching on *Automation Programmer's Reference*. The SIMAN object is created after the simulation has been complied and essentially gives access to the underlying simulation engine. The variables *vPTFactorIndex* and *myPTIndex* will be used to index into SIMAN to access the Arena™ variable *vPTFactor* and the attribute *myPT*.

This example uses VBA forms to obtain input from the user and to display information. Thus, the forms to be used in the example need to be developed. Use Insert > UserForm to create two forms called *Interact* and *UserForm1* as shown in Figure 7-72.

Use the Show Toolbox button to show the VBA controls toolbox. Then, you can select the desired control and place your cursor on the form at the desired location to complete the action. Build the forms as shown in Figures 7-72 and 7-73. To place the labels used in the forms, use the label control (right next to the text box control on the Toolbox). The name of a form can be changed in the Misc > (Name) property as shown in Figure 7-74.

Now that the forms have been built, the controls on the forms can be referenced in other VBA modules. Exhibit 7-2 shows the VBA code for the RunBeginSimulation event.

When the user executes the simulation, the *RunBeginSimulation* event is fired. The first two lines of the routine set the object references to the Model object and to the SIMAN object in order to store the values in the global variables that were previously defined for the VB module. Since the *RunBeginSimulation* method fires at the beginning of the simulation, you do not need to set these global variables in the other functions (e.g., UserFunction1() in Exhibit 7-5. The SIMAN object can then be accessed through the variable *gSIMANObj* to get the indexes of the attribute and variable *myPT* and *vPTFactor*.

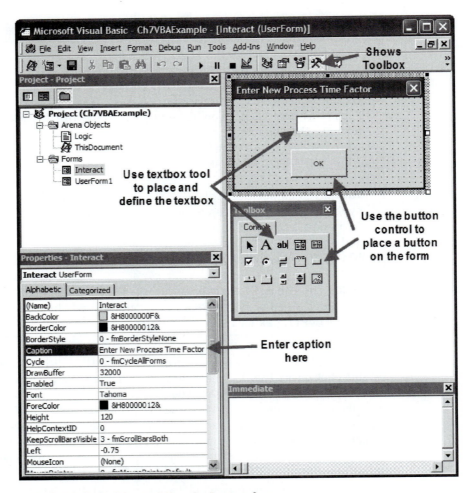

■ FIGURE 7-72 Building the *Interact* form

■ FIGURE 7-73 VBA *UserForm1*

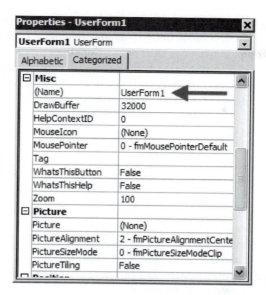

■ **FIGURE 7-74** Properties window

Exhibit 7-2 *RunBeginSimulation* Event Routine

```
Private Sub ModelLogic_RunBeginSimulation()
    '' set object references
    Set gModelObj = ThisDocument.Model
    Set gSIMANObj = ThisDocument.Model.SIMAN

    '' Get the index to the symbol number
    '' if the symbol does not exist there will be an error,
    '' There is no error handling in this example
    '' The SymbolNumber method of the SIMAN object returns
    '' the index associated with the named symbol

    '' get the index for the myPT attribute
    myPTIndex = gSIMANObj.SymbolNumber("myPT")

    '' get the index for the vPTFactor variable
    vPTFactorIndex = gSIMANObj.SymbolNumber("vPTFactor")

    '' set the current value of the text box to the
    '' current value of the vPT variable in Arena
    '' The VariableArrayValue method of the SIMAN object
    '' returns the value of the variable associated with the index
    Interact.TextBox1.value =
            gSIMANObj.VariableArrayValue (vPTFactorIndex)

    '' Display the user form
    Interact.Show
    '' Set the focus to the textbox
    Interact.TextBox1.SetFocus

    '' The code for setting the variable's value is
    '' in the OK button user form code
End Sub
```

When the model is complied each element of the model is numbered. This is called the symbol number. The method, *SymbolNumber("element name")*, on the SIMAN object will return the index symbol number that can be used as an identifier for the named element. To find information on the SIMAN object search the Arena™ help system on SIMAN Object (Automation). There are *many* properties and methods associated with the SIMAN Object. The Arena™ documentation states the following concerning the *SymbolNumber* method.

SymbolNumber Method

Syntax
SymbolNumber (symbolString As String, index1 As Long, index2 As Long) As Long

Description
All defined simulation elements have a unique number. For those constructs that have names, this function may be used to return the number corresponding to the construct name. For example, many of the methods in the SIMAN Object require an argument like resourceNumber or variableNumber, etc. to identify the specific element. Since it is more common to know the name rather than the number of the item, SymbolNumber("MyElementName") is often used for the element-Number type argument.

As indicated in the documentation, an important part of utilizing the SIMAN object's other methods is to have access to an Arena™ element's unique symbol number. The symbol number is an integer and is then used as an index into other method calls to identify the element of interest. In the exhibit, after the indexes to the *myPT* and *vPTFactor* are found, the SIMAN object is used to access the *VariableArrayValue* property. According to Arena's help system the *VariableArrayValue* property can be used either to get or to set the value associated with the variable identified by the index number.

VariableArrayValue Read/Write Property

Syntax
VariableArrayValue (variableNumber As Long) As Double

Description
Gets/Sets the value of the variable, where variableNumber is the instance number of the variable as listed in the VARIABLES Element. For more information on working with variables in automation, please see making variable assignments.

In this code, the value of the variable is accessed and then assigned to the value of the TextBox1 control on the Interact form. Then, the Interact form is told to show itself and the focus is set to the text box, TextBox1, on the form for user entry.

When the Interact form is displayed, the text box will show the current value of the variable. The user can then enter a new value in the text box and press the OK button. Since the user interacted with the OK button, the button press event in the VBA form code can be used to change the value of the variable in Arena™ to what was entered by the user. To enter the code to react to the button press, go to the Interact form and select the OK button on the form. Right-click on the button and from the context menu, select View Code. This will open the form's module and create a VBA event to react to the button click. The code provided in Exhibit 7-3 shows the use of the VariableArrayValue property to assign the value entered into the text box to the *vPTFactor* variable. To access a public variable in the *ThisDocument* module, you precede the variable with *ThisDocument.* (e.g., *ThisDocument.gSIMANObj*). Once the value of the text box has been assigned to the variable, the form is hidden and input focus is given back to the model via the *ThisDocument.gModelObj.Activate* method.

Exhibit 7-3   VBA Code for OK Button

```
Private Sub CommandButton2_Click()
'' when the user clicks the ok button
'' Set the current value of vPT to the value
'' currently in the text box
'' uses the global variable vPTFactorIndex
'' defined in module ThisDocument
'' uses the global variable for the SIMAN object
'' defined in ThisDocument

ThisDocument.gSIMANObj.VariableArrayValue(ThisDocument.vPTFactorIndex)
= TextBox1.value

'' hide the form
Interact.hide
'' Tell the model to resume running
ThisDocument.gModelObj.Activate
End Sub
```

Since the setting of the variable vPTFactor occurs as a result of the code initiated by the *RunBeginSimulation* event, the value supplied by the user will be used for the entire run (unless changed again in the model or in VBA).

The model will now begin to create entities and execute as a normal Arena™ model. Each entity that is created will go through the ASSIGN module and have its *myPT* attribute set to a randomly drawn U(10,20) number. Then the entity proceeds to the PROCESS module where it invokes the SEIZE, DELAY, and RELEASE logic. In this module, the processing time is specified by a user defined function via Arena's UF(fid) function. The UF(fid) function will call associated VBA code from directly in an Arena™ model. You should think of the UF function as a mechanism by which you can easily call VBA functions. The *fid* argument is an integer that will be passed into VBA. This argument can be used to select from other user written functions as necessary. Exhibit 7-4 shows the code for the UF function for this example. When you use the UF function, you must write your own code. To write the UF function, use the drop down box that lists the available VBA events on the ThisDocument module and select the UserFunction event. The UserFunction event routine that is created has two arguments. The first argument is an identifier that represents the current active entity and the second argument is what was supplied by the call to UF(fid) in Arena™. When you create the function it will not have any code. You should organize your code in a similar fashion as shown in Exhibit 7-4. As seen in the exhibit, the supplied function identifier is used in a VBA select-case statement to select from other available user-written functions.

By using this select-case construct, you can easily define a variety of your own functions and call whichever function you need from Arena™ by supplying the appropriate function identifier. The VBA function *UserFunction1()* has been implemented, which accesses the value of the variable *vPTFactor* and the attribute *myPT* to calculate the product of their values. Admittedly, this calculation can be more readily accomplished directly in Arena™ without VBA, but the point is you can implement any complicated calculation using this architecture.

Exhibit 7-5 shows how to access the value of an attribute associated with the active entity. First, the *ActiveEntity* property of the SIMAN object is used to get an identifier for the current active entity (i.e., the entity that entered the VBA block). Next, the *AttributeValue* method of the SIMAN object is used to get the value of the attribute. Arena's help system describes the use of these methods.

Exhibit 7-4 Arena's VBA UserFunction() Event Routine

```
'' This Function allows you to pass a user
'' defined value back to the
'' module which called on the UF(functionID) function in Arena.
'' Use the functionID to select the function that you want via
'' the case statement, add additional functions as necessary
'' The functions can obviously be named something
'' more useful than UserFunctionX()
'' The functions must return a double
Private Function ModelLogic_UserFunction(ByVal entityID As Long, ByVal
functionID As Long) As Double
    '' entityID is the active entity
    '' functionID is supplied when the user calls UF(functionID)

    Select Case functionID
        Case 1
            ModelLogic_UserFunction = UserFunction1()
        Case 2
            ModelLogic_UserFunction = UserFunction2()
    End Select

End Function
```

Exhibit 7-5 Implementing a User-Defined Function

```
Private Function UserFunction1() As Double
    '' each entity has a unique id
    Dim activeEntityID As Long
    '' get the number of the active entity
    '' this could have been passed from ModelLogic_UserFunction active
    EntityID = gSIMANObj.ActiveEntity

    Dim PTvalue As Double
    '' get the value of the myPT attribute
    PTvalue = gSIMANObj.AttributeValue(activeEntityID, myPTIndex, 0, 0)

    Dim factor As Double
    factor = gSIMANObj.VariableArrayValue(vPTFactorIndex)

    '' this could be complicated function of the attribute/variables
    '' here it is very simple (and could have been done in Arena itself
    '' rather than VBA
    UserFunction1 = PTvalue * factor
End Function
```

Next, the *AttributeValue* method of the SIMAN object is used to get the value of the attribute. Arena's help system describes the use of these methods.

ActiveEntity Method

Syntax

ActiveEntity () As Long

Description

Returns the record location (the entity pointer) of the currently active entity, or 0 if there is not one. This is particularly useful in a VBA block Fire event to access attributes of the entity that entered the VBA block in the model.

AttributeValue Method

Syntax

AttributeValue (entityLocation As Long, attributeNumber As Long, index1 As Long, index2 As Long) As Double

Description

Returns the value of general-purpose attribute attributeNumber with associated indices index1 and index2. The number of indices specified must match the number defined for the attribute.

In Arena™, attributes can be defined as multidimensional via the ATTRIBUTES element in the Elements panel. Multidimensional attributes are not discussed in this text, but the *AttributeValue* method allows for this possibility. The two indices that can be supplied indicate to Arena™ how to access the attribute. In the example, these two values are zero, which indicates that this is not a multidimensional attribute. Once the values of the *myPT* attribute and the *vPTFactor* variable are retrieved from the SIMAN object, they are used to compute the value that is returned by the UF function.

With the UF function, you can easily invoke VBA code from essentially any place in your Arena™ model. The UF function is especially useful for returning a value back to Arena™. Using the VBA block, you can also invoke specific VBA code when the entity passes through the block. In this example, after the entity exits the PROCESS module, the entity enters a VBA block. When you use a VBA block in your Arena™ model, each VBA block is given a unique number. Then, in the ThisDocument module, an individual event routine can be created that is associated with each VBA block. In the example, the VBA block is used to open a form and display some information about the entity when it reaches the VBA block. This might be useful to do when running an Arena™ model in order to stop the model execution each time the entity goes through the VBA block to allow the user to interact with the form. However, most of the time the VBA block is used to execute complicated code that depends on the state of the system when the entity passes through the VBA block. Exhibit 7-6 shows the code for our VBA block event.

By now, this code should start to look familiar. The SIMAN object is used to access the values of the attribute and the variable that are of interest and then set them equal to the values for the text boxes that are being used on the form. Then, the form is shown. This brings up the form so that the user can see the values. In *UserForm1*, a command button was defined to close the form. The logic for hiding the form is shown in Exhibit 7-7.

The show-and-hide functionality of VBA forms is used in this example so that new instances of the forms did not have to be created each time they are used. This keeps the form in memory so that the controls on the form can be readily accessed. This is not necessarily the best practice for managing the forms, but allows the discussion to be simplified. As long as you don't have a large number of forms, this approach is reasonable. In addition, in the examples, there is

Exhibit 7-6 Example VBA Block Event

```
Private Sub VBA_Block_1_Fire()
    '' set the values of the text boxes

    UserForm1.TextBox1.value =
gSIMANObj.VariableArrayValue(vPTFactorIndex)
    UserForm1.TextBox2.value =
gSIMANObj.AttributeValue(gSIMANObj.ActiveEntity, myPTIndex, 0, 0)
    UserForm1.TextBox3.value = UserFunction1()
    '' Display the user form
    UserForm1.Show

End Sub
```

Exhibit 7-7 Hiding the User Form

```
Private Sub CommandButton1_Click()
    '' hide the form
    UserForm1.hide
    '' Tell the model to resume running
    ThisDocument.gModelObj.Activate
End Sub
```

no error catching logic. VBA has a very useful error catching mechanism and production code should check for and catch errors. For more on VBA error catching, see the On Error statement in the VBA help file or chapter 13 in Albright (2001).

7.5.9 Generating Correlated Random Variates

The final example involves how to explicitly model dependence (correlation) in the input models discussed in Chapter 3. In the fitting of input models, it was assumed and tested that the sample observations did not have correlation. But, what do you do if the data do contain correlation? For example, let S_i be the service time of the ith customer. What if the S_i have significant correlation? That is, the service times are correlated. Input processes may also be correlated; that is, the time between arrivals might be correlated. Research has shown (see Patuwo et al., 1993, and Livny et al., 1993) that ignoring the correlation when it is in fact present can lead to gross underestimation of the actual performance estimates for the system.

The discussion here is based on the Normal-to-Anything Transformation as discussed in Banks et al. (2006), Cario and Nelson (1996, 1998), and Biller and Nelson (2003). Suppose you have a N(0,1) random variable, Z_i, and a way to compute the CDF, $\Phi(z)$, of the normal distribution. Then, from the discussion of the inverse transform technique in Chapter 3, the random variable, $\Phi(Z_i)$, will have a U(0,1) distribution. Suppose that you wanted to generate a random variable X_i with CDF $F(x)$, then you can use $\Phi(Z_i)$ as the source of uniform random numbers in an inverse transform technique.

$$X_i = F^{-1}(U_i) = F^{-1}(\Phi(Z_i))$$

This transform is called the normal-to-anything (NORTA) transformation. It can be shown that even if the Z_i are correlated (and thus so are the $\Phi(Z_i)$), then the X_i will have the correct CDF and will also be correlated. Unfortunately, the correlation is not directly preserved in the transformation, so that if the Z_i has correlation ρ_Z, the X_i will have ρ_X (not necessarily the same). Thus, in order to induce correlation in the X_i, you must have a method to induce correlation in the Z_i.

One method to induce correlation in the Z_i is to generate the Z_i from a stationary autoregressive time-series model of order 1, that is, AR(1). A stationary AR(1) model with N(0,1) marginal distributions has the following form:

$$Z_i = \phi Z_{i-1} + \varepsilon_i$$

where $i = 2, 3, \ldots$ and $\varepsilon_i \sim iid\, N(0, 1 - \phi^2), i = 2, 3, \ldots, -1 < \phi < 1$, and $Z_1 \sim N(0, 1)$. Thus, it is relatively straightforward to generate this process by first generating Z_1, and then generating $\varepsilon_i \sim iid\, N(0, 1 - \phi^2)$ and using $Z_i = \phi Z_{i-1} + \varepsilon_i$ to compute the next Z_i. It can be shown that this AR(1) process will have lag-1 correlation:

$$\phi = \rho^1 = corr(Z_i, Z_{i+1})$$

Therefore, you can generate a random variable Z_i that has a desired correlation and through the NORTA transformation produce X_i that are correlated with the correlation being functionally related to ρ_Z. By changing ϕ through trial and error, you can get the correlation for the desired X_i. Procedures for accomplishing this are given in the previously mentioned references. The spreadsheet *NORTAExample.xls* can also be used to perform this trial-and-error process.

Implementation of this technique cannot be readily achieved in a general way in Arena™ through the use of standard modules. In order to implement this method, you can make use of Arena's ability to utilize VBA. The Arena™ model, *NORTA-VBA.doe*, shows how to implement the NORTA algorithm with an Arena™ user function. Just as illustrated in the last section, a user function can be written as shown in Exhibit 7-8.

Exhibit 7-8   VBA User Function Event

```
Private Function ModelLogic_UserFunction(ByVal entityID As Long,
ByVal functionID As Long) As Double

    Select Case functionID
        Case 1
            ModelLogic_UserFunction = CorrelatedUniform()
        Case 2
            Dim u As Double
            u = CorrelatedUniform()
            Dim x As Double
            x = expoInvCDF(u, 2)
            ModelLogic_UserFunction = x
    End Select

End Function

Private Function expoInvCDF(u As Double, mean As Double) As Double
    expoInvCDF = -mean * Log(1-u)
End Function
```

Exhibit 7.9   VBA Function to Generate Correlated Uniforms

```
Private Function CorrelatedUniform() As Double
    ''
    Dim phi As Double
    Dim e As Double
    Dim u As Double
    '' change phi inside this function to get a different correlation
    phi = 0.8

    '' generate error
    e = gSimanObj.SampleNormal(0, 1-(phi * phi), 1)
    '' compute next AR(1) Z
    mZ = phi * mZ + e
    '' use normal cdf to get uniform
    u = NORMDIST(mZ)

    CorrelatedUniform = u
End Function
```

When you use UF(1), a correlated U(0,1) random variable will be returned. If the user uses UF(2), then a correlated exponential distribution with a mean of 2 will be returned. Note that in generating the correlated exponential random variable, a correlated uniform is first generated and then used in the inverse transform method to transform the uniform to the proper distribution. Thus, provided that you have functions that implement the inverse transform for the desired distributions, you can generate correlated random variables. Exhibit 7-9 illustrates how the NORTA technique is used to generate the correlated uniform random variables. The variable, phi, is used to control the resulting correlation in the AR(1) process. The function, SampleNormal, associated with the SIMAN object is used to generate a N(0,1) random variable that represents the error in the AR(1) process.

The variable, mZ, has been defined at the VBA module level, and thus it retains its value between invocations of the CorrelatedUniform function. The variable, mZ, represents the AR(1) process. The current value of mZ is multiplied with phi and the error is added to obtain the next value of mZ. The initial value of mZ was determined by implementing the RunBeginReplication event in the ThisDocument module. Because of the NORTA transformation, mZ, is N(0,1), and the function NORMDIST is used to compute the probability associated with the supplied z-value. The NORMDIST function (not shown here) is an implementation of the CDF for the standard normal distribution.

With minor changes, the code supplied in *NORTA-VBA.doe* can be easily adapted to generate other correlated random variables. In addition, the use of the UF function can be expanded to generate other distributions that Arena™ does not have built in (e.g., binomial).

Arena™ is a very capable language, but occasionally you will need to access the full capabilities of a standard programming language. This section illustrated how to utilize VBA in the Arena™ environment by either using the predefined automation events or by defining your own functions or events via the UF function and the VBA block. With these constructs you can make Arena™ into a powerful tool for end users. There is one caveat with respect to VBA integration that must be mentioned. Since VBA is an interpreted language, its execution speed can be slower than compiled code. The use of VBA in Arena™ as indicated in this section can increase the execution time of your models. If execution time is a key factor for your application, then you should consider using C/C++ rather than VBA for your advanced integration needs. Arena^{TM} allows access to the SIMAN runtime engine via C language functions. Essentially, you write your C functions and bundle them into a dynamic linked library which can be linked to Arena^{TM}. For more information on this topic, you should search the Arena^{TM} help system under *Introduction to C Support.*

7.6   Summary

This chapter provided a discussion of miscellaneous modeling constructs that can improve your Arena^{TM} modeling. Advanced resource modeling was covered to allow for resources to be subjected to either scheduled or random-capacity changes. Then, how Arena^{TM} tabulates entity and resource costs were presented. The PICKSTATION allows for an entity to use criteria to select from a set of listed stations. In addition, the PICKUP and DROPOFF modules provide another mechanism by which entities can be added to or removed from the active entity's group. Finally, some programming aspects of Arena^{TM} (including generic stations and VBA) were presented that can make your modeling more productive. In the next chapter, a comprehensive example is presented so that you can experience the full range of applying Arena^{TM} to a simulation problem.

Exercises

7.1 Customers arrive at a one-window drive-in bank according to a Poisson distribution with a mean of 10 per hour. The service time for each customer is exponentially distributed with a mean of 5 minutes. There are three spaces in front of the window including that for the car being served. Other arriving cars can wait outside these three spaces. Use the frequency option of the STATISTIC module to estimate the probability that an arriving customer can enter one of the three spaces in front of the window. Compare your results with your answers to exercise 5-5.

7.2 This problem analyzes the cost associated with the LOTR, Inc. configurations described in Section 4.5.1.2 of Chapter 4. In particular, the master ring maker costs $30 per hour whether she is busy or idle. Her time is considered value added. The rework craftsman earns $22 per hour; however, his time is considered non–value added since it could be avoided if there were no rework. Both the master ring maker and the rework craftsman earn $0.50 for each ring that they work on. The inspector earns $15 per hour without any piecework considerations, and the packer earns $10 per hour (again with no piecework). The inspection time is considered non–value added; however, the packing time is considered value-added. The work rules for the workers include that they should get a 10 minute break after 2 hours into their shift, a 30 minute lunch after 4 hours into their shift, a 10 minute break 2 hours after the completion of lunch, and a 15 minute clean up allowance before the end of the shift.

While many would consider a magic ring priceless, LOTR, Inc. has priced a pair of rings at approximately $10,000. The holding charge for a ring is about 15% per year. Assume that there are 365 days in a year and that each day has 24 hours when determining the holding cost per hour for the pair of rings. Simulate configuration 1 and configuration 2 as given in Chapter 4 and report the entity and resource costs for the configurations. Which configuration would you recommend?

7.3 Reconsider the machine interference problem from Chapter 5. Use Arena's costing functionality and find the best assignment of operators in terms of cost while maintaining an average machine utilization of over 70%.

7.4 A domestic violence and abuse center operates a 24-hour phone center.[1] The calls arrive to the center in a very random fashion. Table 7-6 indicates the number of calls received, recorded on an hourly basis over a period of 7 days. The table does not include any lost calls that resulted because the caller reneged due to being put on hold. The time to complete 100 calls was recorded and is given in Table 7-7.

■ **TABLE 7-6**
Total Calls for each Hour

Starting Hour	Mon.	Tues.	Wed.	Thurs.	Fri.	Sat.	Sun.
0:00	2	3	1	2	1	3	1
1:00	1	2	2	3	3	5	3
2:00	1	0	0	0	1	1	0
3:00	0	1	1	0	1	0	1
4:00	0	0	0	1	1	0	0
5:00	1	0	0	0	1	1	0
6:00	3	4	6	2	3	4	5
7:00	4	3	6	2	2	3	4
8:00	2	5	4	4	2	3	5
9:00	4	6	8	4	5	3	4
10:00	6	5	5	3	6	4	7
11:00	3	9	6	8	4	7	5
12:00	8	11	10	5	15	12	9
13:00	10	9	8	7	10	16	6
14:00	8	6	10	12	12	11	10
15:00	10	9	12	4	10	6	8
16:00	8	6	9	14	12	10	7
17:00	5	10	10	8	10	10	9
18:00	5	4	6	5	6	7	5
19:00	3	4	6	2	3	4	5
20:00	4	3	6	2	2	3	4
21:00	1	2	2	3	3	5	3
22:00	1	2	2	3	3	5	3
23:00	1	0	0	0	1	1	0

■ **TABLE 7-7**
Call Service Times in Minutes

20.48	13.64	17.84	16.86	17.73	12.07	11.22	8.69	16.53	20.66
22.81	10.97	12.17	20.75	8.55	16.93	10.34	12.01	18.85	12.77
15.11	9.53	13.71	16.12	19.50	12.14	15.84	16.35	16.14	13.69
11.54	19.40	9.87	17.10	19.99	15.68	15.92	17.46	20.96	21.02
17.29	13.99	14.53	16.41	14.09	10.89	20.57	12.65	12.33	17.39
13.50	19.06	22.52	12.81	10.21	22.57	12.07	19.94	19.75	13.21
19.11	21.78	13.10	12.09	19.73	10.11	18.85	15.78	10.90	10.99
14.31	9.25	12.46	20.82	14.07	15.66	21.14	13.53	13.34	17.19
15.46	15.19	9.31	12.52	13.11	12.16	14.64	15.32	9.05	18.44
10.72	12.56	16.48	12.18	9.94	20.77	8.54	15.51	10.66	12.16

---

[1] This problem is based in part on Taha (23), problem 17.2, p. 635. Reprinted with permission by permission of Pearson Education, Inc., Upper Saddle River, NJ.

Historic data indicate that the center has been experiencing a 10% annual rate of increase in telephone calls as information about the center has been publicized. The center operates with a staff consisting of both trained volunteers and trained staff members that are dedicated to answering the phone calls. The volunteers work for free, but when scheduled to work they require that at least one trained staff member also be present for every 3 volunteers. Staff members cost $16 per hour (including fringe benefits). Obviously, it is very important that a caller get to speak to a real person and that there is a very low probability that a caller will be put on hold.

(a) Develop staffing plans for the next 3 years that involve only trained staff members that minimizes the total cost while ensuring that the probability that a caller will be put on hold is less than 0.05.

(b) Develop staffing plans for the next 3 years that involve a mix of trained staff and volunteers that minimizes the total cost, while ensuring that the probability a caller will be put on hold is less than 0.05.

(c) Consider how you would handle the following work rules. The staff and volunteers should get a 10-minute break every 2 hours and a 30-minute food/beverage break every 4 hours. Discuss how you would staff the center under these conditions ensuring that there is always someone present (i.e., they are not all on break at the same time).

7.5 The test-and-repair system described in Chapter 5 has three testing stations, a diagnostic station, and a repair station. Suppose that the machines at the test station are subject to random failures. The time between failures is distributed according to an exponential distribution with a mean of 240 minutes and is based on busy time. Whenever a test station fails, the testing software must be reloaded, which takes between 10 and 15 minutes uniformly distributed. The diagnostic machine is also subject to usage failures. The number of uses to failure is distributed according to a geometric distribution with a mean of 100. The time to repair the diagnostic machine is uniformly distributed in the range of 15 to 25 minutes. Examine the effect of explicitly modeling the machine failures for this system in terms of risks associated with meeting the terms of the contract.

7.6 A proposal for a remodeled Sly's Convenience Store has 6 gas pumps each having their own waiting line. The interarrival time distribution of the arriving cars is exponential with a mean of 1 minute. The time to fill up and pay for the purchase is exponentially distributed with a mean of 6 minutes. Arriving customers choose the pump that has the least number of cars (waiting or using the pump). Each pump has room to handle three cars (one in service and two waiting). Cars that cannot get a pump wait in an overall line to enter a line for a pump. The long-run performance of the system is of interest.

(a) Estimate the average time in the system for a customer, the average number of customers waiting in each line, the overall line, and in the system, and estimate the percentage of time that there are {1, 2, 3} customers waiting in each line. In addition, the percentage of time that there are {1, 2, 3, 4 or more} customers waiting to enter a pump line should be estimated. Assume that each car is between 16 and 18 feet in length. About how much space in the overall line would you recommend for the convenience store.

(b) Occasionally, a pump will fail. The time between pump failures is distributed according to an exponential distribution with a mean of 40 hours and is based on clock time (not busy time). The time to service a pump is exponentially distributed with a mean of 4 hours. For simplicity assume that when a pump fails the customer using the pump is able to complete service before the pump is unusable and ignore the fact that some customers may be in line when the pump fails. Simulate the system under these new conditions and estimate the same quantities as requested in part (a). In addition, estimate the percentage of time that each pump spends idle, busy, or failed. Make sure that newly arriving customers do not decide to enter the line of a pump that has failed.

(c) The assumption that a person waiting in line when a pump fails stays in line is unrealistic. Still assume that a customer receiving service when the pump fails can complete service, but now allow waiting customers to exit their line and rejoin the overall line if a pump fails while they are in line. Compare the results of this new model with those of part (b).

7.7 Apply the concepts of generic station modeling to the system described in exercise 6.14. Be sure to show that your results are essentially the same with and without generic modeling.

7.8 A single automatic guided vehicle (AGV) is used to pick up finished parts from three machines and drop them off at a storeroom. The AGV is designed to carry five totes. Each tote can hold up to 10 parts. The machines produce individual parts according to a Poisson process with the rates indicated in the table below. The machines are designed to directly drop the parts into a tote. When a tote is full, it is released down a gravity conveyor to the AGV loading area. It takes 2 seconds for the tote to move along the conveyor to the machine's loading area.

Station	Production Rate	Tote Loading Time	Travel distance to Next station
Machine 1	2 per minute	uniform(3,5) seconds	40 meters
Machine 2	1 per minute	uniform(2,5) seconds	28 meters
Machine 3	3 per minute	uniform(4,6) seconds	55 meters

The AGV moves from station to station in the sequence (Machine 1, Machine 2, Machine 3, storeroom). If there are totes waiting at the machine's loading area, the AGV loads the totes until reaching its capacity. After the loading is completed, the AGV moves to the next station. If there are no totes waiting at the station, the AGV delays for 5 seconds, if any totes arrive during the delay they are loaded; otherwise, the AGV proceeds to the next station. After visiting the third machine's station, the AGV proceeds to the storeroom. At the storeroom any totes that the AGV is carrying are dropped off. The AGV is designed to release all totes at once, which takes between 8 and 12 seconds uniformly distributed. After completing its visit to the storeroom, it proceeds to the first machine. The distance from the storeroom to the first station is 30 meters. The AGV travels at a velocity of 30 meters per minute. Simulate the performance of this system over a period of 10,000 minutes.

(a) Estimate the average system time for a part. What is the average number of totes carried by the AGV? What is the average route time? That is, what is the average time that it takes for the AGV to complete one circuit of its route?

(b) Suppose the sizing of the loading area for the totes is of interest. What is the required size of the loading area (in terms of the number of totes) such that the probability that all totes can wait is virtually 100%.

7.9 Suppose the shuttle bus system described in Section 7.4.3 now has 3 bus stops. Passengers arrive to bus stop 1 according to a Poisson arrival process with a mean rate of 6 per hour. Passengers arrive to bus stop 2 according to a Poisson arrival process with a mean rate of 10 per hour. Passengers arrive to bus stop 3 according to a Poisson arrival process with a mean rate of 12 per hour. Thirty percent of customers arrive to stop 1 desire to go to stop 2 with the remaining going to stop 3. For those passengers arriving to stop 2, 75% want to go to stop 1 and 25% want to go to stop 3. Finally, for those passengers that originate at stop 3, 40% want to go to stop 1 and 60% want to go to stop 2. The shuttle bus now has 20 seats. The loading and unloading times per passenger are still the same as in the example and the travel time between stops is still the same. Simulate this system for 8 hours of operation and estimate the average waiting time of the passengers, the average system time of the passengers, and the average number of passengers left at the stop because the bus was full.

7.10 SQL queries arrive to a database server according to a Poisson process with a rate of 1 query every minute. The time that it takes to execute the query on the server is typically between 0.6 and 0.8 minutes uniformly distributed. The server can only execute 1 query at a time.

(a) Develop a simulation model to estimate the average delay time for a query and the probability that there are 1, 2, 3 or more queries waiting.

(b) Suppose that data on the interarrival times for the queries shows that the time between arrivals are still exponentially distributed with a mean of 1 minute but the arrival process is correlated with a lag-1 correlation of 0.85. Simulate this situation and compare your results to the results of part (a).

7.11 Section 7.5.5 discusses how to store the parameter values of replications in a Microsoft Access database and to read in those parameter values when running replications in Arena^{TM}. In addition, it indicates how to write out the results to a Microsoft Access database. The discussion suggests that if each experimental parameter setting requires multiple replications to simply repeat those parameter settings in the Access data input table. This can be tedious for large experimental designs. This problem considers how to modify the basic logic to allow multiple replications for each design point without duplicating them in the input table. Consider the following pseudo-code.

```
CREATE file reading/writing entity
IF ((MOD(NREP-1,vMaxReps) == 0)
       READ in and ASSIGN parameter values
ENDIF
DELAY until end of replication
WRITE out replication results
DISPOSE file reading/writing entity
```

The variable, *vMaxReps*, is the number of replications that you want each design point repeated. Implement and test this pseudo-code on the *PharmacyModelRWv1.doe* file with each design point being repeated three times.

7.12 Develop a spreadsheet that uses the following set of uniform random number to generate 18 correlated random variables using the NORTA technique with an AR(1) process with $\phi = 0.7$ and a gamma distribution with alpha (shape) parameter equal to 2 and a beta (scale) parameter equal to 5. What is the resulting lag-1 correlation of the gamma random variables?

0.943	0.398	0.372	0.943	0.204	0.794
0.498	0.528	0.272	0.899	0.294	0.156
0.102	0.057	0.409	0.398	0.400	0.997

CHAPTER 8

Application of Simulation Modeling

LEARNING OBJECTIVES

After completing this chapter, you should be able to:

- Understand the issues in developing and applying simulation to real systems.
- Perform experiments and analysis on practical simulation models.

Chapter 1 presented a set of general steps for problem solving called DEGREE. Those steps were expanded into a methodology for problem solving in the context of simulation. As a reminder, the primary steps for a simulation study can be summarized as follows:

1. Problem formulation
   (a) Define the system and the problem
   (b) Establish performance metrics
   (c) Build conceptual model
   (d) Document modeling assumptions
2. Simulation model building
   (a) Model translation
   (b) Input data modeling
   (c) Verification and validation
3. Experimental design and analysis
   (a) Preliminary runs
   (b) Final experiments
   (c) Analysis of results

4. Evaluate and iterate

5. Documentation

6. Implementation of results

To some extent, our study of simulation has followed these general steps. Chapter 1 introduced a basic set of modeling questions and approaches that are designed to assist with step 1:

- What is the system? What information is known by the system?
- What are the required performance measures?
- What are the entities? What information must be recorded or remembered for each entity? How should the entities be introduced into the system?
- What are the resources that are used by the entities? What entities use which resources and how?
- What are the process flows? Sketch the process or make activity flow diagrams.
- Develop pseudo-code for the model.

Answering these questions can be extremely helpful in developing a conceptual understanding of a problem in preparation for the simulation model building steps of the methodology.

Chapter 2 provided the primary modeling constructs to enable the translation of a conceptual model to a simulation model in Arena^{TM}. Since conceptual models often involve randomness, Chapter 3 showed how input distributions can be formed and presented the major methods for obtaining random values to be used in a simulation. Since random inputs imply that the outputs from the simulation will also be random, Chapter 4 showed how to properly analyze the statistical aspects of simulation. This allows appropriate experimental analysis techniques to be successfully employed during a simulation project. Finally, Chapters 5 through 7 delved deeper into a variety of modeling situations to round out the tool set of modeling concepts and Arena^{TM} constructs that you can bring to bear on a problem.

At this point you have learned a great deal concerning the fundamentals of simulation modeling and analysis. The purpose of this chapter is to help you to put your new knowledge into practice by demonstrating the modeling and analysis of a system in its entirety. Ideally, experience in simulating a real system would maximize your understanding of what you have learned; however, a realistic case study should provide this experience given the limitations of a textbook. During the past decade, Rockwell Software (the makers of Arena^{TM}) and the Institute of Industrial Engineers (IIE) have sponsored a student contest involving the use of Arena^{TM} to model a realistic situation. The contest problems have been released to the public and can be found at *www.arenasimulation.com* and are also available on the CD that accompanies this text. This chapter will solve one of the previous contest problems. This process will illustrate the simulation modeling steps from conceptualization to final recommendations. After solving the case study, you should have a better appreciation for and a better understanding of the entire simulation process.

8.1   Problem Description

This section reproduces in its entirety the 7th Annual Contest Problem entitled "SM Testing." Subsequent sections present a detailed solution to the problem. Read through the following section as if you are going to be required to solve the problem. That is, try to apply simulation modeling steps by jotting down your own initial solution. Then, as you study the detailed solution, you can compare your approach to the solution presented here.

8.1.1 Rockwell Software/IIE 7th Annual Contest Problem: SM Testing

SM Testing is the parent company for a series of small medical-laboratory testing facilities. These facilities are often located in or near hospitals or clinics. Many of them are new and came into being as a direct result of cost-cutting measures undertaken by the medical community. In many cases, the hospital or clinic bids its testing out to an external contractor, but provides space for the laboratory in its own facility.

SM Testing provides a wide variety of testing services and has long-term plans to increase its presence in this emerging market. Recently, the company concentrated on a specific type of testing laboratory and has decided to automate most of these facilities. Several pilot facilities have been constructed and have proven to be effective not only in providing the desired service, but are also profitable. The current roadblock to a mass offering of these types of services is the inability to size the automated system properly to specific site requirements. Although a method was developed for the pilot projects, it dramatically underestimated the size of the required system. Thus, additional equipment was required when capacity problems were uncovered. Although this trial-and-error approach eventually provided systems that were able to meet customer requirements, it is not an acceptable approach for the mass introduction of automated systems. The elapsed time from when the systems were initially installed to when they were finally able to meet customer demands ranged from 8 to 14 months. During that time, manual testing supplemented the capacity of the automated system. This proved to be extremely costly.

The company could intentionally oversize the systems as a way of always meeting the projected customer demand, but it is understood that a design that always meets demand may result in a very expensive system with reduced profitability. Managers would like to be able to size these systems easily so that customer requirements are met or surpassed while also minimizing investment. SM Testing has explored several options to resolve this problem, concluding that computer simulation may well provide the technology necessary to size these systems properly.

Before releasing this request for recommendations, the engineering staff developed a standard physical configuration that will be used in all future systems. A schematic of this standard configuration appears in Figure 8-1.

The standard configuration consists of a transportation loop or racetrack joining six different primary locations: one load/unload area and five different test cells. Each testing cell will contain one or more automated testing devices that perform a test specific to that cell. The load/unload area provides the means for entering new samples into the system and removing completed samples from the system.

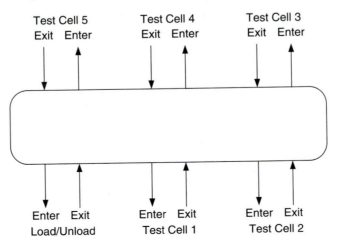

■ **FIGURE 8-1**
Schematic of standard configuration

The transportation loop or racetrack can be visualized as a bucket conveyor or a simple power-and-free conveyor. Samples are transported through the system on special sample holders that can be thought of as small pallets or carts that hold the samples. The number of sample holders depends on the individual system. The transportation loop is 48 feet long, and it can accommodate a maximum of 48 sample holders, spaced at 1-foot increments. Note that because sample holders can also be in a test cell or the load/unload area, the total number of sample holders can exceed 48. The distance between the enter and exit points at the load/unload area and at each of the test cells is 3 feet. The distance between the exit point of one cell and the enter point of the next cell is 5 feet.

Let's walk through the movement of a typical sample through the system. Samples arriving at the laboratory initially undergo a manual preparation for the automated system. The sample is then placed in the input queue to the load/unload area. This manual preparation typically requires about 5 minutes. When the sample reaches the front of the queue, it waits until an empty sample holder is available. At that point, the sample is automatically loaded onto the sample holder, and the unit (sample on a sample holder) enters the transportation loop. The process of a unit entering the transportation loop is much like a car entering a freeway from an on-ramp. As soon as a vacant space is available, the unit (or car) merges into the flow of traffic. The transportation loop moves the units in a counterclockwise direction at a constant speed of 1 foot per second. There is no passing allowed.

Each sample is bar-coded with a reference to the patient file as well as the sequence of tests that need to be performed. A sample follows a specific sequence. For example, one sequence requires that the sample visit test cells 5, 3, and 1 (in that order). Let's follow one of these samples and sample holders (units) through the entire sequence. It leaves load/unload at the position marked exit and moves in a counterclockwise direction past test cells 1 through 4 until it arrives at the enter point for test cell 5. As the unit moves through the system, the bar code is read at a series of points in order for the system to direct the units to the correct area automatically. When it reaches test cell 5, the system checks to see how many units are currently waiting for testing at test cell 5. There is only capacity for three units in front of the testers, regardless of the number of testers in the cell. This capacity is the same for all five test cells. The capacity (three) does not include any units currently being tested or units that have completed testing and are waiting to merge back onto the transportation loop. If room is not available, the unit moves on and will make a complete loop until it returns to the desired cell. If capacity or room is available, the unit will automatically divert into the cell (much like exiting from a freeway). The time to merge onto or exit from the loop is negligible. A schematic of a typical test cell is shown in Figure 8-2.

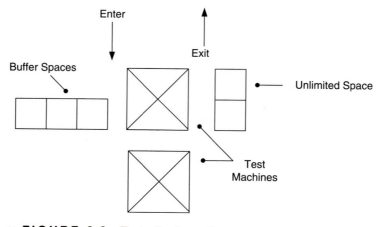

■ **FIGURE 8-2** **Test cell schematic**

As soon as a tester becomes available, the unit is tested, the results are recorded, and the unit attempts to merge back onto the loop. Next it travels to the enter point for test cell 3, where the same logic is applied that was used for test cell 5. Once that test is complete, it is directed to test cell 1 for the last test. When all of the steps in the test sequence are complete, the unit is directed to the enter point for the Unload area.

The data-collection system has been programmed to check the statistical validity of each test. This check is not performed until the sample leaves a tester. If the test results do not fall into accepted statistical norms, the sample is immediately sent back for the test to be performed a second time.

Although there can be a variable number of test machines at each of the test cells, there is only one device at the load/unload area. This area provides two functions: loading of newly arrived samples and unloading of completed samples. Current system logic at this area attempts to ensure that a newly arrived sample never has to wait for a sample holder. Thus, as a sample holder on the loop approaches the enter point for this area, the system checks to see whether the holder is empty or if it contains a sample that has completed its sequence. If the check satisfies either of these conditions, the system then checks to see if there is room for the sample holder in the load/unload area. This area has room for five sample holders, not including the sample holder on the load/unload device or any holders waiting to merge back onto the loop. If there is room, the sample holder enters the area. A schematic for the load/unload area is shown in Figure 8-3.

As long as there are sample holders in front of the load/unload device, it will continue to operate or cycle. It only stops or pauses if there are no available sample holders to process. The specific action of this device depends on the status of the sample holder and the availability of a new sample. There are four possible actions.

- The sample holder is empty and the new sample queue is empty. In this case, there is no action required, and the sample holder is sent back to the loop.
- The sample holder is empty and a new sample is available. In this case, the new sample is loaded onto the sample holder and sent to the loop.
- The sample holder contains a completed sample, and the new sample queue is empty. In this case, the completed sample is unloaded, and the empty sample holder is sent back to the system.
- The sample holder contains a completed sample, and a new sample is available. In this case, the completed sample is unloaded, and the new sample is loaded onto the sample holder and sent to the loop.

The time for the device to cycle depends on many factors, but the staff has performed an analysis and concluded that the cycle time follows a triangular distribution with parameters

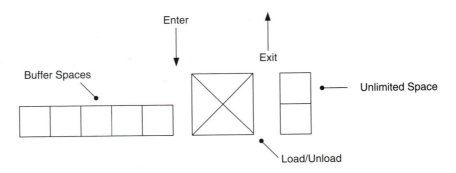

■ **FIGURE 8-3**   **Load/unload cell schematic**

■ **TABLE 8-1**

Tester Cycle Times in Minutes

Tester	Cycle time (minutes)
1	0.77
2	0.85
3	1.03
4	1.24
5	1.7

■ **TABLE 8-2**

Tester Failure and Repair Times

Tester	Mean time between failures (hours)	Mean time to repair (minutes)
1	14	11
3	9	7
4	15	14
5	16	13

0.18, 0.23, and 0.45 (minutes). A sample is not considered complete until it is unloaded from its sample holder. At that time, the system will collect the results from its database and forward them to the individual or area requesting the test.

The time for an individual test is constant but depends on the testing cell. These cycle times are provided in Table 8-1.

Each test performed at Tester 3 requires 1.6 oz of reagent, and 38% of the tests at Tester 4 require 0.6 oz of a different reagent. These are standard reagents and are fed to the testers automatically. Testers periodically fail or require cleaning. The staff has collected data on these activities for four of the testers, summarized in Table 8-2.

The testers for test cell 2 rarely fail, but they do require cleaning after performing 300 tests. The clean time follows a triangular distribution with parameters 5.0, 6.0, and 10.0 (minutes).

The next pilot-testing laboratory will be open 24 hours a day, 7 days a week. The staff has projected demand data for the site, which are provided in Table 8-3. Hour 1 represents the time between midnight and 1 a.m. Hour 24 represents the time between 11 p.m. and midnight. The rate is expressed in average arrivals per hour. The samples arrive without interruption throughout the day at these rates.

Each arriving sample requires a specific sequence of tests, always in the order listed. There are nine possible test sequences, shown with corresponding data in Table 8-4.

SM Testing contracts generally require that test results are provided in 1 hour of sample receipt. For this pilot run, we also need to accommodate rush samples for which test results must be available in 30 minutes. It's estimated that 7% of incoming samples will be labeled "rush." Rush samples are given preference at the load area.

■ **TABLE 8-3**

Sample Arrival Rates

Hour	Rate	Hour	Rate	Hour	Rate
1	119	9	131	17	134
2	107	10	152	18	147
3	100	11	171	19	165
4	113	12	191	20	155
5	123	13	200	21	149
6	116	14	178	22	134
7	107	15	171	23	119
8	121	16	152	24	116

■ **TABLE 8-4**

Test Sequences

Sequence	Sequence steps	Percentage (%)
1	1–2–4–5	9
2	3–4–5	13
3	1–2–3–4	15
4	4–3–2	12
5	2–5–1	7
6	4–5–2–3	11
7	1–5–3–4	14
8	5–3–1	6
9	2–4–5	13

■ **TABLE 8-5**

Equipment Costs for Testers

Equipment	Cost per month ($)
Tester type 1	10,000
Tester type 2	12,400
Tester type 3	8,500
Tester type 4	9,800
Tester type 5	11,200
Sample holder	387

We requested and received cost figures from the equipment manufacturers. These costs include initial capital, operating, and maintenance costs for the projected life of each unit. Monthly costs per unit are provided in Table 8-5.

Management would like to know which equipment configuration would provide the most cost-effective solution while achieving high customer satisfaction. Ideally, the company would always like to provide results in less time than the contract requires. However, management does not believe that the system should include extra equipment just to handle the rare occurrence of a late report.

During a recent SM Testing meeting, a report was presented on the observations of a previous pilot system. The report indicated that completed samples had difficulty entering the load/unload area when the system was loaded lightly. This often caused the completed samples to make numerous loops before they were finally able to exit. A concern was raised that longer-than-necessary test times potentially might cause a system to be configured with excess equipment.

With this in mind, SM Testing approached its equipment vendor and requested a quote to implement alternate logic at the exit point for the load/unload area only. The proposal gives priority to completed samples exiting the loop. When a sample holder on the loop reaches the enter point for this area, the system checks the holder to see whether it is empty or contains a sample that has completed its sequence. If the sample holder contains a completed sample and there is room at load/unload, it leaves the loop and enters the area. If the sample holder is empty, it checks to see how many sample holders are waiting in the area. If that number is less than a given "suggested number," say 2, the sample holder leaves the loop and enters the area. Otherwise, it continues around the loop. The idea is to always attempt to keep a sample holder available for a new sample, but not to fill the area with empty sample holders. The equipment vendor has agreed to provide this new logic at a one-time cost of $85,000. As part of your proposal, evaluate this new logic, including determining the best value of the suggested number.

Your report should include a recommendation for the most cost-effective system configuration. This should include the number of testers at each cell, the number of sample holders, and a decision on the proposed logic. Please provide all cost estimates on a per-month basis.

The company is currently proceeding with the construction of the new facility and will not require a solution for 2 months. Since there are several groups competing for this contract, additional information will not be provided by the company during the analysis period. However, you are encouraged to make additional reasonable, documented assumptions.

8.2 Detailed Solution

From your reading of the contest problem, you should have some initial ideas for how to approach this modeling effort. The following sections will walk you through the simulation modeling steps in order to develop a detailed solution to the contest problem. As you proceed through the sections, you might want to jot down your own ideas and attempt your own analysis. While your approach may be different from what is presented here, your efforts and engagement in the problem will be important to how much you get out of this process. Let's begin the modeling effort with a quick iteration through the basic modeling questions.

8.2.1 Answering the Basic Modeling Questions

- *What is the system? What information is known by the system?*

The purpose of the system is to process test samples in test cells carried by sample holders on a conveyor. The system is well described by the contest problem. Thus, there is no need to repeat the system description here. In summary, the information known by the system includes the following:

- Conveyor speed, total length, and spacing around the conveyor of entry and exit points for the test cells and the load/unload area
- Load/unload machine cycle time, TRIA(0.18, 0.23, 0.45) minutes
- Manual preparation time
- Cycle times for each tester (Table 8-1)
- Time to failure and repair distributions for testers 1, 3, 4, and 5 (Table 8-2)
- Test 2 usage count to cleaning and cleaning time distribution, TRIA(5.0, 6.0, 10.0) minutes
- Sample arrival mean arrival rate by hour of the day (Table 8-3)
- Nine different test sequences for the samples along with the probability associated with each sequence
- A distribution governing the probability of rush samples in the system and a criteria for rush samples to meet (30-minute testing time)
- Equipment costs for testers and sample holders, as well as the cost of implementing additional logic in the model

From this initial review of the information and the contest problem description, you should be able to develop an initial list of the Arena™ modeling constructs that may be required in the model.

The initial list should concentrate on identifying data modules. This will help in organizing the data related to the problem. From the list, you can begin to plan (even at this initial modeling stage) on making the model more "data driven." That is, by thinking about the input parameters in a more general sense (e.g., expressions and variables), you can build the model in a manner that can improve its usability during experimentation. Based on the current information, a basic list of Arena™ modeling constructs might contain the following modules.

CONVEYOR and SEGMENT: To model the conveyor.

STATION: To model the entry and exit points on the conveyor for the testers and the load/unload cell.

SEQUENCE: To model the test sequences for the samples.

EXPRESSION: To hold the load/unload cycle time and for an array of the cycle times for each tester. Can also be used to hold an expression for the manual preparation time, the sequence probability distribution and the rush sample distribution.

FAILURE: To model both count-based (for tester 2) and time-based failures.

ARRIVAL Schedule: To model the nonstationary arrival pattern for the samples.

VARIABLE: To include various system-wide information such as cost information, number of sample holders, load/unload machine buffer capacity, test cell buffer capacity, test result criteria, and so on.

Once you have ideas about the data associated with the problem, you can proceed with the identification of the performance measures that will be the major focus of the modeling effort.

■ *What are the required performance measures?*

From the contest problem description, the monthly cost of the system should be considered a performance measure for the system. The cost of the system will certainly affect its operational performance. In terms of meeting the customer's expectations, it appears that the system time for a sample is important. In addition, the contract specifies that test results must be available within 60 minutes for regular samples and within 30 minutes for rush samples. While late reports do not have to be prevented, they should be rare. Thus, the major performance measures follow:

■ Monthly cost
■ Average system time
■ Probability of meeting contract system time criteria

In addition to these primary performance measures, it would be useful to keep track of resource utilization, queue times, and size of queues, among other system elements, all of which are natural outputs of a typical Arena™ model.

■ *What are the entities? What information must be recorded or remembered for each entity? How should the entities be introduced into the system?*

Recalling the definition of an entity (an object of interest in the system whose movement or operation in the system may cause the occurrence of events), there appear to be two natural candidates for entities in the model.

Samples: Arrive according to a pattern and travel through the system

Sample holders: Move through the system (e.g., on the conveyor)

To further solidify your understanding of these candidate entities, you should consider their possible attributes. If attributes can be identified, then it is highly likely that these are entities. Based on the problem description, every sample must have a priority (rush or not), an arrival time (to compute total system time), and a sequence. Thus, priority, arrival time, and sequence appear to be natural attributes for a sample. Since a sample holder needs to know whether it has a sample, an attribute should be considered to keep track of this for each sample holder. Because both samples and sample holders have clear attributes, there is no reason to drop them from the list of candidate entities.

Thinking about how the candidate entities can enter the model will also assist in helping to understand their modeling. Since the samples arrive according to a nonstationary arrival process (with mean rates that vary by hour of the day), the use of a CREATE module with an ARRIVAL schedule appears to be appropriate. The introduction of sample holders is more problematic. Sample holders don't arrive at the system. They are just in the system when it starts operating. Thus, it is not immediately clear how to introduce the sample holders. Ask yourself, how can a CREATE module be used to introduce sample holders so that they are always in the system? If all the sample holders "arrive" at time 0.0, then they will be available when the system starts up. Thus, a CREATE module can be used to create sample holders.

At this point in the modeling effort, a first iteration on entity modeling has been accomplished; however, it is important to understand that this is just a first attempt at modeling and to remember to be open to future revisions. As you go through the rest of the modeling effort, you should be prepared to revise your current conceptual model as understanding deepens.

To build on the entity modeling, it is natural to ask what resources are used by the entities. This is the next modeling question.

- *What resources are used by the entities? Which entities use which resources and how?*

Again, it is useful to reconsider the definition of a resource.

Resource: A limited quantity of items that are used (e.g., seized and released) by entities as they proceed through the system. A resource has a capacity that governs the total quantity of items that may be available. All the items in the resource are homogeneous, meaning that they are indistinguishable. If an entity attempts to use a resource that does not have units available, it must wait in a queue.

Thus, the natural place to look for resources is where a queue of entities may form. Sample holders (with a sample) wait for testers. In addition, sample holders, with or without a sample, wait for the load/unload machine. Thus, the testers and the load/unload machine are natural candidates for resources in the model. In addition, the problem states the following:

Samples arriving at the laboratory initially undergo a manual preparation for the automated system. The sample is then placed in the input queue to the load/ unload area. This manual preparation typically requires about 5 minutes. When the sample reaches the front of the queue, it waits until an empty sample holder is available.

It certainly appears from this wording that the availability of sample holders can constrain the movement of samples in the model. Thus, sample holders are a candidate for resource modeling.

A couple of remarks about this sort of modeling are in order. First, it should be more evident from the things identified as waiting in the queues that samples and sample holders are even more likely to be entities. For example, sample holders wait in queue to get on the conveyor. Now, since sample holders wait to get on a conveyor, does that mean that the conveyor is a candidate resource? Absolutely, yes! In this modeling, you are identifying resources with a little "r." You are not identifying RESOURCES (i.e., things to put in the RESOURCE module)! Be aware that there are many ways to model resources in a simulation (e.g., inventory as a resource with WAIT/SIGNAL). The RESOURCE module is just one very specific method. Just because you identify something as a potential resource, it does not mean you have to use the RESOURCE module to model how it constrains the flow of entities.

The next step in modeling is to try to better understand the flow of entities by attempting to give a process description.

- *What are the process flows? Sketch the process or make an activity flow diagram.*

The first thing to remember when addressing process flow modeling is that you are building a *conceptual* model for the process flow. Recall that one way to do this is to consider an activity diagram (or some sort of augmented flow chart). Although in many of the previous modeling examples there was almost a direct mapping from the conceptual model to the Arena^{TM} model, you should not expect this to occur for every modeling situation. In fact, you

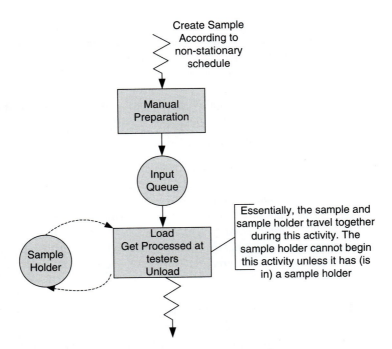

■ **FIGURE 8-4**   **Activity diagram for samples**

should try to build up your conceptual understanding independent of the modeling language. In addition, the level of detail included in conceptual modeling is entirely up to you! Thus, if you don't know how to model a particular detail then you can just "black box it" or just omit it to be handled later.

Suppose an entity goes into an area for which a lot of detail is required. Just put a box on your diagram and indicate in a general manner what should happen in the box. If necessary, you can come back to it later. If you have complex decision logic that would really clutter up the diagram, then omit it in favor of first understanding the big picture. The complicated control logic for sample holders accessing the load/unload device and being combined with samples is an excellent candidate for this technique.

Before proceeding you might want to try to sketch out activity diagrams for the samples and the sample holders. Figure 8-4 presents a simplified activity diagram for the samples. They are created, go through a manual preparation activity, and then flow into the input queue. Whether they wait in the input queue depends on the availability of the sample holder. Once they have a sample holder, they proceed through the system. The sample must have a sample holder to move through the system. Figure 8-5 illustrates a high level activity cycle diagram for the sample holders. Based on a sequence, they convey (with the sample) to the next appropriate tester. After each test is completed, the sample and holder are conveyed to the next tester in the sequence until they have completed the sequence. At that time, the sample holder and sample are conveyed to the load/unload machine where they experience the load/unload cycle time once they have the load/unload machine. If a sample is not available, the sample holder is conveyed back to the load/unload area.

One thing to note in the diagram is that the sequence does not have to be determined until the sample and holder are being conveyed to the first appropriate test cell. Also, it is apparent from the diagram that the load/unload machine is the final location visited by the sample and sample holder. Thus, the enter load/unload station can be used as the last location when using Arena's SEQUENCE module. This diagram does not contain the extra logic for testing if the

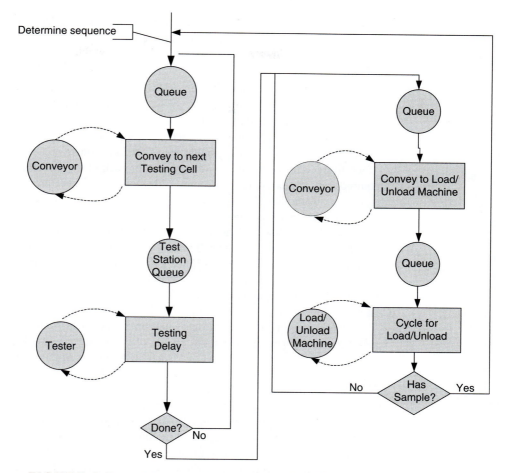

■ FIGURE 8-5   Activity cycle diagram for sample holders

queue in front of a tester is full. It also does not contain the logic to control the buffer at the load/unload machine.

8.2.2   Detailed Modeling

Given your conceptual understanding of the problem, you should now be ready to perform more detailed modeling in order to prepare for implementing the model in Arena^{TM}. When doing detailed modeling, it is often useful to break up the problem into components that can be more easily addressed. This is the natural problem-solving technique called "divide and conquer." Here you must think of the major system components, tasks, or modeling issues that need to be addressed and then work on them first separately and then concurrently in order to have the solution eventually come together as a whole. The key modeling issues for the contest problem appear to be the following:

- Conveyor and station modeling (i.e., the physical modeling)
- Modeling samples
- Test-cell modeling including failures
- Modeling sample holders
- Modeling the load/unload area

- Performance measure modeling (cost and statistical collection)
- Simulation horizon and run parameters

The following sections will examine each of these issues in turn.

8.2.2.1 Conveyor and Station Modeling.

Recall that Arena's conveyor constructs require that each segment of a conveyor be associated with stations in the model. From the problem, the conveyor should be 48 total feet in length. In addition, the problem gives 5-foot and 3-foot distances between the respective enter and exit points. Since the samples access the conveyor 3 feet from the location that they exited the conveyor, a segment is required for this 3-foot distance for the load/unload area and for each of the test cells. Thus, a station will be required for each exit and enter point on the conveyor. Therefore, there are 12 total stations required in the model. The exit point for the load/unload machine can be arbitrarily selected as the first station on the loop conveyor. Figure 8-6 shows the segments for the conveyor. The test cells as well as the load/unload machine have exit and enter stations that define the segments. The exit point is for exiting the cell to access the conveyor. The enter point is for entering the cell (getting off of the conveyor). Since this is a loop conveyor, the last station listed on the segments should be the same as the first station used to start the segments. The problem also indicates that the sample holder takes up 1 foot when riding on the conveyor. Thus, it seems natural to model the cell size for the conveyor as 1 foot.

The conveyor module for this situation is shown in Figure 8-7. The velocity of the conveyor is 1 foot per second with a cell size of 1 foot. Since a sample holder requires 1 foot of space on the conveyor and the cell size is 1 foot, the maximum number of cells occupied by an entity on the conveyor is simply one cell.

Since the stations are defined, the sequences in the model can now be defined. How you implement the sequences depends on how the conveyor is modeled and on how the physical locations of the work cells are mapped to the concept of stations in Arena™. As in any Arena™ model, there are a number of different methods to achieve the same objective. With stations defined for each entry and exit point, the sequences may use each of the entry and exit points. It should be clear that the entry points must be on the sequences; however, because the exit points

Segment - Advanced Transfer

	Name	Beginning Station	Next Stations
1	Loop Conveyer.Segment	ExitLoadUnloadStation	12 rows

Next Stations ☒

	Next Station	Length
1	TestCell1EnterStation	5
2	TestCell1ExitStation	3
3	TestCell2EnterStation	5
4	TestCell2ExitStation	3
5	TestCell3EnterStation	5
6	TestCell3ExitStation	3
7	TestCell4EnterStation	5
8	TestCell4ExitStation	3
9	TestCell5EnterStation	5
10	TestCell5ExitStation	3
11	EnterLoadUnloadStation	5
12	ExitLoadUnloadStation	3

■ **FIGURE 8-6** Conveyor segments

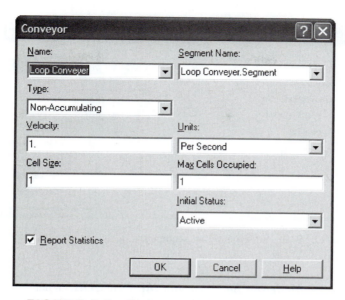

■ **FIGURE 8-7** Conveyor module

are also stations you have the option of using them on the sequences as well. By using the exit point on the sequence, a ROUTE module can be used with the By Sequence option to send a sample/sample holder to the appropriate exit station after processing. This could also be achieved by a direct connection, in which case it would be unnecessary to have the exit stations in the sequences. Figure 8-8 illustrates the sequence of stations for test sequence 1 for the problem. Note that the stations alternate (enter then exit) and that the last station is the station representing the entry point for the load/unload area.

In order to randomly assign a sequence, the sequences can be placed into a set, and then the index into the set randomly generated via the appropriate probability distribution. Figure 8-9 shows the Advanced Set module used to define the set of sequences for the model. In addition, Figure 8-10 shows the implementation of the distribution across the sequences as an expression, called *eSeqIndex*, using the DISC() function in the EXPRESSION module.

Sequence - Advanced Transfer				Steps				
	Name	Steps			Station Name	Step Name	Next Step	Assignments
1	Sequence1	9 rows		1	TestCell1EnterStation			0 rows
2	Sequence2	7 rows		2	TestCell1ExitStation			0 rows
3	Sequence3	9 rows		3	TestCell2EnterStation			0 rows
4	Sequence4	7 rows		4	TestCell2ExitStation			0 rows
5	Sequence5	7 rows		5	TestCell4EnterStation			0 rows
6	Sequence6	9 rows		6	TestCell4ExitStation			0 rows
7	Sequence7	9 rows		7	TestCell5EnterStation			0 rows
8	Sequence8	7 rows		8	TestCell5ExitStation			0 rows
9	Sequence9	7 rows		9	EnterLoadUnloadStation			0 rows

■ **FIGURE 8-8** Test sequence 1

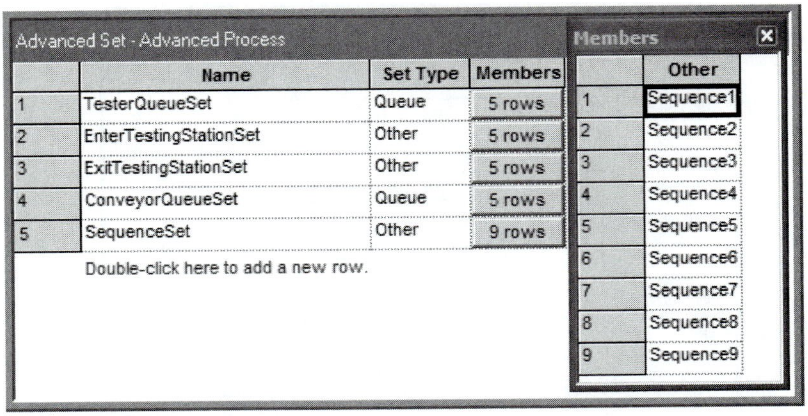

■ **FIGURE 8-9** Set for holding sequences

8.2.2.2 Modeling Samples and the Test Cells. When modeling a large complex problem, you should try to start with a simplified situation. This allows a working model to be developed without unnecessarily complicating the modeling effort. Then, as confidence in the basic model improves, enhancements to the basic model can take place. It will be useful to do just this when addressing the modeling of samples and sample holders.

As indicated in the activity diagrams, once a sample has a sample holder, the sample holder and sample essentially become "one" as they move through the system. Thus, a sample also follows the basic flow as laid out in Figure 8-5. This is essentially what happens to the sample when it is in the "black box" of Figure 8-4. Thus, to simplify the modeling, it is useful to assume that a sample holder is always available whenever a sample arrives. Since sample holders are not being modeled, there is no need to model the details of the load/unload machine. This assumption implies that once a sample completes its sequence it can simply exit at the load/unload area, and that any newly arriving samples simply get on at the load/unload area. With these assumptions, the modeling of the sample holders and the load/unload stations (including its complicated rules) can be bypassed for the time being.

From the conceptual model (activity diagram), it should be clear that the testers can be easily modeled with the RESOURCE module in Arena™. In addition, the diagram indicates that the logic at any test cell is essentially the same. This could lead to the use of generic stations. Based on all these ideas, you should be able to develop pseudo-code for this situation.

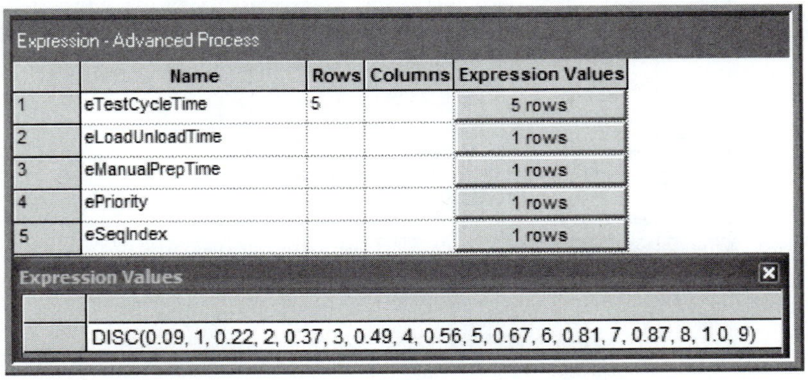

■ **FIGURE 8-10** Discrete distribution for assigning random sequence

Exhibit 8-1   Pseudo-Code Samples

```
CREATE sample according to nonstationary pattern
ASSIGN myEnterSystemTime = TNOW
    myPriorityType ~ Priority CDF
DELAY for manual preparation time
ROUTE to ExitLoadUnloadStation

STATION ExitLoadUnLoadStation
ASSIGN mySeqIndex ~ Sequence CDF
    Entity.Sequence = MEMBER(SequenceSet, mySeqIndex)
    Entity.JobStep = 0
ACCESS Loop Conveyor
CONVEY by Sequence

STATION Generic Testing Cell Enter
DECIDE If NQ at testing cell < cell waiting capacity
    EXIT Loop Conveyor
    SEIZE 1 unit of tester
    DELAY for testing time
    RELEASE 1 unit of tester
    ROUTE to Generic Testing Cell Exit
Else
    CONVEY back to Generic Testing Cell Enter

STATION Generic Testing Cell Exit
ACCESS Loop Conveyor
CONVEY by Sequence

STATION EnterLoadUnloadStation
EXIT Loop Conveyor
RECORD System time
DISPOSE
```

If you are following along, you might want to pause and sketch out your own pseudo-code for the simplified modeling situation.

Exhibit 8-1 shows possible pseudo-code for this initial modeling. Samples are created according to a nonstationary arrival pattern and then are routed to the exit point for the load/unload station. Then, the samples access the conveyor and are conveyed to the appropriate test cell according to their sequence. Once at a test cell, they test if there is room to get off the conveyor. If so, they exit the conveyor and then use the appropriate tester (SEIZE, DELAY, RELEASE). After using the tester, they are routed to the exit point for the test cell, where they access the conveyor and are conveyed to the next appropriate station. If space is not available at the test cell, they do not exit the conveyor, but rather are conveyed back to the test cell to try again. Once the sample has completed its sequence, the sample will be conveyed to the enter load/unload area, where it exits the conveyor and is disposed.

Given the pseudo-code and the previous modeling, you should be able to develop an initial Arena^{TM} model for this simplified situation. For additional practice, you might try to implement the ideas that have been discussed before proceeding.

The Arena^{TM} model (with animation) representing this initial modeling is given in the file, *SMTestingInitialModeling.doe*. The flow chart modules corresponding to the pseudo-code of Exhibit 8-2 are shown in Figure 8-11.

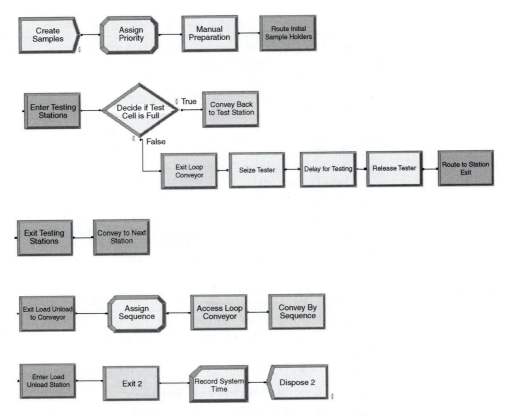

■ **FIGURE 8-11** Initial Arena model for samples and test cells

The following approach was taken when developing the initial Arena™ model for samples and test cells:

Variables: A variable array, *vTestCellCapacity(5)*, was defined to hold the capacity of each test cell. While for this particular problem, the cell capacity was the same for each cell, by making a variable array, the cell capacity can be easily varied if needed when testing the various design configurations.

Expressions: An arrayed expression, *eTestCycleTime(5)*, was defined to hold the cycle times for the testing machines. By making these expressions, they can be easily changed from one location in the model. In addition, expressions were defined for the manual preparation time (*eManualPrepTime*), priority distribution (*ePriority*), and sequence distribution (*eSeqIndex*).

Resources: Five separate resources were defined to represent the testers (*Cell1Tester, Cell2Tester, Cell3Tester, Cell4Tester, Cell5Tester*).

Sets: A resource set, *TestCellResourceSet*, was defined to hold the test cell resources for use in generic modeling. An entity picture set (*SamplePictSet*) with 9 members was defined to be able to change the picture of the sample according to the sequence it was following. A queue set, *TesterQueueSet*, was defined to hold the queues in front of each tester for use in generic modeling. In addition, a queue set, *ConveyorQueueSet*, was defined to hold the access queue for the conveyor at each test cell. Two station sets were defined to hold the enter (*EnterTestingStationSet*) and the exit (*ExitTesting-StationSet*) stations for generic station modeling. Finally, a sequence set, *Sequence-Set*, was use to hold the nine sequences followed by the samples.

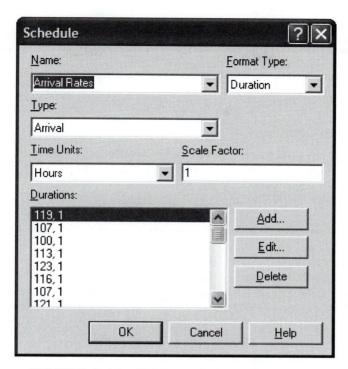

■ **FIGURE 8-12**   ARRIVAL schedule for samples

Schedules: An ARRIVAL schedule (Figure 8-12) was defined to hold the arrival rates by hour to represent the nonstationary arrival pattern for the samples.

Failures: Five failure modules were used to model the failure and cleaning associated with the test cells. Four failure modules (*Tester 1 Failure*, *Tester 3 Failure*, *Tester 4 Failure*, and *Tester 5 Failure*) defined time-based failures and the appropriate repair times. A count-based failure was used to model Tester 2's cleaning after 300 uses. The failures are illustrated in Figure 8-13. An important assumption with the use of the FAILURE module is that the entire resource is categorized as failed when a failure occurs. It is not clear from the contest problem specification what would happen at a test cell that has more than one tester when a failure occurs. Thus, for the sake of simplicity it will be useful to assume that units in a multiple unit tester do not fail individually.

As a first step toward verifying that the model is working as intended, the initial model, *SMTestingInitialModeling.doe*, can be modified so that only one entity is created. For each of the nine different sequences the total distance traveled on the sequence and the total time for testing can be calculated. For example, for the first sequence, the distance from load/unload exit

Failure - Advanced Process

	Name	Type	Up Time	Up Time Units	Count	Down Time	Down Time Units	Uptime in this State only
1	Tester 1 Failure	Time	EXPO(14)	Hours	Hours	EXPO(11)	Minutes	
2	Tester 3 Failure	Time	EXPO(9)	Hours	Hours	EXPO(7)	Minutes	
3	Tester 4 Failure	Time	EXPO(15)	Hours	Hours	EXPO(14)	Minutes	
4	Tester 5 Failure	Time	EXPO(16)	Hours	Hours	EXPO(13)	Minutes	
5	Tester 2 Failure	Count	1.0	Hours	300	TRIA(5 , 6 , 10)	Minutes	

■ **FIGURE 8-13**   Failure modules for SM Testing

■ **TABLE 8-6**

Single-Sample System Time

Sequence	Sequence steps	Distance (ft)	Travel time	Test time	Prep time	Total time
1	1–2–4–5	33	0.55	5.11	5.0	10.11
2	3–4–5	36	0.60	3.97	5.0	9.57
3	1–2–3–4	33	0.55	3.89	5.0	9.44
4	4–3–2	132	2.20	3.12	5.0	10.32
5	2–5–1	84	1.40	3.32	5.0	9.72
6	4–5–2–3	81	1.35	4.82	5.0	11.17
7	1–5–3–4	81	1.35	4.74	5.0	11.09
8	5–3–1	132	2.20	3.50	5.0	10.70
9	2–4–5	36	0.6	3.79	5.0	9.39

to the enter location of cell 1 is 5 feet, the distance from cell 1 exit to cell 2 enter is 5 feet, the distance from cell 2 exit to cell 4 enter is 13 feet, the distance from cell 4 exit to cell 5 enter is 5 feet, and finally, the distance from cell 5 exit to the enter point for load/unload is 5 feet. This totals 33 feet as shown in Table 8-6. The time to travel 33 feet at 1 foot/second is 0.55 minutes. The total testing time for this sequence is $(0.77 + 0.85 + 1.24 + 1.77 = 5.11)$ minutes. Finally, the preparation time is 5 minutes for all sequences. Thus, the total time that it should take a sample following sequence 1 should be 10.11 minutes, assuming no waiting and no failures. To check that the model can reproduce this quantity, nine different model runs were made such that the single entity followed each of the nine different sequences. The total system time for the single entity was recorded and matched exactly those figures given in Table 8-6. This should give you confidence that the current implementation is working correctly, especially that there are no problems with the sequences. The modified model for testing sequence 9 is given in Arena^{TM} file, *SMTestingInitialModelingTest1Entity.doe*. This testing also provides a lower bound on the expected system times for each of the sequences.

Now that a preliminary working model is available, an initial investigation of the resources required by the system and further verification can be accomplished. From the given sequences, the total percentage of the samples that will visit each of the test cells can be tabulated. For example, test cell 1 is visited in sequences 1, 3, 5, 7, and 8. Thus, the total percentage of the arrivals that must visit test cell 1 is $(0.09 + 0.15 + 0.07 + 0.14 + 0.06 = 0.51)$. The total percentage for each of the test cells is provided in Table 8-7.

Using the total percentages given in Table 8-7, the mean arrival rate to each of the test cells for each hour of the day can be computed. From the data given in Table 8-3, you can see that the minimum hourly arrival rate is 100 and that it occurs in the 3rd hour. The maximum hourly arrival rate of 200 customers per hour occurs in the 13th hour. From the minimum and

■ **TABLE 8-7**

Percentage of Arrivals Visiting Each Test Cell

Cell	Percentage from sequences							Total %
1	0.09	0.15	0.07	0.14	0.06			0.51
2	0.09	0.15	0.12	0.07	0.11	0.13		0.67
3	0.13	0.15	0.12	0.07	0.11	0.13		0.71
4	0.09	0.13	0.15	0.12	0.11	0.14	0.13	0.87
5	0.09	0.13	0.07	0.11	0.14	0.06	0.13	0.73

■ **TABLE 8-8**

Offered Load Calculation for Peak Hourly Rate

Test cell	Arrival rate (per hour)	Testing time (minutes)	Testing time (hours)	Service rate (per hour)	Offered load
1	102	0.77	0.01283	77.92	1.31
2	134	0.85	0.01417	70.59	1.90
3	142	1.03	0.01717	58.25	2.44
4	174	1.24	0.02067	48.39	3.60
5	146	1.70	0.02833	35.29	4.14

maximum hourly arrival rates, you can get an understanding of the range of resource requirements for this system. This will not only help when solving the problem, but will also help in verifying that the model is working as intended.

Let's look at the peak arrival rate first. From the discussion in Chapter 5, the offered load is a dimensionless quantity that gives the average amount of work offered per time unit to the c servers in a queuing system. The offered load is defined as $r = \lambda/\mu$. This can be interpreted as each customer arriving with $1/\mu$ average units of work to be performed. It can also be thought of as the expected number of busy servers if there are an infinite number of servers. Thus, the offered load can give a good idea of about how many servers might be utilized. Table 8-8 calculates the offered load for each of the test cells under the peak arrival rate.

The arrival rate for the first test cell is $0.51 \times 200 = 102$. Thus, you can see that a little more than one server can be expected to be busy at the first test cell under peak loading conditions.

You can confirm these numbers by modifying the Arena™ model so that the resource capacities of the test cells are infinite and by removing the failure modules from the resources. In addition, the CREATE module will need to be modified so that the arrival rate is 200 samples per hour. If you run the model for 10 days, 24 hours per day, the results shown in Figure 8-14 will be produced. As seen in the figure, the results very closely match the results of the offered load analysis.

These results provide additional evidence that the initial model is working properly and indicate how many units of each test cell might be needed under peak conditions. The model is provided in the Arena™ file, *SMTestingInitialModelingOnResources.doe*. Table 8-9 indicates the offered load under the minimum arrival rate.

Based on the analysis of the offered load, a preliminary estimate of the resource requirements for each test cell can be determined as in Table 8-10. The requirements were determined by rounding up the computed offered load for each test cell. This provides a range of values, since it is not clear that designing to the maximum is necessary given the nonstationary behavior of the arrival of samples.

Number Busy	Average	Half Width
Cell1Tester	1.3132	0.017101725
Cell2Tester	1.8917	0.018252301
Cell3Tester	2.4236	0.024626714
Cell4Tester	3.5740	0.034768944
Cell5Tester	4.1244	0.047537947

■ **FIGURE 8-14**   **Results from Arena offered load experiment**

■ **TABLE 8-9**
Offered Load Calculation for Minimum Hourly Rate

Test cell	Arrival rate (per hour)	Testing time (minutes)	Testing time (hours)	Service rate (per hour)	Offered load
1	51	0.77	0.01283	77.92	0.65
2	67	0.85	0.01417	70.59	0.95
3	71	1.03	0.01717	58.25	1.22
4	87	1.24	0.02067	48.39	1.80
5	73	1.70	0.02833	35.29	2.07

■ **TABLE 8-10**
Preliminary Test Cell Resource Requirements

Test cell	Resource requirements	
	Low	High
1	1	2
2	1	2
3	2	3
4	2	4
5	3	5

8.2.2.3 Modeling Sample Holders and the Load/Unload Area. Now that a basic working model has been developed and tested, you should feel comfortable with more detailed modeling involving the sample holders and the load/unload area. From the previous discussion, the sample holders appeared to be an excellent candidate for being modeled as an entity. In addition, it also appeared that the sample holders could be modeled as a resource because they also constrain the flow of the sample if a sample holder is not available. There are a number of modeling approaches possible for addressing this situation. By considering the functionality of the load/unload machine, the situation may become more clear.

The problem states that:

As long as there are sample holders in front of the load/unload device, it will continue to operate or cycle. It only stops or pauses if there are no available sample holders to process.

Since the load/unload machine is clearly a resource, the fact that sample holders wait in front of it and that it operates on sample holders indicates that sample holders should be entities. Further consideration of the four possible actions of the load/unload machine can provide some insight on how samples and sample holders interact. As a reminder, the four actions follow:

■ The sample holder is empty and the new sample queue is empty. In this case, there is no action required, and the sample holder is sent back to the loop.

■ The sample holder is empty and a new sample is available. In this case, the new sample is loaded onto the sample holder and sent to the loop.

■ The sample holder contains a completed sample, and the new sample queue is empty. In this case, the completed sample is unloaded, and the empty sample holder is sent back to the system.

■ The sample holder contains a completed sample, and a new sample is available. In this case, the completed sample is unloaded, and the new sample is loaded onto the sample holder and sent to the loop.

Thus, if a sample holder contains a sample, it is unloaded during the load/unload cycle. In addition, if a sample is available, it is loaded onto the sample holder during the load/unload cycle. The cycle time for the load/unload machine varies according to a triangular distribution, but it appears as if both the loading and/or the unloading can happen during the cycle time. Thus, for load, unload, or both, the time using the load/unload machine is triangularly distributed. If the sample holder has a sample, then during the load/unload cycle, they are separated from each other. If a new sample is waiting, then the sample holder and the new sample are combined together during the processing. This indicates two possible approaches to modeling the interaction between sample holders and samples:

BATCH and SEPARATE: The BATCH module can be used to batch the sample and the sample holder together. Then the SEPARATE module can be used to split them apart.

PICKUP and DROPOFF: The PICKUP module can be used to have the sample holder pick up a sample, and the DROPOFF module can be used to have the sample holder drop off a completed sample.

Of the two methods, the PICKUP/DROPOFF approach is potentially easier since it puts the sample holder in charge. In the BATCH/SEPARATE approach, it will be necessary to properly coordinate the movement of the sample and the sample holder (perhaps through MATCH, WAIT, and SIGNAL modules). In addition, the BATCH module requires careful thought about how to handle the formation of the representative entity and its attributes. In what follows the PICKUP/DROPOFF approach will be used. As an exercise, you might attempt the BATCH/SEPARATE approach.

To begin the modeling of sample holders and samples, you need to think about how they are created and introduced into the model. The samples should continue to be created according to the nonstationary arrival pattern, but after preparation occurs they need to wait until they are picked up by a sample holder. This type of situation can be modeled very effectively with a HOLD module with the infinite hold option specified. In a sense, this is just like the bus system situation of Chapter 7. Now you should be able to sketch out the pseudo-code for this situation.

The pseudo-code for this situation has been developed as shown in Exhibit 8-2. As seen in the exhibit, as a sample holder arrives at the enter point for the load/unload station it first checks to see if there is space in the load/unload buffer. If not, it is simply conveyed back around. If there is space, it checks to see if it is empty and there are no waiting samples. If so, it can simply convey back around the conveyor. Finally, if it is not empty or if there is a sample waiting, it will exit the conveyor to try to use the load/unload machine. If the machine is busy, the sample holder must wait in a queue until the machine is available. Once the machine is seized, it can drop off the sample (if it has one) and pick up a new sample (if one is available). After releasing the load/unload machine, the sample holder possibly with a picked up sample, goes to the exit load/unload station to try to access the conveyor. If it has a sample, it conveys by sequence; otherwise, it conveys back around to the entry point of the load/unload area.

One final issue concerning the sample holders needs to be addressed. How should the sample holders be introduced into the model? As mentioned previously, a CREATE module can be used to create the required number of sample holders at time 0.0. Then, the newly created sample holders can be immediately sent to the *ExitLoadUnloadStation*. This will cause the newly created and empty sample holders to try to access the conveyor. They will ride around on the conveyor empty until samples start to arrive, at which time, the logic will cause them to exit the conveyor to pick up samples.

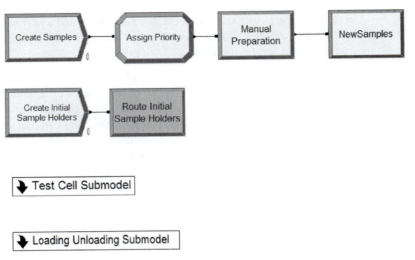

■ **FIGURE 8-15** Overview of entire model

Figure 8-15 provides an overview of the entire model. The samples are created, have their priority assigned, experience manual preparation, and then wait in an infinite hold until picked up by the sample holders. The hold queue for the samples is ranked by the priority of the sample, in order to give preference to rush samples. The initial sample holders are created at time 0.0 and routed to the *ExitLoadUnloadStation*. Two Arena™ submodels were used to model the test cells and the load/unload area. The logic for the test cell submodel is exactly the same as described during the initial model development.

The Arena™ file, *SMTesting.doe*, contains the completed model. The logic described in Exhibit 8-2 is implemented in the Loading Unloading Submodel. The modeling of the alternate logic is also included in this model.

Now that the basic model is completed, you can concentrate on issues related to running and using the results of the model.

8.2.2.4 Performance Measure Modeling. Of the modeling issues identified at the beginning of this section, the only remaining issues are the collection of the performance statistics and the simulation run parameters. The key performance statistics are related to the system time and the probability of meeting the contract time requirements. From your study of Chapters 4 and 5, these statistics can be readily captured using RECORD modules. The submodel for the load/unload area has an additional submodel that implements the collection of these statistics using RECORD modules. Figure 8-16 illustrates this portion of the model. A variable that keeps track of the number of sample holders in use is decremented each time a sample holder drops off a sample (and incremented any time a sample holder picks up a sample). After the sample is dropped off the number of samples completed is recorded with a counter and the system time is tallied with a RECORD module. The overall probability of meeting the 60-minute limit is tallied using a Boolean expression. Then, the probability is recorded according to whether the sample was a rush or not.

The cost of each configuration can also be tabulated using output statistics (Figure 8-17). Even though there is no randomness in these cost values, these values can be captured to facilitate the use of the Process Analyzer and OptQuest.

For example, Table 8-11 shows the total monthly cost calculation for the high-resource case assuming the use of 16 sample holders.

Now that the model has been set up to collect the appropriate statistical quantities, the simulation time horizon and run parameters must be determined for the experiments.

Exhibit 8-2   Pseudo-Code Load/Unload Cell

```
CREATE sample according to nonstationary pattern
ASSIGN myEnterSystemTime = TNOW
    myPriorityType ~ Priority CDF
DELAY for manual preparation time
HOLD until picked up by sample holder

STATION EnterLoadUnloadStation
DECIDE IF NQ at load/unload cell >= buffer capacity
    CONVEY back to EnterLoadUnloadStation
  ELSE//might be able to enter
    If sample holder is empty and
      new sample queue is empty//no need to enter
      CONVEY back to EnterLoadUnloadStation
    ELSE
      EXIT Loop Conveyor
      SEIZE load/unload machine
      DROPOFF sample if loaded
      DELAY for load/unload cycle time
      If new sample queue is not empty
        PICKUP new sample
      End If
      RELEASE load/unload machine
      ROUTE to ExitLoadUnLoadStation
        End If
      End If
STATION ExitLoadUnLoadStation
DECIDE IF the sample holder has a sample
      ASSIGN mySeqIndex ~ Sequence CDF
      Entity.Sequence = MEMBER(SequenceSet, mySeqIndex)
      Entity.JobStep = 0
      End If
ACCESS Loop Conveyor
DECIDE IF the sample holder has a sample
      CONVEY by Sequence
      Else
        CONVEY to EnterLoadUnloadStation
```

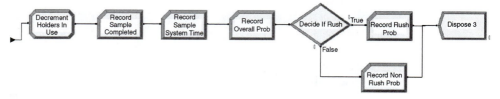

■ **FIGURE 8-16**   System time statistical collection

Statistic - Advanced Process

	Name	Type	Expression	Report Label
1	Tester1Cost	Output	MR(Cell1Tester) * vTesterCostPerMonth (1)	Tester1Cost
2	Tester2Cost	Output	MR(Cell2Tester) * vTesterCostPerMonth (2)	Tester2Cost
3	Tester3Cost	Output	MR(Cell3Tester) * vTesterCostPerMonth (3)	Tester3Cost
4	Tester4Cost	Output	MR(Cell4Tester) * vTesterCostPerMonth (4)	Tester4Cost
5	Tester5Cost	Output	MR(Cell5Tester) * vTesterCostPerMonth (5)	Tester5Cost
6	TotTesterCost	Output	OVALUE(Tester1Cost) + OVALUE(Tester2Cost) + OVALUE(Tester3Cost) + OVALUE(Tester4Cost) + OVALUE(Tester5Cost)	TotTesterCost
7	HolderCost	Output	vNumHolders * vCostPerHolderPerMonth	HolderCost
8	LogicCost	Output	vNewLogicCost * vNewLogicOption	LogicCost
9	TotalCost	Output	OVALUE(HolderCost)+OVALUE(TotTesterCost) + OVALUE(LogicCost)	TotalCost

■ **FIGURE 8-17** Cost collection with output statistics

8.2.2.5 Simulation Horizon and Run Parameters. The setting of the simulation horizon and the run parameters is challenging in this problem because of lack of guidance on this point from the problem description. Without specific guidance you will have to infer from the problem description how to proceed.

It can be inferred from the problem description that SM Testing is interested in using the new design over a period of time, which may be many months or years. In addition, the problem states that SM Testing wants all cost estimates on a per month basis. In addition, the problem provides that during each day the arrival rate varies by hour of the day, but in the absence of any other information it appears that each day repeats this same nonstationary behavior. Because of the nonstationary behavior during the day, steady-state analysis is inappropriate based solely on data collected *during* a day. However, since each day repeats, the *daily* performance of the system may approach steady state over many days. If the system starts empty and idle on the first day, the performance from the initial days of the simulation may be affected by the initial conditions. This type of situation has been termed steady-state cyclical parameter estimation by Law (2007), and is a special case of steady state simulation analysis. Thus, this problem requires an investigation of the effect of initialization bias.

To make this more concrete, let X_i be the average performance from the ith day. For example, the average system time from the ith day. Then, each X_i is an observation in a time series representing the performance of the system for a sequence of days. If the simulation is executed with a run length of, say 30 days, then the X_i constitute *within replication* statistical data, as described in Chapter 4. Thus, when considering the effect of initialization bias, the X_i need to be examined over multiple replications. Therefore, the warm-up period should be in terms of days. The challenge then becomes collecting the daily performance.

■ **TABLE 8-11**

High-Resource Case Cost Example

Equipment	Cost per month ($)	# Units	Total cost
Tester type 1	10,000	2	$20,000
Tester type 2	12,400	2	$24,800
Tester type 3	8,500	3	$25,500
Tester type 4	9,800	4	$39,200
Tester type 5	11,200	5	$56,000
Sample holder	387	16	$6,192
		Grand total	$171,692

There are a number of methods to collect the required daily performance in order to perform an initialization bias analysis. For example, you could create an entity each day and use that entity to record the appropriate performance; however, this would entail special care during the tabulation and would require you to implement the capturing logic for each performance measure of interest. The simplest method is to take advantage of the settings on the Run Setup dialog concerning how the statistics and the system are initialized for each replication.

Let's review the implication of these initialization options. Since there are two check-boxes, one for whether the statistics are cleared at the end of the replication and one for whether the system is reset at the beginning of each replication, there are four possible combinations to consider. The four options and their implications are as follows:

Statistics checked and System checked: This is the default setting. With this option, the statistics are reset (cleared) at the end of each replication and the system is set back to empty and idle. The net effect is that each replication has independent statistics collected on it and each replication starts in the same (empty and idle) state. If a warm-up period is specified, the statistics that are reported are only based on the time after the warm-up period. This option has been used thus far throughout the book.

Statistics unchecked and System checked: With this option, the system is reset to empty and idle at the end of each replication so that each replication starts in the same state. Since the statistics are unchecked, the statistics will *accumulate* across the replica-tions. That is, the average performance for the jth replication includes the perform-ance for all previous replications and the jth replication. If you were to write out the statistics for the jth replication, it would be the cumulative average up to and includ-ing that replication. If a warm-up period is specified, the statistics are cleared at the warm-up time and then captured for each replication. Since this clears the accumu-lation, the net effect of using a warm-up period in this case is that the option functions like option 1.

Statistics checked and System unchecked: With this option, the statistics are reset (cleared) at the end of each replication, but the system is not reset. That is, each subsequent replication begins its initial state using the final state from the previous replication. The value of TNOW is not reset to zero at the end of each replication. In this situation, the average performance for each replication does not include the perform-ance from the previous replication; however, they are not independent observations since the state of the system was not reset. If a warm-up period is specified, a special warm-up summary report appears on the standard text based Arena[TM] report at the warm-up time. After the warm-up, the simulation is then run for the specified number of replications for each run length. Based on how option 3 works, an analyst can effectively implement her or his own batch-means method based on time based batch intervals by specifying the replication length as the batching interval. The number of batches is based on the number of replications (after the warm-up period).

Statistics unchecked and System unchecked: With this option, the statistics are not reset and the system state is not reset. The statistics accumulate across the replications and subsequent replications use the ending state from the previous replication. If a warm-up period is specified, a special warm-up summary report appears on the standard text based Arena[TM] report at the warm-up time. After the warm-up, the simulation is then run for the specified number of replications for each run length.

According to these options, the current simulation can be set up with option 3 with each replication lasting 24 hours and use the number of replications to get the desired number of days of simulation. Since the system is not reset, the variable TNOW is not set back to zero. Thus, each replication causes additional time to evolve during the simulation. The effective run length of the simulation will be determined by the number of days specified for the number of replications. For example, a specification of 60 replications will result in 60 days of simulated time. If the statistics are captured to a file for each replication (day), then the performance by day will be available to analyze for initialization bias.

One additional problem in determining the warm-up period is the fact that there will be many design configurations to be examined. Thus, a warm-up period must be selected that will work on the range of design configurations that will be considered; otherwise, a warm-up analysis will have to be performed for every design configuration. In what follows, the effect of initialization bias will be examined to try to find a warm-up period that will work well across a wide variety of design configurations.

The model was modified to run the high-resource case from Table 8-10 using 48 sample holders. In addition, the model was modified so that the distributions were parameterized by a stream number. This was done so that separate invocations of the program would generate different sample paths. An output statistic was defined to capture the average daily system time for each of the replications and the model was executed 10 different times after adjusting the stream numbers to get 10 independent sample paths of 60 days in length. The 10 sample paths were averaged to produce a Welch plot as shown in Figure 8-18.

From the plot, there is no discernible initialization bias. Thus, it appears that if the system has enough resources, a warm-up period does not seem necessary. However, if the low-resource case of Table 8-10 is run under the same conditions, an execution error will occur that indicates that the maximum number of entities for the professional version of Arena™ is exceeded. In this case, the overall arrival rate is too high for the available resources in the model. The samples build up in the queues (especially the input queue), and unprocessed samples will be carried forward each day, eventually building up to a point that exceeds the number of entities permitted by Arena™. Thus, it should be clear that in the under-resourced case, the performance measures cannot reach steady state. Regardless of whether a warm-up period is set for these cases, they will naturally be excluded as design alternatives because of exceptionally poor performance. Therefore, setting a warm-up period for low-capacity design configurations appears to be unnecessary. Based on this analysis, a 1-day warm-up period will be used to be very conservative across all of the design configurations.

The run setup parameter settings for the initialization bias investigation do not necessarily apply to further experiments. In particular, with the third option, the days are technically *within replication* data. Option 1 appears to be applicable to future experiments. For option 1,

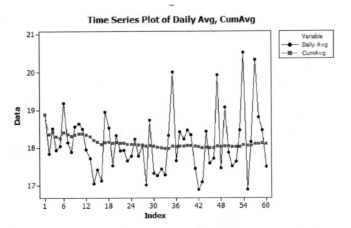

■ **FIGURE 8-18** Welch plot for high-resource configuration

```
Tally

Expression
                                     Average        Half Width

Record Non Rush Prob                  0.9983           0.00
Record Overall Prob                   0.9984           0.00
Record Rush Prob                      0.9600           0.00
SampleSystemTime                     17.8112           0.37

Counter

Count
                                     Average        Half Width

SamplesCompleted                    33688.00         150.40
```

■ **FIGURE 8-19**   **User-defined results for pilot run**

the run length and the number of replications for the experiments need to be determined. The batch means method based on one long replication of many days or the method of replication deletion can be utilized in this situation. Because you might want to use the Process Analyzer and possibly OptQuest during the analysis, the replication deletion method may be more appropriate. Based on the rule of thumb in Banks et al. (2005), the run length should be at least 10 times the amount of data deleted. Because data for 1 day are being deleted, the run length should be set to 11 days so that a total of 10 days of data will be collected. Now, the number of replications needs to be determined.

Using the high-resource case with 48 sample holders, a pilot run was made of five replications of 11 days with 1 day warm-up. The user-defined results of this run are shown in Figure 8-19.

As indicated in the figure, there were a large number of samples completed during each 10-day period and the half-widths of the performance measures are already very acceptable with only five replications. Thus, five replications will be used in all the experiments that follow. Because other configurations may have more variability and thus require more than five replications, a risk is being taken that adequate decisions will be able to be made across all design configurations. Thus, a trade-off is being made between the additional time that pilot runs would entail and the risks associated with making a valid decision.

8.2.3   Preliminary Experimental Analysis

Before proceeding with a full-scale investigation, it will be useful to do two things. First, an assessment for the number of required sample holders is needed. This will provide a range of values to consider during the design alternative search. The second issue is to examine the lower bounds on the required number testers at each test cell. The desire is to prevent a woefully under capacitated system from being examined (either in the Process Analyzer or in an OptQuest run). The execution of a woefully under-capacitated system may cause an execution error that may require the experiments to have to be re-done.

Based on Table 8-10, in order to keep the testers busy in the high-resource case, the system would need to have at least $(2 + 2 + 3 + 4 + 5 = 16)$ sample holders, one sample holder to keep each unit of each test cell busy. To investigate the effect of sample holders on the system, a set of experiments was run using the Process Analyzer on the high-resource case with the number of sample holders steadily decreasing down to 16 sample holders. The results from this initial analysis are shown in Table 8-12. As seen in the table, the performance of the system degrades significantly if the number of sample holders goes below 16. In addition, it should be

■ **TABLE 8-12**
Initial Results

Cell 1	Cell 2	Cell 3	Cell 4	Cell 5	# Holders	Non-rush probability	Rush probability	System time
2	2	3	4	5	48	0.998	0.96	17.811
2	2	3	4	5	44	0.997	0.971	17.182
2	2	3	4	5	40	0.994	0.976	17.046
2	2	3	4	5	36	0.994	0.983	16.588
2	2	3	4	5	32	0.996	0.982	16.577
2	2	3	4	5	28	0.995	0.986	16.345
2	2	3	4	5	24	0.988	0.99	16.355
2	2	3	4	5	20	0.961	0.991	21.431
2	2	3	4	5	18	0.822	0.987	32.839
2	2	3	4	5	17	0.721	0.99	39.989
2	2	3	4	5	16	0.494	0.989	57.565
2	2	3	4	5	12	0	0.988	1309.683
1	1	2	2	3	16	0	0.985	2525.434
1	1	2	2	3	48	0	0.749	2234.676

clear that for the high-resource settings the system is easily capable of meeting the design requirements. In addition, it is clear that too many sample holders can be detrimental to the system time. Thus, a good range for the number of sample holders appears to be from 20 to 36.

The low-resource case was also explored for the cases of 16 and 48 sample holders. The system appears to be highly under-capacitated at the low-resource setting. Since the models were able to execute for the entire run length of 10 days, this provides evidence that any scenarios that have more resources will also be able to execute without any problems.

Based on the preliminary results, you should feel confident to proceed with designing experiments and analyzing the problem in order to make a recommendation. These experiments are described in the next section.

8.2.4   Final Experimental Analysis and Results

While the high-capacity system has no problem achieving good system time performance, it will also be the most expensive configuration. Therefore, there is a clear trade-off between increased cost and improved customer service. Thus, a system configuration that has the minimum cost while also obtaining desired performance needs to be recommended. This situation can be analyzed in a number of ways using the Process Analyzer and OptQuest in the Arena™ environment. Often both tools are used to solve the problem. Typically, in an OptQuest analysis, an initial set of screening experiments may be run using the Process Analyzer to get a good idea on the bounds for the problem. Some of this analysis has already been done. See, for example, Table 8-12. Then, after an OptQuest analysis has been performed, additional experiments might be made via the Process Analyzer to hone the solution space. If only the Process Analyzer is to be used, then a carefully thought out set of experiments can do the job. For teaching purposes, this section will illustrate how to use both the Process Analyzer and OptQuest on this problem.

The contest problem also proposes the use of additional logic to improve the system's performance on rush samples. In a strict sense, this provides another factor (or design parameter) to consider during the analysis. The inclusion of another factor can dramatically increase the number of experiments to be examined. Because of this, an assumption will be made that the additional logic does not significantly interact with the other factors in the model. If this assumption is true, the original system can be optimized to find a basic design configuration. Then, the additional logic can be analyzed to check if it adds value to the basic solution.

In order to recommend a solution, a trade-off between cost and system performance must be established, given the following problem specification:

From this simulation study, management would like to know which configuration would provide the most cost-effective solution while achieving high customer satisfaction. Ideally, the company would always like to provide results in less time than the contract requires. However, management does not believe that the system should include extra equipment just to handle the rare occurrence of a late report.

The objective is clear here: minimize total cost. In order to make the occurrence of a late report "rare," limits can be set on the probability that the rush and non-rush sample's system times exceed the contract limits. This can be done by arbitrarily defining rare as 3 out of 100 samples being late. Thus, at least 97% of the non-rush and rush samples must meet the contract requirements.

8.2.4.1 Using the Process Analyzer on the Problem.

This section uses the Process Analyzer on the problem in an experimental design paradigm. Further information on experimental design methods can be found in Montgomery and Runger (2006). For a discussion of experimental design in a simulation context, please see Law (2007) or Kleijnen (1998). There are six factors (number of units for each of the five testing cells and the number of sample holders). To initiate the analysis, a 2^k factorial experiment can be used. As shown in Table 8-13, the previously determined resource requirements can be readily used as the low and high settings for the test cells. In addition, the previously determined lower- and upper-range values for the number of sample holders can be used as the levels for the sample holders.

This amounts to $2^6 = 64$ experiments; however, because sample holders have a cost, it makes sense to first run the half-fraction of the experiment associated with the lower level of the number of holders to see how the test-cell resource levels interact. The results for the first half-fraction are shown in Table 8-14. The first readily apparent conclusion that can be drawn from this table is that test cells 1 and 2 must have at least two units of resource capacity. Now, considering cases (2, 2, 2, 4, 5) and (2, 2, 3, 4, 5), it is very likely that test cell 3 requires 3 units of resource capacity in order to meet the contract requirements. Lastly, it is very clear that the low levels for cells 4 and 5 are not acceptable. Thus, the search range can be narrowed down to 3 and 4 units of capacity for cell 4 and to 4 and 5 units of capacity for cell 5.

■ TABLE 8-13
First Experiment Factors and Levels

Factor	Levels	
	Low	High
Cell 1 # units	1	2
Cell 2 # units	1	2
Cell 3 # units	2	3
Cell 4 # units	2	4
Cell 5 # units	3	5
# Holders	20	36

■ **TABLE 8-14**
Half-Fraction with # Sample Holders = 20

Cell resource units					Non-rush probability	Rush probability	System time	Total cost
1	2	3	4	5				
1	1	2	2	3	0	0.963	2410.517	100340
2	1	2	2	3	0	0.966	2333.135	110340
1	2	2	2	3	0	0.98	1913.694	112740
2	2	2	2	3	0	0.973	1904.499	122740
1	1	3	2	3	0	0.966	2373.574	108840
2	1	3	2	3	0	0.962	2407.404	118840
1	2	3	2	3	0	0.973	1902.445	121240
2	2	3	2	3	0	0.978	1865.821	131240
1	1	2	4	3	0	0.951	2332.412	119940
2	1	2	4	3	0	0.949	2365.047	129940
1	2	2	4	3	0.003	0.985	496.292	132340
2	2	2	4	3	0.025	0.986	284.205	142340
1	1	3	4	3	0	0.956	2347.05	128440
2	1	3	4	3	0	0.956	2316.931	138440
1	2	3	4	3	0.019	0.984	364.081	140840
2	2	3	4	3	0.122	0.987	157.798	150840
1	1	2	2	5	0	0.968	2394.53	122740
2	1	2	2	5	0	0.965	2360.751	132740
1	2	2	2	5	0	0.98	1873.405	135140
2	2	2	2	5	0	0.982	1865.02	145140
1	1	3	2	5	0	0.963	2391.649	131240
2	1	3	2	5	0	0.965	2361.708	141240
1	2	3	2	5	0	0.974	1926.889	143640
2	2	3	2	5	0	0.98	1841.387	153640
1	1	2	4	5	0	0.951	2387.81	142340
2	1	2	4	5	0	0.955	2352.933	152340
1	2	2	4	5	0.202	0.986	121.199	154740
2	2	2	4	5	0.683	0.991	43.297	164740
1	1	3	4	5	0	0.954	2291.88	150840
2	1	3	4	5	0	0.947	2332.031	160840
1	2	3	4	5	0.439	0.985	69.218	163240
2	2	3	4	5	0.961	0.991	21.431	173240

The other half-fraction with the number of samples at 36 is presented in Table 8-15. The same basic conclusions concerning the resources at the test cells can be made by examining the results. In addition, it is apparent that the number of sample holder can have a significant effect on the performance of the system. It appears that more sample holders hurt the performance of the under capacitated configurations. The effect of the number of sample holders for the highly capacitated systems is some what mixed, but generally the results indicate that 36 sample holders are probably too many for this system. Of course, since this has all been done in an experimental design context, more rigorous statistical tests can be performed to fully test these conclusions. Tests based on the current experimental design will not be done here, but rather the results will be used to set up another experimental design that can help to better examine the system's response.

■ **TABLE 8-15**
Half-Fraction with # Sample Holders = 36

Cell resource units					Non-rush probability	Rush probability	System time	Total cost
1	**2**	**3**	**4**	**5**				
1	1	2	2	3	0	0.776	2319.323	106532
2	1	2	2	3	0	0.764	2260.829	116532
1	2	2	2	3	0	0.729	1816.568	118932
2	2	2	2	3	0	0.714	1779.53	128932
1	1	3	2	3	0	0.768	2238.654	115032
2	1	3	2	3	0	0.775	2243.56	125032
1	2	3	2	3	0	0.73	1762.579	127432
2	2	3	2	3	0	0.717	1763.306	137432
1	1	2	4	3	0	0.796	2244.243	126132
2	1	2	4	3	0	0.79	2268.549	136132
1	2	2	4	3	0.202	0.9	111.232	138532
2	2	2	4	3	0.385	0.915	72.86	148532
1	1	3	4	3	0	0.806	2255.19	134632
2	1	3	4	3	0	0.795	2251.548	144632
1	2	3	4	3	0.36	0.899	76.398	147032
2	2	3	4	3	0.405	0.911	68.676	157032
1	1	2	2	5	0	0.771	2304.784	128932
2	1	2	2	5	0	0.771	2250.198	138932
1	2	2	2	5	0	0.738	1753.819	141332
2	2	2	2	5	0	0.717	1751.676	151332
1	1	3	2	5	0	0.78	2251.174	137432
2	1	3	2	5	0	0.771	2252.106	147432
1	2	3	2	5	0	0.726	1781.024	149832
2	2	3	2	5	0	0.716	1794.93	159832
1	1	2	4	5	0	0.787	2243.338	148532
2	1	2	4	5	0	0.794	2243.558	158532
1	2	2	4	5	0.541	0.897	54.742	160932
2	2	2	4	5	0.898	0.932	28.974	170932
1	1	3	4	5	0	0.791	2263.484	157032
2	1	3	4	5	0	0.799	2278.707	167032
1	2	3	4	5	0.604	0.884	49.457	169432
2	2	3	4	5	0.994	0.983	16.588	179432

Using the initial results in Table 8-12 and the analysis of the two half-fraction experiments, another set of experiments were designed to focus in on the capacities for cells 3, 4, and 5. The experiments are given in Table 8-16. This set of experiments is a $2^4 = 16$ factorial experiment. The results are shown in Table 8-17. The results have been sorted such that the systems that have the higher chance of meeting the contract requirements are at the top of the table. From the results, it should be clear that cell 3 requires at least 3 units of capacity for the system to be able to meet the requirements. It is also very likely that cell 4 requires 4 testers to meet the requirements. Thus, the search space has been narrowed to either 4 or 5 testers at cell 5 and between 20 and 24 holders.

■ **TABLE 8-16**
Second Experiment Factors and Levels

Factor	Levels	
	Low	High
Cell 3 # units	2	3
Cell 4 # units	3	4
Cell 5 # units	4	5
# Holders	20	24

To finalize the selection of the best configuration for the current situation, 10 experimental combinations of cell 5 resource capacity (4, 5) and number of sample holders (20, 21, 22, 23, 24) were run in the Process Analyzer using 10 replications for each scenario. The results of the experiments are shown in Table 8-18. In addition, the multiple comparison procedure was applied to the results based on picking the solution with the best (highest) non-rush probability. The results of that comparison are shown in Figure 8-20. The multiple comparison procedure indicates with 95% confidence that scenarios 3, 4, 5, 7, 8, 9, and 10 are the best. Since there is essentially no difference between them, the cheapest configuration can be chosen, which is scenario 3 with factor levels of (2, 3, 3, 4, 4, and 22) and a total cost of $162,814.

Based on the results of this section, a solid solution for the problem has been determined; however, to give you experience applying OptQuest to a problem, let us examine its application to this situation.

■ **TABLE 8-17**
Results for the Second Set of Experiments

Cell resource units					# Sample holders	Non-rush probability	Rush probability	System time	Total cost
1	2	3	4	5					
2	2	3	4	5	24	0.988	0.99	16.355	174788
2	2	3	4	4	24	0.97	0.988	19.41	163588
2	2	3	4	5	20	0.961	0.991	21.431	173240
2	2	3	4	4	20	0.945	0.989	25.27	162040
2	2	3	3	4	24	0.877	0.987	30.345	153788
2	2	3	3	5	24	0.871	0.988	31.244	164988
2	2	2	4	5	24	0.812	0.984	34.065	166288
2	2	2	4	4	24	0.782	0.986	35.289	155088
2	2	3	3	5	20	0.734	0.991	39.328	163440
2	2	3	3	4	20	0.731	0.992	40.37	152240
2	2	2	3	4	24	0.687	0.987	42.891	145288
2	2	2	4	5	20	0.683	0.991	43.297	164740
2	2	2	3	5	24	0.656	0.98	46.049	156488
2	2	2	4	4	20	0.627	0.989	45.708	153540
2	2	2	3	5	20	0.561	0.987	52.71	154940
2	2	2	3	4	20	0.492	0.985	59.671	143740

■ **TABLE 8-18**
Results for the Final Set of Experiments

Scenario	Cell resource units					# Sample holders	Non-rush probability	Rush probability	System time	Total cost
	1	2	3	4	5					
1	2	2	3	4	4	20	0.917	0.987	26.539	162040
2	2	2	3	4	4	21	0.942	0.987	23.273	162427
3	2	2	3	4	4	22	0.976	0.989	20.125	162814
4	2	2	3	4	4	23	0.974	0.988	19.983	163201
5	2	2	3	4	4	24	0.979	0.989	18.512	163588
6	2	2	3	4	5	20	0.958	0.988	21.792	173240
7	2	2	3	4	5	21	0.967	0.987	19.362	173627
8	2	2	3	4	5	22	0.984	0.988	18.022	174014
9	2	2	3	4	5	23	0.986	0.988	16.971	174401
10	2	2	3	4	5	24	0.98	0.987	16.996	174788

8.2.4.2 Using OptQuest on the Problem. As shown in Figure 8-21, OptQuest for Arena^{TM} can be used to define controls and responses and to setup an optimization model to minimize total cost subject to the probability of non-rush and rush samples meeting the contract requirements being greater than or equal to 0.97. When running OptQuest, it is a good idea to try to give OptQuest a good starting point. Since the high-resource case is already known to be feasible based on the pilot experiments, the optimization can be started using the high-resource case and the number of sample holders in the middle of their range (e.g., 27 sample holders).

The simulation optimization was executed over a period of about 6 hours (wall clock time) until it was manually stopped after 79 total simulations. From Figure 8-22, the best solution was two testers at cells 1 and 2, 3 testers at cell 3, 4 testers at cells 4 and 5, and 23 sample holders. The total cost of this configuration is $163,201. This solution has 1 more sample holder as compared to the solution based on the Process Analyzer. Of course, you would not know this if the analysis was based solely on OptQuest. OptQuest allows top solutions to be saved to a file. Upon looking at those solutions, it becomes clear that the heuristic in OptQuest has found the basic configuration for the test cells (2, 2, 3, 4, 4) and is investigating the number of sample holders. If the optimization had been allowed to run longer, it is very likely that it would have recommended 22 sample holders. At this stage, the Process Analyzer can be used to

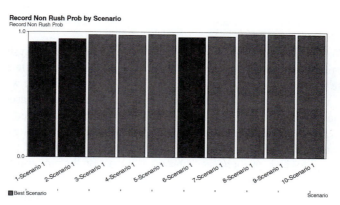

■ **FIGURE 8-20**   Multiple comparison results for non-rush probability

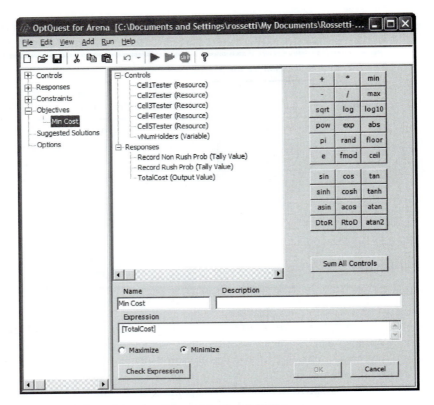

■ **FIGURE 8-21** OptQuest objective definition

hone in on the recommended OptQuest solution by more closely examining the final set of "best" solutions (Figure 8-22). Since that has already been done in the previous section, it will not be included here.

8.2.4.3 Investigating the New Logic Alternative.

Now that a recommended basic configuration is available, the alternative logic needs to be checked to see if it can improve performance for the additional cost. In addition, the "suggested number" for the control logic should be determined. The basic model can be easily modified to contain the alternative logic. This was done in the load/unload submodel. In addition, a flag variable was used to be able to turn on or turn off the use of the new logic so that the Process Analyzer can control whether the logic is included in the model. The implementation can be found in the file *SMTesting.doe*.

To reduce the number of factors involved in the previous analysis, it was assumed that the new logic did not interact significantly with the allocation of the test cell capacity; however, because the suggested number checks for how many samples are waiting at the load/unload station, there may be some interaction between these factors. Based on these assumptions, 10 scenarios with the new logic were designed for the recommended tester configuration (2, 2, 3, 4, 4) varying the number of holders between 20 and 24 with the suggested number (SN) set at both 2 and 3. Table 8-19 presents the results of these experiments. From the results, it is clear that the new logic does not have a substantial impact on the probability of meeting the contract for the non-rush and rush samples. In fact, looking at scenario 3, the new logic may actually hurt the non-rush samples. To complete the analysis, a more rigorous statistical comparison should be performed; however, that task will be skipped for the sake of brevity.

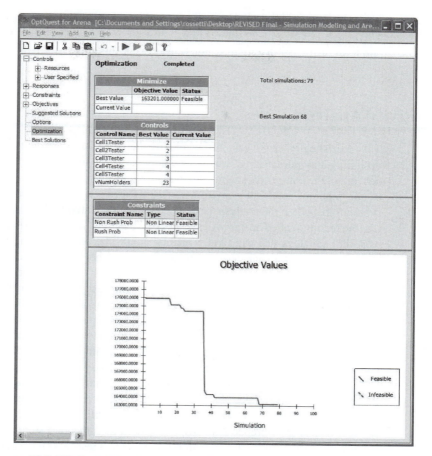

■ **FIGURE 8-22**   OptQuest result

■ **TABLE 8-19**

Results for Analyzing the New Logic

Scenario	Cell resource units					# Sample holders	SN	Non-rush probability	Rush probability	System time
	1	2	3	4	5					
1	2	2	3	4	4	20	2	0.937	0.989	23.937
2	2	2	3	4	4	21	2	0.956	0.988	21.461
3	2	2	3	4	4	22	2	0.957	0.989	21.159
4	2	2	3	4	4	23	2	0.972	0.989	19.011
5	2	2	3	4	4	24	2	0.975	0.986	18.665
6	2	2	3	4	4	20	3	0.926	0.989	25.305
7	2	2	3	4	4	21	3	0.948	0.988	22.449
8	2	2	3	4	4	22	3	0.968	0.989	20.536
9	2	2	3	4	4	23	3	0.983	0.988	18.591
10	2	2	3	4	4	24	3	0.986	0.987	18.17

After all the analysis, the results indicate that SM Testing should proceed with the (2, 2, 3, 4, 4) configuration with 22 sample holders for a total cost of $162,814. The additional one-time purchase of the control logic is not warranted at this time.

8.3   Sensitivity Analysis

This realistically sized case study clearly demonstrates the application of simulation to designing a manufacturing system. The application of simulation was crucial in this analysis because of the complex, dynamic relationships between the elements of the system and because of the inherent stochastic environment (e.g., arrivals, breakdowns, etc.) that were in the problem. Based on the analysis performed for the case study, SM Testing should be confident that the recommended design will suffice for the current situation. However, there are a number of additional steps that can (and probably should) be performed for the problem.

In particular, to have even more confidence in the design, simulation offers the ability to perform a sensitivity analysis on the "uncontrollable" factors in the environment. While the analysis in the previous section concentrated on the design parameters that can be controlled, it is very useful to examine the sensitivity of other factors and assumptions on the recommended solution. Now that a verified and validated model is in place, the arrival rates, breakdown rates, repair times, buffer sizes, and so on, can all be varied to see how they affect the recommendation. In addition, a number of explicit and implicit assumptions were made during the modeling and analysis of the contest problem. For example, when modeling tester breakdowns, Arena's FAILURE module was used. This module assumes that when a failure occurs that all units of the resource become failed. This may not actually be the case. In order to model individual unit failures, the model must be modified. After the modification, the assumption can be tested to see if it makes a difference with respect to the recommended solution. Other aspects of the situation were also omitted. For example, the fact that tests may have to be repeated at test cells was left out due to the fact that no information was given as to the likelihood for repeating the test. This situation could be modeled and a re-test probability assumed. The sensitivity to this assumption could then be examined.

In addition to modeling assumptions, assumptions were made during the experimental analysis. For example, basic resource configuration of (2, 2, 3, 4, 4) was assumed to not change if the control logic was introduced. This can also be tested in a sensitivity analysis. Since performing a sensitivity analysis was discussed in Chapter 4, the mechanics of performing sensitivity analysis will not be discussed here. A number of exercises will ask you to explore these issues.

8.4   Completing the Project

A major aspect of the use of simulation is convincing the key decision makers of the value of the study recommendations. Developing an animation for the Arena™ model is especially useful in this regard. For the SM Testing model, an animation is available in the Arena™ *doe* file. The animation (Figure 8-23) animates the conveyor, usage of resources, and queues, and has a number of variables for monitoring the system. In addition, user-defined views were created for easily viewing the animation and model logic.

A written report is also a key requirement of a simulation project. Chung (2004) provides some guidelines for the report and a presentation associated with a project. An example outline for a simulation project report follows.

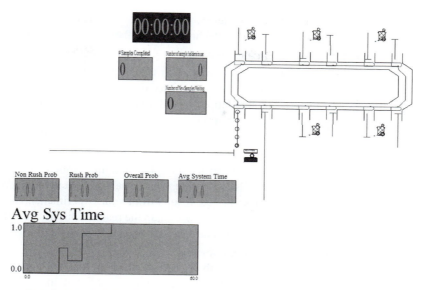

■ FIGURE 8-23   Overview of the animation

Executive Summary

Provide a brief summary of the problem, method, results, and recommendation. It should be oriented toward the decision maker.

1. Introduction

Provide a brief overview of the system and problem context. Orient the reader as to what should be expected from the report.

2. System and Problem Description

Describe the system under study. The system should be described in such a way that an analyst could develop a model from the description. Describe the problem to be solved, indicate the key performance metrics to be used, and the major goals/objectives to be achieved. For example, the contest problem description given in this chapter is an excellent example of describing the system and the problem.

3. Solution and Modeling Approach

(a) Describe the modeling issues that needed to be solved. Describe the inputs for the model and how they were developed. For each major modeling issue, briefly describe your conceptual approach. Relate your description to the system description. Provide activity diagrams or pseudo-code and describe how each relates to your conceptual model. Describe the major outputs from the model and how they are collected.

(b) Describe the overall flow of the Arena™ model and relate it to your conceptual model. Provide an accounting of important modules where the detail is necessary to help the reader understand a tricky or important modeling issue.

(c) State relevant and important assumptions. Discuss their potential effect on your answers and any procedures that you adopted to test your assumptions.

(d) Describe any verification/validation procedures. Describe your up-front analysis. You might approximate parts of your model or perform basic calculations that provide a ballpark idea of the range of your outputs. If you have data from the system concerning expected outputs, then you should compare the simulation output to the system output using statistical techniques. For a more detailed discussion of these techniques, see the excellent discussion in Balci (1998).

4. Experimentation and Analysis

   (a) Describe a typical simulation run. Is the model a finite-horizon or steady-state model? Is the analysis conducted by the replication method or batch means? If it is a steady-state simulation, how did you handle the initial transient period? How did you determine the run length? The number of replications? Describe or reference any special statistical methods that you used so that the reader can understand your results.

   (b) Describe your experimental plan and the results. What are your factors and respective levels? What are your variables in the model? What are you capturing as output? Describe the results from your experiments. Use figures and tables and refer to them in the text to explain the results. Discuss the effect of assumptions and sensitivity analysis and their effect on model results.

5. Conclusions and Recommendations

   Provide the answers to the problem. Back up your recommendations based on your results and analyses. If possible, give the solution in the form of alternatives. For example: "Assuming X, we recommend Y because of Z," or "Given results X, Y, we recommend Z."

References

Appendix

Supporting materials, only if absolutely necessary.

In addition to the above material, a simulation report might also contain plans for how to implement the recommendations. A presentation covering the major issues and recommendations from the report should also be prepared. You can think of the presentation as a somewhat expanded version of the executive summary of the report. It should concentrate on building credibility for your analysis, results, and recommendations. Consider using your animation during the presentation as a tool for building credibility and understanding of the modeled process.

In addition to the model, written report, presentation, and animation, providing model/ user documentation is extremely useful. To assist with some of the documentation, you might investigate Arena's Model Documentation Report available on the Tools menu. This tool produces an organized listing of the modules in the model. However, the tool is no substitute for fully describing the model in a language-independent manner in your report. Even if non–simulation-oriented users will not be interacting with the model, it is very useful to have a user's guide to the model of some sort. This document should describe how to run the model, what inputs to modify, what outputs to examine, and so on. Often a model built for a single use is "re-discovered" and used for additional analysis. A user's guide becomes extremely handy in these situations.

8.5   Some Final Thoughts

This chapter focused on applying simulation to a realistic problem. The solution involved many of the simulation modeling techniques discussed in previous chapters. In addition, Arena™ and its supporting tools were applied to develop a model for the problem and to analyze the system under study. Finally, experimental design concepts were applied in order to recommend design alternatives for the problem. This comprehensive example should give you a good understanding of how to apply simulation in the real world.

A number of other issues in simulation have not been fully described in this text. However, based on the material in this text, you should be well prepared to apply simulation to problems in engineering and business. To continue your education in simulation, consider studying Chung (2004) and Banks (1998). Chapter 13 of Chung (2004) provides a number of examples of the application simulation, and chapters 14 to 21 of Banks (1998) illustrate simulation in areas such as health care, the military, logistics, and manufacturing. In addition, Chapters 22 and 23 of Banks (1998) provide excellent practical advice on how to perform and manage a successful simulation project.

Finally, get involved in professional societies and organizations that support simulation modeling and analysis, such as the following groups.

- American Statistical Association (ASA)
- Association for Computing Machinery: Special Interest Group on Simulation (ACM/SIGSIM)
- Institute of Electrical and Electronics Engineers: Systems, Man, and Cybernetics Society (IEEE/SMC)
- Institute for Operations Research and the Management Sciences: Simulation Society (INFORMS-SIM)
- Institute of Industrial Engineers (IIE)
- National Institute of Standards and Technology (NIST)
- The Society for Modeling and Simulation International (SCS)

An excellent opportunity to get involved is the Winter Simulation Conference. You can easily find information concerning these organizations and the Winter Simulation Conference on the Web.

As a final thought you should try to adhere to the following rules during your practice of simulation:

- Perform only simulation studies that have clearly defined objectives and performance metrics.
- Never perform a simulation study when the problem can be solved through simpler and more direct techniques.
- Always carefully plan, collect, and analyze the data necessary to develop good input distributions for simulation models.
- Never accept simulation results or decisions based on a sample size of one.
- Always use statistically valid methods to determine the sample size for simulation experiments.
- Always check that statistical assumptions (e.g., independence, normality, etc.) are reasonably valid before reporting simulation results.
- Always provide simulation results by reporting the sample size, the sample average, and a measure of sample variation (e.g., sample variance, sample standard deviation, standard error, sample half-width for a specified confidence level, etc.).
- Always perform simulation experiments with a random number generator that has been tested and proven to be reliable.
- Always take steps to mitigate the effects of initialization bias when performing simulations involving steady-state performance.
- Always verify simulation models through such techniques as debugging, event tracing, animation, and so on.

■ Always validate simulation models through common sense analytical thinking techniques, discussions with domain experts, real-data, and sensitivity analysis techniques.

■ Always carefully plan, perform, and analyze simulation experiments using appropriate experimental design techniques.

By following these rules, you should have a successful and professional simulation career.

Exercises

8.1 Perform a sensitivity analysis on the recommendation for the SM Testing contest problem by varying the arrival rate for each hour by plus or minus 10%. In addition, examine the effect of decreasing or increasing the time between failures for test cells 1, 3, 4, and 5 by plus or minus 10% as well as the mean time for repairs.

8.2 What solution should be recommended if the probability of meeting the contract requirements is specified as 95% for non-rush and rush samples?

8.3 Revise the model to handle the independent failure of individual testers at each test cell. Does the more detailed modeling change the recommended solution?

8.4 Revise the model to handle the situation of re-testing at each test cell. Assume that a re-test occurs with a 3% chance and then test the sensitivity of this assumption.

8.5 How does changing the buffer capacities for the load/unload area and the test cells affect the recommended solution?

8.6 Develop a model that separates the load/unload area into two stations, one for loading and one for unloading with each station having its own machine. Analyze where to place the stations on the conveyor and compare your design to the recommended solution for the contest problem.

8.7 Revise the model so that it uses the BATCH/SEPARATE approach to sample holder and sample interaction rather than the PICKUP/DROPOFF approach. Compare the results of your model to the results presented in this chapter.

8.8 Solve one of the Rockwell Software/IIE contest problems available on the CD that accompanies this text.

Bibliography

1. Ahrens J, Dieter V. Computer methods for sampling from the exponential and normal distributions. Commun Assoc Computing Mach 1972;15:873–882.

2. Air Force Systems Command, ASD Directorate of Systems Engineering and DSMC Technical Management Department. Draft Military Standard Systems Engineering, MIL-STD-499B. HQ AFMC/ENS, 4375 Chidlaw Rd Suite 6, WPAFB OH 45433-5006;1994.

3. Albright SC. VBA for Modelers: Developing Decision Support Systems with Microsoft Excel. Belmont, CA: Duxbury Thomson Learning; 2001.

4. Alexopoulos C, Seila AF. Output data analysis. In: Banks J, editor. Handbook of Simulation. New York: John Wiley and Sons; 1998. pp. 225–272.

5. April J, Glover F, Kelly J, Laguna M. Simulation/optimization using ''real-world'' applications. In: Peters BA, Smith JS, Medeiros DJ, Rohrer MW, editors. The Proceedings of the 2001 Winter Simulation Conference. Piscataway, NJ: Institute of Electrical and Electronic Engineers; 2001. pp. 134–138.

6. Askin RG, Goldberg JB. Design and Analysis of Lean Production Systems. New York: John Wiley and Sons; 2002.

7. Askin RG, Standridge CR. Modeling and Analysis of Manufacturing Systems. New York: John Wiley and Sons; 1993.

8. Axsäter S. Inventory Control. New York: Springer Science + Business Media, LLC; 2006.

9. Balci O. Principles of simulation model validation, verification, and testing. Trans Soc Comput Simul Int 1997;14(1):3–12.

10. Balci O. Verification, validation, and testing. In: Banks J, editor. The Handbook of Simulation. New York: John Wiley and Sons; 1998. pp. 335–393.

11. Banks J, Burnette B, Kozloski H, Rose J. Introduction to SIMAN V and CINEMA V. New York: John Wiley and Sons; 1995.

12. Banks J, Carson J, Nelson B, Nicol D. Discrete-Event System Simulation, 4th, ed. Prentice Hall.

13. Ballou RH. Business Logistics/Supply Chain Management: Planning, Organizing, and Controlling the Supply Chain. 5th ed. Englewood Cliffs, NJ: Prentice Hall; 2004.

14. Biller B, Nelson BL. Modeling and generating multivarate time-series input processes using a vector autogressive technique. Assoc Comput Mach Trans Modeling Comput Simul 2003;13:211–237.

15. Biller B, Nelson BL. Fitting time-series input processes for simulation. Operations Res 2005;53:549–559.

16. Blanchard BS, and Fabrycky WJ. Systems Engineering and Analysis. Englewood Cliffs, NJ: Prentice Hall; 1990.

17. Bolch G, Greiner S, de Meer H, Trivedi K. Queueing Networks and Markov Chains. 2nd ed. New York: John Wiley and Sons; 2006.

18. Box GEP, Jenkins GM, Reinsel GC. Time Series Analysis, Forecasting and Control. 3rd ed. Englewood Cliffs, NJ: Prentice Hall; 1994.

19. Box GEP, Muller MF. Note on the generation of random normal deviates. Ann Math Stat 1958;29:610–611.

20. Buzacott JA, Shanthikumar JG. Stochastic Models of Manufacturing Systems. New York: Prentice Hall; 1993.

21. Cario MC, Nelson BL. Autoregressive to anything: time series input processes for simulation. Operations Res Lett 1996;19:51–58.

22. Cario MC, Nelson BL. Numerical methods for fitting and simulating autoregressive-to-anything processes. INFORMS J Comput 1998;10:72–81.

23. Casella G, Berger R. Statistical Inference. Pacific Grove, CA: Wadsworth & Brooks/Cole; 1990.

24. Cash CR, Dippold DG, Long JM, Nelson BL, Pollard WP. Evaluation of tests for initial conditions bias. In: Swain JJ, Goldsman D, Crain RC, Wilson JR, editors. Proceedings of the 1992 Winter Simulation Conference. Piscataway, NJ: Institute of Electrical and Electronics Engineers; 1992. pp. 577–585.

25. Cheng RC. The generation of gamma variables with nonintegral shape parameters. Appl Stat 1977;26(1):71–75.

26. Chopra S, Meindl P. Supply Chain Management: Strategy, Planning, and Operations. 3rd ed. New York: Prentice Hall; 2007.

27. Chung CA. Simulation Modeling Handbook: A Practical Approach. Boca Raton, FL: CRC Press; 2004.

28. Cooper RB. Introduction to Queueing Theory. 3rd ed. Washington, DC: CEEPress Books; 1990.

29. Devroye L. Non-Uniform Random Variate Generation. New York: Springer-Verlag; 1986.

30. Fishman GS. Discrete-Event Simulation: Modeling, Programming, and Analysis. New York: Springer; 2001.

31. Fishman GS. A First Course in Monte Carlo. Pacific Grove, CA: Thomson Brooks/Cole; 2006.

32. Fishman GS, Yarberry LS. An implementation of the batch means method, INFORMS J Comput 1997;9(3):296–310.

33. Goldsman D, Nelson BL. Comparing Systems via Simulation. In: Banks J, editor. Handbook of Simulation. New York: John Wiley and Sons; 1998. pp. 273–306.

34. Glover F, Kelly JP, Laguna M. New Advances for Wedding Optimization and Simulation. In Farrington F, Nembhard HB, Sturrock DT, Evans GW, editors. The Proceedings of the

1999 Winter Simulation Conference. Piscataway, NJ: Institute of Electrical and Electronic Engineers; 1999. pp. 255–260.

35. Glover F, Laguna, M. Tabu Search. Norwell, MA: Kluwer Academic Publishers; 1997.

36. Glynn PW, Whitt W. Extensions of the queueing relation $L = \lambda W$ and $H = \lambda G$, Operations Res 1989;37:634–644.

37. Gross D, Harris CM. Fundamentals of Queueing Theory. 3rd ed. New York: John Wiley and Sons; 1998.

38. Hadley G, Whitin TM. Analysis of Inventory Systems. Englewood Cliffs, NJ: Prentice Hall; 1963.

39. Henriksen J, Schriber T. Simplified approaches to modeling accumulating and non-accumulating conveyor systems. In: Wilson J, Hendricksen J, Roberts S, editors. Proceedings of the 1986 Winter Simulation Conference. Piscataway, NJ: Institute of Electrical and Electronics Engineers; 1986.

40. Hopp WJ, Spearman L. Factory Physics: Foundations of Manufacturing Management. 2nd ed. New York: McGraw-Hill; 2001.

41. Hull TE, Dobell AR. Random number generators. SIAM Rev 1962;4:230–254.

42. Jensen PA, Bard JF. Operations Research Models and Methods. New York: John Wiley and Sons; 2003.

43. Kelly FP. Reversibility and Stochastic Networks. New York: John Wiley and Sons; 1979.

44. Kelton WD, Sadowski RP, Sturrock DT. Simulation with Arena. 3rd ed. New York: McGraw-Hill; 2004.

45. Kendall DG. Stochastic processes occurring in the theory of queues and their analysis by the method of imbedded Markov chains. Ann Math Stat 1953;24:338–354.

46. Kingman JFC. The heavy traffic approximation in the theory of queues. In: Smith WL, Wilkinson RI, editors. Proceedings of the Symposium on Congestion Theory. Chapel Hill, NC: The University of North Carolina Press; 1964. pp. 137–169.

47. Kleijnen JPC. Analyzing simulation experiments with common random numbers. Manag Sci 1988;34:65–74.

48. Kleijnen JPC. Experimental design for sensitivity analysis, optimization, and validation of simulation models. In: Banks J, editor. Handbook of Simulation. New York: John Wiley and Sons; 1998. pp. 173–223.

49. Kleinrock L. Queueing Systems. Volume 1, Theory. New York: John Wiley and Sons; 1975.

50. Lada EK, Wilson JR, Steiger NM. A wavelet-based spectral method for steady-state simulation analysis. In: Chick S, Sànchez PJ, Ferrin D, Morrice DJ, editors. Proceedings of the 2003 Winter Simulation Conference. Piscataway, NJ: Institute of Electrical and Electronics Engineers; 2003. pp. 422–430.

51. Law A. Simulation Modeling and Analysis. 4th ed. New York: McGraw-Hill; 2007.

52. L'Ecuyer P. Testing random number generators. In: Swain JJ, Goldsman D, Crain RC, Wilson JR, editors. Proceedings of the 1992 Winter Simulation Conference. IEEE Press; 1992. pp. 305–313.

53. L'Ecuyer P, Simard R, Kelton WD. An object-oriented random number package with many long streams and substreams. Operations Res 2002;50:1073–1075.

54. Leemis LM, Park SK. Discrete-Event Simulation: A First Course. New York: Prentice Hall; 2006.

55. Little JDC. A proof for the queuing formula $L = \lambda W$. Operations Res 1961;9:383–387.

56. Litton JR, Harmonosky CH. A comparison of selective initialization bias elimination methods. In: Yucesan E, Chen C-H, Snowdon JL, Charnes JM, editors. Proceeding of the 2002 Winter Simulation Conference. Piscataway, NJ: Institute of Electrical and Electronics Engineers; 2002. pp. 1951–1957.

57. Livny M, Melamed B, Tsiolis AK. The impact of autocorrelation on queueing systems. Manag Sci 1993;39(3):322–339.

58. Montgomery DC, Runger GC. Applied Statistics and Probability for Engineers. 4th ed. New York: John Wiley and Sons; 2006.

59. Nahmias S. Production and Operations Analysis. 4th ed. New York: McGraw-Hill; 2001.

60. Patuwo BE, Disney RL, Mcnickle DC. The effect of correlated arrivals on queues. IIE Trans 1993;25(3):105–110.

61. Pegden CD, Shannon RE, Sadowski RP. Introduction to Simulation Using SIMAN. 2nd ed. New York: McGraw-Hill; 1995.

62. Ravindran A, Phillips D, Solberg J. Operations Research Principles and Practice. 2nd ed. New York: John Wiley and Sons; 1987.

63. Ripley BD. Stochastic Simulation. New York: John Wiley and Sons; 1987.

64. Robinson S. Automated analysis of simulation output data. In: Kuhl ME, Steiger NM, Armstrong FB, Joines JA, editors. Proceedings of the 2005 Winter Simulation Conference. Piscataway, NJ: Institute of Electrical and Electronics Engineers; 2005. pp. 763–770.

65. Ross S. Introduction to Probability Models. 6th ed. Washington, DC: Academic Press; 1997.

66. Rossetti MD, Delaney PJ. Control of initialization bias in queueing simulations using queueing approximations. In: Alexopoulos C, Kang K, Lilegdon WR, Goldsman D, editors. Proceedings of the 1995 Winter Simulation Conference. Piscataway, NJ: Institute of Electrical and Electronics Engineers; 1995. pp. 322–329.

67. Rossetti MD, Delaney PJ. Exploring exponentially weighted moving average control charts to determine the warm-up period. In: Kuhl ME, Steiger NM, Armstrong FB, Joines JA, editors. Proceedings of the 2005 Winter Simulation Conference. Piscataway, NJ: Institute of Electrical and Electronics Engineers; 2005. pp. 771–780.

68. Rossetti, M. D, Seldanari F. Multi-objective analysis of hospital delivery systems. Comput Ind Eng 2001;41:309–333.

69. Schruben L. Detecting initialization bias in simulation output. Operations Res 1982;30:569–590.

70. Schruben L, Singh H, Tierney L. Optimal tests for initialization bias in simulation output. Operations Res 1983;31:1167–1178.

71. Schmeiser BW. Batch Size effects in the analysis of simulation output. Operations Res 1982;30:556–568.

72. Silver EA, Pyke DF, Peterson R. Inventory Management and Production Planning and Scheduling. 3rd ed. New York: John Wiley and Sons; 1998.

73. Singh N. Systems Approach to Computer-Integrated Design and Manufacturing, New York: John Wiley and Sons; 1996.

74. Soto J. Statistical testing of random numbers. Proceedings of the 22nd National Information Systems Security Conference; 1999.

75. Stecke KE. Machine interference: assignment of machines to operators. In: Salvendy G, editor. Handbook of Industrial Engineering. New York: John Wiley and Sons; 1992. pp. 1460–1494.

76. Steiger NM, Wilson JR. An improved batch means procedure for simulation output analysis. Manag Sci 2002;48(12):1569–1586.

77. Taha HA. Operations Research: An Introduction. 7th ed. New York: Prentice Hall; 2003.

78. Tijms HC. A First Course in Stochastic Models. New York: John Wiley and Sons; 2003.

79. Welch PD. The statistical analysis of simulation results. In: Lavenberg, S, editor. Computer Performance Modeling Handbook. New York: Academic Press; 1983. pp. 268–328.

80. White KP, Cobb MJ, Spratt SC. A comparison of five steady-state truncation heuristics for simulation. In: Joines JA, Barton RR, Kang K, Fishwick PA, editors. Proceedings of the 2000 Winter Simulation Conference. Piscataway, NJ: Institute of Electrical and Electronics Engineers; 2000. pp. 755–760.

81. Whitt W. The queueing network analyzer. Bell Syst Tech J 1983;62(9):2779–2815.

82. Whitt W. Planning queueing simulations. Manag Sci 1989;35(11):1341–1366.

83. Whitt W. Approximations for the GI/G/m queue. Prod Oper Manag 1993;2(2):114–161.

84. Wilson JR, Pritsker AAB. A survey of re-search on the simulation startup problem. Simulation 1978;31(2):55–58.

85. Zipkin PH. Foundations of Inventory Management. New York: McGraw-Hill; 2000.

APPENDICES

Distributions

A.1   Common Discrete Distributions

Distribution, Random Variable X	PMF	E[X] & V[X]
Bernoulli(p) The number of successes in one trial	$P(X = 1) = p$ $P(X = 0) = 1 - p$	$E[X] = p$ $V[X] = p(1 - p)$
Binomial(n,p) The number of successes in n Bernoulli trials	$P(X = x) = \binom{n}{x} p^x (1 - p)^{n-x}$ $0 \le p \le 1 \quad x = 0, 1, \ldots, n$	$E[X] = np$ $V[X] = np(1 - p)$
Geometric(p) The number of trials until the first success in a sequence of Bernoulli trials	$P(X = x) = p(1 - p)^{x-1}$ $0 \le p \le 1 \quad x = 1, 2, \ldots$	$E[X] = 1/p$ $V[X] = (1 - p)/p^2$
Negative Binomial(r,p), Defn. 1 The number of trials until the rth success in a sequence of Bernoulli trials	$P(X = x) = \binom{x - 1}{r - 1} p^r (1 - p)^{x-r}$ $0 \le p \le 1 \quad x = r, r + 1, \ldots,$	$E[X] = r/p$ $V[X] = r(1 - p)/p^2$
Negative Binomial, Defn. 2 The number of failures prior to the rth success in a sequence of Bernoulli trials	$P(Y = y) = \binom{y + r - 1}{r - 1} p^r (1 - p)^y$ $0 \le p \le 1 \quad y = 0, 1, \ldots,$	$E[Y] = r(1 - p)/p$ $V[Y] = r(1 - p)/p^2$
Poisson(λ) The number of event occurrences during a specified period of time	$P(X = x) = \dfrac{e^{-\lambda} \lambda^x}{x!} \quad x = 0, 1, \ldots,$ $\lambda > 0$	$E[X] = \lambda$ $V[X] = \lambda$
Discrete Uniform(a,b)	$P(X = x) = \dfrac{1}{b - a + 1}$ $a \le b \quad x = a, a + 1, \ldots, b$	$E[X] = (b + a)/2$ $V[X] = \left((b - a + 1)^2 - 1\right)/12$
Discrete Uniform	$P(X = x) = \dfrac{1}{n}$ $x = x_1, x_2, x_3, \ldots, x_n; n \in \aleph$	$E[X] = \dfrac{1}{n} \sum_{i=1}^{n} x_i$ $V[X] = \dfrac{1}{n} \sum_{i=1}^{n} x_i^2 - (E[X])^2$

Note: $\binom{n}{x} = \frac{n!}{(n-x)!x!}$

A.2   Common Continuous Distributions

Distribution	$F(x)$	$E[X]$	$V[X]$
Uniform(a,b)	$\dfrac{x-a}{b-a} \quad a < x < b$	$\dfrac{a+b}{2}$	$\dfrac{(b-a)^2}{12}$
Normal(μ,σ^2)	No Closed Form	μ	σ^2
Exponential(θ)	$1 - e^{-x/\theta}$	θ	θ^2
Erlang(θ, r)	$1 - \sum\limits_{n=0}^{r-1} \dfrac{e^{-x/\theta}(x/\theta)^n}{n!}$	$r\theta$	$r\theta^2$
Gamma(β, α)	If r is a positive integer, see Erlang; otherwise, no closed form	$\alpha\beta$	$\alpha\beta^2$
Weibull(β, α) beta = *scale* alpha = *shape*	$1 - e^{-(x/\beta)^{\alpha}}$	$\dfrac{\beta}{\alpha}\Gamma\left(\dfrac{1}{\alpha}\right)$	$\dfrac{\beta^2}{\alpha}\left\{2\Gamma\left(\dfrac{2}{\alpha}\right) - \dfrac{1}{\alpha}\left(\Gamma\left(\dfrac{1}{\alpha}\right)\right)^2\right\}$
Lognormal (μ_l, σ_l^2)	No Closed Form $\mu = \ln\left(\mu_l / \sqrt{\sigma_l^2 + \mu_l^2}\right)$ $\sigma^2 = \ln\left((\sigma_l^2 + \mu_l^2)/\mu_l^2\right)$	$e^{\mu + \frac{\sigma^2}{2}}$	$e^{2\mu + \sigma^2}\left(e^{\sigma^2} - 1\right)$
Beta(α_1,α_2)	No Closed Form	$\dfrac{\alpha_1}{\alpha_1 + \alpha_2}$	$\dfrac{\alpha_1\alpha_2}{(\alpha_1 + \alpha_2)^2(\alpha_1 + \alpha_2 + 1)}$
Triangular(a, m, b) a = minimum b = maximum m = mode	$\begin{cases} \dfrac{(x-a)^2}{(b-a)(m-a)} & a \leq x \leq m \\ 1 - \dfrac{(b-x)^2}{(b-a)(b-m)} & m \leq x \leq b \end{cases}$	$\dfrac{a+b+m}{3}$	$\dfrac{a^2 + b^2 + m^2 - ab - am - bm}{18}$

A.3   Cumulative Standard Normal Distribution

z	−0.09	−0.08	−0.07	−0.06	−0.05	−0.04	−0.03	−0.02	−0.01	0
−3.9	0.000033	0.000034	0.000036	0.000037	0.000039	0.000041	0.000042	0.000044	0.000046	0.000048
−3.8	0.000050	0.000052	0.000054	0.000057	0.000059	0.000062	0.000064	0.000067	0.000069	0.000072
−3.7	0.000075	0.000078	0.000082	0.000085	0.000088	0.000092	0.000096	0.000100	0.000104	0.000108
−3.6	0.000112	0.000117	0.000121	0.000126	0.000131	0.000136	0.000142	0.000147	0.000153	0.000159
−3.5	0.000165	0.000172	0.000178	0.000185	0.000193	0.000200	0.000208	0.000216	0.000224	0.000233
−3.4	0.000242	0.000251	0.000260	0.000270	0.000280	0.000291	0.000302	0.000313	0.000325	0.000337
−3.3	0.000349	0.000362	0.000376	0.000390	0.000404	0.000419	0.000434	0.000450	0.000466	0.000483
−3.2	0.000501	0.000519	0.000538	0.000557	0.000577	0.000598	0.000619	0.000641	0.000664	0.000687
−3.1	0.000711	0.000736	0.000762	0.000789	0.000816	0.000845	0.000874	0.000904	0.000935	0.000968
−3	0.001001	0.001035	0.001070	0.001107	0.001144	0.001183	0.001223	0.001264	0.001306	0.001350
−2.9	0.001395	0.001441	0.001489	0.001538	0.001589	0.001641	0.001695	0.001750	0.001807	0.001866
−2.8	0.001926	0.001988	0.002052	0.002118	0.002186	0.002256	0.002327	0.002401	0.002477	0.002555
−2.7	0.002635	0.002718	0.002803	0.002890	0.002980	0.003072	0.003167	0.003264	0.003364	0.003467
−2.6	0.003573	0.003681	0.003793	0.003907	0.004025	0.004145	0.004269	0.004396	0.004527	0.004661
−2.5	0.004799	0.004940	0.005085	0.005234	0.005386	0.005543	0.005703	0.005868	0.006037	0.006210
−2.4	0.006387	0.006569	0.006756	0.006947	0.007143	0.007344	0.007549	0.007760	0.007976	0.008198
−2.3	0.008424	0.008656	0.008894	0.009137	0.009387	0.009642	0.009903	0.010170	0.010444	0.010724
−2.2	0.011011	0.011304	0.011604	0.011911	0.012224	0.012545	0.012874	0.013209	0.013553	0.013903
−2.1	0.014262	0.014629	0.015003	0.015386	0.015778	0.016177	0.016586	0.017003	0.017429	0.017864
−2	0.018309	0.018763	0.019226	0.019699	0.020182	0.020675	0.021178	0.021692	0.022216	0.022750
−1.9	0.023295	0.023852	0.024419	0.024998	0.025588	0.026190	0.026803	0.027429	0.028067	0.028717
−1.8	0.029379	0.030054	0.030742	0.031443	0.032157	0.032884	0.033625	0.034380	0.035148	0.035930
−1.7	0.036727	0.037538	0.038364	0.039204	0.040059	0.040930	0.041815	0.042716	0.043633	0.044565
−1.6	0.045514	0.046479	0.047460	0.048457	0.049471	0.050503	0.051551	0.052616	0.053699	0.054799
−1.5	0.055917	0.057053	0.058208	0.059380	0.060571	0.061780	0.063008	0.064255	0.065522	0.066807
−1.4	0.068112	0.069437	0.070781	0.072145	0.073529	0.074934	0.076359	0.077804	0.079270	0.080757
−1.3	0.082264	0.083793	0.085343	0.086915	0.088508	0.090123	0.091759	0.093418	0.095098	0.096800
−1.2	0.098525	0.100273	0.102042	0.103835	0.105650	0.107488	0.109349	0.111232	0.113139	0.115070
−1.1	0.117023	0.119000	0.121000	0.123024	0.125072	0.127143	0.129238	0.131357	0.133500	0.135666
−1	0.137857	0.140071	0.142310	0.144572	0.146859	0.149170	0.151505	0.153864	0.156248	0.158655
−0.9	0.161087	0.163543	0.166023	0.168528	0.171056	0.173609	0.176186	0.178786	0.181411	0.184060
−0.8	0.186733	0.189430	0.192150	0.194895	0.197663	0.200454	0.203269	0.206108	0.208970	0.211855
−0.7	0.214764	0.217695	0.220650	0.223627	0.226627	0.229650	0.232695	0.235762	0.238852	0.241964
−0.6	0.245097	0.248252	0.251429	0.254627	0.257846	0.261086	0.264347	0.267629	0.270931	0.274253
−0.5	0.277595	0.280957	0.284339	0.287740	0.291160	0.294599	0.298056	0.301532	0.305026	0.308538
−0.4	0.312067	0.315614	0.319178	0.322758	0.326355	0.329969	0.333598	0.337243	0.340903	0.344578
−0.3	0.348268	0.351973	0.355691	0.359424	0.363169	0.366928	0.370700	0.374484	0.378280	0.382089
−0.2	0.385908	0.389739	0.393580	0.397432	0.401294	0.405165	0.409046	0.412936	0.416834	0.420740
−0.1	0.424655	0.428576	0.432505	0.436441	0.440382	0.444330	0.448283	0.452242	0.456205	0.460172
0	0.464144	0.468119	0.472097	0.476078	0.480061	0.484047	0.488034	0.492022	0.496011	0.500000

(Continued)

z	0	0.01	0.02	0.03	0.04	0.05	0.06	0.07	0.08	0.09
0	0.500000	0.503989	0.507978	0.511966	0.515953	0.519939	0.523922	0.527903	0.531881	0.535856
0.1	0.539828	0.543795	0.547758	0.551717	0.555670	0.559618	0.563559	0.567495	0.571424	0.575345
0.2	0.579260	0.583166	0.587064	0.590954	0.594835	0.598706	0.602568	0.606420	0.610261	0.614092
0.3	0.617911	0.621720	0.625516	0.629300	0.633072	0.636831	0.640576	0.644309	0.648027	0.651732
0.4	0.655422	0.659097	0.662757	0.666402	0.670031	0.673645	0.677242	0.680822	0.684386	0.687933
0.5	0.691462	0.694974	0.698468	0.701944	0.705401	0.708840	0.712260	0.715661	0.719043	0.722405
0.6	0.725747	0.729069	0.732371	0.735653	0.738914	0.742154	0.745373	0.748571	0.751748	0.754903
0.7	0.758036	0.761148	0.764238	0.767305	0.770350	0.773373	0.776373	0.779350	0.782305	0.785236
0.8	0.788145	0.791030	0.793892	0.796731	0.799546	0.802337	0.805105	0.807850	0.810570	0.813267
0.9	0.815940	0.818589	0.821214	0.823814	0.826391	0.828944	0.831472	0.833977	0.836457	0.838913
1	0.841345	0.843752	0.846136	0.848495	0.850830	0.853141	0.855428	0.857690	0.859929	0.862143
1.1	0.864334	0.866500	0.868643	0.870762	0.872857	0.874928	0.876976	0.879000	0.881000	0.882977
1.2	0.884930	0.886861	0.888768	0.890651	0.892512	0.894350	0.896165	0.897958	0.899727	0.901475
1.3	0.903200	0.904902	0.906582	0.908241	0.909877	0.911492	0.913085	0.914657	0.916207	0.917736
1.4	0.919243	0.920730	0.922196	0.923641	0.925066	0.926471	0.927855	0.929219	0.930563	0.931888
1.5	0.933193	0.934478	0.935745	0.936992	0.938220	0.939429	0.940620	0.941792	0.942947	0.944083
1.6	0.945201	0.946301	0.947384	0.948449	0.949497	0.950529	0.951543	0.952540	0.953521	0.954486
1.7	0.955435	0.956367	0.957284	0.958185	0.959070	0.959941	0.960796	0.961636	0.962462	0.963273
1.8	0.964070	0.964852	0.965620	0.966375	0.967116	0.967843	0.968557	0.969258	0.969946	0.970621
1.9	0.971283	0.971933	0.972571	0.973197	0.973810	0.974412	0.975002	0.975581	0.976148	0.976705
2	0.977250	0.977784	0.978308	0.978822	0.979325	0.979818	0.980301	0.980774	0.981237	0.981691
2.1	0.982136	0.982571	0.982997	0.983414	0.983823	0.984222	0.984614	0.984997	0.985371	0.985738
2.2	0.986097	0.986447	0.986791	0.987126	0.987455	0.987776	0.988089	0.988396	0.988696	0.988989
2.3	0.989276	0.989556	0.989830	0.990097	0.990358	0.990613	0.990863	0.991106	0.991344	0.991576
2.4	0.991802	0.992024	0.992240	0.992451	0.992656	0.992857	0.993053	0.993244	0.993431	0.993613
2.5	0.993790	0.993963	0.994132	0.994297	0.994457	0.994614	0.994766	0.994915	0.995060	0.995201
2.6	0.995339	0.995473	0.995604	0.995731	0.995855	0.995975	0.996093	0.996207	0.996319	0.996427
2.7	0.996533	0.996636	0.996736	0.996833	0.996928	0.997020	0.997110	0.997197	0.997282	0.997365
2.8	0.997445	0.997523	0.997599	0.997673	0.997744	0.997814	0.997882	0.997948	0.998012	0.998074
2.9	0.998134	0.998193	0.998250	0.998305	0.998359	0.998411	0.998462	0.998511	0.998559	0.998605
3	0.998650	0.998694	0.998736	0.998777	0.998817	0.998856	0.998893	0.998930	0.998965	0.998999
3.1	0.999032	0.999065	0.999096	0.999126	0.999155	0.999184	0.999211	0.999238	0.999264	0.999289
3.2	0.999313	0.999336	0.999359	0.999381	0.999402	0.999423	0.999443	0.999462	0.999481	0.999499
3.3	0.999517	0.999534	0.999550	0.999566	0.999581	0.999596	0.999610	0.999624	0.999638	0.999651
3.4	0.999663	0.999675	0.999687	0.999698	0.999709	0.999720	0.999730	0.999740	0.999749	0.999758
3.5	0.999767	0.999776	0.999784	0.999792	0.999800	0.999807	0.999815	0.999822	0.999828	0.999835
3.6	0.999841	0.999847	0.999853	0.999858	0.999864	0.999869	0.999874	0.999879	0.999883	0.999888
3.7	0.999892	0.999896	0.999900	0.999904	0.999908	0.999912	0.999915	0.999918	0.999922	0.999925
3.8	0.999928	0.999931	0.999933	0.999936	0.999938	0.999941	0.999943	0.999946	0.999948	0.999950
3.9	0.999952	0.999954	0.999956	0.999958	0.999959	0.999961	0.999963	0.999964	0.999966	0.999967

A.4 Percentage Points $t_{\alpha,v}$ of the Student t-Distribution (v degrees of freedom)

v \ α	0.1	0.05	0.025	0.01	0.005	0.0025	0.001	0.0005
1	3.078	6.314	12.706	31.821	63.657	127.321	318.309	636.619
2	1.886	2.920	4.303	6.965	9.925	14.089	22.327	31.599
3	1.638	2.353	3.182	4.541	5.841	7.453	10.215	12.924
4	1.533	2.132	2.776	3.747	4.604	5.598	7.173	8.610
5	1.476	2.015	2.571	3.365	4.032	4.773	5.893	6.869
6	1.440	1.943	2.447	3.143	3.707	4.317	5.208	5.959
7	1.415	1.895	2.365	2.998	3.499	4.029	4.785	5.408
8	1.397	1.860	2.306	2.896	3.355	3.833	4.501	5.041
9	1.383	1.833	2.262	2.821	3.250	3.690	4.297	4.781
10	1.372	1.812	2.228	2.764	3.169	3.581	4.144	4.587
11	1.363	1.796	2.201	2.718	3.106	3.497	4.025	4.437
12	1.356	1.782	2.179	2.681	3.055	3.428	3.930	4.318
13	1.350	1.771	2.160	2.650	3.012	3.372	3.852	4.221
14	1.345	1.761	2.145	2.624	2.977	3.326	3.787	4.140
15	1.341	1.753	2.131	2.602	2.947	3.286	3.733	4.073
16	1.337	1.746	2.120	2.583	2.921	3.252	3.686	4.015
17	1.333	1.740	2.110	2.567	2.898	3.222	3.646	3.965
18	1.330	1.734	2.101	2.552	2.878	3.197	3.610	3.922
19	1.328	1.729	2.093	2.539	2.861	3.174	3.579	3.883
20	1.325	1.725	2.086	2.528	2.845	3.153	3.552	3.850
21	1.323	1.721	2.080	2.518	2.831	3.135	3.527	3.819
22	1.321	1.717	2.074	2.508	2.819	3.119	3.505	3.792
23	1.319	1.714	2.069	2.500	2.807	3.104	3.485	3.768
24	1.318	1.711	2.064	2.492	2.797	3.091	3.467	3.745
25	1.316	1.708	2.060	2.485	2.787	3.078	3.450	3.725
26	1.315	1.706	2.056	2.479	2.779	3.067	3.435	3.707
27	1.314	1.703	2.052	2.473	2.771	3.057	3.421	3.690
28	1.313	1.701	2.048	2.467	2.763	3.047	3.408	3.674
29	1.311	1.699	2.045	2.462	2.756	3.038	3.396	3.659
30	1.310	1.697	2.042	2.457	2.750	3.030	3.385	3.646
31	1.309	1.696	2.040	2.453	2.744	3.022	3.375	3.633
32	1.309	1.694	2.037	2.449	2.738	3.015	3.365	3.622
33	1.308	1.692	2.035	2.445	2.733	3.008	3.356	3.611
34	1.307	1.691	2.032	2.441	2.728	3.002	3.348	3.601
35	1.306	1.690	2.030	2.438	2.724	2.996	3.340	3.591
40	1.303	1.684	2.021	2.423	2.704	2.971	3.307	3.551
45	1.301	1.679	2.014	2.412	2.690	2.952	3.281	3.520
50	1.299	1.676	2.009	2.403	2.678	2.937	3.261	3.496
∞	1.282	1.645	1.960	2.326	2.576	2.807	3.090	3.291

A.5 Percentage Points $\chi^2_{\alpha,\nu}$ of the Student t-Distribution (ν degrees of freedom)

ν \ α	0.1	0.05	0.025	0.01	0.005	0.0025	0.001	0.0005
1	2.706	5.991	12.833	26.217	48.290	80.097	124.839	182.445
2	4.605	9.488	19.023	36.191	61.581	96.599	144.567	206.463
3	6.251	12.592	23.337	41.638	68.053	105.330	155.528	219.555
4	7.779	14.067	26.119	45.642	73.166	111.513	162.788	227.847
5	9.236	16.919	28.845	48.278	76.969	115.203	167.610	233.753
6	10.645	18.307	31.526	52.191	82.001	122.543	176.014	244.351
7	12.017	21.026	35.479	57.342	88.236	129.835	184.379	253.737
8	13.362	22.362	36.781	58.619	89.477	131.046	186.762	256.079
9	14.684	23.685	38.076	61.162	93.186	135.878	191.520	261.926
10	15.987	24.996	39.364	62.428	94.419	137.083	193.894	264.262
11	17.275	27.587	43.195	67.459	100.554	144.293	202.184	274.751
12	18.549	28.869	44.461	68.710	101.776	145.491	203.366	275.914
13	19.812	30.144	46.979	71.201	105.432	150.273	209.265	282.886
14	21.064	32.671	49.480	74.919	109.074	155.041	215.149	289.844
15	22.307	33.924	50.725	76.154	111.495	157.419	217.499	292.160
16	23.542	35.172	53.203	79.843	115.117	162.166	223.363	299.101
17	24.769	36.415	54.437	81.069	117.524	164.535	225.705	301.412
18	25.989	37.652	55.668	82.292	118.726	165.718	226.876	302.567
19	27.204	40.113	59.342	87.166	124.718	172.801	235.053	312.945
20	28.412	41.337	60.561	88.379	125.913	173.978	236.220	314.096
21	29.615	42.557	61.777	89.591	127.106	176.332	239.716	317.549
22	30.813	43.773	62.990	90.802	128.299	177.508	240.880	318.699
23	32.007	46.194	66.617	95.626	134.247	184.548	249.018	329.038
24	33.196	47.400	67.821	96.828	135.433	185.720	250.179	330.185
25	34.382	48.602	69.023	99.228	138.987	189.229	254.818	334.771
26	35.563	49.802	70.222	100.425	140.169	191.566	257.135	338.207
27	36.741	50.998	71.420	101.621	141.351	192.733	258.292	339.352
28	37.916	52.192	73.810	104.010	144.891	196.232	262.920	343.929
29	39.087	54.572	76.192	107.583	148.424	200.889	267.541	349.644
30	40.256	55.758	77.380	108.771	149.599	202.052	269.849	351.928
31	41.422	56.942	78.567	109.958	150.774	203.214	271.002	354.211
32	42.585	58.124	80.936	112.329	154.294	207.858	275.612	358.774
33	43.745	59.304	82.117	114.695	156.637	210.177	279.066	363.333
34	44.903	60.481	83.298	115.876	157.808	211.336	280.217	364.472
35	46.059	62.830	85.654	118.236	161.314	215.965	284.815	369.026
40	51.805	68.669	92.689	126.462	170.634	226.354	297.433	383.800
50	63.167	82.529	108.937	145.099	192.610	251.601	325.970	415.493
60	74.397	95.081	123.858	162.398	212.111	274.395	352.072	445.906
70	85.527	107.522	137.517	178.421	230.347	294.799	374.665	470.592
80	96.578	119.871	151.084	194.342	248.485	315.110	398.294	497.434
90	107.565	132.144	165.696	210.176	266.537	335.339	420.718	521.964
100	118.498	144.354	179.114	225.933	283.390	354.376	441.954	545.318

APPENDIX B

Kolmogorov-Smirnov Test Critical Values

Sample size	Level of Significance		
n	$D_{(0.1)}$	$D_{(0.05)}$	$D_{(0.01)}$
10	0.36866	0.40925	0.48893
11	0.35242	0.39122	0.46770
12	0.33815	0.37543	0.44905
13	0.32549	0.36143	0.43247
14	0.31417	0.34890	0.41762
15	0.30397	0.33760	0.40420
16	0.29472	0.32733	0.39201
17	0.28627	0.31796	0.38086
18	0.27851	0.30936	0.37062
19	0.27136	0.30143	0.36117
20	0.26473	0.29408	0.35241
25	0.23768	0.26404	0.31657
30	0.21756	0.24170	0.28987
35	0.20185	0.22425	0.26897
Approximation			
Over 35	$1.22/\sqrt{n}$	$1.36/\sqrt{n}$	$1.63/\sqrt{n}$

See Miller (1956). Additional values can be computed at: http://www.ciphersbyritter.com/JAVASCRP/NORM-CHIK.HTM#KolSmir

A P P E N D I X   C

Arena™

C.1   Arena's Mathematical and Logical Operators

Operator	Operation	Priority
Math Operators		
**	Exponentiation	1 (highest)
/	Division	2
*	Multiplication	2
−	Subtraction	3
+	Addition	3
Logical Operators		
.EQ., ==	Equality comparison	4
.NE., < >	Non-equality comparison	4
.LT., <	Less than comparison	4
.GT., >	Greater than comparison	4
.LE., <=	Less than or equal to comparison	4
.GE., > =	Greater than or equal to comparison	4
.AND., &&	Conjunction (and)	5
.OR., ||	Inclusive disjunction (or)	5

C.2   Arena's Distributions

Beta distribution	BETA(Alpha1,Alpha2[,Stream])
Normal distribution	NORM(Mean,SD[,Stream])
Empirical Continuous distribution	CONT(Prob1,Value1,Prob2,Value2, . . . [,Stream])
NSExpo distribution	Non-homogeneous Poisson process
Empirical DISCRETE distribution	DISC(Prob1,Value1,Prob2,Value2, . . . [,Stream])
Poisson distribution	POIS(Mean[,Stream])
k-Erlang distribution	ERLA(Mean, k[,Stream])
Lognormal distribution	LOGN(LogMean,LogStd[,Stream])
Random Number Between 0 and 1	RA
Exponential distribution	EXPO(Mean[,Stream])
Triangular distribution	TRIA(Min,Mode,Max[,Stream])
Gamma distribution	GAMM(scale,shape[,Stream])
Uniform distribution	UNIF(Min,Max[,Stream])
Johnson distribution	JOHN(shape1,shape2,scale,location[,Stream])
Weibull distribution	WEIBull(scale,shape[,Stream])

C.3   Arena's Mathematical Functions

Function	Description
ABS(a)	Absolute value
ACOS(a)	Arc cosine
AINT(a)	Truncate
AMOD(a1, a2)	Real remainder, returns (a1-(AINT(a1/a2)*a2))
ANINT(a)	Round to nearest integer
ASIN(a)	Arc sine
ATAN(a)	Arc tangent
COS(a)	Cosine
EP(a)	Exponential (e^a)
HCOS(a)	Hyperbolic cosine
HSIN(a)	Hyperbolic sine
HTAN(a)	Hyperbolic tangent
MN(a1, a2, . . .)	Minimum value
MOD(a1, a2)	Integer remainder, same as AMOD except the arguments are truncated to integer values first
MX(a1, a2, . . .)	Maximum value
LN(a)	Natural logarithm
LOG(a)	Common logarithm
SIN(a)	Sine
SQRT(a)	Square root
TAN(a)	Tangent

Analytical Results for Various Queuing Systems

Results for P_0 and P_n for various Queueing Systems

Notation	Parameters	P_0	P_n
M/M/1	$\lambda_n = \lambda; \mu_n = \mu;$ $c = 1; \rho = \dfrac{\lambda}{c\mu} = r$	$P_o = 1 - r$	$P_n = P_o r^n$
M/M/c	$\lambda_n = \lambda;$ $\mu_n = \begin{cases} n\mu & 0 \le n < c \\ c\mu & n \ge c \end{cases};$ $\rho = \lambda/c\mu;\ r = \lambda/\mu$	$P_0 = \left[\displaystyle\sum_{n=0}^{c-1} \dfrac{r^n}{n!} + \dfrac{r^c}{c!(1-\rho)} \right]^{-1}$	$P_n = \begin{cases} \dfrac{r^n}{n!} P_0 & 1 \le n < c \\ \dfrac{r^n}{c! c^{n-c}} P_0 & n \ge c \end{cases}$
M/M/1/k	$\lambda_n = \begin{cases} \lambda & n < k \\ 0 & n \ge k \end{cases};$ $\mu_n = \begin{cases} \mu & 0 \le n \le k \\ 0 & n > k \end{cases}$ $c = 1;\ \rho = \dfrac{\lambda}{c\mu} = r;$ $\lambda_e = \lambda(1 - P_k)$	$P_0 = \begin{cases} \dfrac{1-r}{1-r^{k+1}} & r \ne 1 \\ \dfrac{1}{k+1} & r = 1 \end{cases}$	$P_n = \begin{cases} P_0 r^n & r \ne 1 \\ \dfrac{1}{k+1} & r = 1 \end{cases}$
M/M/c/k	$\lambda_n = \begin{cases} \lambda & n < k \\ 0 & n \ge k \end{cases};$ $\mu_n = \begin{cases} n\mu & 0 \le n < c \\ c\mu & c \le n \le k \end{cases}$ $\rho = \lambda/c\mu;$ $r = \lambda/\mu;\ \lambda_e = \lambda(1 - P_k)$	$P_0 \begin{cases} \left[\displaystyle\sum_{n=0}^{c-1} \dfrac{r^n}{n!} + \dfrac{r^c}{c!} \dfrac{1 - \rho^{k-c+1}}{1-\rho} \right]^{-1} & \rho \ne 1 \\ \left[\displaystyle\sum_{n=0}^{c-1} \dfrac{r^n}{n!} + \dfrac{r^c}{c!}(k-c+1) \right]^{-1} & \rho = 1 \end{cases}$	$P_n = \begin{cases} \dfrac{r^n}{n!} P_0 & 1 \le n < c \\ \dfrac{r^n}{c! c^{n-c}} P_0 & c \le n \le k \end{cases}$
M/M/c/c M/G/c/c	$\lambda_n = \begin{cases} \lambda & n < c \\ 0 & n \ge c \end{cases};$ $\mu_n = \begin{cases} n\mu & 0 \le n \le c \\ 0 & n > c \end{cases}$ $\rho = \lambda/c\mu;$ $r = \lambda/\mu;\ \lambda_e = \lambda(1 - P_c)$	$P_0 = \left[\displaystyle\sum_{n=0}^{c} \dfrac{r^n}{n!} \right]^{-1}$	$P_n = \dfrac{r^n}{n!} P_0 \quad 0 \le n \le c$
M/M/1/k/k	$\lambda_n = \begin{cases} (k-n)\lambda & 0 \le n < k \\ 0 & n \ge k \end{cases}$ $\mu_n = \begin{cases} \mu & 0 \le n \le k \\ 0 & n > k \end{cases}$ $r = \lambda/\mu;\ \lambda_e = \lambda(k - L)$	$P_0 = \left[\displaystyle\sum_{n=0}^{k} \prod_{j=0}^{n-1} \left(\dfrac{\lambda_j}{\mu_{j+1}} \right) \right]^{-1}$	$P_n = \dbinom{k}{n} n! r^n P_0 \quad 0 \le n \le k$
M/M/c/k/k	$\lambda_n = \begin{cases} (k-n)\lambda & 0 \le n < k \\ 0 & n \ge k \end{cases}$ $\mu_n = \begin{cases} n\mu & 0 \le n < c \\ c\mu & n \ge c \end{cases}$ $r = \lambda/\mu;\ \lambda_e = \lambda(k - L)$	$P_0 = \left[\displaystyle\sum_{n=0}^{k} \prod_{j=0}^{n-1} \left(\dfrac{\lambda_j}{\mu_{j+1}} \right) \right]^{-1}$	$P_n = \begin{cases} \dbinom{k}{n} r^n P_0 & 1 \le n < c \\ \dbinom{k}{n} \dfrac{n!}{c^{n-c} c!} r^n P_0 & c \le n \le k \end{cases}$

Results for L_q for various Queueing Systems

Notation	L_q
M/M/1	$L_q = \dfrac{r^2}{1-r}$
M/M/c	$L_q = \left(\dfrac{r^c \rho}{c!(1-\rho)^2} \right) P_0$
M/M/1/k	$L_q = \begin{cases} \dfrac{\rho}{1-\rho} - \dfrac{\rho(k\rho^k + 1)}{1-\rho^{k+1}} & \rho \neq 1 \\[2ex] \dfrac{k(k-1)}{2(k+1)} & \rho = 1 \end{cases}$
M/M/c/k	$L_q = \begin{cases} \dfrac{P_0 r^c \rho}{c!(1-\rho)^2}\left[1 - \rho^{k-c} - (k-c)\rho^{k-c}(1-\rho) \right] & \rho < 1 \\[2ex] \dfrac{r^c(k-c)(k-c+1)}{2c!}P_0 & \rho = 1 \end{cases}$
M/M/c/c M/G/c/c	$L_q = 0$
M/M/1/k/k	$L_q = k - \left(\dfrac{\lambda + \mu}{\lambda} \right)(1 - P_0)$
M/M/c/k/k	$L_q = \displaystyle\sum_{n=c}^{k} (n-c)P_n$

Results M/M/c $\rho = \lambda / c\mu$

c	P_0	L_q
1	$1 - \rho$	$\dfrac{\rho^2}{1-\rho}$
2	$\dfrac{1-\rho}{1+\rho}$	$\dfrac{2\rho^3}{1-\rho^2}$
3	$\dfrac{2(1-\rho)}{2+4\rho+3\rho^2}$	$\dfrac{9\rho^4}{2+2\rho-\rho^2-3\rho^3}$

Results M/G/1 and M/D/1

	Parameters	L_q
M/G/1	$E[ST] = \dfrac{1}{\mu};\ Var[ST] = \sigma^2;\ r = \lambda/\mu$	$L_q = \dfrac{\lambda^2\sigma^2 + r^2}{2(1-r)}$
M/D/1	$E[ST] = \dfrac{1}{\mu};\ Var[ST] = 0;\ r = \lambda/\mu$	$L_q = \dfrac{r^2}{2(1-r)}$

Analytical Results for (r,Q) Inventory Model

r = reorder point Q = reorder quantity

λ = mean customer demand rate in units/time

L = constant lead time (measured in time-units)

$g(x; t)$ = probability mass function for Poisson distribution with rate λ, representing the number of events that occur in an interval of length t

$$g(x; t) = \frac{(\lambda t)^x e^{-\lambda t}}{x!}$$

$G(x; t) = \Pr\{X \leq x\} = \sum_{i=0}^{x} g(i; t) = $ the Poisson cumulative distribution function

$G^0(x; t) = 1 - G(x; t) = $ the Poisson complementary cumulative distribution function

$G^1(x; t) = -(x - \lambda t)G^0(x; t) + (\lambda t)g(x; t) = $ Poisson first order loss function

$G^2(x; t) = (1/2)\left\{ \left[(x - \lambda t)^2 + x\right]G^0(x; t) - (\lambda t)(x - \lambda t)g(x; t) \right\} = $ Poisson second order loss function

$$\overline{SO} = \frac{1}{Q}\left[G^1(r; L) - G^1(r + Q; L)\right]$$

$$\overline{B} = \frac{1}{Q}\left[G^2(r; L) - G^2(r + Q; L)\right]$$

$$\overline{I} = (1/2)(Q + 1) + r - \lambda L + \overline{B}$$

$$\overline{OF} = \frac{\lambda}{Q}$$

h = holding cost for the item in units($/unit/time)

b = back ordering cost for the item in units ($/unit/time)

k = order preparation cost in units ($/order)

TC = total cost of the policy per time, $TC = k\overline{OF} + h\overline{I} + b\overline{B}$ in units ($/time)

Panel Modules

Basic Process Panel Modules

CREATE	Used to create and introduce entities into the model according to a pattern.
DISPOSE	Used to dispose of entities once they have completed their activities within the model.
PROCESS	Used to allow an entity to experience an activity with the possible use of a resource.
ASSIGN	Used to make assignments to variables and attributes within the model
RECORD	Used to capture and tabulate statistics within the model.
BATCH	Used to combine entities into a permanent or temporary representative entity.
SEPARATE	Used to create duplicates of an existing entity or to split a batched group of entities.
DECIDE	Used to provide alternative flow paths for an entity based on probabilistic or condition based branching.
VARIABLE	Used to define variables for use within the model.
RESOURCE	Used to define a quantity of units of a resource that can be seized and released by entities.
QUEUE	Used to define a waiting line for entities whose flow is currently stopped within the model.
ENTITY	Used to define different entity types for use within the model.
SET	Used to define a list of elements within Arena that can be indexed by the location in the list.
SCHEDULE	Used to define a staffing schedule for resources or a time-based arrival pattern.

Advanced Process Panel Modules

SEIZE	Allows an entity to request a number of units of a resource. If the units are not available, the entity waits in a queue.
DELAY	Allows an entity to experience a delay in movement via the scheduling of an event.
RELEASE	Releases the units of a resource seized by an entity.
MATCH	Allows entities to wait in queues until a user specified matching criteria occurs.
HOLD	Holds entities in a queue until a signal is given or until a condition in the model is met.
SIGNAL	Signals entities in a HOLD queue to proceed.
PICKUP	Allows an entity to pick up and place other entities into a group associated with the entity.
DROPOFF	Allows an entity to drop off entities from its entity group.
SEARCH	Allows an entity to search a queue for entities that match search criteria.
REMOVE	Allows an entity to remove other entities directly from a queue.
STORE	Indicates that the entity is in a STORAGE
UNSTORE	Indicates that the entity is no longer in a STORAGE
ADJUST VARIABLE	Adjusts a variable to a target value at a specified rate.
READWRITE	Allows input and output to occur within the model.
FILE	Defines the characteristics of the operating system file used within a READWRITE module.
EXPRESSION	Allows the user to define named logical/mathematical expressions that can be used throughout the model.
STORAGE	Demarks a location/concept that may contain entities.
ADVANCED SET	Used to define a list of elements within Arena that can be indexed by the location in the list.
FAILURE	Used to define unscheduled capacity changes for resources according to a time pattern or a usage indicator.
STATISTIC	Used to define and manage time-based, observation based, and replication statistics.

Advanced Transfer Panel Modules

STATION	Allows the marking in the model for a location to which entities can be directed for processing.
ROUTE	Facilitates the movement between stations with a time delay.
ENTER	Represents STATION, DELAY, and/or EXIT/FREE
LEAVE	Represents ROUTE or TRANSPORT or CONVEY with DELAY option
PICKSTATION	Allows entity to decide on its next station based on conditions.
ACCESS	Requests space on a conveyor. Entity waits if no space is available.
CONVEY	After obtaining space on a conveyor, causes the entity to be conveyed to its destination station.
EXIT	Releases space on a conveyor
START	Causes a stopped conveyor to start transferring entities.
STOP	Causes a conveyor to stop transferring entities.
ALLOCATE	Assigns a transporter to an entity without moving the transporter. The entity now has control of the transporter. The entity may wait in queue if no transporters are available.
MOVE	Moves an allocated transporter to a station destination.
REQUEST	Asks a transporter for a pick up. The requesting entity waits until a transporter is allocated and moves to the pick up location.
TRANSPORT	Same as REQUEST followed by MOVE to the entities desired location.
FREE	Causes the entity to release an allocated transporter.
HALT	Changes the state of the transporter to inactive.
ACTIVATE	Changes the state of the transporter to active.
SEQUENCE	Allows for pre-specified routes of stations to be defined and attributes to be assigned when entities are transferred to the stations.
CONVEYOR	Defines a conveyor as a list of segments and provides the velocity and space characteristics of the conveyor.
SEGMENT	Defines the distance between two stations as a segment on a conveyor.
TRANSPORTER	Defines a mobile resource and its characteristics, capable of free path or guided path movement.
DISTANCE	Defines the from-to distances between stations for free path transporters.
NETWORK	Defines a set of transporter links between intersections that represents a guided path network for transporters.
NETWORKLINK	Defines the characteristics of a space constraining path between intersections within guided path networks.
ACTIVITY AREA	Defines stations that are part of an area to facilitate the collection of aggregate statistics on the group of stations.

Miscellaneous Useful Blocks, Attributes, and Variables

IF-ELSEIF-ELSE-ENDIF	From the Blocks panel, these modules allow standard logic based flow of control.
WHILE-ENDWHILE	From the Blocks panel, these modules allow for iterative looping.
BRANCH	From the Blocks panel, this module allows probabilistic and condition based path determination along with cloning of entities.
TNOW	The current simulation time.
NREP	Current replication number.
MREP	Maximum number of replications.
J	Index in SEARCH module
NQ(queue name)	Number of entities in the named queue.
MR(resource name)	The current capacity of the named resource
NR(resource name)	The current number of busy units of the named resource
NE(station)	The number of entities transferring to the named station.
Entity.SerialNumber	A number assigned to an entity upon creation. Duplicates will have this same number.
Enity.Jobstep	The entity's current position in its squence.
Entity.Sequence	The entity's sequence when using a transfer option.
Entity.Station	The entity's location or destination.
Entity.CurrentStation	The entity's location.
Enity.CreateTime	The value of TNOW when the entity was created.
IDENT	A unique number assigned to an entity while in the model. No entities have the same IDENT number.

APPENDIX G

Useful Equations

Cumulative Distribution Function

$$F(x) = P(X \le x) = \int\limits_{-\infty}^{x} f(u)du$$

Variance $Var[X] = E[X^2] - (E[X])^2$

Covariance $cov(X, Y) = E[XY] - E[X]E[Y]$

Correlation $cor[X, Y] = \dfrac{cov[X, Y]}{\sqrt{V[X] \times V[Y]}}$

Sample average $\overline{X}(n) = \dfrac{1}{n}\sum\limits_{i=1}^{n} X_i$

Time Average $\overline{Y}(n) = \dfrac{\int_{t_0}^{t_n} Y(t)dt}{t_n - t_0}$

Sample variance

$$S^2(n) = \dfrac{1}{n-1}\sum\limits_{i=1}^{n}(X_i - \overline{X})^2$$

$(1 - \alpha)\%$ Confidence Interval

$$\overline{x} \pm t_{\alpha/2, n-1}\dfrac{s}{\sqrt{n}}$$

Normal Sample Size Approximation

$n \ge \left(\dfrac{z_{\alpha/2}s}{h}\right)^2 h$ is desired half-width

Half-width Ratio Sample Size

Approximation $n \cong n_0\left(\dfrac{h_0}{h}\right)^2$

Sample Size for Proportions

$$n = \left(\dfrac{z_{\alpha/2}}{E}\right)\hat{p}(1 - \hat{p})$$

LCG Equation
$R_{i+1} = (aR_i + c)\bmod m \quad for\ i = 0, 1, 2, \ldots$

$z = y \bmod m \Leftrightarrow z = y - m\left\lfloor \dfrac{y}{m} \right\rfloor$

Chi-Squared Test Statistic

$$\chi^2 = \sum\limits_{j=1}^{k} \dfrac{(N_j - np_j)}{np_j}$$

Reject the null hypothesis if $\chi^2 > \chi^2_{\alpha, k-s-1}$, where (s) is the number of estimated parameters

Kolmogorov-Smirnov Test Statistic
$\hat{F}(x)$ hypothesized distribution

$$D_n^+ = \max_{1 \le i \le n}\left\{\dfrac{i}{n} - \hat{F}(x_{(i)})\right\}$$

$$D_n^- = \max_{1 \le i \le n}\left\{\hat{F}(x_{(i)}) - \dfrac{i-1}{n}\right\} \quad D_n = \max\{D_n^+, D_n^-\}$$

Little's Formula $L = \lambda W \quad L_q = \lambda W_q$

$B = \lambda E[ST] = \dfrac{\lambda}{\mu} \quad \rho = \dfrac{B}{c} = \dfrac{\lambda}{c\mu}$

Bonferroni Inequality

$$P\{\cap_{i=1}^{k} E_i\} \ge 1 - \sum\limits_{i=1}^{k}\alpha_i$$

Index